WIRELESS SENSOR NETWORKS: APPLICATION-CENTRIC DESIGN

Edited by **Dr. Geoff V Merret and**
Dr. Yen Kheng Tan (Editor-in-Chief)

Wireless Sensor Networks: Application-Centric Design
Edited by Dr. Geoff V Merret and Dr. Yen Kheng Tan (Editor-in-Chief)

Published by InTech

Janeza Trdine 9, 51000 Rijeka, Croatia

Publishing Process Manager Jelena Marusic

Technical Editor Teodora Smiljanic

Cover Designer InTech Design Team

Additional hard copies can be obtained from orders@intechopen.com

Wireless Sensor Networks: Application-Centric Design, Edited by Dr. Geoff V Merret and Dr. Yen Kheng Tan (Editor-in-Chief)
p. cm.
ISBN 978-953-307-321-7

For EU product safety concerns:
IN TECH d.o.o., Prolaz Marije Krucifikse Kozulić 3, 51000 Rijeka, Croatia,
info@intechopen.com or visit our website at intechopen.com.

Meet the editors

Yen Kheng Tan received his B.Eng degree in Electrical and Computer Engineering (ECE) from National University of Singapore (NUS) in 2003 and the Master of Technological Design (Mechatronics Engineering) degree jointly offered by NUS and the Eindhoven University of Technology (TU/e) in 2006. Since 2006, he has been working towards his Ph.D degree in the Department of ECE of NUS. Presently, he is leading a team of research engineers at the Energy Research Institute @NTU (ERI@N). His research interests are micro-power generations from solar, wind, thermal and vibration energy sources, energy harvesting wireless sensor network, nonlinear control and motor drives, electric vehicles, wireless power transfer, rehabilitation engineering and assistive technology. Yen Kheng is on the Technical Program Committee of numerous conferences and reviewer of papers for a number of prestigious international journals. He is also the book editor of several books as well. [http://www.ece.nus.edu.sg/emdl/yenkheng.htm]

Geoff Merrett received the BEng (Hons) Electronic Engineering (2004) and PhD Electronic and Electrical Engineering (2009) degrees from the University of Southampton, where he is presently a lecturer in the School of Electronics and Computer Science. His PhD thesis was entitled "Energy- and Information-Managed Wireless Sensor Networks: Modelling and Simulation," and he has research interests in energy-aware wireless sensing and sensor networks, energy harvesting, and modelling and simulation; he is particularly interested in the novel application of technology to wearable and pervasive healthcare. Dr Merrett is a Co-Investigator on two EPSRC funded projects in the areas of energy harvesting. He is a member of the IET and IEEE, has authored a number of journal and conference publications in the area of wireless sensor networks, and is an active reviewer for a number of prestigious international journals.

Contents

Preface

About this Book

Over the past decade, there has been a prolific increase in the research, development and commercialisation of Wireless Sensor Networks (WSNs) and their associated technologies (see Figure 1). This rise has been a result of a number of contributing factors, including continued miniaturisation (leading towards an era of truly 'pervasive' and 'invisible' computing); low-power circuits, devices and computation (for example, the ultra-low-power sleep states now found in microcontrollers); and efficient short-range communication (such as ZigBee and Bluetooth). The dramatic rise in WSN activity, fuelled by the prospect of a new computing paradigm, has resulted in the topic being researched (and taught) in the electronics and computer science departments of Universities around the world.

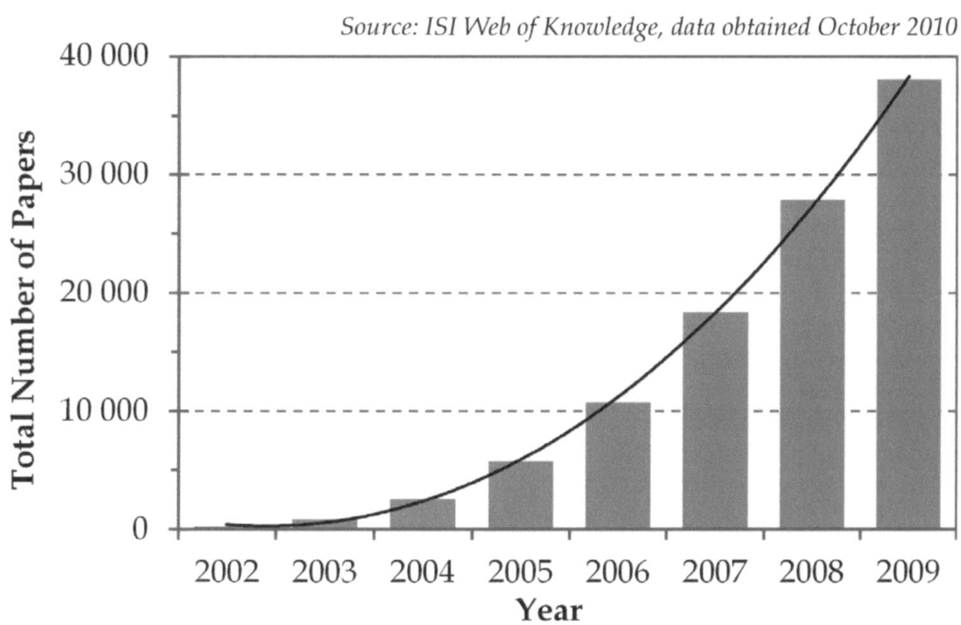

Figure 1. The increase in research into WSNs, shown by the total number of published papers (as catalogued on the ISI Web of Knowledge) matching the topic (sensor network*).

While enabling technologies such as low-power circuitry have permitted the conception and growth of WSNs (for example a microcontroller's ultra-low-power sleep states enable a vast reduction in the average power consumption obtained through duty cycled operation, a technique which underpins the operation of most implementations), the principal reason for the explosion of research is, in my opinion, due to the volume of WSN applications that can be conceived and realised. To name a few, they have found use in healthcare, defence and security, environmental monitoring, process control, structural health monitoring, condition monitoring, building automation, multimedia provision and advertising. However, as a result of the broad array of pertinent applications, WSN researchers have also realised the application specificity of the domain; it is incredibly difficult, if not impossible, to find an application-independent solution to most WSN problems (be it a routing algorithm, MAC protocol, energy harvesting architecture, or data processing algorithm). Hence, research into WSNs dictates the adoption of an application-centric design process.

Research into WSN applications not only concerns the technical issues and system integration strategies of deployment, but also the communication and processing of data, alongside the analysis, understanding and modelling of the application parameters that are of interest. As such, this book is not intended to be a comprehensive review of all WSN applications and deployments to date. Instead, it is a collection of state-of-the-art research papers discussing current applications and deployment experiences, but also the communication and data processing technologies that are fundamental in further developing solutions to applications. Whilst a common foundation is retained through all chapters, this book contains a broad array of often differing interpretations, configurations and limitations of WSNs. I believe that these aid to highlight the rich diversity and sheer scale of this ever-changing research area.

Organisation

The chapters of this book have been categorised into three distinct sections: applications and case studies (section A), communication and networking technologies (section B), and information and data processing technologies (section C). These are described below:

Applications and Case Studies: The first chapter of this section serves as an introduction to the book, providing a concise overview of WSNs, discussing their history, platforms and architectures, research challenges, and application. The remainder of the section discusses current applications and their implementation, from experiences of monitoring agricultural processes to the different methods by which accidental falls of the elderly can be detected and classified.

Communication and Networking Technologies: Alongside the predominant theme of this book, this section contains a collection of state-of-the-art technical papers on application-specific communication and networking problems. These include modelling wireless propagation through water, routing strategies for hybrid radio frequency/fibre optic WSNs, through to an overview of radio transceiver design.

Information and Data Processing Technologies: The final section of this book provides an insight into current research on information and data processing, including data fusion, target tracking, fault tolerance and multimedia provision in WSNs.

The readership of this book is intended to be postgraduate/postdoctoral researchers, and professional engineers. Some of the chapters may also be of interest to master's level students that are undertaking modules that are particularly relevant to this field.

Dr. Geoff V Merret and Dr. Yen Kheng Tan (Editor-in-Chief)

Part 1

Applications and Case Studies

1

Wireless Sensor Networks - An Introduction

Qinghua Wang
Dept. of Electronics and Telecommunications
Norwegian University of Science and Technology
Norway

Ilangko Balasingham
Dept. of Electronics and Telecommunications
Norwegian University of Science and Technology
The Interventional Centre, Oslo University Hospital
Institute of Clinical Medicine, University of Oslo
Norway

This chapter provides a detailed introduction to the history and current state of the art with regard to wireless sensor networks (WSNs).

1. History

The origins of the research on WSNs can be traced back to the Distributed Sensor Networks (DSN) program at the Defense Advanced Research Projects Agency (DARPA) at around 1980. By this time, the ARPANET (Advanced Research Projects Agency Network) had been operational for a number of years, with about 200 hosts at universities and research institutes (Chong & Kumar, 2003). DSNs were assumed to have many spatially distributed low-cost sensing nodes that collaborated with each other but operated autonomously, with information being routed to whichever node was best able to use the information. At that time, this was actually an ambitious program. There were no personal computers and workstations; processing was mainly performed on minicomputers and the Ethernet was just becoming popular (Chong & Kumar, 2003). Technology components for a DSN were identified in a Distributed Sensor Nets workshop in 1978 (*Proceedings of the Distributed Sensor Nets Workshop*, 1978). These included sensors (acoustic), communication and processing modules, and distributed software. Researchers at Carnegie Mellon University (CMU) even developed a communication-oriented operating system called Accent (Rashid & Robertson, 1981), which allowed flexible, transparent accesss to distributed resources required for a fault-tolerant DSN. A demonstrative application of DSN was a helicopter tracking system (Myers et al., 1984), using a distributed array of acoustic microphones by means of signal abstractions and matching techniques, developed at the Massachusetts Institute of Technology (MIT).
Even though early researchers on sensor networks had in mind the vision of a DSN, the technology was not quite ready. More specifically, the sensors were rather large (i.e. shoe box and

This work was carried out during the tenure of an ERCIM "Alain Bensoussan" Fellowship Programme and is part of the MELODY Project, which is funded by the Research Council of Norway under the contract number 187857/S10.

up) which limited the number of potential applications. Further, the earliest DSNs were not tightly associated with wireless connectivity. Recent advances in computing, communication and microelectromechanical technology have caused a significant shift in WSN research and brought it closer to achieving the original vision. The new wave of research in WSNs started in around 1998 and has been attracting more and more attention and international involvement. In the new wave of sensor network research, networking techniques and networked information processing suitable for highly dynamic ad hoc environments and resource-constrained sensor nodes have been the focus. Further, the sensor nodes have been much smaller in size (i.e. pack of cards to dust particle) and much cheaper in price, and thus many new civilian applications of sensor networks such as environment monitoring, vehicular sensor network and body sensor network have emerged. Again, DARPA acted as a pioneer in the new wave of sensor network research by launching an initiative research program called SensIT (Kumar & Shepherd, 2001) which provided the present sensor networks with new capabilities such as ad hoc networking, dynamic querying and tasking, reprogramming and multi-tasking. At the same time, the IEEE noticed the low expense and high capabilities that sensor networks offer. The organization has defined the IEEE 802.15.4 standard (*IEEE 802.15 WPAN Task Group 4*, n.d.) for low data rate wireless personal area networks. Based on IEEE 802.15.4, ZigBee Alliance (*ZigBee Alliance*, n.d.) has published the ZigBee standard which specifies a suite of high level communication protocols which can be used by WSNs. Currently, WSN has been viewed as one of the most important technologies for the 21st century (*21 Ideas for the 21st Century*, 1999). Countries such as China have involved WSNs in their national strategic research programmes (Ni, 2008). The commercialization of WSNs are also being accelerated by new formed companies like Crossbow Technology (*Crossbow Technology*, n.d.) and Dust Networks (*Dust Networks, Inc.*, n.d.).

2. Hardware Platform

A WSN consists of spatially distributed sensor nodes. In a WSN, each sensor node is able to independently perform some processing and sensing tasks. Furthermore, sensor nodes communicate with each other in order to forward their sensed information to a central processing unit or conduct some local coordination such as data fusion. One widely used sensor node platform is the Mica2 Mote developed by Crossbow Technology (*Crossbow Technology*, n.d.). The usual hardware components of a sensor node include a radio transceiver, an embedded processor, internal and external memories, a power source and one or more sensors.

2.1 Embedded Processor

In a sensor node, the functionality of an embedded processor is to schedule tasks, process data and control the functionality of other hardware components. The types of embedded processors that can be used in a sensor node include Microcontroller, Digital Signal Processor (DSP), Field Programmable Gate Array (FPGA) and Application-Specific Integrated Circuit (ASIC). Among all these alternatives, the Microcontroller has been the most used embedded processor for sensor nodes because of its flexibility to connect to other devices and its cheap price. For example, the newest CC2531 development board provided by Chipcon (acquired by Texas Instruments) uses 8051 microcontroller, and the Mica2 Mote platform provided by Crossbow uses ATMega128L microcontroller.

2.2 Transceiver

A transceiver is responsible for the wireless communication of a sensor node. The various choices of wireless transmission media include Radio Frequency (RF), Laser and Infrared. RF based communication fits to most of WSN applications. The operational states of a transceiver are Transmit, Receive, Idle and Sleep. Mica2 Mote uses two kinds of RF radios: RFM TR1000 and Chipcon CC1000. The outdoor transmission range of Mica2 Mote is about 150 meters.

2.3 Memory

Memories in a sensor node include in-chip flash memory and RAM of a microcontroller and external flash memory. For example, the ATMega128L microcontroller running on Mica2 Mote has 128-Kbyte flash program memory and 4-Kbyte static RAM. Further, a 4-Mbit Atemel AT45DB041B serial flash chip can provide external memories for Mica and Mica2 Motes (Hill, 2003).

2.4 Power Source

In a sensor node, power is consumed by sensing, communication and data processing. More energy is required for data communication than for sensing and data processing. Power can be stored in batteries or capacitors. Batteries are the main source of power supply for sensor nodes. For example, Mica2 Mote runs on 2 AA batteries. Due to the limited capacity of batteries, minimizing the energy consumption is always a key concern during WSN operations. To remove the energy constraint, some preliminary research working on energy-harvesting techniques for WSNs has also been conducted. Energy-harvesting techniques convert ambient energy (e.g. solar, wind) to electrical energy and the aim is to revolutionize the power supply on sensor nodes. A survey about the energy-harvesting sensor nodes is provided by (Sudevalayam & Kulkarni, 2008).

2.5 Sensors

A sensor is a hardware device that produces a measurable response signal to a change in a physical condition such as temperature, pressure and humidity. The continual analog signal sensed by the sensors is digitized by an analog-to-digital converter and sent to the embedded processor for further processing. Because a sensor node is a micro-electronic device powered by a limited power source, the attached sensors should also be small in size and consume extremely low energy. A sensor node can have one or several types of sensors integrated in or connected to the node.

3. Operating System

The role of any operating system (OS) is to promote the development of reliable application software by providing a convenient and safe abstraction of hardware resources. OSs for WSN nodes are typically less complex than general-purpose OSs both because of the special requirements of WSN applications and because of the resource constraints in WSN hardware platforms.

TinyOS (*TinyOS Community Forum*, n.d.) is perhaps the first operating system specifically designed for WSNs. It features a component-based architecture which enables rapid innovation and implementation while minimizing code size as required by the severe memory constraints inherent in WSNs. TinyOS's component library includes network protocols, distributed services, sensor drivers, and data acquisition tools - all of which can be further refined for a

custom application. Unlike most other OSs, TinyOS is based on an event-driven programming model instead of multithreading. TinyOS programs are composed into event handlers and tasks with run-to-completion semantics. When an external event occurs, such as an incoming data packet or a sensor reading, TinyOS calls the appropriate event handler to handle the event. Event handlers can post tasks that are scheduled by the TinyOS kernel at a later stage. Both the TinyOS system and programs written for TinyOS are written in a special programming language called nesC which is an extension of the C programming language. NesC is designed to detect race conditions between tasks and event handlers. Currently, TinyOS has been ported to over a dozen platforms and numerous sensor boards. A wide community uses it in simulation to develop and test various algorithms and protocols. According to the figure published on TinyOS forum, over 500 research groups and companies are using TinyOS on the Berkeley/Crossbow Motes. Because TinyOS is open source, numerous groups are actively contributing code to the development of TinyOS and thus making it even more competitive.
Contiki (*Contiki*, n.d.) is another open source OS specifically designed for WSNs. The Contiki kernel is event-driven, like TinyOS, but the system supports multithreading on a per-application basis. Furthermore, Contiki includes protothreads that provide a thread-like programming abstraction but with a very small memory overhead. Contiki provides IP communication, both for IPv4 and IPv6. Many key mechanisms and ideas from Contiki have been widely adopted within the industry. The uIP embedded IP stack, originally released in 2001, is today used by hundreds of companies in systems such as freighter ships, satellites and oil drilling equipment. Contiki's protothreads, first released in 2005, have been used in many different embedded systems, ranging from digital TV decoders to wireless vibration sensors. Contiki's idea of using IP communication in low-power WSNs has lead to an IETF standard and an international industry alliance - IP for Smart Objects (IPSO) Alliance (*IPSO Alliance - promoting the use of IP for Smart Objects*, n.d.).
There are also other OSs that can be used by WSNs. For example, SOS (*SOS Embedded Operating System*, n.d.) is an event-driven OS for mote-class sensor nodes that adopts a more dynamic point on the design spectrum. The prime feature of SOS is its support for loadable modules. A complete system is built from smaller modules, possibly at run-time. To support the inherent dynamism in its module interface, SOS also focuses on supporting dynamic memory management. Unfortunately, SOS is no longer under active development due to the graduation of the core developers. LiteOS (*LiteOS*, n.d.) is an open source, interactive, UNIX-like operating system designed for WSNs. With the tools that come from LiteOS, it is possible to operate one or more WSNs in a Unix-like manner. It is also possible to develop programs for nodes, and wirelessly distribute such programs to sensor nodes.

4. Networking

4.1 Network Architecture

A WSN is a network consisting of numerous sensor nodes with sensing, wireless communications and computing capabilities. These sensor nodes are scattered in an unattended environment (i.e. sensing field) to sense the physical world. The sensed data can be collected by a few sink nodes which have accesses to infrastructured networks like the Internet. Finally, an end user can remotely fetch the sensed data by accessing infrastructured networks. Fig. 1 shows the operation sketch map of WSNs.
In Fig. 1, two kinds of network topologies are shown. The sensor nodes either form a flat network topology where sensor nodes also act as routers and trasfer data to a sink through

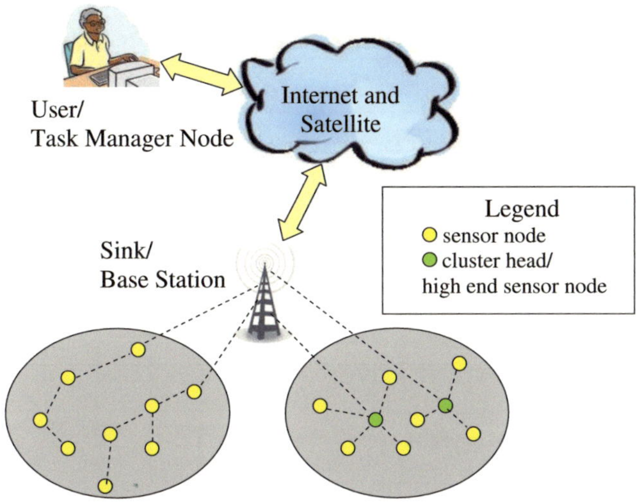

(a) Flat sensor network; (b) Clustered sensor network.

Fig. 1. The Operation of WSNs.

multi-hop routing, or a hierarchical network topology where more powerful fixed or mobile relays are used to collect and route the sensor data to a sink.

4.2 Protocol Stack of WSNs

The protocol stack used by the sink, cluster head and sensor nodes are shown in Fig. 2. According to (Akyildiz et al., 2002), the sensor network protocol stack is much like the traditional protocol stack, with the following layers: application, transport, network, data link, and physical. The physical layer is responsible for frequency selection, carrier frequency generation, signal detection, modulation and data encryption. The data link layer is responsible for the multiplexing of data streams, data frame detection, medium access and error control. It ensures reliable point-to-point and point-to-multipoint connections in a communication network. The network layer takes care of routing the data supplied by the transport layer. The network layer design in WSNs must consider the power efficiency, data-centric communication, data aggregation, etc. The transportation layer helps to maintain the data flow and may be important if WSNs are planned to be accessed through the Internet or other external networks. Depending on the sensing tasks, different types of application software can be set up and used on the application layer.

WSNs must also be aware of the following management planes in order to function efficiently: mobility, power, task, quality of service (QoS) and security management planes. Among them, the functions of task, mobility and power management planes have been elaborated in (Akyildiz et al., 2002). The power management plane is responsible for minimizing power consumption and may turn off functionality in order to preserve energy. The mobility management plane detects and registers movement of nodes so a data route to the sink is always maintained. The task management plane balances and schedules the sensing tasks assigned to the sensing field and thus only the necessary nodes are assigned with sensing tasks and the remainder are able to focus on routing and data aggregation. QoS management in WSNs (Howitt et al., 2006) can be very important if there is a real-time requirement with regard to

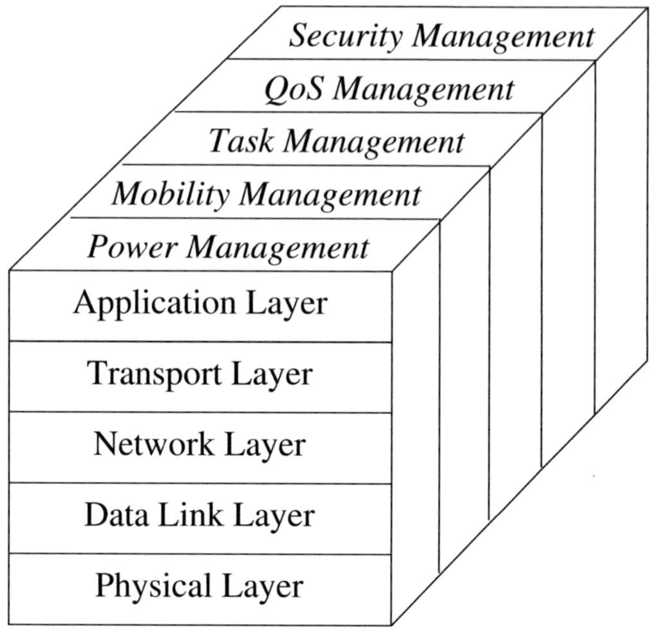

Fig. 2. The Protocol Stack of WSNs (This figure is an extended version of Figure 3 in (Akyildiz et al., 2002)).

the data services. QoS management also deals with fault tolerance, error control and performance optimization in terms of certain QoS metrics. Security management is the process of managing, monitoring, and controlling the security related behavior of a network. The primary function of security management is in controlling access points to critical or sensitive data. Security management also includes the seamless integration of different security function modules, including encryption, authentication and intrusion detection. Please refer to the author's publications (Wang & Zhang, 2008; 2009) for more information about security management in WSNs. It is obvious that networking protocols developed for WSNs must address all five of these management planes.

5. Applications

The original motivation behind the research into WSNs was military application. Examples of military sensor networks include large-scale acoustic ocean surveillance systems for the detection of submarines, self-organized and randomly deployed WSNs for battlefield surveillance and attaching microsensors to weapons for stockpile surveillance (Pister, 2000). As the costs for sensor nodes and communication networks have been reduced, many other potential applications including those for civilian purposes have emerged. The following are a few examples.

5.1 Environmental Monitoring

Environmental monitoring (Steere et al., 2000) can be used for animal tracking, forest surveillance, flood detection, and weather forecasting. It is a natural candidate for applying WSNs (Chong & Kumar, 2003), because the variables to be monitored, e.g. temperature, are usually distributed over a large region. One example is that researchers from the University of Southampton have built a glacial environment monitoring system using WSNs in Norway (Martinez et al., 2005). They collect data from sensor nodes installed within the ice and the sub-glacial sediment without the use of wires which could disturb the environment. Another example is that researchers from EPFL have performed outdoor WSN deployments on a rugged high mountain path located between Switzerland and Italy (Barrenetxea et al., 2008). Their WSN deployment is used to provide spatially dense measures to the Swiss authorities in charge of risk management, and the resulting model will assist in the prevention of avalanches and accidental deaths.

5.2 Health Monitoring

WSNs can be embedded into a hospital building to track and monitor patients and all medical resources. Special kinds of sensors which can measure blood pressure, body temperature and electrocardiograph (ECG) can even be knitted into clothes to provide remote nursing for the elderly. When the sensors are worn or implanted for healthcare purposes, they form a special kind of sensor network called a body sensor network (BSN). BSN is a rich interdisciplinary area which revolutionizes the healthcare system by allowing inexpensive, continuous and ambulatory health monitoring with real-time updates of medical records via the Internet. One of the earliest researches on BSNs was conducted in Imperial College London, where a specialized BSN sensor node and BSN Development Kit have been developed (*BSN Research in Imperial College London*, n.d.).

5.3 Traffic Control

Sensor networks have been used for vehicle traffic monitoring and control for some time. At many crossroads, there are either overhead or buried sensors to detect vehicles and to control the traffic lights. Furthermore, video cameras are also frequently used to monitor road segments with heavy traffic. However, the traditional communication networks used to connect these sensors are costly, and thus traffic monitoring is usually only available at a few critical points in a city (Chong & Kumar, 2003). WSNs will completely change the landscape of traffic monitoring and control by installing cheap sensor nodes in the car, at the parking lots, along the roadside, etc. Streetline, Inc. (*Streetline, Inc.*, n.d.) is a company which uses sensor network technology to help drivers find unoccupied parking places and avoid traffic jams. The solutions provided by Streetline can significantly improve the city traffic management and reduce the emission of carbon dioxide.

5.4 Industrial Sensing

As plant infrastructure ages, equipment failures cause more and more unplanned downtime. The ARC Advisory Group estimates that 5% of production in North America is lost to unplanned downtime. Because sensor nodes can be deeply embedded into machines and there is no infrastructure, WSNs make it economically feasible to monitor the "health" of machines and to ensure safe operation. Aging pipelines and tanks have become a major problem in the oil and gas industry. Monitoring corrosion using manual processes is extremely costly, time consuming, and unreliable. A network of wireless corrosion sensors can be economically deployed to reliably identify issues before they become catastrophic failures. Rohrback Cosasco Systems (RCS) (*Rohrback Cosasco Systems*, n.d.) is the world leader in corrosion monitoring technology and is applying WSNs in their corrosion monitoring. WSNs have also been suggested for use in the food industry to prevent the incidents of contaminating the food supply chain (Connolly & O'Reilly, 2005).

5.5 Infrastructure Security

WSNs can be used for infrastructure security and counterterrorism applications. Critical buildings and facilities such as power plants, airports, and military bases have to be protected from potential invasions. Networks of video, acoustic, and other sensors can be deployed around these facilities (Chong & Kumar, 2003). An initiative in Shanghai Pudong International Airport has involved the installation of a WSN-aided intrusion prevention system on its periphery to deter any unexpected intrusions. The Expo 2010 Shanghai China (*Expo 2010 Shanghai China*, n.d.) has also secured its expo sites with the same intrusion prevention system.

6. Security

While the future of WSNs is very prospective, WSNs will not be successfully deployed if security, dependability and privacy issues are not addressed adequately. These issues become more important because WSNs are usually used for very critical applications. Furthermore, WSNs are very vulnerable and thus attractive to attacks because of their limited prices and human-unattended deployment (Wang & Zhang, 2009).

6.1 Security Threats in WSNs

A typical WSN consists of hundreds or even thousands of tiny and resource-constrained sensor nodes. These sensor nodes are distributedly deployed in uncontrollable environment for the collection of security-sensitive information. Individual sensor nodes rely on multi-hop wireless communication to deliver the sensed data to a remote base station. In a basic WSN scenario, resource constraint, wireless communication, security-sensitive data, uncontrollable environment, and even distributed deployment are all vulnerabilities. These vulnerabilities make WSNs suffer from an amazing number of security threats. WSNs can only be used in the critical applications after the potential security threats are eliminated.

6.1.1 Physical Layer Threats

Comparing WSNs with traditional networks, there are more threats to WSNs in the physical layer, due to the non-tamper-resistant WSN nodes and the broadcasting nature of wireless transmission. Typical types of attacks in the physical layer include physical layer jamming and the subversion of a node.

6.1.2 Link Layer Threats

The data link layer is responsible for the multiplexing of data streams, data frame detection, medium access, and error control. The following attacks can happen in the link layer of WSNs: Link layer jamming; Eavesdropping; Resource exhaustion and traffic analysis.

6.1.3 Network Layer Threats

Threats in the network layer mostly aim at disturbing data-centric and energy efficient multi-hop routing, which is the main design principle in WSNs. The following threats and attacks in this layer are identified in (Wang & Zhang, 2009): Spoofed, altered, or replayed routing information; Sybil attack; Selective forwarding; Sinkhole attack; Wormhole attack and flooding.

6.1.4 Application Layer Threats

Many WSNs' applications heavily rely on coordinated services such as localization, time synchronization, and in-network data processing to collaboratively process data (Sabbah et al., 2006). Unfortunately, these services represent unique vulnerabilities such as: False data filtering; Clock un-synchronization; False data injection.

6.2 Countermeasures

WSN Threats presented above either violate network secrecy and authentication, such as packet spoofing, or violate network availability, such as jamming attack, or violate some other network functionalities. Generally, countermeasures to the threats in WSNs should fulfill the following security requirements (Wang & Zhang, 2009):

- Availability, which ensures that the desired network services are available whenever required.
- Authentication, which ensures that the communication from one node to another node is genuine.
- Confidentiality, which provides the privacy of the wireless communication channels.
- Integrity, which ensures that the message or the entity under consideration is not altered.
- Non-reputation, which prevents malicious nodes to hide or deny their activities.
- Freshness, which implies that the data is recent and ensures that no adversary can replay old messages.
- Survivability, which ensures the acceptable level of network services even in the presence of node failures and malicious attacks.
- Self-security, countermeasures may introduce additional hardware and software infrastructures into the network, which must themselves be secure enough to withstand attacks.

Depending on applications, countermeasures should also fulfill appropriate performance requirements.

6.2.1 Key Management

When setting up a sensor network, one of the first security requirements is to establish cryptographic keys for later secure communication. The established keys should be resilient to attacks and flexible to dynamic update. The task that supports the establishment and maintenance of key relationships between valid parties according to a security policy is called key management. Desired features of key management in sensor networks include energy awareness, localized impact of attacks, and scaling to a large number of nodes.

6.2.2 Authentication

As sensor networks are mostly deployed in human-unattended environments for critical sensing measurements, the authentication of the data source as well as the data are critical concerns. Proper authentication mechanisms can provide WSNs with both sensor and user identification ability, can protect the integrity and freshness of critical data, and can prohibit and identify impersonating attack. Traditionally, authentication can be provided by public-key schemes as digital signature and by symmetric-key schemes as message authentication code (MAC). Besides, key-chain schemes using symmetric keys determined by asymmetric key-exchange protocols are also popular for broadcast authentication in WSNs.

6.2.3 Intrusion Detection

Security technologies, such as authentication and cryptography, can enhance the security of sensor networks. Nevertheless, these preventive mechanisms alone cannot deter all possible attacks (e.g., insider attackers possessing the key). Intrusion detection, which has been successfully used in Internet, can provide a second line of defense.

6.2.4 Privacy Protection

As WSN applications expand to include increasingly sensitive measurements in both military tasks and everyday life, privacy protection becomes an increasingly important concern. For example, few people may enjoy the benefits of a body area WSN, if they know that their personal data such as heart rate, blood pressure, etc., are regularly transmitted without proper privacy protection. Also, the important data sink in a battlefield surveillance WSN may be firstly destroyed, if its location can be traced by analyzing the volume of radio activities.

7. Standardization

In the area of WSNs, several standards are currently either ratified or under development. The major standardization bodies are the Institute of Electrical and Electronics Engineers (IEEE), the Internet Engineering Task Force (IETF), the International Society for Automation (ISA) and the HART Communication Foundation, etc. These standardization bodies have different focuses and they provide global, open standards for interoperable, low-power wireless sensor devices. Table 1 provides the comparisons of different standards currently available for the communication protocols of WSNs.

7.1 IEEE 802.15.4

IEEE 802.15.4 is a standard which specifies the physical layer and MAC layer for low-rate wireless personal area networks. It is the basis for the ZigBee and WirelessHART specification, each of which further attempts to offer a complete networking solution by developing the upper layers which are not covered by the standard.
The features of IEEE 802.15.4 include (*IEEE 802.15 WPAN Task Group 4*, n.d.):

- Data rates of 250 kbps, 40 kbps, and 20 kbps.
- Two addressing modes; 16-bit short and 64-bit IEEE addressing.
- Support for critical latency devices, such as joysticks.
- CSMA-CA channel access.
- Automatic network establishment by the coordinator.
- Fully handshaked protocol for transfer reliability.
- Power management to ensure low power consumption.
- 16 channels in the 2.4GHz ISM band, 10 channels in the 915MHz ISM band and one channel in the 868MHz band.

7.2 Zigbee

ZigBee is a standard for a suite of high level communication protocols based on the IEEE 802.15.4 standard for low power and low data rate radio communications. Zigbee is initiated and maintained by the Zigbee Alliance - a large consortium of industry players. The typical application areas of Zigbee include: Smart energy monitoring; Health care monitoring; Remote control; Building automation and home automation, etc.

7.3 WirelessHART

WirelessHART is an open-standard wireless mesh network communications protocol designed to meet the needs for process automation applications. The protocol utilizes IEEE 802.15.4 compatible DSSS radios and it is operating in the 2.4GHz ISM radio band. On the data link layer, the protocol uses TDMA technology to arbitrate and coordinate communications between devices. WirelessHART provides highly secure communications by using AES-128 block ciphers with individual Join and Session Keys and Data-Link level Network Key. WirelessHART supports the standard HART Application Layer and is compatible with existing HART tools, applications and system integration technology. The other outstanding features of WirelessHART include reliability and scalability. Typically, the communication reliability for a well-formed WirelessHART network is greater than 3σ and normally greater than 6σ. Adding new devices can further improve the network and its communication reliability.

8. Summary

WSNs have been identified as one of the most prospective technologies in this century. This chapter provides information concerning both its history and current state of the art. In concrete terms, the authors provide an overview about the hardware, software and networking protocol design of this important technology. The authors also discuss the security and ongoing standardization of this technology. Depending on applications, many other techniques such as localization, synchronization and in-network processing can be important, which are not discussed in this chapter.

9. References

21 Ideas for the 21st Century (1999). *Business Week* pp. 78–167.

Akyildiz, I. F., Su, W., Sankarasubramaniam, Y. & Cayirci, E. (2002). A survey on sensor networks, *IEEE Communications Magazine* **40**(8): 102–114.

Barrenetxea, G., Ingelrest, F., Schaefer, G. & Vetterli, M. (2008). Wireless sensor networks for environmental monitoring: The sensorscope experience, *Proc. of 20th IEEE International Zurich Seminar on Communications (IZS'08)*.

BSN Research in Imperial College London (n.d.). http://ubimon.doc.ic.ac.uk/bsn/m621.html.

Chong, C.-Y. & Kumar, S. P. (2003). Sensor networks: Evolution, opportunities, and challenges, *Proceedings of the IEEE* **91**(8): 1247–1256.

Connolly, M. & O'Reilly, F. (2005). Sensor networks and the food industry, *Proc. of Workshop on Real-World Wireless Sensor Networks (REALWSN'05)*.

Contiki (n.d.). http://www.sics.se/contiki.

Crossbow Technology (n.d.). http://www.xbow.com.

Dust Networks, Inc. (n.d.). http://www.dustnetworks.com.

Expo 2010 Shanghai China (n.d.). http://www.expo2010.cn.

Hill, J. L. (2003). *System Architecture for Wireless Sensor Networks*, PhD thesis, Doctor of Philosophy in Computer Science, University of California at Berkeley, USA.

Howitt, I., Manges, W. W., Kuruganti, P. T., Allgood, G., Gutierrez, J. A. & Conrad, J. M. (2006). Wireless industrial sensor networks: Framework for qos assessment and qos management, *ISA Transactions* **45**(3): 347–359.

IEEE 802.15 WPAN Task Group 4 (n.d.). http://www.ieee802.org/15/pub/TG4.html.

IPSO Alliance - promoting the use of IP for Smart Objects (n.d.). http://www.ipso-alliance.org.

Kumar, S. & Shepherd, D. (2001). Sensit: Sensor information technology for the warfighter, *Proc. of the 4th International Conference on Information Fusion (FUSION'01)*, pp. 3–9 (TuC1).

LiteOS (n.d.). http://www.liteos.net.

Martinez, K., Padhy, P., Riddoch, A., Ong, H. L. R. & Hart, J. K. (2005). Glacial environment monitoring using sensor networks, *Proc. of Workshop on Real-World Wireless Sensor Networks (REALWSN'05)*.

Myers, C., Oppenheim, A., Davis, R. & Dove, W. (1984). Knowledge-based speech analysis and enhancement, *Proc. of the International Conference on Acoustics, Speech and Signal Processing*.

Ni, L. M. (2008). China's national research project on wireless sensor networks, *Proc. of the 2008 IEEE International Conference on Sensor Networks, Ubiquitous, and Trustworthy Computing (SUTC'08)*, p. 19.

Pister, K. S. J. (2000). Military applications of sensor networks, *of Institute for Defense Analyses Paper P-3531, Defense Science Study Group*.

Proceedings of the Distributed Sensor Nets Workshop (1978). Pittsburgh, USA. Department of Computer Science, Carnegie Mellon University.

Rashid, R. & Robertson, G. (1981). Accent: A communication oriented network operating system kernel, *Proc. of the 8th Symposium on Operating System Principles*, pp. 64–75.

Rohrback Cosasco Systems (n.d.). http://www.cosasco.com.

Sabbah, E., Majeed, A., Kang, K., Liu, K. & AbuGhazaleh, N. (2006). An application driven perspective on wireless sensor network security, *Proc. of the 2nd ACM Workshop on QoS and Security for Wireless and Mobile Networks*.

SOS Embedded Operating System (n.d.). https://projects.nesl.ucla.edu/public/sos-2x/doc/.

Steere, D., Baptista, A., McNamee, D., Pu, C. & Walpole, J. (2000). Research challenges in environmental observation and forecasting systems, *Proc. of 6th International Conference on Mobile Computing and Networking (MOBICOMM'00)*, pp. 292–299.

Streetline, Inc. (n.d.). http://www.streetlinenetworks.com.

Sudevalayam, S. & Kulkarni, P. (2008). Energy Harvesting Sensor Nodes: Survey and Implications, *Technical Report TR-CSE-2008-19*, Department of Computer Science and Engineering, Indian Institute of Technology Bombay.

TinyOS Community Forum (n.d.). http://www.tinyos.net.

Wang, Q. & Zhang, T. (2008). Sec-snmp: Policy-based security management for sensor networks, *Proc. of the International Conference on Security and Cryptography (SECRYPT'08), in conjunction with ICETE 2008*.

Wang, Q. & Zhang, T. (2009). A survey on security in wireless sensor networks, *in* Y. Zhang & P. Kitsos (eds), *Security in RFID and Sensor Networks*, CRC Press, Taylor & Francis Group, chapter 14, pp. 293–320.

ZigBee Alliance (n.d.). http://www.zigbee.org.

Type	IEEE 802.11b/g	Bluetooth	UWB-IR	IEEE 802.15.4	Zigbee	Wireless HART	IEEE 802.15.6[a]
Band	ISM 2.4GHz	ISM 2.4GHz	3.1GHz-10.6GHz	ISM 2.4GHz, 915MHz, 868MHz	ISM 2.4GHz	ISM 2.4GHz	400MHz, ISM 2.4GHz, 3.1-10.6GHz
Spreading	DSSS	FHSS/TDD	Baseband	DSSS	DSSS	DSSS	DSSS
Modulation	BPSK, QPSK, CCK, OFDM	GFSK	Impulse radio, time-domain	O-QPSK	O-QPSK	O-QPSK	Group PPM
Range	100m	10m	<5m	10m	10m	<250m	3-10m
Rate	<54Mbps	1Mbps	20Kbps, 250Kbps, 10Mbps	250Kbps	250Kbps	250Kbps	<10Mbps
Power	High	Low	Ultra low	Very low	Very low	Battery or line power	Ultra low
Roaming	Yes	No	Yes	Yes	Yes	Yes	Yes
No. of nodes	32 (per access point)	8 (per piconet)	10-1000	<65536	<65536	5-65536	<256
Power consumption	Medium	Low	Ultra low	Very low	Very low	Very low	Ultra low
Complexity	Complex	Very complex	Simple transmitter but complex receiver	Simple	Simple	Simple	Simple
Security	WEP, WPA	64bit or 128 bit	128 bit	NULL, 32bit, 64bit or 128bit	128 bit	128 bit	Scrambled mapping code
Target cost	High	Medium	Very low	Low	Low	Low	Low

Table 1. Comparisons of different standards available for the communication protocols of WSNs.

[a] All values in this column are not official. IEEE 802.15.6 is a communication standard optimized for wireless body area networks and it is still under active development.

2

Wireless Sensor Networks for On-field Agricultural Management Process

Luca Bencini, Davide Di Palma, Giovanni Collodi and Gianfranco Manes
Department of Electronics and Telecommunications - University of Florence
Italy

Antonio Manes
Netsens S.r.l.
Italy

1. Introduction

Agriculture is one of the most ancient activities of man in which innovation and technology are usually accepted with difficulty, unless real and immediate solutions are found for specific problems or for improving production and quality. Nevertheless, a new approach of gathering information from the environment could represent an important step towards high quality and eco-sustainable agriculture.

Nowadays, irrigation, fertilization and pesticides management are often left to the farmer's and agronomist's discretion: common criteria used to guarantee safe culture and plant growth are often giving a greater amount of chemicals and water than necessary. There is no direct feedback between the decision of treating or irrigating plants and the real effects in the field. Plant conditions are usually committed to sporadic and faraway weather stations that cannot provide accurate and local measurements of the fundamental parameters in each zone of the field. Also, agronomic models, based on these monitored data, cannot provide reliable information. On the contrary, agriculture needs detailed monitoring in order to obtain real time feedback between plants, local climate conditions and man's decisions.

The concept of precision agriculture has been around for some time now. Blackmore et al., in 1994 (Blackmore, 1994) defined it as a comprehensive system designed to optimize agricultural production by carefully tailoring soil and crop management to correspond to the unique condition found in each field while maintaining environmental quality. The early adopters during that time found precision agriculture to be unprofitable and the instances in which it was implemented were few and far between. Further, the high initial investment in the form of electronic equipment for sensing and communication meant that only large farms could afford it. The technologies proposed at this point comprised three aspects: *Remote Sensing* (RS), *Global Positioning System* (GPS) and *Geographical Information System* (GIS). RS coupled with GPS coordinates produced accurate maps and models of the agricultural fields. The sampling was typically through electronic sensors such as soil probes and remote optical scanners from satellites. The collection of such data in the form of electronic computer databases gave birth to the GIS. Statistical analyses were then conducted on the data and the variability of agricultural land was charted with respect to its properties. The technology, apart from being

non-real-time, involved the use of expensive technologies like satellite sensing and was labor intensive where the maps charting the agricultural fields were mostly manually done.
Over the last seven years, the advancement in sensing and communication technologies has significantly brought down the cost of deployment and running of a feasible precision agriculture framework. Emerging wireless technologies with low power needs and low data rate capabilities have been developed which perfectly suit precision agriculture (Wang et al., 2006). The sensing and communication can now be done on a real-time basis leading to better response times. The wireless sensors are cheap enough for wide spread deployment and offer robust communication through redundant propagation paths (Akyildiz & Xudong, 2005). Thanks to these features, the *Wireless Sensor Networks* (WSNs) (Akyildiz & Xudong, 2005) have become the most suitable technology to fit an invasive method of monitoring the agricultural environment.
In this chapter, an end-to-end monitoring WSN technology-based solution is presented, joining hardware optimization with communications protocols design and a suitable interface. In particular Section 2 discusses the system requirements and illustrates the overall system characterization in conjunction with related work. Sections 3, 4 and 5 deal respectively with the system in terms of hardware, protocol and software design. Section 6 describes the actual experiences, focusing on several case study analyses for highlighting the effectiveness and accurateness of the developed system. Sections 7 and 8 describe respectively the commercial system "VineSense", born from the experimental solution, and some agronomic results. Finally, in Section 9 some conclusions are drawn in order to explain the future direction of the current research study.

2. System Requirement and Architecture

The requirements that adopting a WSN are expected to satisfy in effective agricultural monitoring concern both *system level* issues (i.e., unattended operation, maximum network life time, adaptability or even functionality and protocol self-reconfigurability) and *final user* needs (i.e., communication reliability and robustness, user friendliness, versatile and powerful graphical user interfaces). The most relevant mainly concerns the supply of *stand-alone* operations. To this end, the system must be able to run unattended for a long period, as nodes are expected to be deployed in zones that are difficult to maintain. This calls for optimal energy management ensuring that the energy spent is directly related to the amount of traffic handled and not to the overall working time. In fact, energy is nevertheless a limited resource and the failure of a node may compromise WSN connectivity as the network gets partitioned. Other issues to be addressed are the capabilities of quickly setting-up an end-to-end communication infrastructure, supporting both synchronous and asynchronous queries, and of dynamically reconfiguring it. An additional requirement is *robust* operative conditions, which need fault management since a node may fail for several reasons. Other important properties are *scalability* and *adaptability* of the network's topology, in terms of the number of nodes and their density in unexpected events with a higher degree of responsiveness and reconfigurability. This also implies the development of a *plug and play* sensor interface and the provisioning of remote connectivity. Finally, several user-oriented attributes, including *fairness, latency, throughput* and enhanced data querying schemes (i.e., *time-driven* (Al-Karaki & Kamal, 2004) or *query-driven*) need to be taken into account even if they could be considered secondary with respect to our application purposes because the WSN's cost/performance trade-off (Langendoen & Halkes, 2004).

A WSN system was developed according to the afore mentioned requirements. The system, shown in Fig. 1, comprises a self-organizing mesh WSN endowed with sensing capabilities, a GPRS Gateway, which gathers data and provides a TCP-IP based connection toward a Remote Server, and a Web Application, which manages information and makes the final user capable of monitoring and interacting with the instrumented environment.

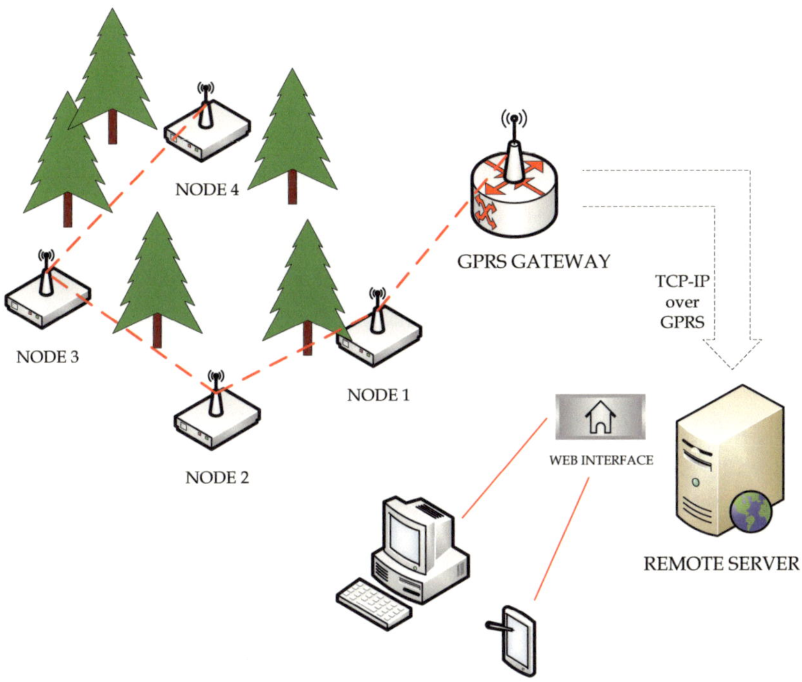

Fig. 1. Wireless Sensor Network System

3. Hardware Design

Focusing on an end-to-end system architecture, every constitutive element has to be selected according to application requirements and scenario issues, especially regarding the hardware platform. Many details have to be considered, involving the energetic consumption of the sensor readings, the power-on and power-save status management and a good trade-off between the maximum radio coverage and the transmitted power. After an accurate investigation of the out-of-the-shelf solutions, 868 MHz *Mica2 motes* (Mica2 Series, 2002) were adopted according to these constraints and to the reference scenarios. The *Tiny Operative System* (TinyOS) running on this platform ensures full control of mote communication capabilities to attain optimized power management and provides necessary system portability towards future hardware advancements or changes. Nevertheless, Mica2 motes are far from perfection, especially in the RF section, since the power provided by the transceiver (*Chipcon CC1000*) is not completely available for transmission. However, it is lost to imperfect coupling with the antenna, thus reducing the radio coverage area. An improvement of this section was performed, using more suitable antennas and coupling circuits and increasing the transmitting power with a power amplifier, thus increasing the output power up to 15 dBm while respecting international restrictions and standards. These optimizations allow for greater radio coverage (about

200 m) and better power management. In order to manage different kinds of sensors, a compliant sensor board was adopted, allowing up to 16 sensor plugs on the same node;, this makes a single mote capable of sensing many environmental parameters at a time (Mattoli et al., 2005). Sensor boards recognize the sensors and send *Transducer Electronic Datasheets* (TEDS) through the network up to the server, making it possible for the system to recognize an automatic sensor. The overall node stack architecture is shown in Fig. 2.

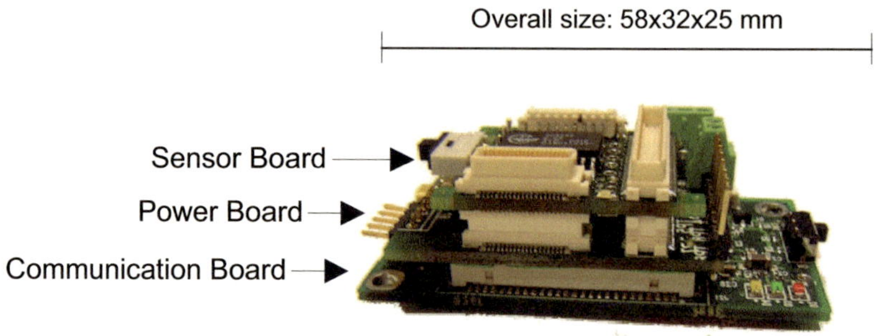

Fig. 2. Node Stack Architecture

The GPRS embedded Gateway, shown in Fig. 3, is a stand-alone communication platform designed to provide transparent, bi-directional wireless TCP-IP connectivity for remote monitoring. In conjunction with *Remote Data Acquisition* (RDA) equipment, such as WSN, it acts when connected with a Master node or when directly connected to sensors and transducers (i.e., Stand-Alone weather station, Stand-Alone monitoring camera).

Fig. 3. GPRS Gateway

The main hardware components that characterize the gateway are:

- a miniaturized GSM/GPRS modem, with embedded TCP/IP stack (Sveda et al., 2005), (Jain et al., 1990);
- a powerful 50 MHz clock microcontroller responsible for coordinating the bidirectional data exchange between the modem and the master node to handle communication with the Remote Server;

- an additional 128 KB SRAM memory added in order to allow for data buffering, even if the wide area link is lost;
- several A/D channels available for connecting additional analog sensors and a battery voltage monitor.

Since there is usually no access to a power supply infrastructure, the hardware design has also been oriented to implement low power operating modalities, using a 12 V rechargeable battery and a 20 W solar panel.
Data between the Gateway and Protocol Handler are carried out over TCP-IP communication and encapsulated in a custom protocol; from both local and remote interfaces it is also possible to access part of the Gateway's configuration settings. The low-level firmware implementation of communication protocol also focuses on facing wide area link failures. Since the gateway is always connected with the Remote Server, preliminary connectivity experiments demonstrated a number of possible inconveniences, most of them involving the *Service Provider Access Point Name* (APN) and *Gateway GPRS Support Node* (GGSN) subsystems. In order to deal with these drawbacks, custom procedures called *Dynamic Session Re-negotiation* (DSR) and *Forced Session Re-negotiation* (FSR), were implemented both on the gateway and on the CMS server. This led to a significant improvement in terms of disconnection periods and packet loss rates.
The DSR procedure consists in a periodical bi-directional control packet exchange, aimed at verifying the status of uplink and downlink channels on both sides (gateway and CMS). This approach makes facing potential deadlocks possible if there is asymmetric socket failure, which is when one device (acting as client or server) can correctly deliver data packets on the TCP/IP connection but is unable to receive any. Once this event occurs (it has been observed during long GPRS client connections, and is probably due to Service Provider Access Point failures), the DSR procedure makes the client unit to restart the TCP socket connection with the CMS.
Instead, the FSR procedure is operated on the server side when no data or service packets are received from a gateway unit and a fixed timeout elapses: in this case, the CMS closes the TCP socket with that unit and waits for a new reconnection. On the other side, the gateway unit should catch the close event exception and start a recovery procedure, after which a new connection is re-established. If the close event should not be signaled to the gateway (for example, the FSR procedure is started during an asymmetric socket failure), the gateway would anyway enter the DSR recovery procedure.
In any case, once the link is lost, the gateway unit tries to reconnect with the CMS until a connection is re-established.

4. Protocol Design

The most relevant system requirements, which lead the design of an efficient Medium Access Control (MAC) and routing protocol for an environmental monitoring WSN, mainly concern power consumption issues and the possiblity of a quick set-up and end-to-end communication infrastructure that supports both synchronous and asynchronous queries. The most relevant challenge is to make a system capable of running unattended for a long period, as nodes are expected to be deployed in zones that are difficult to maintain. This calls for optimal energy management since a limited resource and node failure may compromise WSN connectivity. Therefore, the MAC and the network layer must be perfected ensuring that the energy used is directly related to the amount of handled traffic and not to the overall working time.

Other important properties are scalability and adaptability of network topology, in terms of number of nodes and their density. As a matter of fact, some nodes may either be turned off may join the network afterward.
Taking these requirements into account, a MAC protocol and a routing protocol were implemented.

4.1 MAC Layer Protocol

Taking the IEEE 802.11 Distributed Coordination Function (DCF) (IEEE St. 802.11, 1999) as a starting point, several more energy efficient techniques have been proposed in literature to avoid excessive power waste due to so called idle listening. They are based on periodical preamble sampling performed at the receiver side in order to leave a low power state and receive the incoming messages, as in the WiseMAC protocol (El-Hoiydi et al., 2003). Deriving from the classical contention-based scheme, several protocols (S-MAC (Ye et al., 2002), TMAC (Dam & Langendoen, 2003) and DMAC (Lu et al., 2004)) have been proposed to address the overhead idle listening by synchronizing the nodes and implementing a duty cycle within each slot.
Resorting to the above considerations, a class of MAC protocols was derived, named *Synchronous Transmission Asynchronous Reception* (STAR) which is particularly suited for a flat network topology and benefits from both WiseMAC and S-MAC schemes. More specifically, due to the introduction of a duty-cycle, it joins the power saving capability together with the advantages provided by the offset scheduling, without excessive overhead signaling. According to the STAR MAC protocol, each node might be either in an idle mode, in which it remains for a time interval T_l (*listening time*), or in an energy saving sleeping state for a T_s (*sleeping time*). The transitions between states are synchronous with a period frame equal to $T_f = T_l + T_s$ partitioned in two sub-intervals; as a consequence, a duty-cycle function can also be introduced:

$$d = \frac{T_l}{T_l + T_s} \tag{1}$$

To provide the network with full communication capabilities, all the nodes need to be *weakly synchronized*, meaning that they are aware at least of the awaking time of all their neighbors. To this end, as Fig. 4 shows, a node sends a *synchronization message* (SYNC) frame by frame to each of its neighbor nodes known to be in the listening mode (Synchronous Transmission), whereas, during the set-up phase in which each node discovers the network topology, the control messages are asynchronously broadcasted. On the other hand, its neighbors periodically awake and enter the listening state independently (Asynchronous Reception). The header of the synchronization message contains the following fields: a unique node identifier, the message sequence number and the *phase*, or the time interval after which the sender claims to be in the listening status waiting for both synchronization and data messages from its neighbors. If the node is in the sleeping status, the phase ϕ is evaluated according to the following rule:

$$\phi_1 = \tau - T_l \tag{2}$$

where τ is the time remaining to the next frame beginning. Conversely, if the mote is in the listening status, ϕ is computed as:

$$\phi_2 = \tau + T_s \tag{3}$$

In order to fully characterize the STAR MAC approach, the related energy cost normalized can be evaluated as it follows:

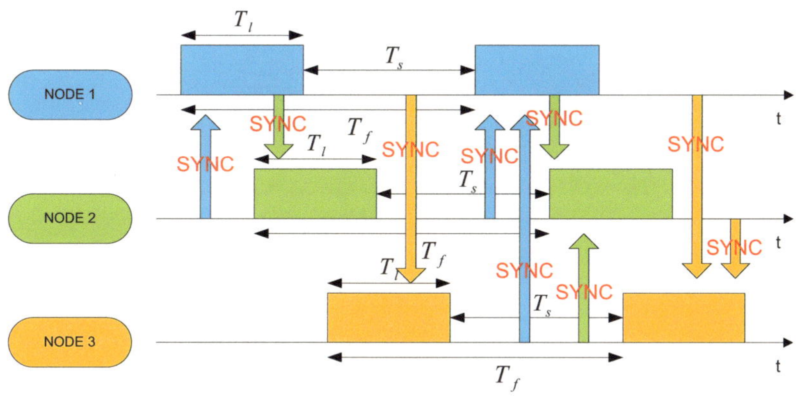

Fig. 4. STAR MAC Protocol Synchronization Messages Exchange

$$C = c_{rx} d T_f + c_{sleep}[T_f(1-d) - NT_{pkt}] + NC_{tx} \quad [mAh] \tag{4}$$

where c_{sleep} and c_{rx} represent the sleeping and the receiving costs $[mA]$ and C_{tx} is the single packet transmission costs $[mAh]$, T_f is the frame interval $[s]$, d is the duty cycle, T_{pkt} is the synchronization packet time length $[s]$ and finally N is the number of neighbors. When the following inequality is hold:

$$NT_{pkt} \ll T_f \tag{5}$$

then:

$$C \simeq c_{rx} d T_f + c_{sleep} T_f(1-d) + NC_{tx} \quad [mAh] \tag{6}$$

The protocol cost normalized to the synchronization time is finally:

$$\frac{C}{T_f} = c_{rx} d + c_{sleep}(1-d) + \frac{NC_{tx}}{Tf} \quad [mA] \tag{7}$$

As highlighted in Table 1, it usually happens that $c_{tx} \ll c_{sleep} \ll c_{rx}$, where $c_{tx} = C_{tx}/T_{pkt}$ and T_{pkt} is the packet transmission time $[s]$ assumed equal to 100 ms as worst case. This means that the major contribution to the overall cost is represented by the listening period that the STAR MAC protocol tries to suitably minimize.

c_{rx}	12 mA
c_{sleep}	0.01 mA
C_{tx}	30 mAh
c_{tx}	0.001 mA

Table 1. Power Consumption Parameters for the Considered Platform.

In Fig. 5(a) the normalized cost versus the number of neighbor nodes is shown for the S-MAC and STAR MAC schemes. It is worth noticing that the performance of the proposed protocol is better with respect to the existing approach for a number of neighbor nodes greater than 7. In Fig. 5(b) the normalized costs of S-MAC and STAR MAC approaches are compared with

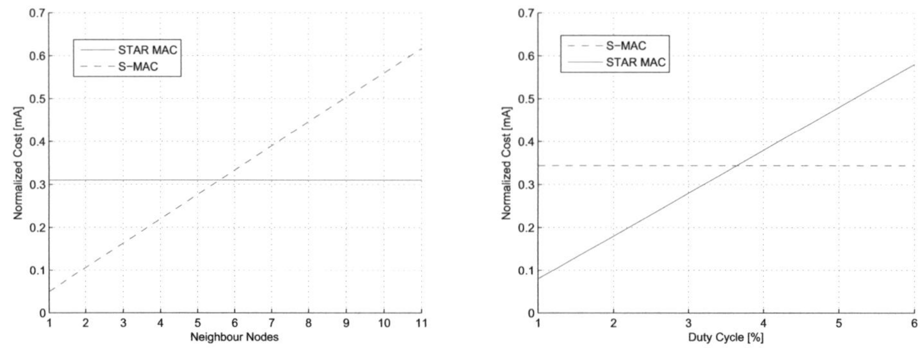

(a) Normalized Cost vs. Neighbor Nodes (b) Normalized Cost vs. Duty Cycle Duration

Fig. 5. STAR MAC Performance

respect to the duty cycle duration for a number of neighbor nodes equal to 8. It is possible to notice that for $d < 3.5\%$ the proposed protocol provide a significant gain.
Nevertheless, for densely deployed or high traffic loaded WSN, STAR MAC approach might suffers the shortcoming of cost increasing due to the large number of unicasted messages. To limit this effect, an enhanced approach, named STAR+, was introduced, aiming at minimizing also the packet transmission cost. According to it, only one synchronization packet is multicasted to all the neigh- bor nodes belonging to a subset, i.e., such that they are jointly awake for a time interval greater than T_l. This leads to an additional advantage, as the number of neighbors increases allowing better performance with respect to scalability and a power saving too. Besides, the synchronization overhead is reduced with a consequent collisions lowering. Under this hypothesis the normalized cost might be expressed as:

$$\frac{C}{T_f} = c_{rx}d + c_{sleep}(1-d) + \frac{KC_{tx}}{Tf} \quad [mA] \tag{8}$$

where K is the number of subsets. Since $K \leq N$, the normalized cost results to be remarkably lowered, especially if number of nodes and duty-cycle get higher, even if the latter case is inherently power consuming.

4.2 Network Layer Protocol

In order to evaluate the capability of the proposed MAC scheme in establishing effective end-to-end communications within a WSN, a routing protocol was introduced and integrated according to the *cross layer* design principle (Shakkottai et al., 2003). In particular, we refer to a proactive algorithm belonging to the class link-state protocol that enhance the capabilities of the *Link Estimation Parent Selection* (LEPS) protocol. It is based on periodically information needed for building and maintaining the local routing table, depicted in Table 2. However, our approach resorts both to the signaling introduced by the MAC layer (i.e., synchronization message) and by the Network layer (i.e., ping message), with the aim of minimizing the overhead and make the system more adaptive in a cross layer fashion. In particular, the parameters transmitted along a MAC synchronization message, with period T_f, are the following:

- *next hop* (NH) to reach the gateway, that is, the MAC address of the one hop neighbor;

- *distance* (HC) to the gateway in terms of number of needed hops;
- *phase* (PH) that is the schedule time at which the neighbor enter in listening mode according to Equation (2) and Equation (3);
- *link quality* (LQ) estimation as the ratio of correctly received and the expected synchronization messages from a certain neighbor.

Target	NH	HC	PH	LQ	BL	CL
Sink 1	A	N_A	ϕ_A	η_A	B_A	C_A
	B	N_B	ϕ_B	η_B	B_B	C_B
Sink 2	C	N_C	ϕ_C	η_C	B_C	C_C
	D	N_D	ϕ_D	η_D	B_D	C_D

Table 2. Routing Table General Structure

On the other hand, the parameters related to long-term phenomena are carried out by the ping messages, with period $T_p \gg T_f$, in order to avoid unnecessary control traffics and, thus, reducing congestion. Particularly, they are:

- *battery level* (BL) (i.e., an estimation of the energy available at that node);
- *congestion level* (CL) in terms of the ratio between the number of packets present in the local buffer and the maximum number of packets to be stored in.

Once, the routing table has been filled with these parameters, it is possible to derive the proper metric by means of a weighted summation of them. It is worth mentioning that the routing table might indicate more than one destination (*sink*) thanks to the ping messages that keep trace of the intermediate nodes within the message header.

5. Software and End User Interface Design

The software implementation was developed, considering a node as both a single element in charge of accomplishing prearranged tasks and as a part of a complex network in which each component plays a crucial role in the network's maintenance. As far as the former aspect is concerned, several TinyOS modules were implemented for managing high and low power states and for realizing a finite state machine, querying sensors at fixed intervals and achieving anti-blocking procedures, in order to avoid software failure or deadlocks and provide a robust stand alone system. On the other hand, the node has to interact with neighbors and provide adequate connectivity to carry the messages through the network, regardless of the destination. Consequently, additional modules were developed according to a cross layer approach that are in charge of managing STAR MAC and multihop protocols. Furthermore, other modules are responsible for handling and forwarding messages, coming from other nodes or from the gateway itself. Messages are not only sensing (i.e., measures, battery level) but also control and management messages (i.e., synchronization, node reset). As a result, a full interaction between the final user and the WSN is guaranteed.
The final user may check the system status through graphical user interface (GUI) accessible via web. After the log-in phase, the user can select the proper pilot site. For each site the deployed WSN together with the gateway is schematically represented through an interactive map. In addition to this, the related sensors display individual or aggregate time diagrams

for each node with an adjustable time interval (Start/Stop) for the observation. System monitoring could be performed both at a high level with a user friendly GUI and at a low level by means of message logging.
Fig. 6 shows some friendly Flash Player applications that, based on mathematical models, analyze the entire amount of data in a selectable period and provide ready-to-use information. Fig. 6(a) specifically shows the aggregate data models for three macro-parameters, such as vineyard water management, plant physiological activity and pest management. The application, using cross light colors for each parameter, points out normal (green), mild (yellow) or heavy (red) stress conditions and provides suggestions to the farmer on how to apply pesticides or water in a certain part of the vineyard. Fig. 6(b) shows a graphical representation of the soil moisture measurement. Soil moisture sensors positioned at different depths in the vineyard make it possible to verify whether a summer rain runs off on the soil surface or seeps into the earth and provokes beneficial effects on the plants: this can be appreciated with a rapid look at the soil moisture aggregate report which, shows the moisture sensors at two depths with the moisture differences colore in green tones. Fig. 6(c) highlights stress conditions on plants, due to dry soil and/or to hot weather thanks to the accurate trunk diametric growth sensor that can follow each minimal variation in the trunk giving important information on plant living activity. Finally, Fig. 6(d) shows a vineyard map: the green spots are wireless units, distributed in a vineyard of one hectare.

6. Real World Experiences

The WSN system described above was developed and deployed in three pilot sites and in a greenhouse. Since 2005, an amount of 198 sensors and 50 nodes have continuously sent data to a remote server. The collected data represents a unique database of information on grape growth useful for investigating the differences between cultivation procedures, environments and treatments.

6.1 Pilot Sites Description

The first pilot site was deployed in November 2005 on a sloped vineyard of the Montepaldi farm in Chianti Area (Tuscany - Italy). The vineyard is a wide area where 13 nodes (including the master node) with 24 sensors, running STAR MAC and dynamic routing protocols were successfully deployed. The deployment took place in two different steps: during the first one, 6 nodes (nodes 9,10,14,15,16,17) were placed to perform an exhaustive one week test. The most important result regards the multi-hop routing efficiency, estimated as:

$$\eta_{MHop} = \frac{M_{EU}}{M_{ex}} \tag{9}$$

where η_{MHop} is the efficiency, M_{EU} are the messages correctly received by the remote user and M_{ex} are the expected transmitted messages. For the gateway neighbors, η_{MHop} is very high, over 90%. However, even nodes far from the gateway (i.e., concerning an end-to-end multihop path) show a message delivery rate (MDR) of over 80%. This means that the implemented routing protocol does not affect communication reliability. After the second deployment, in which nodes 11,12,13,18,19,20 were arranged, the increased number of collisions changed the global efficiency, thus decreasing the messages that arrived to the end user, except for nodes 18,19,20, in which an upgraded firmware release was implemented. The related results are detailed in Table 3.

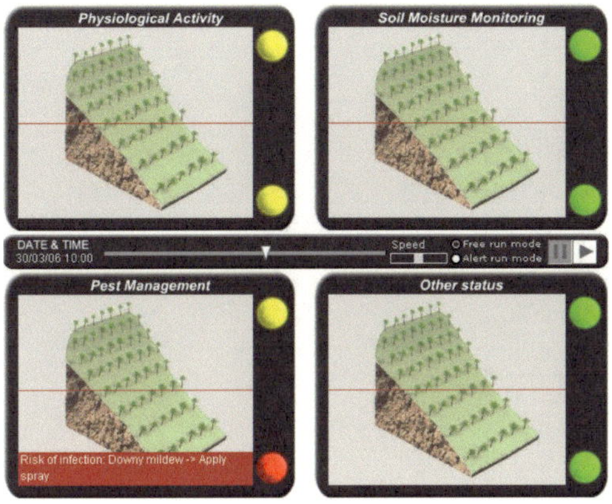

(a) Aggregate Data Models for Vineyard Water Management, Plant Physiological Activity and Pest Management

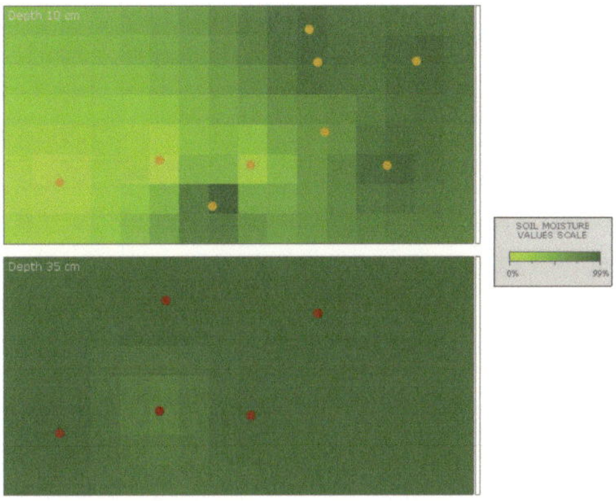

(b) Soil Moisture Aggregation Report: the upper map represent soil moisture @ 10 cm in the soil and the lower map represents soil moisture @ 35cm in the vineyard after a slipping rain

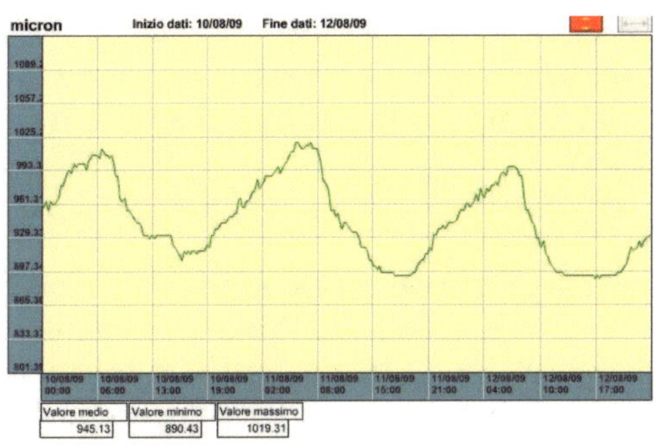

(c) Trunk Diametric Growth Diagram: daily and nightly metabolic phases

(d) Distributed Wireless Nodes in a Vineyard

Fig. 6. Flash Player User Interface

This confirms the robustness of the network installed and the reliability of the adopted communications solution, also considering the power consumption issues: batteries were replaced on March 11th 2006 in order to face the entire farming season. After that, eleven months passed before the first battery replacement occurred on February 11th 2007, confirming our expectations and fully matching the user requirements. The overall Montepaldi system has been running unattended for one year and a half and is going to be a permanent pilot site. So far, nearly 2 million samples from the Montepaldi vineyard have been collected and stored in the server at the University of Florence Information Services Centre (CSIAF), helping agronomist experts improve wine quality through deeper insight on physical phenomena (such as weather and soil) and the relationship with grape growth.
The second pilot site was deployed on a farm in the Chianti Classico with 10 nodes and 50 sensors at about 500 *m* above sea level on a stony hill area of 2.5 hectares. The environmental

Location	MDR
Node 9	72.2%
Node 10	73.7%
Node 11	88.5%
Node 12	71.4%
Node 13	60.4%
Node 14	57.2%
Node 15	45.6%
Node 16	45.4%
Node 17	92.1%
Node 18	87.5%
Node 19	84.1%

Table 3. Message Delivery Rate for the Montepaldi Farm Pilot Site

variations of the the "terroir" have been monitored since July 2007, producing one of the most appreciated wines in the world.
Finally, the third WSN was installed in Southern France in the vineyard of Peach Rouge at Gruissan. High sensor density was established to guarantee measurement redundancy and to provide a deeper knowledge of the phenomena variation in an experimental vineyard where micro-zonation has been applied and where water management experiments have been performed for studying plant reactions and grape quality.

6.2 Greenhouse

An additional deployment at the University of Florence Greenhouse was performed to let the agronomist experts conduct experiments even in seasons like Fall and Winter, where plants are quiescent, thus breaking free from the natural growth trend. This habitat also creates the opportunity to run several experiments on the test plants, in order to evaluate their responses under different stimuli using in situ sensors.
The greenhouse environmental features are completely different from those of the vineyard: as a matter of fact, the multipath propagation effects become relevant, due to the indoor scenario and the presence of a metal infrastructure. A highly dense node deployment, in terms of both nodes and sensors, might imply an increased network traffic load. Nevertheless, the same node firmware and hardware used in the vineyard are herein adopted; this leads to a resulting star topology as far as end-to-end communications are concerned.
Furthermore, 6 nodes have been in the greenhouse since June 2005, and 30 sensors have constantly monitored air temperature and humidity, plants soil moisture and temperature, differential leaf temperature and trunk diametric growth. The sensing period is equal to 10 minutes, less than the climate/plant parameter variations, providing redundant data storage. The WSN message delivery rate is extremely high: the efficiency is over 95%, showing that a low number of messages are lost.

7. VineSense

The fruitful experience of the three pilot sites was gathered by a new Italian company, Netsens, founded as a spinoff of the University of Florence. Netsens has designed a new monitoring

system called VineSense based on WSN technology and oriented towards market and user applications.
VineSense exalts the positive characteristics of the experimental system and overcomes the problems encountered in past experiences, thus achieving an important position in the wireless monitoring market.
The first important outcome of the experimental system, enhanced by VineSense is the idea of an end-to-end system. Sensors deployed in the field constantly monitor and send measurements to a remote server through the WSN. Data can be queried and analyzed by final users thanks to the professional and user-friendly VineSense web interface. Qualified mathematic models are applied to monitoring parameters and provide predictions on diseases and plant growth, increasing agronomists' knowledge and reducing costs while paving the road for new vineyard management.
VineSense improves many aspects of the experimental system, both in electronics and telecommunications.
The MAC and Routing protocol tested in the previous experimental system showed such important and significant results in terms of reliability that the same scheme was also adopted in the VineSense system and minimal changes were introduced: the routing protocol is lighter in terms of data exchange, building the route with different parameters, aimed at increasing the message success rate, such as master node distance and received signal strength.
A more secure data encryption was adopted in data messages to protect customers from malicious sniffing or to discourage possible competitors from decrypting network data.
Furthermore, a unique key-lock sequence was also implemented on each wireless node to prevent stealing, ensuring correct use with only genuine Netsens products and only in combination with its master node, which comes from the factory.
The new wireless nodes are smaller, more economical, more robust and suited for vineyard operations with machines and tractors. The electronics are more fault-tolerant, easier to install and more energy efficient: only a 2200 mAh lithium battery for 2-3 years of continuous running without human intervention. Radio coverage has been improved up to 350 m and nodes deployment can be easily performed by end users who can rely on a smart installation system with instantaneous radio coverage recognition. Some users have also experimented with larger area coverage, measuring a point-to-point communication of about 600 m in the line of sight.
Hypothetically, a VineSense system could be composed of up to 255 wireless nodes and more than 2500 sensors, considering a full sensor set per node, but since it is a commercial system these numbers are much more than necessary to cover farmers' needs.
Sensors used in the VineSense system are low-cost, state-of-the-art devices designed by Netsens for guaranteeing the best accuracy-reliability-price ratio. The choice of Netsens to develop custom and reliable sensors for the VineSense system is not only strategic from a marketing point of view, since it frees VineSense from any kind of external problems, such as external supplying, delays, greater costs and compliancy. It is also a consequence of the "System Vision", where VineSense is not only a wireless communication system product, but an entire system with no "black holes" inside so as to provide the customer with a complete system with better support.
The VineSense wireless-sensor unit is shown in Fig. 7(a).
Recovery strategies and communication capabilities of the stand-alone GPRS gateway have been improved: in fact, data received by wireless nodes are both forwarded in real-time to a remote server and temporarily stored on board in case of abrupt disconnections; moreover,

automatic reset and restart procedures avoid possible software deadlocks or GPRS network failures. Finally, a high-gain antenna guarantees good GPRS coverage almost everywhere.
The GPRS gateway firmware has been implemented for remotely managing of the acquisition settings, relieving users of the necessity of field maintenance.
The GPRS gateway communication has been greatly improved introducing new different communication interfaces, such as Ethernet connection (RJ-45), USB data downloading and the possibility of driving an external Wi-Fi communication system for short-range transmissions.
Since the beginning of 2010, the "Always On" connection started to be fully used and it boosted the VineSense system, enlarging its possible field of application: a complete bidirectional communication was established between the GPRS unit in the field and the remote server at Netsens. The previous "one way" data flow, from the vineyard to the internet, was gone over by a new software release, able to send instantaneous messages from the VineSense web interface to the field: the monitoring system was changed into the monitoring and control system, sending automatic, scheduled or asynchronous commands to the gateway station or to nodes, i.e. to open or close irrigation systems or simply to download a firmware upgrade.
In Fig. 7(b) the GPRS gateway with weather sensors is shown.

(a) VineSense Wireless Unit

(b) VineSense GPRS Gateway with Weather Sensors

Fig. 7. VineSense Hardware Elements

The web interface is the last part of VineSense's end-to-end: the great amount of data gathered by the sensors and stored in the database needs a smart analysis tool to become useful and usable. For this reason different tools are at the disposal of various kinds of users. On one hand, some innovative tools such as control panels for real time monitoring or 2D chromatic maps create a quick and easy approach to the interface. On the other hand, professional plots and data filtering options allow experts or agronomists to study them more closely.

8. Agronomic Results

The use of VineSense in different scenarios with different agronomic aims has brought a large amount of important results.

When VineSense is adopted to monitor soil moisture positive effects can be obtained for plants and saving water, thus optimizing irrigation schedules. Some examples of this application can be found in systems installed in the Egyptian desert where agriculture is successful only through wise irrigation management. In such a terroir, plants suffer continuous hydric stress during daylight due to high air temperature, low air humidity and hot sandy soils with a low water retention capacity. Water is essential for plant survival and growth, an irrigation delay can be fatal for the seasonal harvest therefore, a reliable monitoring system is necessary. The adoption of VineSense in this scenario immediately resulted in continuous monitoring of the irrigating system, providing an early warning whenever pump failure occurred. On the other hand, the possibility to measure soil moisture at different depths allows agronomists to decide on the right amount of water to provide plants; depending on different day temperatures and soil moisture, pipe schedules can be changed in order to reduce water waste and increase water available for plants.
An example of different pipe schedules is shown in Fig. 8.

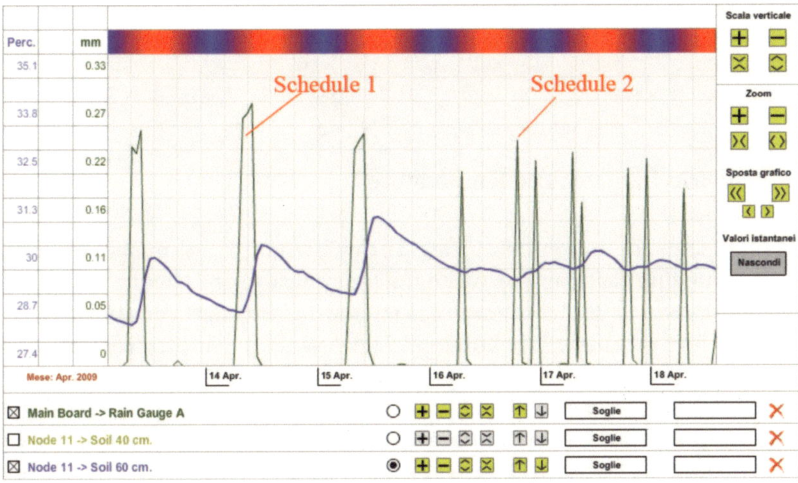

Fig. 8. Different Pipe Schedules in Accordance with Soil Moisture Levels

Originally, the irrigation system was opened once a day for 5 hours giving 20 liters per day (schedule 1); since sandy soils reach saturation very rapidly most of this water was wasted in deeper soil layers; afterwards irrigation schedules were changed (schedule 2), giving the same amount of water in two or more times per day; the water remained in upper soil layers at plant root level, reducing wastes and increasing the amount of available water for plants, as highlighted by soil moisture at 60 *cm* (blue plot).
Another important application of the VineSense system uses the dendrometer to monitor plant physiology. The trunk diametric sensor is a mechanical sensor with $+/-$ 5 microns of accuracy; such an accurate sensor can appreciate stem micro variations occurring during day and night, due to the xilematic flux inside the plant. Wireless nodes measure plant diameter every 15 minutes, an appropriate time interval for following these changes and for creating a plot showing this trend. In normal weather conditions, common physiologic activity can be recognized by agronomists the same as a doctor can do reading an electrocardiogram; when air temperature increases and air humidity falls in combining low soil moisture levels, plants change their activity in order to face water stress, preserve their grapes and especially them-

selves. This changed behavior can be registered by the dendrometer and plotted in the VineSense interface, warning agronomists about incoming risks; as a consequence, new irrigation schedules can be carried into effect.
Fig. 9 shows an example of a plant diametric trend versus air temperature.

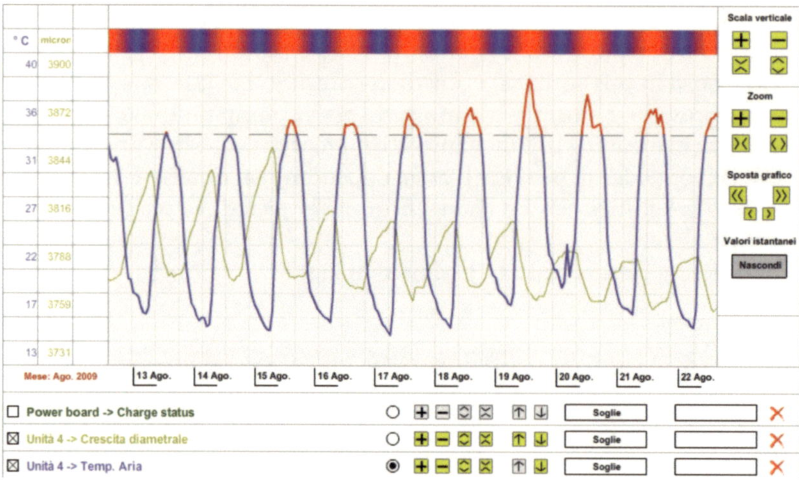

Fig. 9. Plant Diametric Trend vs. Air Temperature

The blue plot represents the air temperature in 10 days, from 13th August until the 22nd August 2009 in Italy; the blue line becomes red when the temperature goes over a 35 degree threshold. During the period in which the temperature is so high, plant stem variations are reduced due to the lower amount of xilematic flux flowing in its vessels, a symptom of water leakage.
WSN in agriculture are also useful for creating new databases with historical data: storing information highlighting peculiarities and differences of vineyards provides agronomists an important archive for better understanding variations in plant production capabilities and grape ripening. Deploying wireless nodes on plants in interesting areas increases the knowledge about a specific vineyard or a specific terroir, thus recording and proving the specificity of a certain wine. I.E., the quality of important wines such CRU, coming from only one specific vineyard, can be easily related to "grape history": data on air temperature and humidity, plant stress, irrigation and rain occurring during the farming season can assess a quality growing process, that can be declared to buyers.
Finally, VineSense can be used to reduce environmental impact thanks to a more optimized management of pesticides in order to reach a sustainable viticulture. Since many of the most virulent vine diseases can grow in wet leaf conditions, it is very important to monitor leaf wetness in a continuous and distributed way. Sensors deployed in different parts of vineyards are a key element for agronomists in monitoring risky conditions: since wetness can change very rapidly during the night in a vineyard and it is not homogeneous in a field, a real time distributed system is the right solution for identifying risky conditions and deciding when and where to apply chemical treatments. As a result, chemicals can be used only when they are strictly necessary and only in small parts of the vineyard where they are really needed, thus reducing the number of treatments per year and decreasing the amount of active substances sprayed in the field and in the environment. In some tests performed in 2009 in Chianti, the amount of pesticides was reduced by 65% compared to the 2008 season.

Leaf wetness sensors on nodes 2 and 3 measure different wetness conditions as shown in Fig. 10. The upper part of the vineyard is usually wetter (brown plot) than the lower part (blue plot) and sometimes leaf wetness persists for many hours, increasing the risk of attacks on plants.

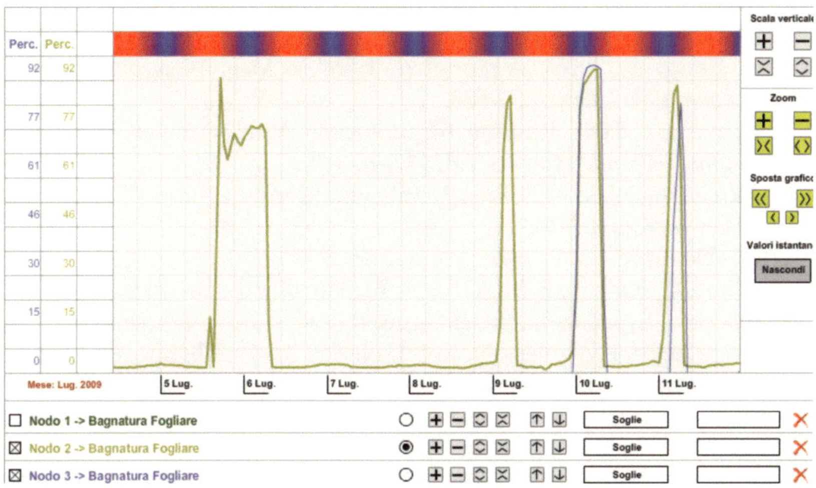

Fig. 10. Different Leaf Wetness Conditions in a Small Vineyard

9. Conclusion

This paper deals with the design, optimization and development of a practical solution for application to the agro-food chain monitoring and control. The overall system was addressed in terms of the experienced platform, network issues related both to communication protocols between nodes and gateway operations up to the suitable remote user interface. Every constitutive element of the system chain was described in detail in order to point out the features and the remarkable advantages in terms of complexity reduction and usability.

To highlight the effectiveness and accurateness of the developed system, several case studies were presented. Moreover, the encouraging and unprecedented results achieved by this approach and supported by several pilot sites into different vineyard in Italy and France were shown.

The fruitful experience of some pilot sites was gathered by a new Italian company, Netsens, founded as a spin off of the University of Florence. Netsens has designed a new monitoring system called VineSense based on WSN technology and oriented towards market and user applications. In order to point out the improvements of the new solution respect to the experimental one, the main features of VineSense were described. Moreover, some important agronomic results achieved by the use of VineSense in different scenarios were sketched out, thus emphasizing the positive effects of the WSN technology in the agricultural environment. Nowadays, the application of the solution described in this paper is under investigation to the more general field of environmental monitoring, due to its flexibility, scalability, adaptability and self-reconfigurability.

10. References

Blackmore, S. (1994). Precision Farming: An Introduction, *Outlook on Agriculture Journal*, **Vol. 23**, pp. 275-280.

Wang, N., Zhang, N. & Wang, M. (2006). Wireless sensors in agriculture and food industry - Recent development and future perspective, *Computers and Electronics in Agriculture Journal*, **Vol. 50**, pp. 114-120.

Akyildiz, I.F. & Xudong, W. (2005). A Survey on Wireless Mesh Networks, *IEEE Communication Magazine*, **Vol. 43**, pp. S23-S30.

Al-Karaki, J. & Kamal, A. (2004). Routing Techniques in Wireless Sensor Networks: a Survey, *IEEE Communication Magazine*, **Vol. 11**, pp. 6-28.

Langendoen, K. & Halkes, G. (2004). Energy-Efficient Medium Access Control, *The Embedded Systems Handbook*, pp. 2-30.

Mica2 Series, Avaiable on `http://www.xbow.com`.

Mattoli, V., Mondini, A., Razeeb, K.M., Oflynn, B., Murphy, F., Bellis, S., Collodi, G., Manes, A., Pennacchia, P., Mazzolai, B., & Dario, P. (2005). Development of a Programmable Sensor Interface for Wireless Network Nodes for Intelligent Agricultural Applications, *Proceedings of IE 2005*, IEEE Computer and Communications Societies, Sydney, pp. 1-6.

Sveda, M., Benes, P., Vrba, R. & Zezulka, F. (2005). Introduction to Industrial Sensor Networking, *Handbook of Sensor Networks: Compact Wireless and Wired Sensing Systems*, pp. 10-24.

Jain, J.N. & Agrawala, A.K. (1990). *Open Systems Interconnection: Its Architecture and Protocols*, Elsevier.

IEEE Standard 802.11 (1999). *Wireless LAN Medium Access Control (MAC) and Physical Layer (PHY) Specifications*, IEEE Computer Society.

El-Hoiydi, A., Decotignie, J., Enz, C. & Le Roux, E. (2003). WiseMAC, an Ultra Low Power MAC Protocol for the WiseNET Wireless Sensor Network, *Proceedings of SENSYS 2003*, Association for Computer Machinery, Los Angeles (CA), pp. 244-251.

Ye, W., Heidemann, J. & Estrin, D. (2002). An Energy-Efficient MAC Protocol for Wireless Sensor Networks, *Proceedings of INFOCOM 2002*, IEEE Computer and Communications Societies, New York (NY), pp. 1567-1576.

Dam, T. & Langendoen, K. (2003). An Adaptive Energy-Efficient MAC Protocol for Wireless Sensor Networks, *Proceedings of SENSYS 2003*, Association for Computer Machinery, Los Angeles (CA), pp. 171-180.

Lu, G., Krishnamachari, B. & Raghavendra, C. (2004). Adaptive Energy-Efficient and Low-Latency MAC for Data Gathering in Sensor Networks, *Proceedings of WMAN 2004*, Institut fur Medien.Informatik, Ulm (Germany), pp. 2440-2443.

Shakkottai, S., Rappaport, T. & Karlsson, P. (2003). Cross-Layer Design for Wireless Networks, *IEEE Communication Magazine*, **Vol. 41**, pp. 77-80.

3

Wildlife Assessment using Wireless Sensor Networks

Harry Gros-Desormeaux, Philippe Hunel and Nicolas Vidot
LAMIA, Université des Antilles et de la Guyane,
Campus de Schœlcher, B.P. 7209, 97275 Schoelcher, French West Indies
France

1. Introduction

The endangered species always drew the attention of the scientific community since their disappearance would cause irreplaceable loss. To help these species to survive, their habitat is protected by the laws of environmental protection. Sometimes this protection is not enough, because their natural evolution is the main cause of their disappearance. However, to save them, it is sometimes possible to transfer them elsewhere that should be similar to their previous habitat to avoid disturbing the balance of wildlife. To model a habitat, several parameters must be of interest and are generally defined by experts. This is the case for the number of singing birds which will be studied in this paper.

Today, advances in sensor technology enable the monitoring of species and their habitat at a very low cost. Indeed, the increasing sophistication of wireless sensors bids opportunities that enable new challenges in a lot of areas, including the surveillance one. Progress in their miniaturization leads to micro-sensors of size of cubic millimeters which, used in large quantity, produce huge amounts of data. This paper promotes the use of sensors for monitoring bird endangered in their habitat. Actual methods for counting endangered birds use mainly human labor and because they are not really comprehensive leads to poor estimation. The use of sensors deployed in critical environments can help the census of these species and even generate new data on their customs.

Among the challenges that the use of the sensor technology enable, energy efficiency is the most critical for these wireless networks since battery depletion totally disables a sensor. In addition, designing algorithms for wireless networks stems from the distributed computer science domain with limited devices. Memory space and computational power are often of a magnitude less than miles than their desktop counterparts. This paper investigate the problem and proposes to approximate the number of birds by geometric means derived in a graph problem.

Our paper is organized as follows. First, Section 2 provides an overview of techniques generally used to estimate the locations of multiple sources with a unknown sensor network. Section 4 details our heuristics used to count birds. Section 5 introduces a distributed algorithm for counting birds. Experimentation confirms the effectiveness of our counting systems in Section 6. Then we conclude in Section 7 and gives an overview of our future work.

2. Previous Work

Source localization is an area of interest that has been widely studied in these recent years. A comprehensive review of incentives techniques and source localization has been written by Krim and Viberg in (Krim & Viberg, 1996) and it is not difficult to understand that problem has been of particular focus for military needs. Indeed, radar and sonars are a direct application of source localization.

Several acoustic parameters such as bandwidth, distance sensors, reverberation and thus change the way the location of the sources are handled. In addition, the algorithms of source localization depends strongly on physics and rely on the sound characteristics of waveform to calculate location sources. Waveform audio is known to be *broadband* (30Hz-15kHz) and sensors usually record the sound from *near-field sources*. The following presents some algorithms of interest which satisfy these two properties. Near-fields algorithms like close-formed ones (Smith. & Abel, 1987) use time delays between sensors location to estimate the source position. However, though they are computationally less expensive than maximum-likelihood parametric algorithms (Chen et al., 2001a), they cannot handle efficiently multiple sources (Chen et al., 2001b). Maximum-likelihood (ML) algorithms are inspired by the fact that source location information is contained in the linear phase shift of the sensor data spectrum obtained through a discrete Fourier Transform applied to the wideband data. However, ML techniques are dominated by low-cost suboptimal techniques like the well-known MUSIC algorithm (Schmidt, 1986) which leverages spectral calculus on signal and noise subspaces to find sources locations.

Unlike these approaches, we do not use the acoustic properties of the song of the bird to find its location. Indeed, we assume that our sensors are simple and only detect songs relevant to the monitored specie. Further, our sensors are wireless and rely on battery power to function. It is important to notice that our algorithms do not try to pinpoint birds, but rather estimate the number of songbirds that inhabit a region. In our case, only approximate geometric information is sufficient to establish this estimate.

3. Recognizing the birdsong

The recognition process of *birdsong* is the first part of our counting systems. Today, it is true that the performance levels made in the treatment of audio signals are high, but this requires large memory and processing power of large size which could exclude limited capacity of devices such as wireless sensors.

Recognition of species based on acoustic analysis has been widely studied in recent years and usually falls within the scope of the classification field. This is particularly the case for recognition of *bird songs*. Indeed, for a particular song, it is necessary to determine if it belongs to a specie. For example, the work of Seppo Fagerlund (Fagerlund, 2007) uses support vector machines to classify the different species of birds based on their songs. Similarly, Jim Cai et al. (Cai et al., 2007) propose a method recognition based on neural networks to find the membership of a song to a bird class. Our recognition process, inspired by the work of Rabiner (Rabiner & Wilpon, 1979), leverages the same mechanics by means of a clustering algorithm to classify the song.

Figure 1 gives an overview of our wireless counting system.

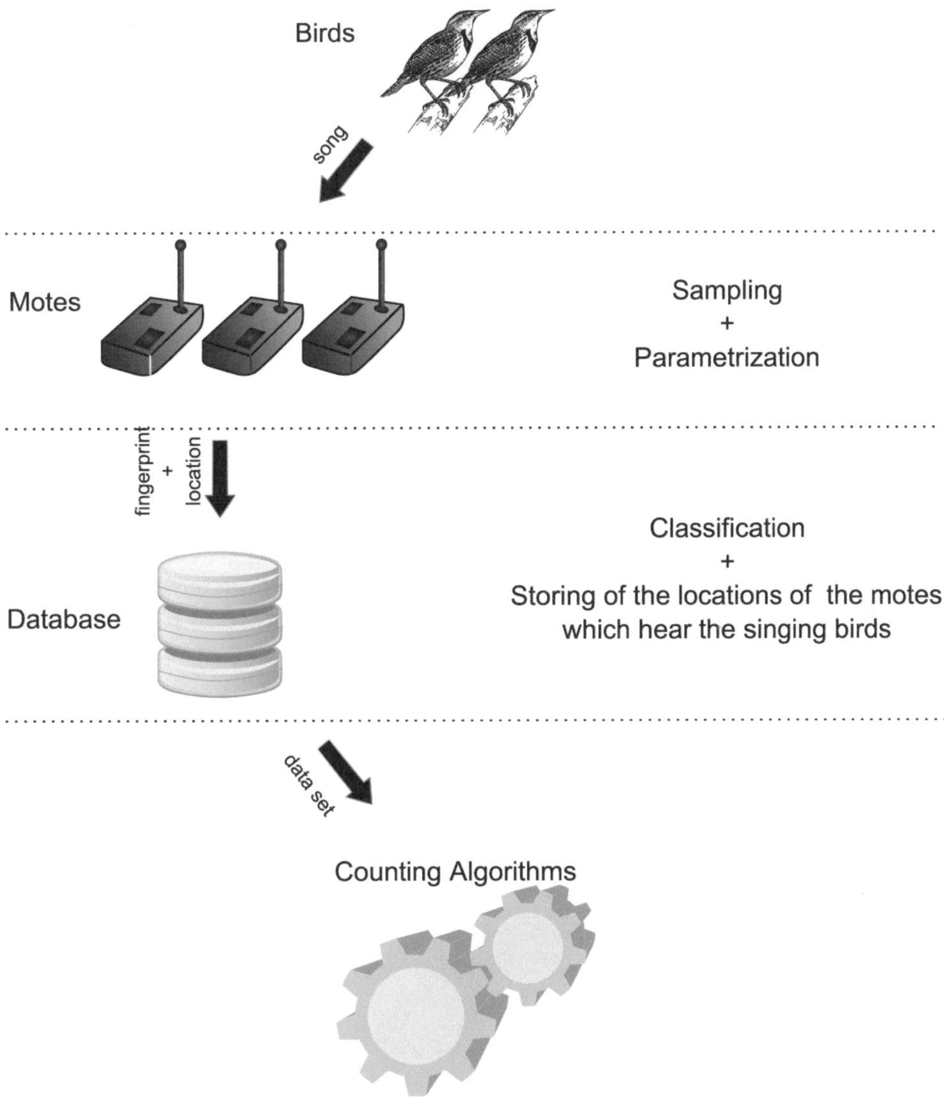

Fig. 1. The Counting System

Bird Species Recognition Using Clustering

Our classification method is twofold : a parameterization transformation process of the song in a certain fingerprint, and clustering process to determine its membership.

The parametrization process uses the songs of the birds to create a series of coefficients that describe the signal. Although various parameterization methods LPC, LPCC, PLP, dots exist, we use the MFCC Mel Frequency Cepstral Coefficient because our analysis is limited to a very limited vocabulary on limited devices. Indeed, Christopher Levy compared in (Lévy et al., 2006) different parameterization methods on small systems such as mobile phones for reduced vocabulary and have showed that the parameterization based on MFCC is much more effective for such systems.

Once the fingerprint is obtained from the parameterization process, it is added in a set with other fingerprints, themselves derived from a database containing a large number of songs of individuals known as the specie. Subsequently, a clustering algorithm (K-Means or EM) is used on all the fingerprints to determine their similarity and to create one or more clusters in which will be the *bird cluster*. For a given footprint, the problem is then to determine its membership to the bird cluster. If that's the case, data location + Mote timestamp is stored in the database for further processing counting algorithms. Our recognition results are compelling because almost all birds are classified correctly in our case.

4. The Counting Algorithm

This section is devoted to our counting heuristics inspired by the triangulation detection used by R. E. Bell to count owls in the forest (Bell, 1964). Our method differs essentially from the fact that we do not use semi-directional devices but omni-directional wireless sensors to loosely locate a birdsong. In our theoretical framework, all motes share the same characteristics building, which means they have the same (processing power, memory, battery, radius of detection, etc). Optimizing routes in wireless sensors networks here are out of concern. We only focus on the manner to detect birds in their habitat viewed as a 2D area. Further, we do not have any assumptions on the number of birds, on their movements or even their customs.
More formally, let denote $M = \{m_1, \ldots, m_n\}$ the set of all the motes which covers the habitat. Each mote has the same detection radius r. All motes can report information to the base station B which holds our counting algorithm, assuming that B is always reachable by every mote. Let $F_t : M \rightarrow \{0, 1\}$, the detection function which returns 1 if a mote m_i detects a bird, 0 otherwise at time t. The base station stores the detection array $D_t = [F_t(m_1), \ldots, F_t(m_n)]$ which reveals the detection state of each mote at time t. Note that the base stores detection arrays at a sampling rate determined empirically, that is, detection arrays D_i are stored in a data set D at the base B. Fig. 2 shows an example of motes placed on a 2D area.

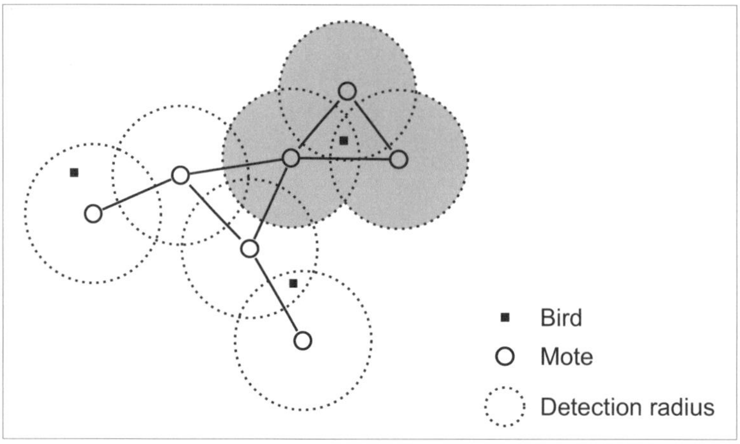

Fig. 2. Motes, birds and the hard underlying unity Chart

We propose to count one bird for all the motes which trigger at time t and for which radius of detection intersect mutually. We call such a set a Maximal Detection Set denoted MDS(N) with $N \subset M$ where N is the set of the motes which trigger at time t. The grayed area in figure 2 is a MDS. Let's denote such a subset $W = \{m \in M \mid \forall m_i, m_j \in M, r(m_i) \wedge r(m_j)\}$.

Finding the Maximum Detection Set is similar to find a maximum clique (Bomze et al., 1999). Let's see why.
A unit disk graph $G(V, E)$ is an intersection graph of disks of unit radius, that is, $\forall ij \in E$, the unit circle of center i intersects the unit circle of center j. The set of each center of these circles is called the model of the unit disk graph. This class of graph is well studied and is extensively used in the field of ad hoc networks (Kuhn et al., 2008). Indeed, UDGs (Unit Disk Graphs) can represent an ideal view of an ad hoc networks and provides strong theoretical result due to the geometric properties of the model. For example, Clark and al. (Clark et al., 1990) show that finding a maximal clique for an UDG is polynomial given its model. More recently, Raghavan and Spinrad (Raghavan & Spinrad, 2003) have shown that it is even possible to compute the maximum clique without the model in polynomial time.
Without loss of generality, let $G(V, E)$ a graph where V is the set of the motes and E, the set of edges where the edge ij exists if and only if the detection radius of mote i intersects the detection radius of mote j. Clearly, G is a unit disk graph. Unfortunately, a clique in G only gives motes which are pairwise adjacent and we are interested in motes which are mutually adjacent, that is motes which intersect mutually. We propose to alter all triangles (clique of size 3) which do not have a mutual intersection in the graph i.e we remove one edge in the triangle. As a consequence, all cliques of more than three vertices will have a mutual intersection.

Theorem 4.1. *If a graph $G(V, E)$ only has triangles formed from motes whose detection radius intersect mutually, then all motes forming a clique in G have detection radii intersecting mutually.*

Proof. By definition, all clique of size three have detection radii which intersect mutually. Now, assume that all motes clique of size n intersect mutually. Let choose such a clique that we call $S = \{m_1, \ldots, m_n\}$ and let's add a new mote m_{n+1} to S. Assume that $S + \{m_{n+1}\}$ form a clique for which some motes do not intersect mutually. Clearly, m_{n+1} form at least two proper intersections with S, and the detection radius of the mote m_{n+1} cannot intersect mutually at least with two other radii detection. But, by definition, all triangles intersect mutually which is a contradiction. □

Reichling (Reichling, 1988) uses convex programming to find the common intersection of a set of disks in $O(k)$ steps where k is the number of constraints of the convex program. Moreover, all the triangles in a graph can be computed in $O(mn)$ steps where m is the number of edges and n, the number of vertices. Thus, we can alter all triangles which do not have a common intersection in $O(kmn)$ steps. Several strategies could be used to alter a triangle. However, removing the longest edge in a triangle seems to be the most relevant one since the number of altered triangles would be reduced. Intuitively, a longest edge in a "bad" triangle is more likely to be common to another "bad" triangle. Unfortunately, the underlying unit disk graph can loose its nature since it might become a quasi-unit disk graph[1] for which the maximum clique problem is known to be NP-complete (Ceroi, 2002).
Algorithm 1 recursively constructs the maximum set of all motes which triggers at time t and removes a MDS built from this set. For each MDS removed, the number of birds iterates. This procedure is run for each detection array and the maximum number found over these detection arrays is an estimation of the number of singing birds. This algorithm complexity is bounded by the MDS search which consists in finding a clique in the unit disk graph underlying our network. Breu (Breu, 1996) has given an algorithm which find a maximum clique in a unit disk graph with complexity $O(n^{3.5} \log n)$. However, the alteration of the underlying unit disk graph leads to a NP-complete algorithm.

[1] Model which takes into account non-circular detection area

begin
  $L \leftarrow \emptyset$;
  foreach $d \in D$ **do**
    NumberOfBirds $\leftarrow 0$;
    Construct the underlying altered unit disk graph $G(V, E)$ from d;
    while $V \neq \emptyset$ **do**
      Search for a maximum clique in G;
      Remove this clique from G;
      NumberOfBirds $\leftarrow$ NumberOfBirds $+ 1$;
    Add NumberOfBirds to L;
  return $\max_{l \in L} l$;
end

Algorithm 1: The Counting Heuristic

Refining the Counting Heuristic

In the following, we suggest a little enhancement of our scheme. Indeed, we partition successive detection arrays pairwise in order to refine our estimation of the number of birds. Intuitively, the habitat is divided in such a manner that birds in a part could not have moved to another one between two instants (for each couple of detection arrays). A threshold is empirically fixed for the flight speed of the birds such that no birds can fly over that value. This leads to the decomposition of the environment in several sub-environments. Then, each sub-environment is processed with algorithm 1. For example, assume that we have 10 birds in an area. Halve this area and put 5 birds in one part, and 5 in the counterpart. Now, assume that the 5 birds in the first part sing together at time t, the other ones sing together at time $t + 1$ and these parts are too distant such that birds in one part can go in the other part between the two time steps. In that case, algorithm 1 outputs 5 birds as estimate. Our next algorithm halves the environment in two parts such that birds in two. As a consequence, we can apply algorithm 1 on each part independently and take the sum of the estimates found on each part, which gives 10 birds.

Data: A list of detection arrays $D = D_1, \ldots, D_m$
Result: An estimation of the number of birds in the habitat
begin
  $L \leftarrow \emptyset$;
  while $|D| > 0$ **do**
    Partition detection arrays D_i and D_{i+1} respectively in $X = \{X_1, \ldots, X_k\}$ and $Y = \{Y_1, \ldots, Y_k\}$;
    $Z = \emptyset$;
    for $i \leftarrow 1$ *to* k **do**
      Process X_i and Y_i with algorithm 1 and put the maximum of the number of birds counted in Z;
    Add $\sum_{z \in Z} z$ to L;
  return $\max_{l \in L} l$;
end

Algorithm 2: The Enhanced Counting Algorithm

For sake of clarity, in algorithm 2, the number of detection arrays is even and only two successive detection arrays are partitioned. The next section presents another way to count the singing birds in their habitat. This next version is designed to be partially distributed on the motes.

5. The Swarm Counting Protocol

Our next counting method can be seen as two levels, a local and a global one. At the local level, motes cooperates sending information to count the number of singing birds in their neighborhood. At the global level, motes aggregates data to find a more accurate estimation of the number of singing birds in the habitat.
Like the technique previously described, we assume that the motes layout forms a unit disk graph. First, motes have to estimate locally how many birds had sang. Then, they send this data to the base station which derives from all the information the estimate for the number of singing birds. In our scheme, motes all have a set of rules which are the following. They are all in a passive state until some songs trigger them. When triggered, they switch to an active state and tell to their neighbors[2] that they detect a bird. Then they listen for their neighborhood during a specified time. Finally, they deduce the number of singing birds in the vicinity from their active answering neighbors, and send this number to the base station.

Local Counting

Our local counting is somewhat similar to the one in section 4. It leverages the trilateration technique to estimate a number of birds in the vicinity. All motes know their neighbors' topology and are in an initial passive state when they are waiting for signals (bird songs). Whenever a mote is triggered , it sends a signal to its neighbors and listen for whose which were triggered too. If two or more neighbors have an intersecting detection area, we assume that only one bird is counted for these motes. In figure 3, the black mote hears a bird song, asks its neighbors if they heard too and waits for their reply. Remark that the number of birds counted is the number of neighbors which are independent mutually in each neighborhood, i.e the cardinal of the maximum independent set[3] in the graph induced by the neighbors.

Global Counting

Now, assume that all motes have counted the birds in their vicinity and have sent their local count to the base station. Now, all these information have to be aggregated accordingly to find an estimate of the number of singing birds at this instant. Because, the neighborhood was used to derive the local counting, obviously, motes which are neighbors will influence each other in the counting process. So, summing up their local count can lead to an over-estimate of the number of singing birds. Note that is also the case for motes which are at distance 2, that is neighbors of neighbors in the unit disk graph, since they can share common neighbors. Therefore, only motes which are more than distant 2 each other will sum up their count. Our estimation will be the maximum number of birds which could be counted over aggregated nodes in the underlying graph.
In figure 4, the black nodes are at distant 3. So a global counting of singing birds could be four. Remark that such a counting has to be done for all set of nodes which are at more than distance 2 each other. If such a technique seems to lead to a combinatorial explosion of the

[2] As previously, neighbors are adjacent nodes in the unit disk graph
[3] The largest set of vertices which are not pairwise adjacent

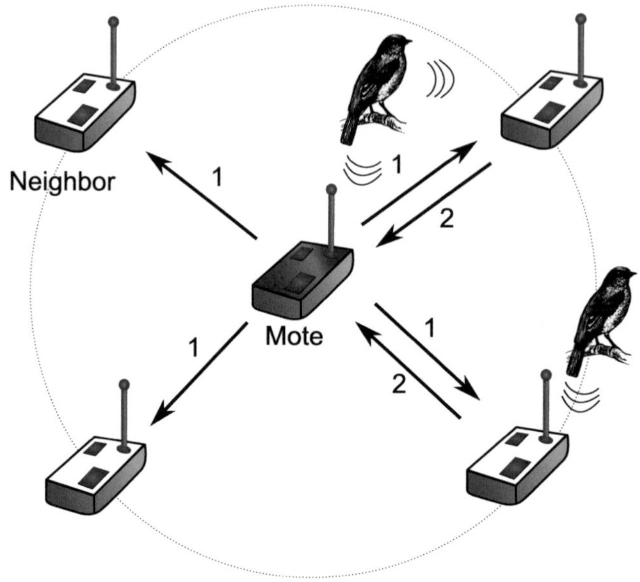

Fig. 3. Motes collaboration at the local level

set of motes which can be aggregated, the underlying graph has some nice properties which allows to find the estimate in linear time.

More formally, let $N(i)$ define the neighbors of a mote i, that is

$$\forall i \in V, \quad N(i) = \{j \in V | \quad ij \in E\}$$

and

$$\forall A \subset V, \quad N(A) = \bigcup_{i \in A} N(i)$$

Let $G_{2+}(V, E_{2+})$ define the graph where

$$\forall i, j \in V^2, \quad ij \in E_{2+} \text{ iff } j \in N(N(i)) \backslash i$$

The graph G_{2+} is the graph of all motes which are at most 2-distant between them. Let $S(.)$ denote the mapping which maps a vertex $v \in V$ to the number of birds counted locally. Let $\mathcal{C}$ be the set of all independent set in graph G_{2+}. Our estimation is the sum of birds counted locally for each motes derived from the maximum weighted independent set of G_{2+}, i.e

$$\max_{c \in \mathcal{C}} \sum_{v \in c} S(v)$$

Lemma 5.1. *G_{2+} is a chordal graph.*

Proof. Proof Remark that in G_{2+}, all vertices are simplicial[4]. Thus, there exists a perfect elimination ordering on its vertices and de facto G_{2+} is chordal. □

[4] Vertices for which neighbors induce a clique in the graph.

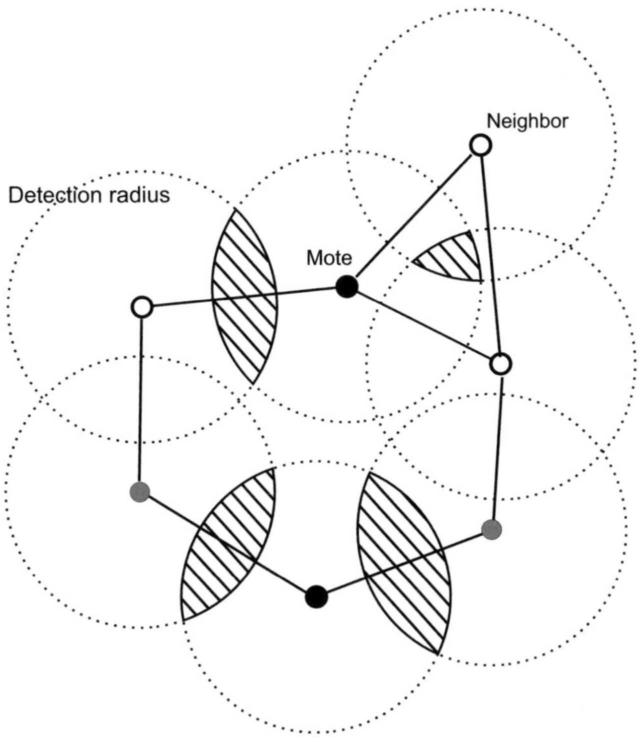

Fig. 4. Example of underlying unit disk graph in local and global detection

Chordal graphs are graph for which vertices do not induce cycles without chord of size more or equal to four. They are perfect graphs and well discussed in (Golumbic, 1980). It is also well known that finding a maximum weighted independent set in chordal graph is linear (Leung, 1984). Thus, our later algorithm finds its estimation of the number of birds in linear time given G_{2+}.

Let's see why and how our algorithm is not so sensible to noise and encompasses non circular detection area. One of the most interesting features of swarm computing (Blum & Merkle, 2008) is that nodes (swarm entities) create mechanisms which tend to be resilient to disruption and failure. Similarly, our last counting technique leverages the swarm intelligence since motes collaborates each other to derive their local count. The more the motes are, better the estimate is. There are two cases where inconsistencies could appear :

1. Motes can have a different status from what it would be. For example, a mote could stay in a passive state while it would have heard "a bird song". However, neighbor motes tend to negate this last effect. Conversely, motes could "wake up" while no birds have sung. This latter case is somewhat less frequent and is easier to correct since this mote could be a one-vertex connected component in the underlying graph, fact which is prone to be an erratic behavior of the mote.
2. Objects can occlude bird songs, that is detection area is no more circular. In that case, the occluded motes would stay in a passive state. Fortunately, the swarm could correct this drawback by multiplicity : other closer motes could hear the birds too.

Therefore, note that the layout of the motes is somewhat important and a simple way to tackle the occlusion problem is to rise the density of the motes on the monitored environment. It is even possible to only increase the number of motes where occlusion problems could occur.

The next section is dedicated to experiments which prove our algorithm efficiency, even in the presence of noise.

6. Experiments

6.1 Context

Endangered species receive attention from the scientific community since their disappearance would lead to irreplaceable losses. To help these species to survive, their habitat is protected by laws of environmental protection. Sometimes, this protection is not sufficient since their habitat evolution is the main cause of their vanishing. In order to save them, they must be transferred elsewhere. Obviously, the new habitat has to be similar to the previous one to minimally disrupt the equilibrium of the wildlife. To model a habitat, several parameters have to be fixed by an expert. This study precedes the MOM project for which wireless sensor networks have to be used to monitor an endangered specie. So these simulations are the first steps to the deployment of WSNs over the Caravelle location in Martinique (a French Caribbean island). Indeed, birds called "White-breasted Thrasher" are a specie which is only known to be in in the Caravelle. They are considered endangered since specialists think that only fifty of them are still alive there.

6.2 Testbed

Environment

For the need of the simulations, we wrote a tool which aims at generating the data necessary to run our counting heuristics described previously. Our test environment comprises :

- an Intel Core 2 Duo E6750 2.67 GHZ,
- 4 Go RAM,
- Windows Vista 64 bits for Operating System,
- and the JDK 1.6 Update 10 (x64) since our tool is written in java.

Parameters

Simulations parameters were calibrated to be the closest to our tested area. The dimension of our habitat is about 1000m$\times$1000m. Birds can fly at 2 meters per second, stay at place, take random directions with uniform probability. They sing with some probability fixed empirically. This latter parameter is fixed at 0.2 for each sample record (detection array). Finally, motes are placed randomly on our area.

6.3 Performance Evaluation

Figure 5 shows our three different algorithms estimation for counting fifty living birds in the habitat. *Algo1* stands for the algorithm which only rely on the underlying UDG. *Longest* is the algorithm which alters the longest edges of "bad triangles". *Swarm* is the algorithm presented in section 5. *True* is the number of birds which really sang. Each test has been driven 50 times and the mean of the estimations was taken as the final result. Error deviation is shown for

Motes	Algo1	Longest	Swarm	True
100.00	10.10	10.72	9.98	17.58
200.00	13.86	16.56	12.62	17.38
300.00	15.16	18.64	12.92	18.02
400.00	15.70	20.46	13.84	17.58
500.00	16.76	22.56	15.04	17.36
600.00	17.28	24.12	15.80	17.80
700.00	17.88	25.74	16.08	17.66
800.00	18.38	27.74	17.04	17.64
900.00	18.40	29.50	17.60	17.86
1000.00	18.30	31.22	17.98	17.80

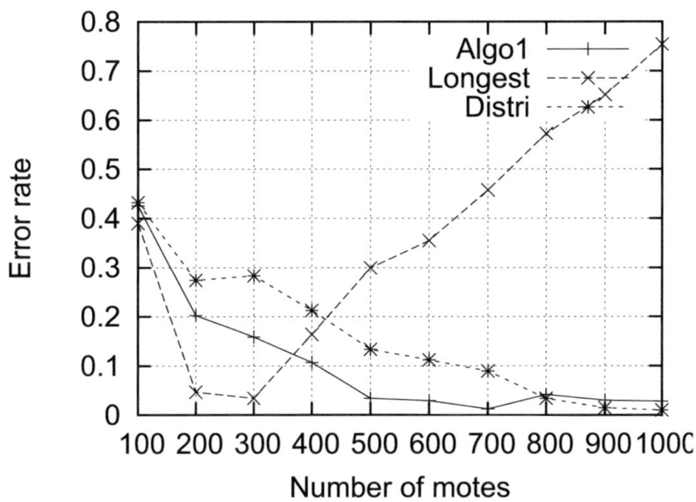

Fig. 5. Number of Birds found by the heuristics and error deviation for 50 Birds in the habitat

each algorithm on the graph near the tables and gives an idea of how the algorithms perform along the parameters.

Experiments show that algorithm *Algo1* performs nearly as well as algorithm *Swarm* whenever the number of motes is high. However, *Algo1* tends to over-count the birds. The outputted number of birds depends on the manner the MDS are removed. Let's sketch a brief example on figure 6. There exists three MDS at first step in each configuration. Grayed areas represent the MDS removed on each configuration at each step. Configuration 1 leads to two grayed area whereas configuration 2 leads to three grayed area. That is how two birds can be counted as three. To reduce this drawback, we could run several times the counting process which would remove the MDS randomly and then take the minimum number of birds over these countings. However, such a scheme does not guarantee that we will not over-count and further, will highly rise the execution time of the whole process.

To validate our schemes, noise is added as a parameter in our simulator. We assume that 20% of the motes malfunction. Table 1 shows the percentage error for *Algo1* and *Swarm* in the presence of noise.

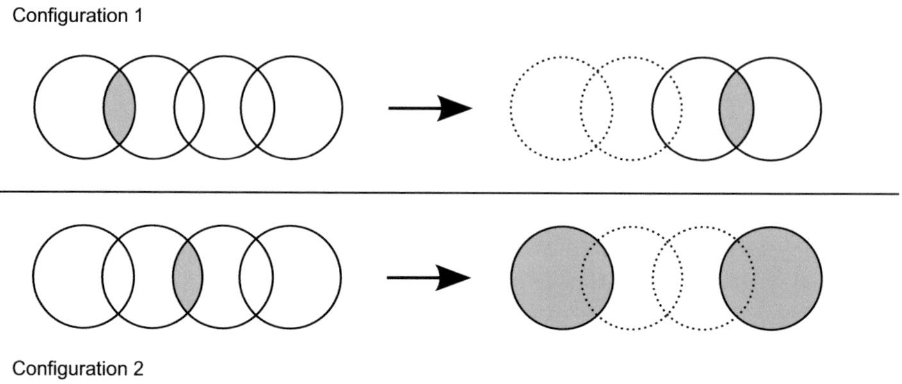

Fig. 6. Example of bad counting

Table 1. Relative error for the counting algorithms for 50 birds

Motes	Algo1	Swarm	Algo1 with noise (%)	Swarm with noise(%)
100	45,08	46,33	50,51	50,85
200	27,24	33,83	32,69	37,34
300	15,27	25,79	20,59	28,62
400	8,85	19,31	13,33	23,22
500	9,32	19,77	13,75	23,86
600	3,31	12,69	6,86	17,49
700	1,93	4,54	1,36	9,53
800	4,41	3,72	0,93	7,67
900	4,30	2,49	2,04	6,23
1000	7,11	1,49	4,47	1,03

Clearly, the algorithm based on the swarm counting protocol seems more sensitive to noise than its counterpart. Note that without noise, algorithm *Algo1* over-counts the number of birds. Therefore, in presence of noise, the approximate of the number of birds tends to be more precise. Conversely, algorithm *Swarm* already undercounts the number of birds originally. So, noise degrades even more the approximate of the number of birds which generally leads to a worse counting.

Finally, we decided to fix the number of motes which will be used on the Caravelle habitat to 1000 and vary the number of birds in our simulator to confirm our estimation. Results are shown on figure 7.
Algorithm *Longest* suffers the same drawback seen in figure 5 when the number of motes is high, so much that estimation are too high. This results from the fact that altering triangles tends to create much more cliques to remove.
Algorithm *Swarm* gives slightly better estimations in configuration using a high number of motes. However, its efficiency lowers whenever less motes are used. Indeed, the lesser the motes you have, the lesser you cover the habitat. Furthermore, our schemes rely on a high number of motes to better estimate the singing birds except algorithm *Longest* which could be used to estimate the songbirds whenever the number of motes are low.

Birds	Algo1	Longest	Swarm	True
10.00	5.94	10.20	5.44	5.34
20.00	9.08	16.08	8.76	8.90
30.00	12.76	20.58	12.20	12.28
40.00	15.90	25.48	15.06	14.76
50.00	18.30	31.22	17.98	17.80
60.00	21.64	34.76	20.62	20.28
70.00	23.92	37.74	22.56	23.08

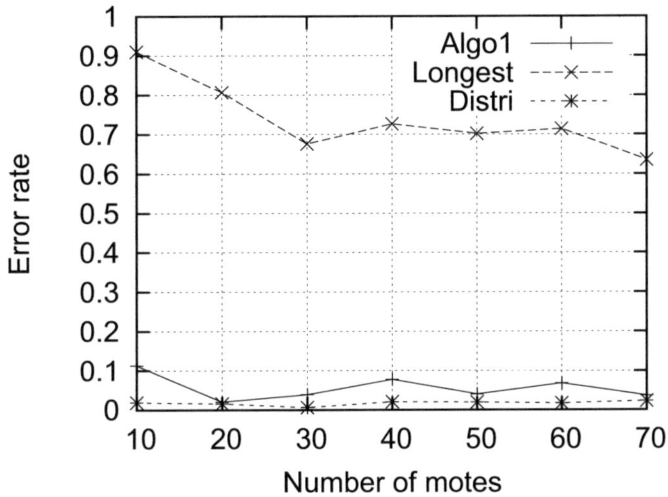

Fig. 7. Number of Birds found by the heuristics and error deviation for 1000 motes in the habitat

These results suggest to design an hybrid algorithm which will switch along determined thresholds. However, remind that the data was generated from simulations and these thresholds could be different from what our experimentations outputted. In our case, using the *Swarm* algorithm for counting the birds seems to be so far the best solution to apply in the *Caravelle* since more motes give accurate estimations.

7. Conclusion

Endangered species are a known problem that drew attention from the community these last years. Habitat monitoring with wireless sensors networks could lead to several improvements in the way to tackle the problem of the survival of these species. We have proposed a first technique to estimate the number of birds using wireless microphone motes scattered in an habitat. Our method derives from the motes layout a unit disk graph and removes continuously maximum cliques to count the number of birds. A limitation of this technique could be the maximum clique problem but simulations have shown that estimations are still suitable if unit disk graphs are used to represent the motes network. We have also proposed a linear algorithm to estimate the singing birds in the habitat and have shown that it is as much as

efficient (quality) as our first one. This scheme can be fully distributed on a suitable wireless sensors network. Such a distributed scheme would deny the need of a powerful base station since the counting process would totally shift from the base to the motes.
Counting singing birds is a first step in our habitat monitoring project and surely is not sufficient to identify specificities of the monitored specie. One major goal of habitat monitoring is the reintroduction of the specie in another environment which will share the same characteristics. We intend to work in this way by monitoring several parameters of interests in an environment to model it and compare it with another ones.

8. References

Bell, R. E. (1964). A Sound Triangulation Method for Counting Barred Owls, *The Wilson Bulletin*, The Wilson Ornithological Society.

Blum, C. & Merkle, D. (eds) (2008). *Swarm Intelligence: Introduction and Applications*, Natural Computing Series, Springer.
URL: *http://dx.doi.org/10.1007/978-3-540-74089-6*

Bomze, I. M., Budinich, M., Pardalos, P. M. & Pelillo, M. (1999). The Maximum Clique Problem, *Research Report CS-99-1*, Dipartimento di Informatica, Univerità Ca' Foscari di Venezia.

Breu, H. (1996). Algorithmic aspects of constrained unit disk graphs, *Technical Report TR-96-15*, Department of Computer Science, University of British Columbia. Tue, 22 Jul 1997 22:20:10 GMT.
URL: *ftp://ftp.cs.ubc.ca/pub/local/techreports/1996/TR-96-15.ps.gz*

Cai, J., Ee, D., Pham, B., Roe, P. & Zhang, J. (2007). Sensor network for the monitoring of ecosystem: Bird species recognition, *Intelligent Sensors, Sensor Networks and Information*, Queensland Univ. of Technol., Brisbane, pp. 293–298.

Ceroi, S. (2002). The clique number of unit quasi-disk graphs, *Rapport*.
URL: *http://hal.inria.fr/inria-00072169/en/; http://hal.ccsd.cnrs.fr/docs/00/07/21/69/PDF/RR-4419.pdf*

Chen, J. C., Hudson, R. E. & Yao, K. (2001a). A maximum-likelihood parametric approach to source localizations, *Proceedings of the Acoustics, Speech, and Signal Processing*, IEEE Computer Society, Washington, DC, USA, pp. 3013–3016.

Chen, J. C., Hudson, R. E. & Yao, K. (2001b). Joint maximum-likelihood source localization and unknown sensor location estimation for near-field wideband signals, *in* F. T. Luk (ed.), *Advanced Signal Processing Algorithms, Architectures, and Implementations XI*, Vol. 4474, SPIE, pp. 521–532.
URL: *http://link.aip.org/link/?PSI/4474/521/1*

Clark, B. N., Colbourn, C. J. & Johnson, D. S. (1990). Unit disk graphs, *Discrete Math.* **86**(1-3): 165–177.

Fagerlund, S. (2007). Bird species recognition using support vector machines, *EURASIP J. Appl. Signal Process.* **2007**(1): 64–64.

Golumbic, M. C. (1980). *Algorithmic Graph Theory and Perfect Graphs*, Academic Press.

Krim, H. & Viberg, M. (1996). Two decades of array signal processing research — the parametric approach, *IEEE Signal Processing Magazine*, Vol. 13(3).

Kuhn, F., Wattenhofer, R. & Zollinger, A. (2008). Ad hoc networks beyond unit disk graphs, *Wireless Networks* **14**(5): 715–729.
URL: *http://dx.doi.org/10.1007/s11276-007-0045-6*

Leung, J. Y. T. (1984). Fast algorithms for generating all maximal independent sets of interval, circular-arc and chordal graphs, *ALGORITHMS: Journal of Algorithms* **5**.

Lévy, C., Linarès, G. & Bonastre, J.-F. (2006). Gmm-based acoustic modeling for embedded speech recognition, *In International Conference on Speech Communication and Technology*.

Rabiner, L. R. & Wilpon, J. G. (1979). Considerations in applying clustering techniques to speaker-independent word recognition, *The Journal of the Acoustical Society of America* **66**(3): 663–673.
URL: *http://link.aip.org/link/?JAS/66/663/1*

Raghavan, V. & Spinrad, J. (2003). Robust algorithms for restricted domains, *ALGORITHMS: Journal of Algorithms* **48**.

Reichling, M. (1988). On the detection of a common intersection of k convex objects in the plane, *Information Processing Letters* **29**(1): 25–29.

Schmidt, R. (1986). Multiple emitter location and signal parameter estimation, *IEEE Transactions on Antennas and Propagation* **34**(3): 276–280.

Smith., J. & Abel, J. (1987). Closed-form least-squares source location estimation from range-difference measurements, *IEEE Transactions on Acoustics, Speech and Signal Processing* **35**(12): 1661–1669.

4

Wireless Sensor Network for Disaster Monitoring

Dr. Maneesha Vinodini Ramesh
Amrita Center for Wireless Networks and Applications,
Amrita Vishwa Vidyapeetham (Amrita University)
India

1. Introduction

This chapter provides a framework of the methodical steps and considerations required when designing and deploying a Wireless Sensor Network (WSN) to a given application. A real example is used to demonstrate WSN deployment in action.

WSN has many possible applications that have not yet been explored. WSN is a fast growing technology however much written about WSN is still theory. 'How to deploy WSNs,' although having much theory written still currently lacks a practical guide.

Using our research experience and the practical real life solutions found when deploying a WSN for the application of Landslide Detection this chapter outlines the steps required when conducting a real world deployment of a WSN.

In this chapter the application for WSN most focused on is for purpose of detecting natural disasters. WSN can be useful to disaster management in two ways. Firstly, WSN has enabled a more convenient early warning system and secondly, WSN provides a system able to learn about the phenomena of natural disasters.

Natural disasters are increasing world wide due to the global warming and climate change. The losses due to these disasters are increasing in an alarming rate. Hence, it is would be beneficial to detect the pre-cursors of these disasters, early warn the population, evacuate them, and save their life. However, these disasters are largely unpredictable and occur within very short spans of time. Therefore technology has to be developed to capture relevant signals with minimum monitoring delay. Wireless Sensors are one of the cutting edge technologies that can quickly respond to rapid changes of data and send the sensed data to a data analysis center in areas where cabling is inappropriate.

WSN technology has the capability of quick capturing, processing, and transmission of critical data in real-time with high resolution. However, it has its own limitations such as relatively low amounts of battery power and low memory availability compared to many existing technologies. It does, though, have the advantage of deploying sensors in hostile environments with a bare minimum of maintenance. This fulfills a very important need for any real time monitoring, especially in hazardous or remote scenarios.

Our researchers are using WSNs in the landslide scenario for estimating the chance occurrence of landslides. India faces landslides every year with a large threat to human life causing annual loss of US $400 million (27). The main goal of this effort is to detect rainfall induced landslides which occur commonly in India.

Many papers have highlighted the need for a better understanding of landslide phenomena and attempted to create systems that gather and analyse that data (1), (14) & (31).

The capacity of sensors and a WSN to collect and collate and analyse valuable worthwhile data, in an ordered manner, for studying landslide phenomena or other natural disasters and has not fully been explored.
Landslide prone-area are usually situated in terrains that are steep, hostile, difficult to access making monitoring landslides a strenuous activity. The wireless sensor network offers itself as an effective, reliable, low maintenance solution.
Using WSN for real-time continuous monitoring has been proven possible as shown the example of (9) who developed a Drought Forecast and Alert System (DFAS) using a WSN. This success in conjunction with (4) who developed a durable wireless sensor node able to remotely monitor soil conditions and (26) who proposed a design for slip surface localization in WSNs motivated our researchers to the design, develop, and deploy a real-time WSN for landslide detection. This system is deployed to monitor and detect landslides, in a landslide prone area of Kerala, India, and is further supported by laboratory setups.
This landslide detection system using a WSN is the first in India, one of the first in the world of its kind. It is also one of the first landslide field deployments backed up by a laboratory setup and modeling software. This system has been operational and collecting data for the last two years, and has issued landslide warnings in July 2009. The current system can be replicated in other rainfall induced landslide prone areas around the world.
One particular advance was the design of a Deep Earth Probe (DEP) to support the deployment of sensors. Previous landslide monitoring procedures have used sensors yet they have not implemented connecting all the sensors to a single wireless sensor node ((29); (28); (14); (1)). We have designed a sensor placement strategy that can be adapted for any landslide prone area and potentially for placing sensors to detect other natural disasters, in other disaster prone areas.
The chapter is arranged as follows: Requirement Analysis consisting of: Analysis of Scenario, Selection of Geophysical Sensors, Placement of Geophysical Sensors, Spatial Distribution of the Deep Earth Probe (DEP), Wireless Sensor Network Requirements, Algorithm Requirements, Network requirements (data transmission requirements/method), Data Analysis Requirements and Data Visualization Requirements;
Followed by sections on: Wireless Sensor Network Architecture; Wireless Network Design and Architecture; Wireless Sensor Network Algorithms; Wireless Software Architecture; Design of Interfacing Sensors and Power Management Methods; Field Deployment Methods and Experiences; Field Selection; Deployment of Deep Earth Probe (DEP); Network Implementation and Integration; Validation of the Complete System - Landslide Warning Issued; and lastly, Conclusion and Future Work.

2. Requirement Analysis

This section will describe in detail how to design a real-time Wireless Sensor Network (WSN), and what are the considerations/requirements that have to be analyzed for designing the network for any scenario. The different processes that will contribute to a WSN design are:

- *Analysis of Scenario*

  Wireless Sensor Networks (WSN) could be useful in a vast and diverse amount of applications. The chosen target scenario must be understood and investigated thoroughly in order to choose the most appropriate sensors and network. A comprehensive analysis of the scenario is one of the first steps to undertake when considering the design of the system. The constraints found (from the analysis of the scenario) determine and govern the overall size and type of network and sensors required.

Understanding the characteristics of a scenario allows logical links to be made about how to detect the occurrence of land movement. The scenario here is landslides and is then further specified to become 'rainfall induced landslides'. The importance of specialization is that landslides would be too generic and there would be too many other factors to consider.

Each landslide behaves differently. Factors playing strong roles in landslide occurrence include slope subsurface factors such as: the type of soil and its properties, soil layer structure, the depth of the soil to bedrock, the presence of quartz or other mineral veins, and the depth of the water table, among others and slope surface factors such as: the types of foliage and vegetation, the topographical geography, human alterations to the landscape, and the amount, intensity, and duration of rainfall.

Landslides are one of the major catastrophic disasters that happen around the world. Their occurrence can be related to several causes such as geological, morphological and physical effects, as well as human activities (30). Basically, landslides are the down-slope movement of soil, rock and organic materials due to the influence of gravity. These movements are short-lived and suddenly occurring phenomena that cause extraordinary landscape changes and destruction of life and property. Some slopes are susceptible to landslides whereas others are more stable. Many factors contribute to the instability of slopes, but the main controlling factors are the nature of the soil and underlying bedrock, the configuration of the slope, the geometry of the slope, and groundwater conditions.

In India, (27) the main landslide triggers are intense rainfall and earthquakes. Landslides can also be triggered by gradual processes such as weathering, or by external mechanisms including:

- Undercutting of a slope by stream erosion, wave action, glaciers, or human activity such as road building,
- Intense or prolonged rainfall, rapid snowmelt, or sharp fluctuations in groundwater levels,
- Shocks or vibrations caused by earthquakes or construction activity,
- Loading on upper slopes, or
- A combination of these and other factors.

Some of the factors that aggravate the incidence of landslides are environmental degradation on account of the heavy pressure of population, decline in forest cover, change in agricultural practices, and the development of industry and infrastructure on unstable hill slopes, among others.

In India, the main landslide triggers are intense rainfall and earthquakes. Under heavy rainfall conditions, rain infiltration on the slope causes instability, a reduction in the factor of safety, transient pore pressure responses, changes in water table height, a reduction in shear strength which holds the soil or rock, an increase in soil weight and a reduction in the angle of repose. When the rainfall intensity is larger than the slope saturated hydraulic conductivity, runoff occurs (12).

The key principal parameters that initiate the rainfall induced landslides are:

1. Rainfall: Rainfall is one of the main triggers for the landslide. The increase in the rainfall rate or its intensity increases the probability of landslide. Hence monitoring rainfall rate is essential for the detection and prediction of landslides. This can be performed by incorporating a rain gauge with the complete system for monitoring landslides.
2. Moisture: The moisture level in the soil will increase as the rainfall increases. Enormous increase in moisture content is considered to be a primary indication for landslide initiation. Hence, it is very important to know the soil moisture at which the soil loses sheer strength and eventually triggers failure.
3. Pore pressure: The pore pressure piezometer is one of the critical sensors needed for the rainfall induced landslide detection. As rainfall increases rainwater accumulates at the pores of the soil. This exerts a negative pressure and also it causes the loosening of soil strength. So the groundwater pore pressure must be measured, as this measurement provides critical information about how much water is in the ground. As the amount of water in the ground is directly related to the soil cohesion strength, this parameter is one of the most important for slope stability and landslide prediction.
4. Tilt: Sliding of soil layers has to be measured for identifying the slope failures. This can be performed by measuring the angular tilt (angular slide) during the slope failure.
5. : Vibrations: Vibrations in the earth can be produced during the initiation of a landslide, as the land mass starts to move, but does not fully slide. These vibrations can be monitored and taken as a precursor to a full landslide.

- *Selection of Geophysical Sensors*

  Landslide detection requires measurement of principal parameters discussed in the above section. The key geophysical sensors such as rain gauge, soil moisture sensors, pore pressure transducers, strain gauges, tiltmeters, and geophones are identified for measuring the principal parameters. These sensors are selected based on their relevance in finding the causative geological factors for inducing landslides under heavy rainfall conditions.

  The details of the selected sensors are:

  - Dielectric moisture sensors: Capacitance-type dielectric moisture sensors are used to monitor the changes experienced in volumetric water content.
  - Pore pressure piezometers: Pore pressure piezometers are used to capture the pore pressure variations, as the rainfall rate varies. Either the vibrating wire piezometer or the strain gauge type piezometer is used for in this deployment.
  - Strain gauges: When attached to a DEP (Deep Earth Probe), a strain gauge can be used to measure the movement of soil layers. Strain gauges of different resistance such as 100Ω, 350Ω, and 1000Ω have been used for deployment, to measure deflections in the DEP of 0.5 mm per meter.
  - Tiltmeters: Tiltmeters are used for measuring the soil layer movements such as very slow creep movements or sudden movements. High accuracy tiltmeters are required for this scenario.

- Geophones: The geophone is used for the analysis of vibrations caused during a landslide. The characteristics of landslides demand the measurement of frequencies up to 250 Hz. The resolution should be within 0.1 Hz and these measurements need to be collected real-time.
- Rain gauges: Maximum rainfall of 5000 mm per year needs to be measured using the tipping bucket. The tipping bucket type of wireless rain gauge, in which the tipping event is counted as .001 inch of rainfall, has been deployed.
- Temperature sensors: The physical properties of soil and water change with temperature. A resolution of 1/10th degree Celsius, measured every 15 minutes, is sufficient. Temperature measurements are collected using the rain gauge.

Cost-Effective Considerations Cost-effective solutions have been explored, e.g. using strain gauges for monitoring slope movement. Investigation into the sensors is a necessary pursuit. Searching for cost effective, yet reliable sensors and accessing their ability to process that data is an issue. When choosing appropriate sensors for your given application it is necessary to access the usefulness of a sensor and its ability to provide the type of worthwhile data required. Developing the ability of sensors effects the applications currently available.

Another factor when considering the most appropriate sensors is how cost-effective the sensor is, for example our team opted to use strain gauges for monitoring slope movement which are significantly cheaper than tiltmeters. Though in choosing to use strain gauges it took a much longer to develop the signal conditioning and electronics to interface the strain gauge to the wireless sensor nodes. It also took a longer time to learn how to accurately interpret the data resulting from the strain gauges since strain gauges capture more noise (and unwanted signals) than other more expensive sensors. Therefore signal conditioning was required to extract the relevant signals that determine slope movements from the strain gauge's raw data.

Nested Dielectric Moisture Sensor is another cost effective choice made by the researchers for monitoring the infiltration rate.

- *Placement of Geophysical Sensors*

The chosen, above mentioned, sensors or a combination of them can be used for detecting landslides. The terrain and type of landslide will determine the group of sensors to be used in a particular location for detecting landslides. All the chosen geophysical sensors are capable of real-time monitoring with bare minimum maintenance. A DEP (Deep Earth Probe) was devised to deploy these many sensors as a stack, attached to a vertical pipe, in different locations of the landslide prone site. This generalized design for the DEP, and the sensor placement procedures at the DEP has been developed and implemented to simplify future deployments. This design can be adapted for any landslide prone area and potentially for placing sensors to detect other natural disasters, in other disaster prone areas. Preparation with an 'eye on the future' is an integral part of the development of a practical system, as this design for a DEP proves. Currently replication of this particular system is being requested across much of India by the Government, the design of the DEP will enable each procedure to be much more organised and simplify deployment.

The ideal depth for the DEP to be deployed would be the same as the depth of the bedrock in that location.

The DEP design uses a heterogeneous structure with different types of geophysical sensors at different positions. The geological and hydrological properties, at the location of each of the DEPs, determine the total number of each of the geophysical sensors needed and their corresponding position on the DEP. These geophysical sensors are deployed or attached inside or outside of the DEP according to each of their specific deployment strategies.

All the geological sensors on the DEP are connected to the wireless sensor node via a data acquisition board as shown in Figure 1. This apparatus, including the DEP with its sensors, the data acquisition board and the wireless sensor node, is conjunctly termed a wireless probe (WP).

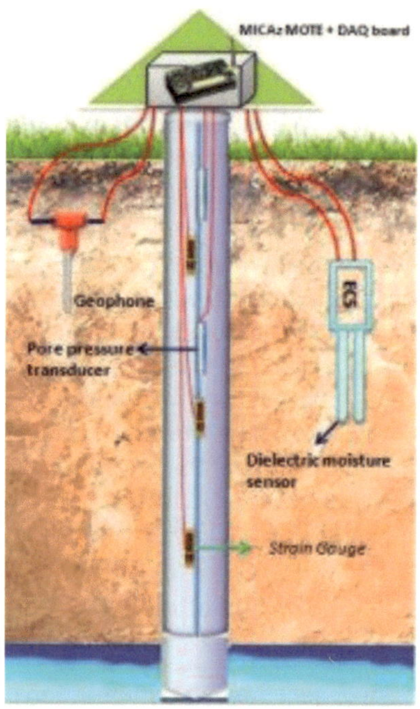

Fig. 1. Multi Sensor Deep Earth Probe

- *Spatial Distribution of the DEP (Deep Earth Probe)*

  Challenges come when wide area monitoring is required. Different approaches can be used for determining the spatial distribution and deployment of Wireless Probes (WPs). The different approaches considered are the Random Approach, the Matrix Approach, the Vulnerability Index Approach, and the Hybrid Approach. In the Random Approach, WPs can be deployed at all possible locations according to the terrain structure of a landslide prone mountain. Whereas in the Matrix Approach, the total area of deployment, A, is sectored into a matrix of NxN size, and one WP is placed in each cell of the matrix. The cell size of the matrix is selected by the smallest value of the maximum range covered by each sensor present with the DEP. In the Vulnerability Index Approach, WPs are deployed in vulnerable regions that have been identified during the site investigation, terrain mapping, and soil testing. The Hybrid Approach incorporates more than one approach stated earlier. After considering these different approaches, a particular approach suitable for the deployment area has to be selected.

- *Wireless Sensor Network Requirements*

  Landslide detection requires wide area monitoring, and real-time, continuous data collection, processing, and aggregation. Wireless Sensor Networks (WSNs) are the key emerging technology that has the capability to real-time, continuous data collection, processing, aggregation with minimum maintenance. Any wide area monitoring must determine the

  - maximum number of wireless sensor nodes,
  - maximum number of relay nodes,
  - maximum frequency of data collection from each node per minute,
  - maximum data rate required,
  - maximum power required for sampling, transmitting, processing, and receiving,
  - maximum tolerance limit of delay,
  - maximum tolerance limit of data packet loss,

- *Algorithm Requirements*

  Wide area monitoring requires efficient algorithm development for data collection, processing, and transmission. The different criteria to be analyzed for designing the algorithms are: the total area of deployment, maximum and minimum transmission range, maximum number of sensor nodes necessary, maximum number of sensor nodes available, maximum amount of power available (in the battery), the corresponding transmission range, data storage capability of each node, availability of constant power source, maximum bandwidth availability, frequency of data collection and transmission specific to the application scenario, and the data aggregation method suitable for the application under consideration.

  Analysis of the above requirements contributes to the development of required algorithms for designing the network topology, data collection algorithm, data aggregation algorithm, data dissemination method, energy optimized network, networks with maximum life time, time synchronized network, localization techniques etc.

- *Network Requirements*

  The design and development of the complete network architecture requires the knowledge and understanding of relevant technologies such as wireless networks, wired networks, cellular networks, satellite networks etc., maximum number of nodes, maximum data rate, available bandwidth, traffic rate, delay, distance between the point of data initiation and its destination, effect of terrain structure, vegetation index, climate variation etc., on data transmission, delay, and data packet loss, accessibility/connectivity of the area, location of DEP (Deep Earth Probe), transmission range, identification of the communication protocol and radio interface technology, integration of the application specific algorithms for data collection and aggregation, routing and fault tolerance etc. These requirements have to be thoroughly analyzed with regard to the conditions of the deployment area, maximum data transmission distance, traffic rate, and the available technologies. Choose the best technologies that can be integrated effectively to achieve minimum data packet loss, delay, minimum power consumption, and fast arrival of data.

- *Data Analysis Requirements* The data received from the deployment area has to be modeled and analyzed according the application scenario requirements. Statistical models and pattern recognition techniques can be used for further data analysis to determine the warning levels. Warning levels are the level of indication (from the sensors) that a landslide maybe becoming possible or about to occur. Along with this data analysis architecture has to be developed for effective and fast data analysis.
- *Data Visualization Requirements* The development of real-time systems requires the design and development of: a data dissemination method, a channel or technology that can be used for data dissemination (within the shortest amount of time), and the data visualization criteria & methods specific to the application scenario. The method of data dissemination, and the allowable delay for data dissemination, and the techniques that should be adopted for data dissemination will depend on the application scenario under consideration. The architecture for data visualization has to be developed with the goal of effective and fast streaming of data.

3. Wireless Sensor Network Architecture

This current deployment used a placement strategy using the Hybrid Approach, by incorporating both the Matrix Approach and the Vulnerability Index Approach. The whole deployment area was initially sectored using Matrix Approach. In each cell, the deployment location of the Wireless Probe (WP) is decided after considering the Vulnerability Index Approach. This has helped to maximize the collection of relevant information from the landslide prone area.

The wide area monitoring using Wireless Sensor Network (WSN) is achieved using a regionalized two-layer hierarchical architecture. Since the geological and hydrological properties of each of the locations, of the landslide prone area, differ with respect to the different regions they belong to they are divided into regions. The data received from each of the sensors cannot be aggregated together due to the variability in soil geological and hydrological properties. So the whole landslide prone area is divided into regions possessing soil geological and hydrological properties unique to their region. In this particular case, the deployment area is divided into three regions such as crown region, middle region, and toe region of the slope as shown in Figure 2, and numerous WPs are deployed in these regions.

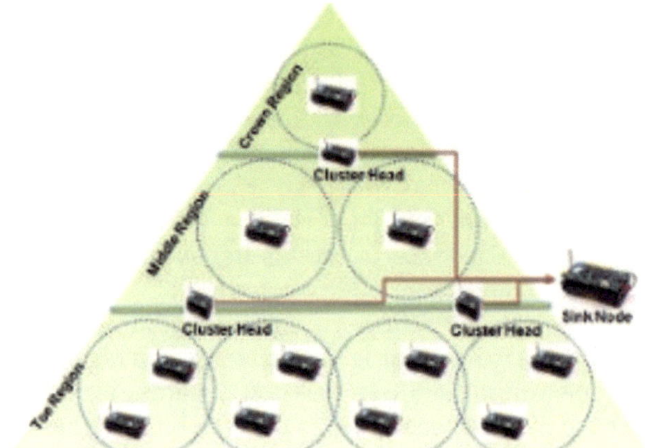

Fig. 2. Regionalized Wireless Sensor Network Architecture for Landslides

4. Wireless Network Design and Architecture

One of the important requirements for any landslide detection system is the efficient delivery of data in near real-time. This requires seamless connectivity with minimum delay in the network. The architecture we have developed for satisfying the above requirements is shown in the Figure 3.

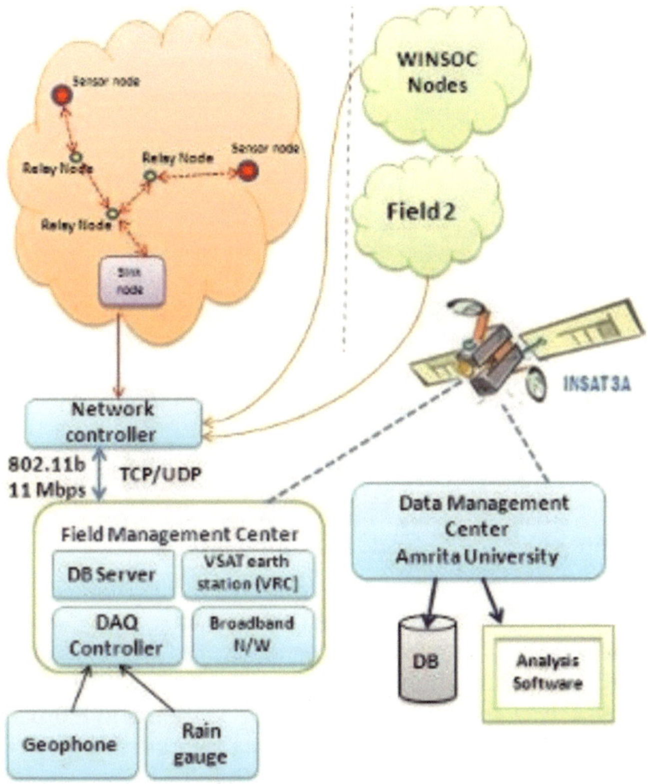

Fig. 3. Wireless Sensor Network Architecture For Landslide Detection

The wireless sensor network follows a two-layer hierarchy, with lower layer wireless sensor nodes, sample and collect the heterogeneous data from the DEP (Deep Earth Probe) and the data packets are transmitted to the upper layer. The upper layer aggregates the data and forwards it to the sink node (gateway) kept at the deployment site.

The current network has 20 wireless sensor nodes spread on two different hardware platforms. The first hardware platform is Crossbow MicaZ. This MicaZ network follows a two-layer hierarchy, with a lower level (wireless probes) and a higher level (cluster head), to reduce the energy consumption in the total network. The wireless probes (lower level nodes) sample and collect the heterogeneous data from the DEP (Deep Earth Probe) and the data packets are transmitted to the higher level. The higher level aggregates the data and forwards it to the probe gateway (sink node) kept at the deployment site.

The second hardware platform, used, is the newly developed WINSOC wireless sensor nodes. One purpose of this WINSOC network is to extensively test and validate the WINSOC nodes, shown in Figure 4, with respect to performance reliability and energy trade-offs between the two hardware platforms in a landslide scenario. WINSOC nodes are endowed with a WINSOC distributed consensus algorithm. Another purpose of this network is to test and validate the performance and scalability of the WINSOC distributed consensus algorithm in a landslide scenario. This network is scalable as it provides the capability to incorporate any new field networks to the current network.

Fig. 4. Field Deployment of WINSOC Node With MiniatureAntenna

Data received at the gateway has to be transmitted to the Field Management Center (FMC) which is approximately 500m away from the gateway. A Wi-Fi network is used between the gateway and FMC to establish the connection. The FMC incorporates facilities such as a VSAT (Very Small Aperture Terminal) satellite earth station and a broadband network for long distant data transmission. The VSAT satellite earth station is used for data transmission from the field deployment site at Munnar, Kerala, South India to the Data Management Center (DMC), situated at our university campus 300 km away.

The DMC consists of the database server and an analysis station, which performs data analysis and landslide modeling and simulation on the field data to determine the landslide probability. The real-time data and the results of the data analysis are real-time streamed on the Internet. Alert services such as E-Mail, SMS and MMS are implemented to alert about: the probability of landslides, status of the network and for monitoring the system components.

Fault tolerance is achieved even during extreme weather conditions. For example, if the VSAT network becomes unavailable, the WAWN adapts by using the broadband or GPRS connectivity at the FMC for uploading the real-time data directly to a web page with minimum delay and thus providing fault tolerance.

The entire system is equipped to remotely monitor the level of battery charges and the level of solar charging rate, and indicate faulty wireless sensor nodes or geological sensors. A feedback loop is used that remotely changes, the sampling rate of the geological sensors, with respect to the real-time climatic variations.

This proposed network architecture is scalable, as any number of nodes and new landslide deployment fields can be incorporated via a Wi-Fi network to the same FMC. In future, this will

provide the capability to monitor many very large areas and also to incorporate the different spatio-temporal analysis to provide an even better understanding of landslides.
The Munnar region experiences frequent landslides and has several landslide prone areas within every 1 sq km, which can be utilized as future extension sites for landslide detection systems. The different deployment sites can connect to the FMC via a Wi-Fi network.

5. Wireless Sensor Network Algorithms

The wireless sensor network designed and deployed for wide area landslide monitoring requires efficient data collection, data aggregation, energy management, and fault tolerant methods.
Regionalized dynamic clustering method is designed and implemented for effective geological and hydrological data collection using the wireless sensor network.
Threshold based temporal data collection and data aggregation method (19)is designed and implemented for effective data aggregation. This algorithm combined with the newly designed state transition algorithm (18) contributes optimum energy consumption by each node and in increasing the life time of the whole network, avoiding unnecessary collection, processing and transmission of redundant data thus achieving increased energy efficiency and the simplification of the data analysis & visualization process.
Fault tolerant methods are designed and integrated in the wireless sensor network for effective handling of node failure, reduced signal strength, high data packet loss, and low balance energy per node.

6. Wireless Software Architecture

Real-time monitoring and detection of landslides require seamless connectivity together with minimum delay for data transmission. The existing Wireless Sensor Network (WSN) system for landslide detection incorporates various heterogeneous wireless networks such as the WSN, Wi-Fi, satellite network, and broadband network. Each of these networks perform at different frequency range, that contributes to different traffic rate, congestion, data packet loss, buffering methods, delay, and different data collection, transmission, and processing methods. Hence to reduce the complexity in dealing with different types of wireless network, generic software architecture was designed and implemented for achieving all the requirements of each of the wireless network. This wireless software architecture includes wireless sensor network software, wireless sensor gateway software, and a middleware for heterogeneous wireless networks.

7. Design of Interfacing Sensors and Power Management Methods

We designed special purpose interfacing circuits, since the commercially available wireless sensor nodes do not include implanted geophysical sensors necessary for landslide monitoring, and also the geophysical sensors cannot be connected directly to the data acquisition board, integrated with the wireless sensor node. The special purpose interfacing circuit act as an intermediary to remove the variance experienced between the required input voltage for a data acquisition board and the output voltage received from the geophysical sensors. Thus design requirements of the interfacing board are described in (17), and the output from the interfacing board is directly fed into the data acquisition board inputs. Later, the signals were software adjusted to obtain the original sensor outputs, and hence the sensor data. The details of the interfacing circuit requirements are shown in Figure 5.

Sensor	Output type	Signal pre-processing
Strain gauge piezometer	Dual wire analog	Level shifting, Amplification
Vibrating wire piezometer	RS-232 from the data logger	None
Dielectric Moisture Sensor	Single wire analog	None
Tiltmeter	Single wire analog	Voltage reduction, Amplification
Geophone	Dual wire analog	Level shifting, Amplification

Fig. 5. Interfacing Circuit Requirements (17)

For any Wireless Sensor Network (WSN), power constraints are one of the major problems faced by wide area deployments, for real-time monitoring and detection. In the current deployment, maximum power is consumed for excitation of geophysical sensors than that of transmission, processing, or reception by a wireless sensor node. Indigenous power circuits are developed to provide constant power for the excitation of the geophysical sensors, wireless sensor nodes, and interfacing circuits, since each of them requires different levels of power. This power circuit board is designed with high efficiency regulator chips to provide multiple outputs from a single power battery input, a non-regulated 6 Volts DC supplied from rechargeable lead acid batteries. To increase the lifetime of the lead acid batteries, they are automatically recharged by the solar recharging unit using the charge controller.
Along with hardware power management methods, software methods are also incorporated. Software power solutions are implemented in the wireless sensor network by integrating switching on-off of geological sensors and by the dynamic adjustment of frequency of sensor measurements, according to the different state transitions of wireless sensor nodes as described in the research paper (18). Efficient use of power and an optimized lifetime has been achieved by these hardware and software solutions.

8. Field Deployment Methods and Experiences

The Wireless Sensor Network (WSN) for landslide detection system is deployed at Anthoniar colony, Munnar, Idukki (Dist), Kerala (State), India, shown in Figure 6. The deployment site has historically experienced several landslides, with the latest one occurring in the year 2005, which caused a death toll of 8 people.
The WSN for landslide detection system is deployed in an area of 7 acres of mountain. The whole area consists of approximately 20 wireless sensor nodes, 20 DEPs consisting of approximately 50 geological sensors. Important research focal points were deciding the DEP locations, designing and constructing the DEPs, DEP deployment methods, interfacing circuitry, WSN, Wi-Fi network, satellite network, and power solutions, soil tests, and data analysis.

Extensive field investigations were conducted for identifying the possible landslide prone areas for the deployment of the system and also for identifying the possible locations for DEP deployment. The borehole locations for the DEPs were chosen so as to measure the cumulative effect of geographically specific parameters that cause landslides.
The field deployment was performed in two phases. The pilot deployment in January 2008 to March 2008 and the main deployment from January 2009 to June 2009. The period in between these phases involved extensive testing and calibration processes.

8.1 Field Selection

Extensive field investigations were conducted for identifying the possible landslide prone areas for the deployment of the system, in the state of Kerala, India. Approximately 15 landslide prone areas have been visited and studied, that had historically experienced landslides. After extensive investigation of the 15 sites, five sites were identified as potential field deployment sites for a Wireless Sensor Network (WSN) in landslide monitoring applications. Other sites were also visited but were not deemed suitable for the field deployment due to various factors, including: difficulty of access, uncertainty about the landslide risk, lack of communication facilities, and the size of the potential landslide, among others.

Fig. 6. Deep Earth Probe Deployment Locations at the Anthoniar Colony Site, Munnar, Kerala, India

From the shortlisted five landslide prone sites, Anthoniar Colony was selected for deployment, which is located 700 meters Northwest of Munnar town. A first slide had occurred many years earlier. On July 25th, 2005, another landslide also occurred in Munnar at the Anthoniar colony. A torrential rainfall of 460mm in the middle of the monsoon period was the primary trigger. Two levels of slide area can be observed at the Anthoniar Colony, as shown in the Figure 6.
There is a high probability for another slide at this location. Some of the factors that indicate a probability of landslide is the seepage flow during the dry season, long vertical and horizontal cracks, soil material has large amount of quartz vein, and the soil type is reddish colored sticky clay. Even now when the rain falls, water will flow down on to the top of the houses that are at the foot of the hill, indicating water saturation and higher pore pressure at the toe region, which can indicate a landslide in future.

8.2 Deployment of DEP (Deep Earth Probe)

One of the important activities required for deploying the landslide detection system is bore hole drilling. Bore hole location and its depth, determines the maximum amount of geological and hydrological properties that can be gathered from the field for the functioning of the landslide detection system. Hence, the most important parameters that determine the bore hole design include the decision of location, depth, and diameter of the planned bore hole, soil sample extraction methods to be adopted, field tests involved, and the bore hole drilling method.

Different types of bore hole drilling are available such as the hand auger method, and the rotary drilling method. The pilot deployment the hand auger method was used. In the main deployment the rotary drilling method was used to drill deeper holes as deep as 23 meters as it consumed less time and labor compared to the hand auger method. Soil sample extraction and field permeability tests were performed to collect the relevant soil properties, geologic and hydrologic properties from each of the locations. Drilling was continued until the bed rock was observed. If bedrock was too deep, drilling was continued until the observation of weathered rock. If even weathered rock was too deep, drilling was stopped at a major soil layer change after the water table. The decision of the bore hole depth was chosen to be dependent on the location of the hole, vulnerability of the location, sensor deployment requirement, water table height, and location of weathered rock or bed rock.

The DEP (Deep Earth Probe) design is influenced by the local geological and hydrological conditions, the terrain structure, and accessibility of that location. The distribution pattern of different types of geophysical sensors at different depths of the DEP is unique depending on the characteristics of the specific location.

The DEPs were designed in a two stage process. Initial DEP designs were made for the pilot deployment, which consists of two DEPs, and in the main deployment, the spatial granularity was increased to 20 DEPs and 20 wireless sensor nodes. Multiple DEPs were installed in six locations (labeled henceforth as either $C1, C2, \cdots, C6$), shown in Figure 6.

In the main deployment DEPs are placed significantly deeper into the ground than in the pilot deployment, on average 2 to 5 times deeper, with a maximum depth of up to 23 meters, penetrating to the weathered rock or bed rock. The geological sensors are attached to the ABS plastic inclinometer casing according to the geological or hydrological parameter that will be measured. The maximum number of eight external sensors can be attached to the data acquisition board of Crossbow's wireless sensor nodes. The details of the connected geological sensors as on June 2009, is detailed in the Figure 7.

External hardware components have been put in two enclosures that are used to protect the data acquisition and transmission equipment for the wireless sensor node - one electronics box and one power box and are attached to poles equipped with solar panels and external antenna. These were designed and then fabricated at the University as shown in Figure 8.

8.3 Network Implementation and Integration

The network consists of a Wireless Sensor Network (WSN), Wi-Fi, a satellite network, a broadband network, a GPRS and GSM network. The network integration of all the components required different software and hardware implementations. The design and development of a WSN for the landslide scenario involves the consideration of different factors such as terrain structure, vegetation index, climate variation, accessibility of the area, location of DEP (Deep Earth Probe), transmission range, identification of the communication protocol and radio interface technology, the application specific algorithms for data collection and aggregation,

DEP Location	Piezometer Depth (m)	Moisture Sensor Depth (m)		Strain Gauge Depth(m)						Geophone (m)	Rain Gauge Height (m)
		1	2	1	2	3	4	5	6		
C1	5.5	0.8	3.05	1.5x	4x	4alpha	7x			0.5	
C2	6.8	1.73		3x	3y	6x	6y	9x	9y		
C3	15.4	1		7.5x	7.5y	10.5x	10.5y	16.5x	19.5x		
C4	18	2		3.5x	3.5y	6.5x	9.5x	11.5x	11.5y		
C5	14	1.37		4.25x	4.25y	10.75x	10.5 alpha	10.5 beta	15x		3
C6	19.75	2		2x	2y	5x	5y	11x	17x		

Fig. 7. Main Deployment - Details of Geological Sensor Deployment, Location and its Depth of Deployment

routing and fault tolerance etc. The wireless sensor nodes used for the deployment are 2.4 GHz MicaZ motes from Crossbow. The MDA 320 from Xbow is the data acquisition board used to interface the sensors with the MicaZ motes. The MicaZ samples and processes the sensor values from the MDA board that has up to 8 channels of 16-bit analog input, logs it, and sends it to the communication routines for packetizing, framing, check sum generation, etc.

Although the manufacturer specified that the MicaZ nodes could transmit up to 100 meters, in ideal field conditions (flat dirt ground, dry weather) the maximum range was around 50 or 60 meters even when the motes were being placed 2 meters above the ground. Due to this shorter transmission range, a number of relay nodes are used to maintain communication between the DEPs. These relay nodes required extensive testing for careful placement such that the network connectivity can be maintained even in the worst case weather conditions. External antennas were also used to maintain network connectivity.

The hardware of the original probe gateway is from Crossbow and is named as Stargate. Later Stargate became unavailable and out of production thereby forcing the use of a new gateway. The new gateway is based on an AMD Geode Mini-ITX Motherboard. The base station listens to the packet transmissions, in the sensor network, and logs the packet transmissions if they are addressed to the base station. After this, the sensor data is stored either accessed through the Wi-Fi network or through the Ethernet interface of the probe gateway.

Data received at the probe gateway is transmitted to the Field Management Center (FMC), using a Wi-Fi network. The Wi-Fi network uses standard, off-the-shelf Wi-Fi components, such as a compact flash Wi-Fi card, at the gateway, and an Ethernet wireless access point, at the Field Management Center. The Wi-Fi network allows us to install the gateway at any scalable distance from the FMC. The FMC incorporates a VSAT (Very Small Aperture Terminal) satellite earth station and a broadband/GPRS network for long distant data transmission.

Data received at the FMC is transmitted to the Data Management Center (DMC) using a satellite network. The data received at the DMC is analyzed using an in-house designed data analysis and visualization software. This software is interfaced with landslide modeling software and data analysis software developed at Amrita University. Landslide modeling software provides the factor of safety of the mountain and the probability of landslide occurrence with respect to the signals received from the deployed sensors. Data analysis software provides

Fig. 8. C3 Sensor Column of Main Deployment

the capability to compare and analyze data from different DEPs, different sensors in the same DEP, the same sensors in different DEPs, selective comparison, etc.
This data analysis and visualization software is also capable of real streaming the data and the results of the data analysis, over the Internet. Which makes it possible for the scientists around the world to analyze the data with very minimal delay and effective warning can be issued on time.

9. Validation of the Complete System - Landslide Warning Issued

A novel and innovative decision support system for landslide warning has been developed using a three level warning (Early, Intermediate and Imminent). The decision for each level depends on the moisture (for an Early warning), pore pressure (for an Intermediate warning), and movement (for an Imminent warning) sensor data values correlating with the rainfall intensity. Along with the three level warning system, the results of the landslide modeling software is compared to avoid false alarms. Landslide modeling software incorporates the raw sensor data from the field deployment site, along with data from soil tests, lab setup , and other terrain information to determine the Factor of Safety (FS) (term used to quantify

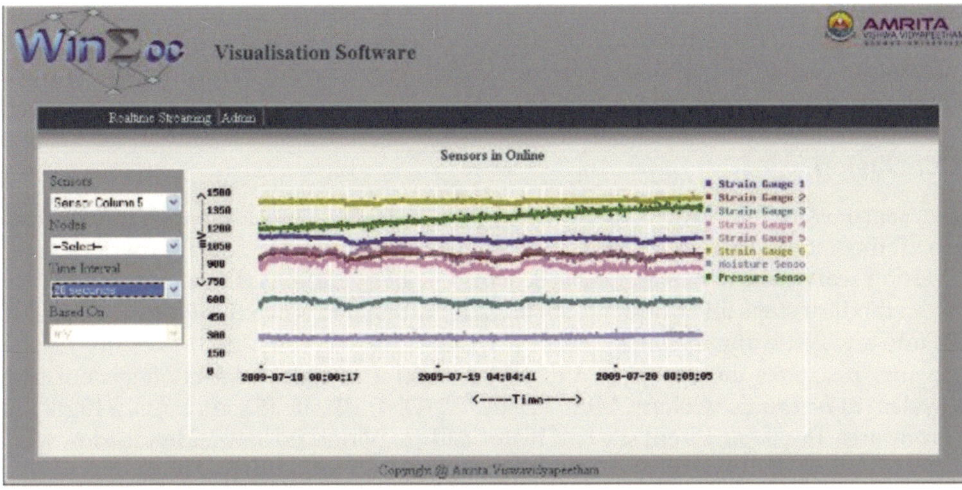

Fig. 9. Snapshot from the real streaming software, for a period of 18 July 2009 (00: 00:17) to 20 July 2009 (08:09:05) for location 5, the middle position of the hill

the slope stability). Dependent on the results reaching a threshold (that is, if $FS \leq 1$), each grid point could be pronounced 'unsafe' or 'safe'. This implementation is incorporated into the data visualization software and the results are real-time streamed to the website.

In July 2009, high rain fall was experienced at our deployment site and multiple landslides occurred all over the state of Kerala, India. The data analysis showed an increase in pore pressure and also noticeable soil movements. The pore pressure transducer deployed 14 meter deep from the surface at location 5 (which is a vulnerable area), showed a gradual increase in pore pressure. The strain gauges deployed at location 5, at various depths such as 4.25 x, 4.25 y, 10.75 x, 10.5 alpha, 10.5 beta and 15 x show noticeable movements of underneath soil. More strain gauge soil movement is shown at position 10.75 x, 10.5 alpha and 10.5. Other sensors at 6 different locations at Anthoniar Colony also showed observable soil movements and increase in pore pressure.

Our real streaming software currently incorporated to www.winsoc.org website can be used to view the pattern.

Figure 9 shows the real-time streaming data, for a period of July 18th, 2009 to July 20th, 2009 for location 5. The figure shows an increase in the pore pressure and also soil movements at the middle position of the hill, which is actually a vulnerable area after the previous landslide of July, 2005. Additionally, the soil moisture sensor readings at location 1, the toe region of the hill, were already saturated. The strain gauges at location 1 and location 4 also showed slight soil movements.All of the above analysis shows the vulnerability of Anthoniar Colony to possible landslides. In this context, we issued a preliminary warning through television channels, and the official Kerala State Government authorities were informed. The government authorities considered the warning seriously. Higher officials made visits to the landslide prone area and the people were asked to evacuate with the warning given below.

We would like to inform you that in case the torrential rainfall prevails, it would be wiser to alert the people of this region and advise them to relocate to another area till the region comes back to normalcy in terms of pore pressure and underneath soil movements.

As the rainfall reduced, the real-time streaming software showed the pore pressure reducing and then stabilizing. This situation helped us to validate the complete system. As a result of the successful warning issuance and system validation, the Indian government now wants to extent the network to all possible landslide areas.

10. Conclusion

Wireless Sensor Networks (WSNs) are still an emerging technology and much literature available is still theoretical, therefore practical deployment guides using actual experience are few if any. Using real practical experience, this overview of operations is one such guide providing the methodical steps and outlining the basic requirements when designing and deploying a WSN into any given application.
This chapter discusses the design and deployment of a landslide detection system using a WSN system at Anthoniar Colony, Munnar, Idukki (Dist), Kerala (State), India, a highly landslide prone area The deployment site had historically experienced several landslides, with the latest one occurring in the year 2005, which caused a death toll of 10 (people).
Our researchers, at Amrita University, designed and deployed a Wireless Sensor Network for the purpose of landslide detection. The complete functional system consists of 50 geological sensors and 20 wireless sensor nodes. This network has the capability to provide real-time data through the Internet and also to issue warnings ahead of time using the innovative three level warning system developed as part of this work. The system incorporates energy efficient data collection methods, fault tolerant clustering approaches, and threshold based data aggregation techniques. This wireless sensor network system is in place. For two years it has been gathering vast amounts of data, providing better understanding of landslide scenario and has been poised to warn of any pertinent landslide disaster in future. The system has proved its validity by delivering real warning to the local community during heavy rains in the last monsoon season (July 2009). This system is scalable to other landslide prone areas and also it can be used for flood, avalanche, and water quality monitoring with minor modifications.
This development describes a real experience and makes apparent the significant advantages of using Wireless Sensor Networks in Disaster Management. The knowledge gained from this actual experience is useful in the development of other systems for continuous monitoring and detection of critical and emergency applications.

Acknowledgments

The author would like to express gratitude for the immense amount of motivation and research solutions provided by Sri. Mata Amritanandamayi Devi, The Chancellor, Amrita University. The authors would also like to acknowledge Dr. P. Venkat Rangan, Dr. H. M. Iyer, Dr. P. V. Ushakumari, Dr. Bharat Jayaraman, Dr. Nirmala Vasudevan, Mr. Sangeeth Kumar, Mr. Joshua (Udar) D. Freeman, Mr. Vijayan Selvan, Mr. Kalainger (Kailash) Thangaraju, Mr. Mukundan T. Raman, Ms Rekha. Prabha, Ms. Thushara. Eranholi, Mr. Manohar. B. Patil, Ms. Erica (Thapasya) S. Fernandes for their valuable contribution to this work.

11. References

[1] Biavati, G.; Godt, W. & McKenna, J. P. (2006). Drainage effects on the transient, near-surface hydrologic response of a steep hillslope to rainfall: implications for slope stability, Edmonds, Washington, USA, *Natural Hazards Earth System Science*, Vol. 6, pp. 343-355, 2006.

[2] Caine, N. (1980). The rainfall intensity-duration control of shallow landslides and debris flows, *Geografiska Annaler*, Vol. 62A, pp. 23-27, 1980.

[3] Christanto, N.; Hadmoko, D.S.; Westen, C. J.; Lavigne, F.; Sartohadi, J. & Setiawan, M. A. (2009). Characteristic and Behavior of Rainfall Induced Landslides in Java Island, Indonesia : an Overview, Geophysical Research Abstracts, Vol. 11, 2009.

[4] Garich, E. A. (2007). Wireless, Automated Monitoring For Potential Landslide Hazards, Masters Thesis, Texas A & M University, 2007.

[5] Hill, C. & Sippel, K. (2002) Modern Deformation Monitoring: A Multi Sensor Approach, *In Proceedings of FIG XXII International Congress*, Washington D.C, USA, April 19-26, 2002.

[6] Iverson, R.M. (2000). Landslide triggering by rain infiltration, *Water Resource Research*, Vol. 36, pp. 1897-1910, 2000.

[7] Kimura, H. & Yamaguchl. Y. (2000). Detection of landslide areas using satellite radar interferometry, *Photogrammetric Engineering & Remote Sensing*, Vol. 6 (3), pp. 337-344, 2000.

[8] Kumar, V.S.; Sampath, S.; Vinayak, P. V. S. S. K. & Harikumar, R. (2007). Rainfall intensity characteristics in at coastal and high altitude stations in Kerala, *Journal of Earth Systems Sciences*, Vol. 116 (5), pp. 451-463, 2007.

[9] Kung, H.; Hua, J. & Chen, C. (2006) Drought Forecast Model and Framework Using Wireless Sensor Networks, *Journal of Information Science and Engineering*, Vol. 22, pp. 751-769, 2006.

[10] Kunnath, A. T. & Ramesh, M. V. (2010). Integrating geophone network to real-time wireless sensor network system for landslide detection, *In Proceedings of The Third International Conference on Sensor Technologies and Applications*, SENSORCOMM 2010, IEEE Xplore, 2010.

[11] Kuriakose, S.; Jetten, V. G.; van Westen, C. J.; Sankar, G. & van Beek, L. P. H. (2008). Pore Water Pressure as a Trigger of Shallow Landslides in the Western Ghats of Kerala, India: Some Preliminary Observations from an Experimental Catchment, *Physical Geography*, Vol. 29 (4), pp. 374-386, 2008.

[12] LAN, Hengxing.; ZHOU, C.; Lee , C. F.; WANG, S. & Faquan, W. U. (2003). Rainfall-induced landslide stability analysis in response to transient pore pressure - A case study of natural terrain landslide in Hong Kong, *Science in China Ser. E Technological Sciences*, Vol. 46, pp. 52-68, 2003.

[13] Liu, H.; Meng, Z. & Cui, S. (2007). A Wireless Sensor Network Prototype for Environmental Monitoring in Greenhouses, IEEE Xplore, 2007.

[14] Marchi, L.; Arattano, M. & Deganutti, A. M. (2002). Ten years of debris-flow monitoring in the Moscardo Torrent (Italian Alphs), *Geomorphology*, Elseveir, Vol. 46, pp. 1-17, 2002.

[15] Martinez, C.; Jorge, E. & Olarte Montero, Juan, eds. (2001). Socioeconomic and Environmental Impacts of Landslides in the Western Hemisphere, *Proceedings of the Third Panamerican Symposium on Landslides*, July 29 to August 3, Cartagena, Colombia, 2001.

[16] Musaloiu-E, R.; Terzis, A.; Szlavecz, K.; Szalay, A.; Cogan, J. & Gray, J. (2006). Life Under your Feet: A Wireless Soil Ecology Sensor Network, 2006.

[17] Ramesh, M. V. (2009). Real-time Wireless Sensor Network for Landslide Detection, *In Proceedings of The Third International Conference on Sensor Technologies and Applications*, SENSORCOMM 2009, IEEE Digital Library, 2009.

[18] Ramesh, M.V.; Raj, R. T.; Freeman, J. D.; Kumar, S. & Rangan, P. V. (2007). Factors and Approaches for Energy Optimized Wireless Sensor Network to Detect Rainfall In-

duced Landslides, *In Proceedings of the 2007 International Conference on Wireless Networks (ICWN'07)*, pp. 435-438, CSREA Press, June, 2007.
[19] Ramesh, M. V. & Ushakumari, P. (2008). Threshold Based Data Aggregation Algorithm To Detect Rainfall Induced Landslides, *In Proceedings of the 2008 International Conference on Wireless Networks (ICWN'08)*, Vol. 1, pp. 255-261, 2008.
[20] Ramesh, M. V.; Vasudevan, N. & Freeman, J. (2009). Real Time Landslide Monitoring via Wireless Sensor Network,*In EGU Geophysical Research Abstracts*, Vol. 11, EGU2009-14061, EGU General Assembly, 2009.
[21] Ramesh, M. V.; Kumar, S. & Rangan, P. V. (2009). Wireless Sensor Network for Landslide Detection, *In the Proceedings of the 2009 International Conference on Wireless Networks (ICWN'09)*, pp. 89-95, 2009.
[22] Ramesh, M. V. & Soman, K. P. (2009). Wireless Sensor Network LocalizationWith Imprecise Measurements Using Only a Quadratic Solver, *In the Proceedings of the 2009 International Conference on Wireless Networks (ICWN'09)*, pp. 109-112, 2009.
[23] Raj, R.; Ramesh, M. V. & Kumar, S. (2008). Fault Tolerant Clustering Approaches in Wireless Sensor Network for Landslide Area Monitoring, *In Proceedings of the 2008 International Conference on Wireless Networks (ICWN'08)*, Vol. 1, pp. 107-113, 2008.
[24] Reid, M. E.; LaHusen, M. E. & Iverson, R.M. (1988). Hydraulic factors triggering a shallow landslide failure, *Bulletin of the Association of Engineering Geologists*, Vol. 25(3), pp. 349-361, 1988.
[25] Takehiko, O. & Yoshitsugu, T. (1998). The Characteristics of Pore Water Pressure Fluctuation on the Slip Surface Based on Drainage Well Works: Study of Tairasawa Landslide Area, Saitama Prefecture, *Landslides*, Vol. 35(3), pp. 24-33, 1998.
[26] Terzis, A.; Anandarajah, A.; Moore, K. & Wang, I. J.(2006). Slip Surface Localization in Wireless Sensor Networks for Landslide Prediction, *In Proceedings of Information Processing in Sensor Networks (IPSN'06)*, IEEE, 2006.
[27] Thampi, P. K.; Mathai, J.; Sankar, G. & Sidharthan, S. (1997). Landslides: Causes, Control and Mitigation, Centre for Earth Science Studies, Trivandrum
[28] Tofani, V.; Dapporto, S.; Vannocci, P. & Casagli, N. (2005). Analysis of infiltration, seepage processes and slope instability mechanisms during the November 2000 storm event in Tuscany, *Advances in Geosciences*, Vol. 2, pp. 301-304, 2005.
[29] Tsaparas, I.; Rahardjo, H.; Toll, D. G. & Leong, E. C. (2002). Controlling parameters for rainfall-induced landslides, *Computers and Geotechnics*, Vol. 29, pp. 1-27, 2002.
[30] Turner, A. K. (1996). Landslides: Investigation and Mitigation, *United States Transportation Research Board, National Research Council*, 1996.
[31] Van Asch, Th. W. J.; Buma, J. & Van Beek, L. P. H. (1999). A view on some hydrological triggering systems in landslides, *Geomorphology*, Vol. 30 (1-2), pp. 25-32, October 1999.
[32] Wang et.al 2003 Wang, G. & Sassa, K. (2003). Pore-pressure generation and movement of rainfall-induced landslide: Effect of grain size and fine-particle content, *Engineering Geology*, Vol. 69, pp. 109-125, 2003.
[33] Wilson, R. C. & Wieczorek, G. F. (1995). Rainfall Thresholds for the Initiation of Debris Flowa at La Honda, California, *Environmental and Engineering Geoscience*, Vol. 1 (1), pp. 11-27, 1995.

5

Urban Microclimate and Traffic Monitoring with Mobile Wireless Sensor Networks

Francesco Chiti and Romano Fantacci
University of Florence
Italy

1. Introduction

Climate is usually defined as the *average* of the atmospheric conditions over both an extended period of time and a large region. Small scale patterns of climate resulting from the combined influence of topography, urban buildings structure, watercourses, vegetation, are known as *microclimates*, which refers to a *specific* site or location. The microclimate scale may be at the level of a settlement (urban or rural), neighborhood, cluster, street or buffer space in between buildings or within the building itself. Specifically, the dispersion and dilution of air pollutants emitted by vehicles is one of the most investigated topics within urban meteorology, for its fundamental impact on the environment affecting cities of all sizes. This issues concern the average and peak values of various air pollutants as well as their temporal trends and spatial variability. The accurate detection of these values might be advantageously exploited by public authorities to better plan the public and private transportation by evaluating the impact on people health, while controlling the greenhouse phenomenon.

As the unpredictable nature of a climate variations requires an incessant and ubiquitous sensing, Wireless Sensor Networks (WSNs) represent a key technology for environmental monitoring, hazard detection and, consequently, for decision making (Martinez et al., 2004). A WSN is designed to be self-configuring and independent from any pre-existing infrastructure, being composed of a large number of elementary Sensor Nodes (SNs) that can be large-scale deployed with small installation and maintenance costs.

Literature contains several examples of frameworks for evaluating the urban air quality with WSNs, as it is reported in (Santini et al., 2008). In addition, in (Cordova-Lopez et al., 2007) it is addressed the monitoring of exhaust and environmental pollution through the use of WSN and GIS technology. As micro-climate monitoring usually requires deploying a large number of measurement tools, in (Shu-Chiung et al., 2009) it is adopted vehicular wireless sensor networks (VWSNs) approach to reduce system complexity, while achieving fine-grained monitoring.

Another aspect strictly correlated with microclimate establishment is represented by the ecologic footprint of traffic congestion due to inefficient traffic management. As a consequence, an increasing number of cities are going to develop intelligent transport system (ITS) as an approach to harmonize roads and vehicles in optimized and green paths. ITSs involves several technologies as advanced informatics, data communications and transmissions, electronics and computer control with the aim of real-time traffic reporting and alerting. Such a framework allows remote operation management and self-configuration of traffic flows, as well as

specific information delivering to vehicles concerning, for instance, traffic congestion or the presence of accidents (Pinart et al., 2009)[1]. Thus, the research on data acquisition scheme has become a key point to enable effective ITSs.
At present, the acquisition of real-time traffic data is by means of installation and use of *wired* monitoring equipment in most cities. However several concerns are associated with this choice: firstly, with the continuous expansion of the city size and the increasing or the traffic roads, the more the number of wired monitoring equipment increases, the more the cost grows (*scalability*). Further, the installation of wired monitoring equipment does not have the flexibility, being difficult to (re)deploy. Finally, as urban traffic congestion has a certain degree of space-time randomness, then it is inappropriate to install monitoring equipment in *fixed* locations. On the other hand, large-scale universal installation will cause larger waste. To solve these problems, a promising approach is currently represented by WSNs applicable to all types of urban environment (Laisheng et al., 2009), as they have no space constraints, flexible distribution, mobile convenience and quick reaction.

2. Background

2.1 ITS Communications Paradigms Overview

ITS services availability relies on the presence of an infrastructure usually comprising fixed devices interconnected by an underlying network, either wired or wireless. Data exchange toward or among mobile terminals is inherently wireless, since information should directly reach the drivers through PDAs or on-board transceivers; in evidence, IEEE 802 committee has activated 11p Task Group to define a Wi-Fi extension for Wireless Access in Vehicular Environments (WAVE) (Jiang & Delgrossi, 2008). Moreover, wireless connections are needed also for data gathering, according to the Wireless Sensor Network (WSN) paradigm, comprising large number of devices in charge of sensing and relay informations to the core network (Tubaishat et al., 2009).
Within the above scenario, several communications paradigms are possible (Yousefi et al., 2006). The case in which fixed Access Points (APs) allow mobile nodes to join the network is usually referred as *infrastructure-to-vehicle* (I2V) communications and can support advanced applications such as web surfing, multimedia streaming, remote vehicle diagnostics, real time navigation, to name a few; on the other side, *vehicle-to-vehicle* (V2V) communications represent the option in which mobile nodes can directly communicate to each other without any need of infrastructure. Although V2V and I2V communications are both prominent research fields, this paper is mainly focused on the latter, as it aims to efficiently exchange short amounts of data, collected and aggregated by an in-field deployed WSN, to nomadic users, while keeping the complexity of on-board circuitry as low as possible. It is worth noticing that a reliable I2V scheme is extremely valuable even for V2V communications since, whenever a direct link among vehicles is not available, message exchange can leverage on the infrastructure instead of being successively relayed by few low-reliable mobile nodes, as addressed in (Gerla et al., 2006).
The urban environment is usually composed of a large number of mobile terminals that are likely to quickly change their reference AP, therefore facing frequent disconnection and reconnection procedures, so that it may be not viable to deliver the total amount of required data within a single session. Moreover, the urban channel is affected by long and short term fading that introduces additional delays for data retransmissions (in the case of TCP traffic)

[1] These goals could be summarized in two main fields as traffic flow *forecast* and traffic *congestion control*.

or sensibly lowers the data reliability (in the case of UDP traffic); these issues are addressed in details in (Bychkovsky et al., 2006) and (Ott & Kutscher, 2004), providing a practical case study involving IEEE 802.11b.

In general, content distribution through overlay networks is more efficient when compared to traditional solutions using multiple unicasts. In order to achieve higher throughput and failure resilience, parallel downloading from multiple overlay nodes represents a typical approach in most recent proposals (Wu & Li, 2007). However, the same content may be unnecessarily supplied by multiple nodes, rising the problem of the so called content reconciliation, which usually is a time and bandwidth consuming operation (Byers, Considine, Mitzenmacher & Rost, 2002).

2.2 Rateless Codes

Rateless (or digital fountain) codes are a recently introduced class of forward error correction codes with universally capacity-approaching behavior over erasure channels with arbitrary erasure statistics. The first practical rateless codes, called LT codes (Luby, 2002), are based on a simple encoding process where the source message of length k information symbols is encoded into a message containing a potentially infinite number of encoded symbols. Each encoded symbol is an independently and randomly created representation of the source message and, as soon as the receiver correctly receives any set of k' encoded symbols, where k' is only slightly larger than k, it is able to recover the source message.

LT encoding process is defined by the LT code degree distribution $\Omega(d)$, which is a probability mass function defined over the interval $[1,k]$. To create a new encoded symbol, the degree d is randomly sampled from the degree distribution $\Omega(d)$, d information symbols are uniformly and randomly selected from the information message, and the encoded symbol is obtained by XOR-ing d selected information symbols. Usually, information and encoded symbols are equal-length binary data packets and the XOR operation is the bit-wise XOR. Encoded symbols are transmitted over an erasure channel and decoded at the receiver using the iterative Belief-Propagation (BP) algorithm. BP algorithm for erasure channel-decoding iteratively recovers information symbols from the degree one encoded packets, and cancels out the recovered information symbols in all the remaining encoded packets (which may result in a new set of degree one encoded packets). The iterations of this simple process can lead to the complete message recovery, otherwise the receiver will have to wait for additional encoded packets in order to complete the decoding process. The key problem of LT code design is the design of the degree distribution $\Omega(d)$ that will enable source message recovery from any slightly more than k received encoded symbols using the iterative BP decoding algorithm. This problem is solved asymptotically in (Luby, 2002), where it is shown that using so called robust soliton degree distribution, it is possible to recover the source message from any k' encoded symbols, where $k' \to k$ asymptotically, with encoding/decoding complexity of the order $O(k \cdot logk)$.

Rateless codes are usually applied in multicast scenarios, where the source message is entirely available to the source node. However, in many practical systems such as wireless ad-hoc networks, WSN or p2p networks, the message of interest might be distributed over many or all network nodes. As shown recently - see (Vukobratovic et al., 2010) and references therein - distributed rateless coding may be performed as efficiently as its centralized counterparts, and may provide a number of benefits in distributed network scenarios for applications such as data gathering, data persistence and distributed data storage.

3. Proposed Approach

3.1 System Requirements and Architecture

The reference system model is derived from a real world case study, inspired by the Tuscany Region project "Metropolitan Mobility Agency Supporting Tools" (SSAMM), devoted to enhance the quality of urban transportation system introducing innovative paradigms. The addressed urban communications scenario is modeled as a two-level network, as illustrated in Fig. 1. In particular, the lower level is composed of a large number of Sensor Nodes (SNs), positioned in such way that suitable and effective sampling of the road traffic is achieved within the area of interest (Tanner, 1957). Whenever possible, SNs are deployed in correspondence with road infrastructures such as posts, lamps and traffic lights, typically arranged in a square grid fashion. Their purpose is to collect traffic flow information[2] and relay it to the higher layer consisting of interconnected network of APs. In addition to this fixed SNs, also mobile sensors are introduced; it could be the case of a public vehicle equipped with gas analyzers for the classical air pollutants NO, NO_2, O_3 and NO, in order to record air pollution and meteorological data within different urban zones. In the meanwhile vehicles can deliver information regarding the interarrival time between adjacent APs, which is useful in estimating the congestion level. Finally, APs deliver gathered data toward Mobile Collector (MC) usually referred to as *data mule*.
As the proposed application scenario is concerned with fast and efficient information retrieval, these drawbacks could be faced by introducing an appropriate *data dissemination* algorithm, enhancing the information *persistence* throughout the network without an excessive overload in terms of total packet transmissions. To face MC inherent mobility, a *distributed data gathering* protocol has been introduced (Stefanovic et al., 2011) to efficiently collect all the sensed data by visiting only an arbitrary subset of the SNs; this general requirement is extremely important in urban scenarios, since path are usually space-time *constrained*. This has been achieved resorting to a distributed implementation of rateless codes (Byers, Luby & Mitzenmacher, 2002), a particular class of erasure correction codes that rely on sparse binary coefficient data combining, being suitable for the envised I2V data dissemination application, in which devices exhibit low computational capabilities. Moreover, it has been introduced an adequate *data dissemination protocol* which has been integrated with a MC *data gathering* scheme specifically designed for a urban wide area monitoring WSN in order to allow reliable and accurate sensing collection.

3.2 Communications Scheme

Each homogeneous subset of SNs is connected in a star-wise or tree topology to an AP; APs encode and exchange packets received from SNs, then broadcast the information to MCs. MCs usually join the network without need of an association with a specific AP by adopting a *passive operation mode* and continuously collecting information regarding the surrounding environment broadcasted by APs[3]. Nevertheless, whenever MCs are involved in disseminating their own information, they explicitly associate to the best AP and operate both in *transmitting* and *receiving* modes. However, inter-vehicle communications are not hereafter considered. Finally, we assume that MCs have on-board capabilities to process the downloaded data according to suitable applications in order to interpret current traffic information. Specifically,

[2] Different types of information, e.g., average crossroad waiting time, presence of roadworks or accidents, could be of interest.

[3] This is adopted in order to lower the implementation complexity and cost of the mobile equipment, minimize the downloading time avoiding access contentions and complex handover procedure.

the collected real-time data provide opportunity for on-board computer to perform optimal route calculation, delay estimates, and present driver with visual map representation of critical locations where accidents, high pollutants concentrations or severe road congestions took place. Although the push mode is possible, data processing is usually implemented in an automatic (i.e., periodic) manner in order to guarantee as much as possible real-time monitoring of the traffic-load conditions. In particular, the latter mode of operation makes it possible for MCs to be informed about dangerous situations (e.g., accidents) in a short time span, hence allowing for increased safety of people and vehicles.

The communication between SNs (even mounted on board of a MC) and APs, and between MCs and APs, is assumed to be based on wireless technology. As recently shown, IEEE 802.11b/g standards demonstrated significant potential for vehicular applications (Bychkovsky et al., 2006). Another candidate could be IEEE 802.11p (still in the draft stage), whose one of the aims is to support efficient data exchange between roadside infrastructure and vehicles.

Regarding the communications between APs, it is accomplished by leveraging on a pre-existing infrastructure deployed in a urban area, i.e., connecting APs to wired Metropolitan Area Network (MAN), such the one adopted by Florence Municipality, called *FI-Net*, comprised of a double fiber optics ring with a 2×2.5 Gbps full-duplex capacity. Full-mesh wireless interconnections are not considered as the adoption of an IEEE 802.11 unique radio interface could pose several limitations in terms of coverage or, equivalently, scalability.

However, the communication scenario described above fits in the infrastructure mode of the IEEE 802.11 standard in which several APs are interconnected using an external distributed system, forming an *Extended Service Set* (ESS).

3.3 Distributed Data Gathering, Encoding and Dissemination

The system application, residing in APs, periodically performs the following three procedures, according to (Stefanovic et al., 2011): (i) data gathering from SNs, (ii) encoding and (iii) disseminating encoded data to MCs. We refer to these three stages as upload, encoding and download phase, respectively, and the period encompassing all of them as *data refreshment period*. According to the IEEE 802.11 standard, the link time in every AP coverage zone is divided in superframes (IEEE, 2007), and the data refreshment period in each zone is aligned with superframe boundaries (see Fig. 2).

During the upload phase, every AP polls all SNs in its domain[4] and collects the most recent measurements. As typically foreseen by most of the IEEE 802.11 standards (IEEE, 2007), superframes are divided into the Contention-Free Period (CFP) and Contention-Based Period (CBP), where the former is used to avoid MAC collisions and deliver prioritized information to MCs. The polling phase can be accomplished within the CFP part of a typical frame. CFP always starts after a beacon with a delivery traffic information map (DTIM) field sent by AP to STAs (that is SNs in our case). STAs associated with AP learn when the CFP should begin and automatically set their NAV to MaxCFPDuration when the CFP is expected to begin. Then AP individually polls each STA with a CF-poll message waiting for DATA and CF-ACK messages from it, where messages are separated by a Short Interframe Space (SIFS) period. We assume

[4] SNs can be either fixed, i.e., infrastructured, or mobile, i.e. on board of MCs. In the latter case it is necessary that the sojourn time of MC is comparable with superframe. This condition is easily satisfied in the case of public transportation means close to a regular or temporary stop where an AP has been placed.

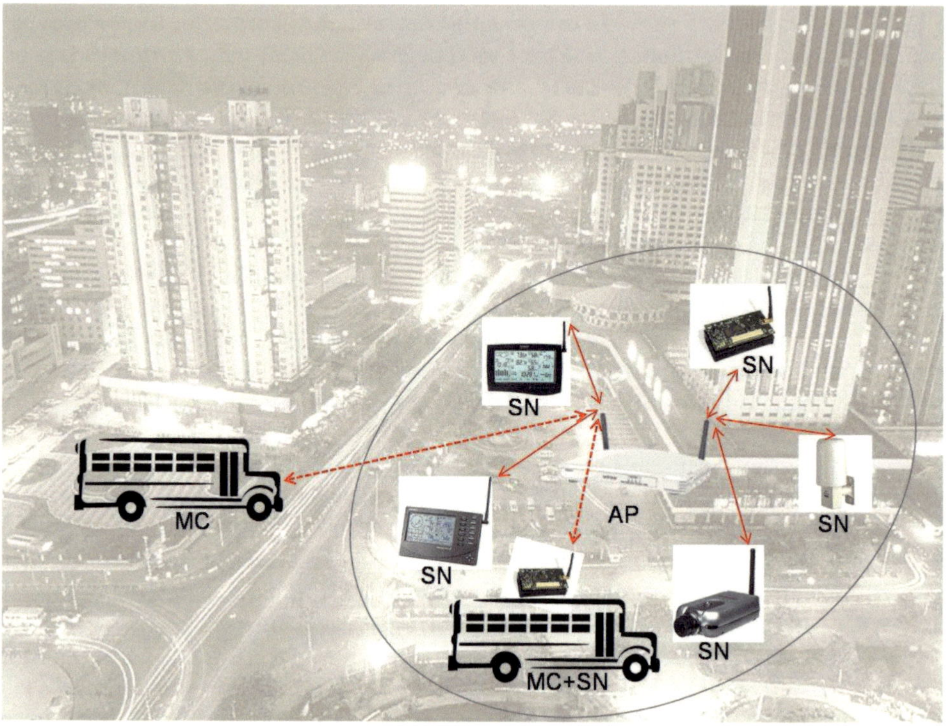

Fig. 1. Reference network topology for general I2V applications.

that APs are globally synchronized[5], so the actual upload takes place in the *first* superframe period following the start of the data refreshment period. Each SN uploads its measurements within a single data packet of length L bits. Since SNs and APs form an *infrastructured* network, it has been supposed that nodes have been previously deployed in line-of-sight (LoS) fashion in order to optimize link quality; however possible packet losses are managed by means Automatic Repeat reQuest (ARQ) scheme, so that from an application point of view data delivering could be considered reliable. In particular, to match the constraint of polling completion within the first frame, a maximum ARQ retransmission persistency equal to N_A attempts, where this parameter is selected in order to yield a negligible *residual* packet error probability.
On reception, AP stores and uniquely indexes each received data packet, where the indexing scheme is known to all APs. The total number of stored data packets in APs network per data refreshment period is k, which is equal to the total number of SNs. These k data packets represent a single data generation, upon which the rateless coding is performed. The differentiation among data generations can be achieved using appropriate field in packet header, allowing MCs to maintain global time-references.
After the upload phase, the system application, distributed over all APs in the network, performs distributed rateless encoding of collected data packets (i.e., rateless coding is used at the

[5] It is easily provided by the preexisting MAN communications infrastructure which arrange a sort of *APs network*.

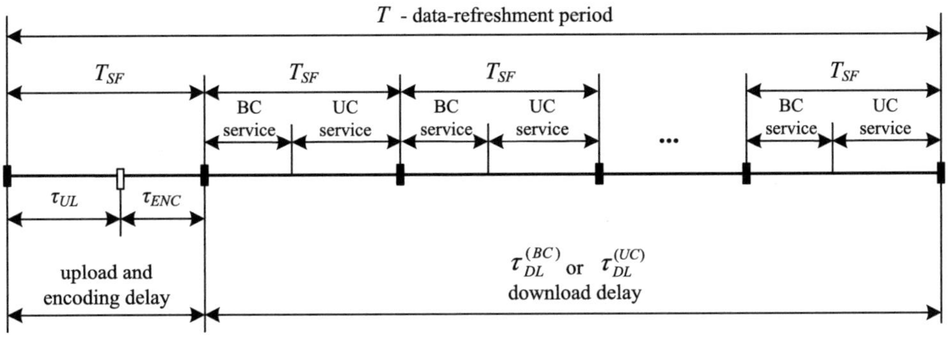

Fig. 2. Data refreshment period of the proposed application.

application level). Each AP independently produces k_{AP} encoded packets, where the actual value of k_{AP} is chosen such that it is sufficient for successful data recovery by all MCs with high probability (w.h.p.). Specifically,

$$k_{AP} \geq \left(\frac{1 + \epsilon^{(\max)}}{1 - P_{PL}} \right) \cdot k$$

where $\epsilon^{(\max)}$ is the reception overhead that allows for decoding w.h.p. and which depends on the properties of the applied rateless codes, while P_{PL} is the estimated link-layer packet loss probability. For each encoded packet, AP draws degree d from the employed degree distribution and then randomly selects d data packets from the pool of all k data packets residing in the AP network during the *current* data refreshment period. In general case, most of the selected data packets are likely to be stored in other APs, so the AP has to request them using the known indexing scheme. After reception of the missing data packets, AP creates encoded packets by simple bitwise XOR of associated data packets.

Finally, in the download phase, each AP disseminates encoded packets by simply *broadcasting* them to MCs currently falling in its coverage area. This approach has been adopted in order to minimize the complexity and the power consumption of MC receiver by always keeping it in a receiving mode. Depending on the provided service, two kinds of dissemination are possible: broadcast (BC) and geocast (GC) ones. The BC service covers simultaneous *global* distribution of the most important data such as key traffic info to all the associated MCs, using the broadcast MAC address. The GC service could be used for an additional (e.g., traffic congestion and air quality) *local* delivering of uncoded packets containing information on the actual hot spot (e.g., context aware information for navigation software enahnced services).

Due to its duration and broadcast nature, the BC service is capable of delivering significantly larger amount of data per superframe to its users as compared to the GC service, which is why the former is preferable for delay-sensitive real-time information delivery. The dissemination starts in the first superframe that follows the encoding phase, and lasts until the next data collection phase (i.e., the next data refreshment period). For the purpose of BC service, the natural choice is to use CFP part of the superframe, as it guarantees delivery of traffic-info updates to all subscribed MCs within the service area.

While traveling within the service area, each MC performs a channel sensing at periodic intervals (say θ), dynamically selects the best carrier and transparently roams among adjacent APs, while downloading encoded packets from APs, until it collects enough for sensor data recovery using the iterative BP algorithm. The number of excessive encoded packets compared to k sensor packets is measured by the reception overhead ϵ[6]; i.e., for successful recovery MC needs in total $k' = (1+\epsilon') \cdot k$ encoded packets, where ϵ' usually is a small positive number. Since each encoded packet is an innovative representation of the original data, *any* subset of $k' = (1+\epsilon') \cdot k$ taken from the set of *all* the encoded packets in the network allows for restoration of the whole original data. This property of rateless codes makes them a perfect candidate to be used at the application level for content delivery in vehicular networks, since packet losses caused by the varying link characteristics are compensated simply by reception of the new packets and there is no need for standard acknowledgment-retransmission mechanisms which can not be supported by a semi-duplex architecture as the one adopted. In other words, the usage of connection-oriented transport protocols like TCP can be avoided, as UDP-like transport provides a satisfactory functionality. Moreover, the loosing of packets caused by channel error or by the receiver deafness during the selection of a different AP does not impact on BC scheme, as MC continues downloading data *without* any need for (de/re)association, session management or content reconciliation.

4. Simulation Results

The simulation setup assumes that the urban area is covered by a regular hexagonal lattice, where each non-overlapping hexagon represents the coverage area of a single AP and the hexagon side length is equal to the AP transmission range. MCs move throughout the lattice using the rectangular grid that models urban road-infrastructure, associating with the nearest AP. The overlay hexagonal AP lattice is independent and arbitrarily aligned with the underlying rectangular road-grid. The MCs move according to the Manhattan mobility model (Bai et al., 2003), a model commonly used for metropolitan traffic. In brief, Manhattan mobility model assumes a regular grid consisting of horizontal and vertical (bidirectional) streets; at each intersection, MC continues in the same direction with probability 0.5 or turns left/right with probability 0.25 in each case. The MC speed is uniformly chosen from a predefined interval and changes on a time-slot basis (time-slot duration is a model parameter), with the speed in the current time-slot being dependent on the value in the previous time-slot. Besides temporal dependencies, Manhattan mobility model also includes spatial dependencies, since the velocity of a MC depends on the velocity of other MCs moving in the same road segment and in the same direction; as we are interested only in I2V communications from the perspective of a single user (i.e., a single MC), spatial dependencies are omitted in our implementation.

The purpose of the simulations is to estimate the duration of the download phase, as the most important and the lengthiest phase of the data refreshment period. In each simulation run, while moving on the road grid, the MC starts receiving the encoded data from the AP in whose coverage zone it is currently located. The reception of the encoded packets continues until the MC collects enough to successfully decode all the original data. If during this process, MC happens to move to another AP zone, it simply associates to a new local AP (i.e., handover takes place) and starts to receive its encoded packets. Also, if the AP has transmitted all of its encoded packets to the MC, but it failed to decode the data (e.g., due to link-layer packet losses), the MC suspends data reception until it enters the new AP coverage zone. The

[6] This takes into account both the decoding overhead as well as the redundancy needed in the presence of erasure channel.

simulation run ends when the decoding is finished and all the original data packets are retrieved. All the presented results are obtained by performing 1000 simulation runs for each set of parameters.

System Parameter	**Value**
AP transmission range	400 m
N_{AP} (no. of APs in the system)	40
N_s (no. of sensor nodes per AP)	50
k (no. of data packets)	2000
L (data packet length)	250 byte
$k \cdot L$ (total amount of original data)	4 Mbit $\approx$ 0.48 Mbyte
c, δ (rateless code parameters)	0.03, 0.5
k_{AP} (no. of encoded packets per AP)	3600
R (bit-rate)	6, 11, 12, 24 Mbit/s
T_{SF} (superframe duration)	100 ms
τ_{HO} (handover time)	0.5 s
P_{PL} (packet-loss probability)	0.3
road-segment length	150 m
velocity	4 - 17 m/s
acceleration	$\pm$ 0.6 m/s^2
mobility model time-slot duration	2 s

Table 1. Simulation Parameters

Table 1 summarizes the values for the communication and mobility model parameters used in simulations. The number of APs is chosen such that it provides a coverage area which is approximately equal to a medium-sized city area. The data packet length is estimated in such way that is sufficient to accommodate single sensor readings and additional headers (i.e., IEEE 802.11 MAC and LLC, network and transport layer). The values for bit-rate and superframe duration are selected as suggested in (Bohm & Jonsson, 2008) and (Eriksson et al., 2008), pessimistic assumption on packet-loss rate and estimate of the mean MC handover time were taken from (Bychkovsky et al., 2006), the average road segment length (i.e., average distance between two intersections) from (Peponis et al., 2007). The number of encoded packets per AP, k_{AP} is chosen such that a MC could decode all original data with probability of 0.99, when downloading from a single AP and considering employed rateless code properties and assumed link-layer packet-loss rate. In other words, $k_{AP} > \left(1 + \epsilon^{(max)}\right) \cdot k \cdot L/(1 - P_{PL})$.

Fig. 3 presents the probability P_{SD} that the MC successfully decodes the sensor data as a function of time, for the BC service and $T_{SF}^{(BC)} = 0.1 \cdot T_{SF}$. The value for T_{SF} is selected such that it leaves enough room for the GC service and other usual best-effort services. As it can be observed from the figure, for higher bit-rates (i.e., $R > 6$ Mbit/s), the MCN is able to

successfully decode w.h.p. all the data in the time span of several seconds. The positive effect of rateless coding is inherent in the fact that, even in the worst case, the data refreshment period is below 15s, a value that still allows for real-time information updates and which could be decreased further by assigning a larger superframe fraction to the BC service. As opposed to rateless encoded data delivery, the uncoded data delivery would result in retransmission feedback implosion for BC service, overwhelming the sender (i.e., AP) with unwanted traffic. The probability of successful decoding for GC service is presented in Fig. 4, where the fraction of the superframe assigned to a single user is assumed to be $T_{SF}^{(GC)}/N_{MN} = 0.01 \cdot T_{SF}$; the values for $T_{SF}^{(UC)}$ and N_{MN} are taken from the realistic analysis given in (Bohm & Jonsson, 2008). Fig. 4 demonstrates that for the standard GC service, the data refreshment period is of the order of minutes rather than seconds, which limits its usage for the applications that tolerate larger update periods. However, this period would be significantly longer if rateless coding was not used, since the link layer retransmissions would make the data delivery process considerably less efficient. Finally, it can be observed that, for the GC service, the differences in transmission bit-rate have a significant impact on the download delay, which makes higher bit-rates desirable.
Fig. 5 presents the duration of the time interval $T_{0.99}$ for which a MN, using the GC service, decodes all the original data with probability $P_{SD} = 0.99$, as a function of the number of users N_{MN} and for the fixed $T_{SF}^{(GC)} = 0.8 \cdot T_{SF}$. The figure shows a linear increase in $T_{0.99}$ as the rate decreases or N_{MN} increases, verifying that the content reconciliation phase is indeed unnecessary, since the change of the AP does not introduce additional delays apart from the handover time. In other words, after a handover, MC seamlessly advances both with the receiving and decoding processes.
Finally, Fig. 6 shows the cumulative distribution function F_T of the number of transmitted packets using the GC service from an AP to any MC within a single AP domain. The $T_{SF}^{(GC)}/N_{MN}$ ratio of the UC service is set to $0.005 \cdot T_{SF}$ or $0.01 \cdot T_{SF}$. As it can be observed, the number of transmitted packets to a MC reaches the threshold value k_{AP} equal to 3600 for the selected parameter values (Table I), in all cases but for $R = 6$ Mbit/s and $T_{SF}^{(GC)}/N_{MN} = 0.005 \cdot T_{SF}$. This means that number of encoded packets per AP (i.e., k_{AP}) is properly dimensioned to allow a single user to collect enough of encoded packets to decode all the data w.h.p. while moving through a single AP coverage zone.

To summarize the benefits provided by the proposed I2V data dissemination based on the rateless codes over traditional methods, it is worth noticing first of all that, by their design, rateless codes are tuned to the changing wireless link conditions and have a close-to-the-minimal reception overhead. Furthermore, each rateless coded packet is an equally important representation of the original data, which makes lengthy TCP-like reliability mechanisms unnecessary. These factors influence the time allocations within the superframe, allowing larger number of mobile nodes to be serviced during designated service-time portion of the superframe, or alternatively, service-time portion shortening, providing larger time allocations for best-effort traffic. Finally, while roaming through the network, mobile users can simply continue with data download from the new local AP after a handover, avoiding the redundant content reconciliation phase.

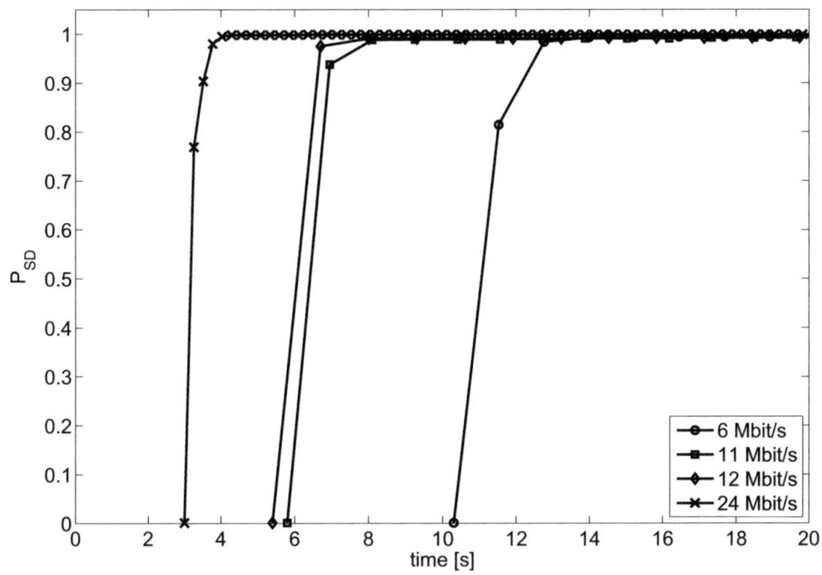

Fig. 3. Probability of successful decoding P_{SD} for BC service, $T_{SF}^{(BC)} = 0.1 \cdot T_{SF}$.

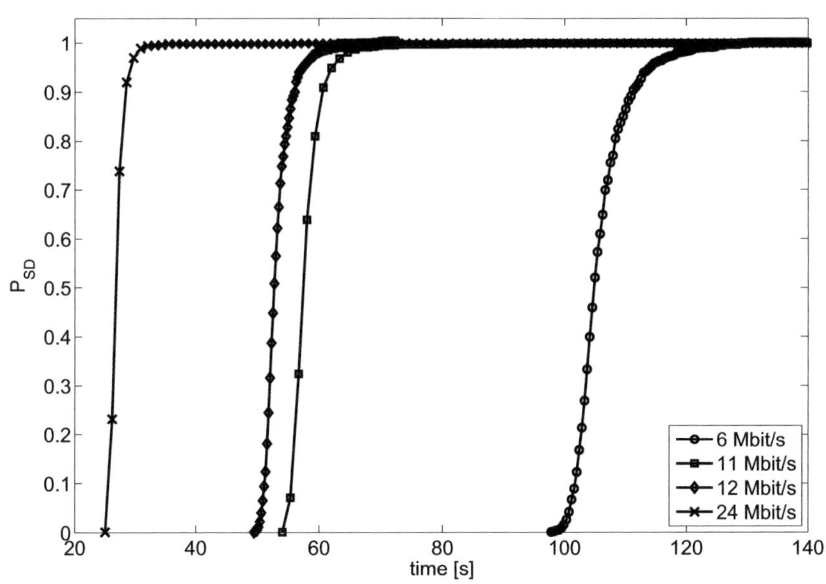

Fig. 4. Probability of successful decoding P_{SD} for GC service, $T_{SF}^{(GC)} / N_{MN} = 0.01 \cdot T_{SF}$.

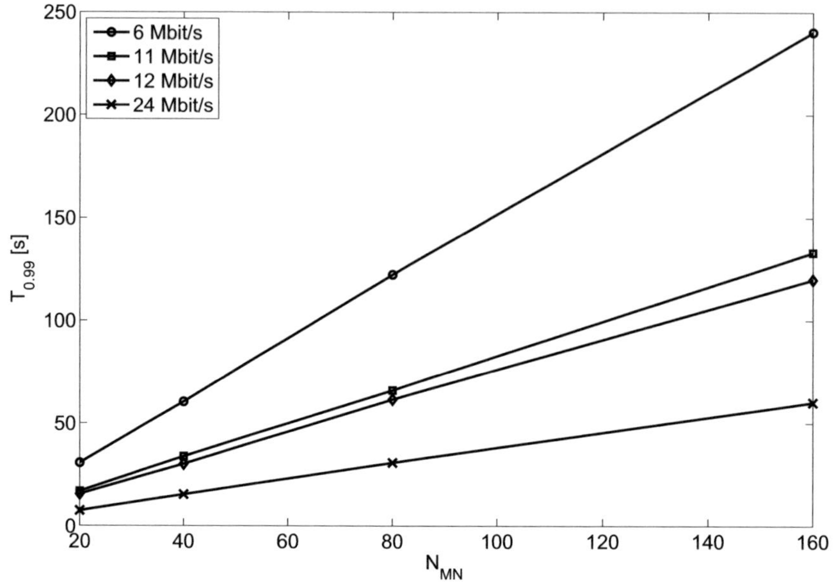

Fig. 5. Duration of time-interval $T_{0.99}$ for which MC decodes all data with $P_{SD} = 0.99$ for GC service.

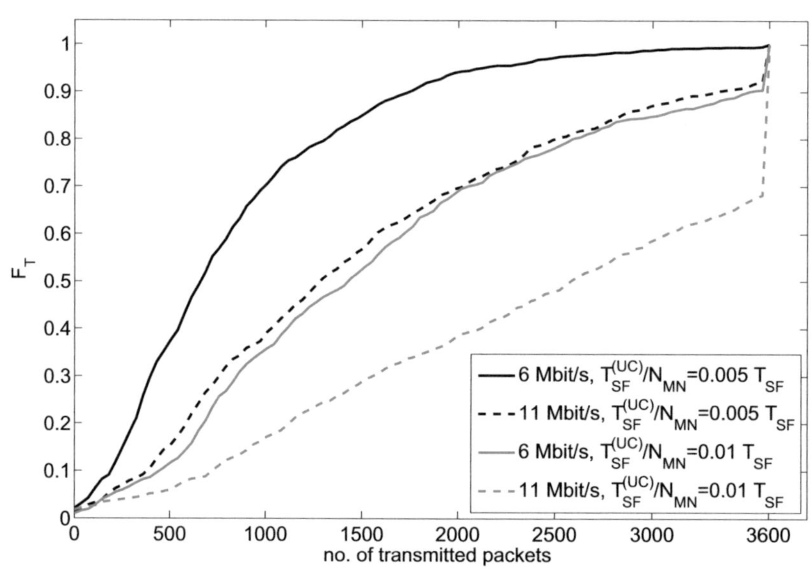

Fig. 6. Cumulative distribution function F_T of number of transmitted packets to MN in single AP cell for GC service.

Acknowledgment

This work was supported in part by by the Italian National Project "Wireless multiplatfOrm mimo active access netwoRks for QoS-demanding muLtimedia Delivery" (WORLD), under grant number 2007R989S, as well as by the Tuscany Region projects "Metropolitan Mobility Agency Supporting Tools" (SSAMM) and "Microparticulate Monitoring via Wireless Sensor Networks" (MAPPS) The authors would like also to thank the partners of EU FP7-REGPOT-2007-3 - "AgroSense" project for their fruitful discussion and comments.

5. References

Bai, F., Sadagopan, N. & Helmy, A. (2003). IMPORTANT: A framework to systematically analyze the Impact of Mobility on Performance of RouTing protocols for Adhoc NeTworks, *Proc. of IEEE INFOCOM 2003*, San Francisco, CA, USA.

Bohm, A. & Jonsson, M. (2008). Supporting real-time data traffic in safety-critical vehicle-to-infrastructure communication, *Proc. of IEEE LCN 2008*, Montreal, QC, Canada.

Bychkovsky, V., Hull, B., Miu, A., Balakrishnan, H. & Madden, S. (2006). A Measurement Study of Vehicular Internet Access Using *in Situ* Wi-Fi Networks, *Proc. of ACM MobiCom 2006*, Los Angeles, CA, USA.

Byers, J., Considine, J., Mitzenmacher, M. & Rost, S. (2002). Informed Content Delivery Across Adaptive Overlay Networks, *Proc. of ACM SIGCOMM 2002*, Pittsburg, PA, USA.

Byers, J., Luby, M. & Mitzenmacher, M. (2002). A Digital Fountain Approach to Asynchronous Reliable Multicast, *IEEE Journal on Selected Areas in Communications* **20**(8): 1528–1540.

Cordova-Lopez, L. E., Mason, A., Cullen, J. D., Shaw, A. & Al-ShammaŠa, A. (2007). Online Vehicle and Atmospheric Pollution Monitoring using GIS and Wireless Sensor Networks, *Proc. of ACM IntŠl Conference on Embedded Networked Sensor Systems (SenSys)*, pp. 87Ű–101.

Eriksson, J., Balakrishnan, H. & Madden, S. (2008). Cabernet: Vehicular Content Delivery Using WiFi, *Proc. of ACM MobiCom 2008*, San Francisco, CA, USA.

Gerla, M., Zhou, B., Lee, Y. Z., Soldo, F., Lee, U. & Marfia, G. (2006). Vehicular Grid Communications: The Role of the Internet Infrastructure, *Proc. of WICON Š06*, Boston, MA, USA.

IEEE (2007). Ieee 802.11-2007 wireless lan medium access control and physical layers specifications.

Jiang, D. & Delgrossi, L. (2008). IEEE 802.11p: Towards an International Standard for Wireless Access in Vehicular Environments, *Proc. of IEEE VTC2008-Spring*, Singapore.

Laisheng, X., Xiaohong, P., Zhengxia, W., Bing, X. & Pengzhi, H. (2009). Research on traffic monitoring network and its traffic flow forecast and congestion control model based on wireless sensor networks, *Measuring Technology and Mechatronics Automation, 2009. ICMTMA '09. International Conference on*, Vol. 1, pp. 142 –147.

Luby, M. (2002). LT Codes, *Proc. of IEEE FOCS 2002*, Vancouver, BC, Canada.

Martinez, K., Hart, J. & Ong, R. (2004). Environmental Sensor Networks, *IEEE Computer Journal* **37**: 50–56.

Ott, J. & Kutscher, D. (2004). Drive-thru Internet: IEEE 802.11b for Automobile Users, *Proc. of IEEE Infocom 2004*, Hong Kong.

Peponis, J., Allen, D., Haynie, D., Scoppa, M. & Zhang, Z. (2007). MEASURING THE CONFIGURATION OF STREET NETWORKS: the Spatial profiles of 118 urban areas in the 12 most populated metropilitan regions in the US, *Proc. of 6th International Space Syntax Symposium*, Istanbul, Turkey.

Pinart, C., Calvo, J. C., Nicholson, L. & Villaverde, J. A. (2009). ECall-compliant early crash notification service for portable and nomadic devices, *Proc. of IEEE VTC2009-Spring*, Barcelona, Spain.

Santini, S., Ostermaier, B. & Vitaletti, A. (2008). First Experiences using Wireless Sensor Network for Noise Pollution Monitoring, *Proc. of 3rd ACM Workshop on Real-World Wireless Sensor Networks (REALWSN'08)*, Glasgow, United Kingdom.

Shu-Chiung, H., You-Chiun, W., Chiuan-Yu, H. & Yu-Chee, T. (2009). A Vehicular Wireless Sensor Network for CO2 Monitoring, *Proc. of IEEE Sensors*, pp. 1498 – 1501.

Stefanovic, C., Crnojevic, V., Vukobratovic, D., Niccolai, L., Chiti, F. & Fantacci, R. (2011). Urban Infrastructure-to-Vehicle Traffic Data Dissemination Using Rateless Codes, *to appear in IEEE Journal on Selected Areas on Communications special issue on Vehicular Communications and Networks* .

Tanner, J. C. (1957). The Sampling of Road Traffic, *Journal of the Royal Statistical Society. Series C (Applied Statistics)* **6**(3): 161–170.

Tubaishat, M., Zhuang, P., Qi, Q. & Shang, Y. (2009). Wireless sensor networks in intelligent transportation systems, *Wireless Communications and Mobile Computing 2009* **9**(3): 287–302.

Vukobratovic, D., Stefanovic, C., Crnojevic, V., Chiti, F. & Fantacci, R. (2010). Rateless Packet Approach for Data Gathering in Wireless Sensor Networks, *to appear in IEEE Journal on Selected Areas in Communications special issue on Simple Wireless Sensor Networking Solutions)* **28**(7).

Wu, C. & Li, B. (2007). Outburst: Efficient Overlay Content Distribution with Rateless Codes, *Proc. of IFIP Networking 2007*, Atlanta, GA, USA.

Yousefi, S., Mousavi, M. S. & Fathy, M. (2006). Vehicular Ad Hoc Networks (VANETs): Challenges and Perspectives, *Proc. of ITST 2006*, Chengdu, China.

6

Improving Greenhouse's Automation and Data Acquisition with Mobile Robot Controlled system via Wireless Sensor Network

István Matijevics* and Simon János**
*University of Szeged, Institute of Informatics
Hungary
**Subotica Tech, Department of Informatics
Serbia

1. Introduction

The function of a greenhouse is to create the optimal growing conditions for the full lifecycle of the plants. Using autonomous measuring stations helps to monitor all the necessary parameters for creating the optimal environment in the greenhouse. The robot equipped with sensors is capable of driving to the end and back along crop rows inside the greenhouse. This chaper deals with the implementation of mobile measuring station in greenhouse environment. It introduces a wireless sensor network that was used for the purpose of measuring and controlling the greenhouse application. Continuous advancements in wireless technology and miniaturization have made the deployment of sensor networks to monitor various aspects of the environment increasingly flexible. Climate monitoring is vitally important to the operation in greenhouses and the quality of the collected information has a great influence on the precision and accuracy of control results. Currently, the agro-alimentary market field incorporates diverse data acquisition techniques. Normally, the type of acquisition system is chosen to be optimal for the control algorithm to be used. For traditional climate monitoring and control systems, all sensors are distributed through the greenhouse and connected to the device performing the control tasks. These equipments use time-based data sampling techniques as a consequence of using time-based controllers. Typical applications of WSNs include monitoring, tracking, and controlling. Some of the specific applications are habitat monitoring, object tracking, etc. In a typical application, a WSN is scattered in a region where it is meant to collect data through its sensor node. The WSN-based controller has allowed a considerable decrease in the number of changes in the control action and made possible a study of the compromise between quantity of transmission and control performance. In modern greenhouses, several measurement points are required to trace down the local climate parameters in different parts of the big greenhouse to make the greenhouse automation system work properly. Cabling would make the measurement system expensive and vulnerable. Moreover, the cabled measurement points are difficult to relocate once they are installed. Thus, a wireless sensor network (WSN) consisting of small-size wireless sensor nodes equipped with radio and one or several sensors, is an attractive and cost-efficient option to build the required measurement system. In this work, we developed a wireless sensor node for greenhouse monitoring by integrating a sensor platform provided SunSPOT by Sun Microsystems with few sensors capable to measure four climate variables. Continuous advancements in wireless technology and miniaturization have made the deployment of sensor networks to monitor various aspects of the environment increasingly flexible.

2. Mobile platform

Mobile robotics is a young field of research. Its roots include many engineering and science disciplines, from mechanical, electrical and electronics engineering to computer, cognitive and social sciences. The Board Of Education is a complete, low-cost development platform equipped with the needed sensors for humidity, temperature, light, etc. As shown in Figure 1, the Boe-Bot is a great tool with which to get started with robotics.

Fig. 1. Assembled Boe-Bot

The SunSPOT WSN module makes it possible for the Boe-Bot robot's BASIC Stamp 2 microcontroller brain to communicate wirelessly with a web based user interface running on a nearby PC. The BASIC Stamp microcontroller runs a small PBASIC program that controls the Boe-Bot robot's servos and optionally monitors sensors while it communicates wirelessly with the web server.

3. Control scheme for mobile robots

A mobile robot needs locomotion mechanisms that enable it to move throughout its known or unknown environment. But there are a large variety of possible ways to move, and so the selection of a robot's approach to locomotion is an important aspect of mobile robot design.

Figure 2, presents the control scheme for mobile robot systems. In the laboratory, there are research robots that can walk, jump, run, slide, skate, swim, fly, and, of course, roll. Any of these activities has its own control algorithm (Gy. Mester, 2009).

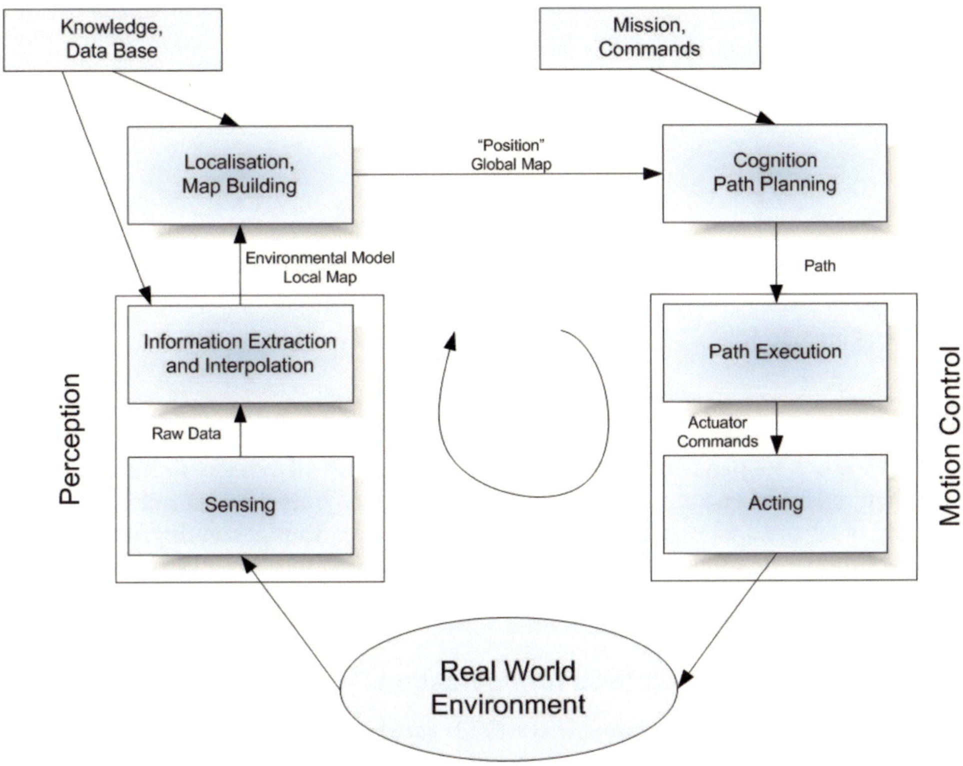

Fig. 2. Reference control scheme for mobile robot systems

Locomotion is the complement of manipulation. In manipulation, the robot arm is fixed but moves objects in the workspace by imparting force to them. In locomotion, the environment is fixed and the robot moves by imparting force to the environment. In both cases, the scientific basis is the study of actuators that generate interaction forces, and mechanisms that implement desired kinematical and dynamic properties. The wheel has been by far the most popular mechanism in mobile robotics and in man-made vehicles in general. It can achieve very good efficiencies, and does so with a relatively simple mechanical implementation. On Figure 3, the kinematics of the mobile robot is depicted. In addition, balance is not usually a research problem in wheeled robot designs, because wheeled robots are almost always designed so that all wheels are in ground contact at all times (Gy. Mester, 2009).

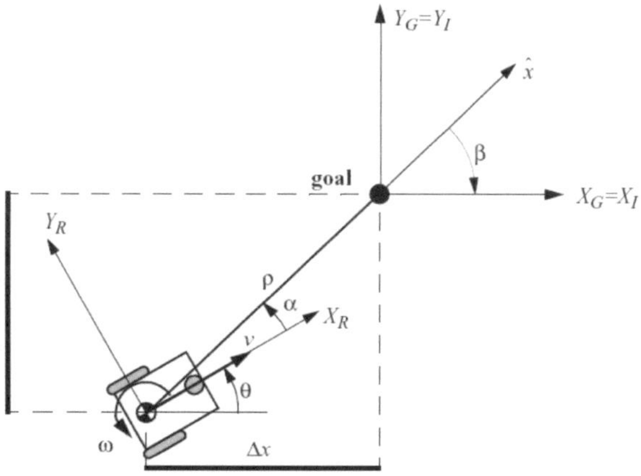

Fig. 3. Robot kinematics and its frames of interests

Thus, three wheels are sufficient to guarantee stable balance, although, as we shall see below, two-wheeled robots can also be stable (R. Siegwart, 2004). When more than three wheels are used, a suspension system is required to allow all wheels to maintain ground contact when the robot encounters uneven terrain. Motion control might not be an easy task for this kind of systems. However, it has been studied by various research groups, and some adequate solutions for motion control of a mobile robot system are available (Gy. Mester, 2009).

4. Using Potential Fields method for navigation

A potential field consists of two imaginary fields (*attractive potential* and *repulsive potential*) and used to avoid a collision with unexpected obstacle while moving in a predetermined path. The Attractive Potential forces the robot to move through a predetermined path and the Repulsive Field, assumed to be generated by obstacles, forces the robot to move a different way to avoid the collision (O. Khatib, 1986). The Artificial Potential Field approach is a local path planner method that was introduced by Khatib. This method defines obstacles as repelling force sources, and goals as attracting force sources. The path is then influenced by the composition of the two forces, which produces a robot motion that moves away from obstacles while moving towards the target goal. The approach is mathematically simple and is able to produce real-time acceptable results for collision avoidance even in dynamic environments. The most known limitation of this approach is the local minima, which refers to locations that trap the robot and prevent it from reaching the target goal location. This main problem has been addressed by many different techniques that try to solve or at least minimize its impact (O. Khatib, 1985).

4.1 Attractive Potential Field

The attractive potential field corresponds to the component responsible for the potentials that attract the robot towards the target goal position. At all locations in the environment the action vector will point to the target goal.

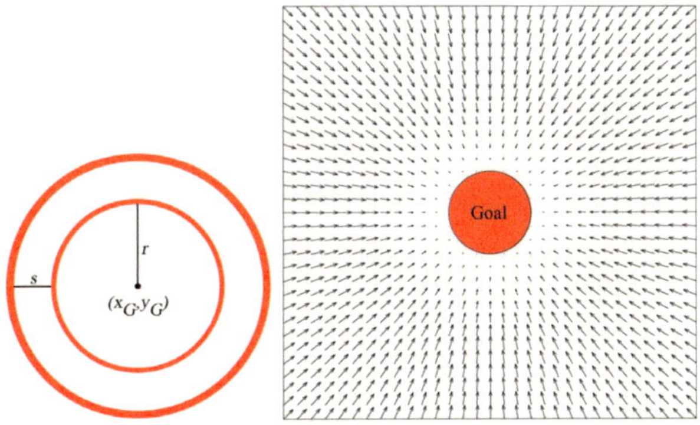

Fig. 4. Attractive potential field action vectors pointing to the goal and goal representation (M. Goodrich, 2002)

Usually, the action vector is found by applying a scalar potential field function to the robot's position and then calculating the gradient of that function.

$$\nabla = [\nabla x, \nabla y] = [\frac{\partial U}{\partial x}, \frac{\partial U}{\partial y}] \tag{1}$$

After defining: (M. Goodrich, 2002)

- $[x_G, y_G]$ as the position of the goal;
- r as the radius of the goal;
- $[x_R, y_R]$ as the position of the robot;
- s as the size of the goal's area of influence;
- α as the strength of the attractive field $(\alpha > 0)$

We can compute ∇x and ∇y using the following steps:

1. Find the distance d between the goal and the robot:

$$d = \sqrt{(x_G - x_R)^2 + (y_R - y_G)^2} \tag{2}$$

2. Find the angle θ between the robot and the goal:

$$\theta = \tan^{-1}\left(\frac{y_G - y_R}{x_G - x_R}\right) \tag{3}$$

3. Set ∇x and ∇y according to the rules:

If $d < r$ then $\nabla x = \nabla y = 0$

$$\text{If } r \le d \le s + r \text{ then } \begin{cases} \nabla x = \alpha(d - r)\cos(\theta) \\ \nabla y = \alpha(d - r)\sin(\theta) \end{cases} \quad (4)$$

$$\text{If } d > s + r \text{ then } \begin{cases} \nabla x = \alpha s\cos(\theta) \\ \nabla y = \alpha s\sin(\theta) \end{cases}$$

The last step presents three simple rules that characterize three different behaviors for the robot according to its relative position towards the goal:

- In the first rule of step 3, $d < r$ means that the robot is in the goal area. In this case, no forces act and ∇x and ∇y are set to zero.
- In the second rule, $r \le d \le s + r$ means that the robot is inside the area of inuence of the goal. The action vector is set using α , d and s .
- In the third and last rule, $d > s + r$ means that the robot is outside the goal area and also outside its area of influence. The action vector is set to with s and α thus reaching higher values.

4.2 Repulsive Potential Field

The repulsive potential field is the component that is responsible for forcing the robot to stay away from the obstacles it encounters on its path. All repulsive action vectors point away from the obstacle surface driving the robot away from the obstacle.

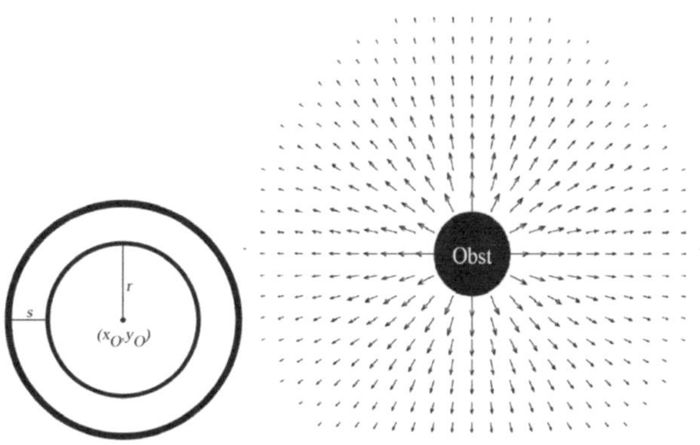

Fig. 5. Repulsive potential field action vectors pointing away from the obstacle and obstacle representation (M. Goodrich, 2002)

Similarly to the Attractive Potential, we calculate the repulsive action vector.

After defining: (M. Goodrich, 2002)

- $[x_O, y_O]$ as the position of the obstacle;
- r as the radius of the obstacle;
- $[x_R, y_R]$ as the position of the robot;
- s as the size of the obstacle's area of influence;
- β as the strength of the repulsive field $(\beta > 0)$

We can compute ∇x and ∇y using the following steps:

1. Find the distance d between the obstacle and the robot:

$$d = \sqrt{(x_O - x_R)^2 + (y_R - y_O)^2} \tag{5}$$

2. Find the angle θ between the robot and the obstacle:

$$\theta = \tan^{-1}\left(\frac{y_O - y_R}{x_O - x_R}\right) \tag{6}$$

3. Set ∇x and ∇y according to the rules:

If $d < r$ then $\begin{cases} \nabla x = -sign(\cos(\theta))\infty \\ \nabla y = -sign(\sin(\theta))\infty \end{cases}$

If $r \le d \le s + r$ then $\begin{cases} \nabla x = -\beta(s + r - d)\cos(\theta) \\ \nabla y = -\beta(s + r - d)\sin(\theta) \end{cases}$ (7)

If $d > s + r$ then $\nabla x = \nabla y = 0$

Similar to the attractive potential rules, these rules are also simple and characterize three different behaviors for the robot according to its position relative to the obstacle. It is important to notice that all action vectors need to point away from the obstacle, hence the need to use negative values (M. Goodrich, 2002).

- In the first rule of step 3, the robot is within the radius of the obstacle, so the action vector needs to be infinite, expressing the need to escape from the robot.

- In the second rule, where the robot is outside the obstacle's radius but inside its area of influence, the action vector is set to a high value in order to express the need to escape the current location.
- In the third rule, where the robot is outside the area of influence of the obstacle, the action vector is set to zero, meaning that no repulsive forces are acting on the robot (M. Goodrich, 2002).

Since the repulsive force only acts when the robot is inside the area of influence of the obstacle, the value of s must be carefully chosen. A small value for s can cause trajectory problems by causing abrupt changes on the path and some constraints on the speed of the robot. A large value for s may cause also problems on the robot's movement since it can constrain movement in small places where the robot could pass.

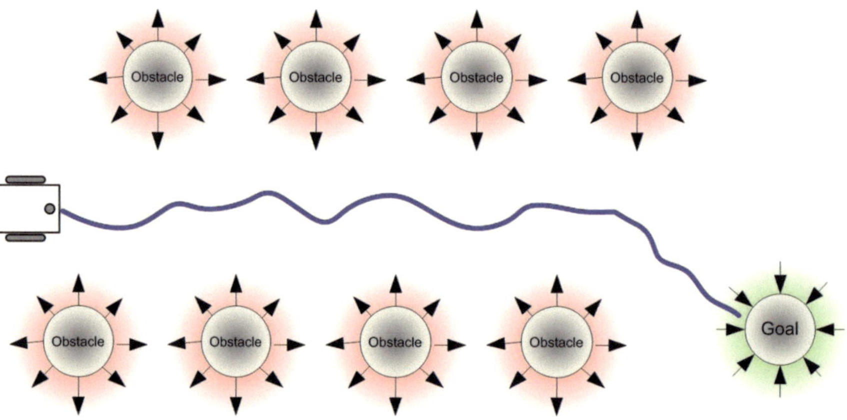

Fig. 6. Potential Fields simulation

The repulsive force has the objective of repelling the robot only if it is close to an obstacle and its velocity points towards that obstacle (M. Goodrich, 2002).

4. WSN and Event-Based System for Greenhouse Climate Control

A wireless sensor network (WSN) is a computer network consisting of spatially distributed autonomous devices using sensors to cooperatively monitor physical or environmental conditions, such as temperature, sound, vibration, pressure, motion or pollutants, at different locations (Sun Microsystems, 2002). The development of wireless sensor networks was originally motivated by military applications such as battlefield surveillance. Figure 7, presents the sensor node architecture. However, wireless sensor networks are now used in many civilian application areas, including environment and habitat monitoring, healthcare applications, home automation, and traffic control.

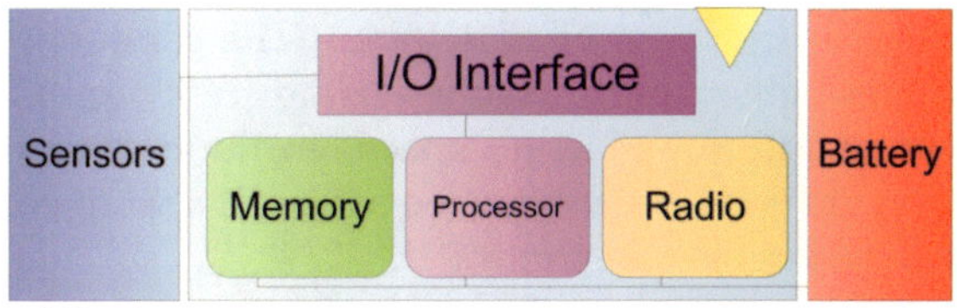

Fig. 7. Sensor Node Architecture

In addition to one or more sensors, each node in a sensor network is typically equipped with a radio transceiver or other wireless communications device, a small microcontroller, and an energy source, usually a battery. Figure 8, shows the typical wireless sensor network.

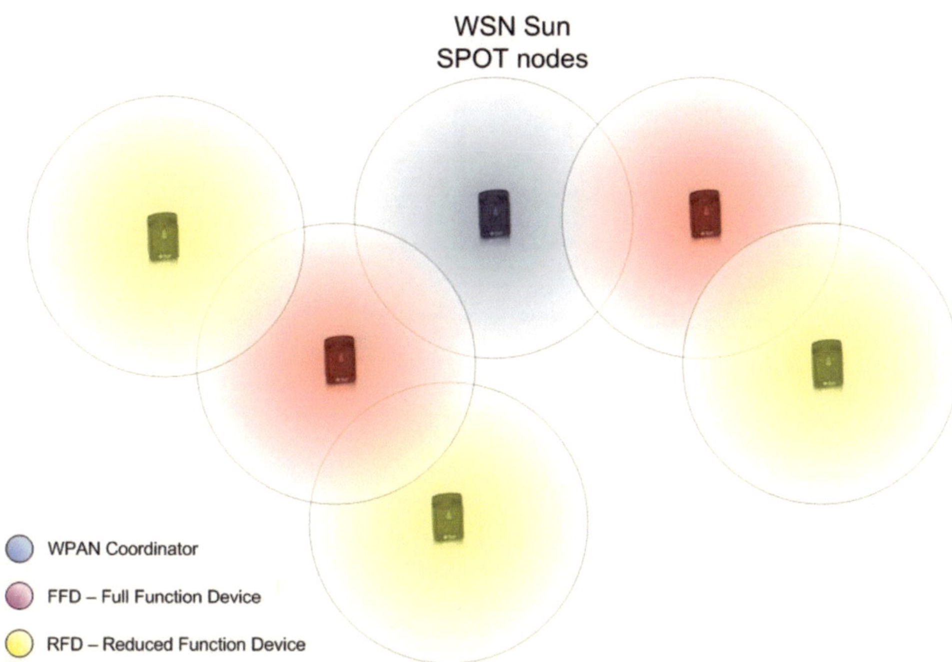

Fig. 8. Typical wireless sensor network (WSN)

The size a single sensor node can vary from shoebox-sized nodes down to devices the size of grain of dust. The cost of sensor nodes is similarly variable, ranging from hundreds of dollars to a few cents, depending on the size of the sensor network and the complexity required of individual sensor nodes (Sun Microsystems, 2005). Size and cost constraints on sensor nodes result in corresponding constraints on resources such as energy, memory, computational speed and bandwidth. In computer science, wireless sensor networks are an active research area with numerous workshops and conferences arranged each year (S. Scaglia, 2008). As commented above, this paper is devoted to analyzing diurnal and nocturnal temperature control with natural ventilation and heating systems, and humidity control as a secondary control objective. Under diurnal conditions, the controlled variable is the inside temperature and the control signal is the vent opening. The use of natural ventilation produces an exchange between the inside and outside air, usually provoking a decrease in the inside temperature of the greenhouse. The controller must calculate the necessary vent opening to reach the desired setpoint. The commonest controller used is a gain scheduling PI scheme where the controller parameters are changed based on some disturbances: outside temperature and wind speed. In the case of nocturnal temperature control, forced-air heaters are used to increase the inside temperature and an on/off control with dead/zone was selected as heating controller.

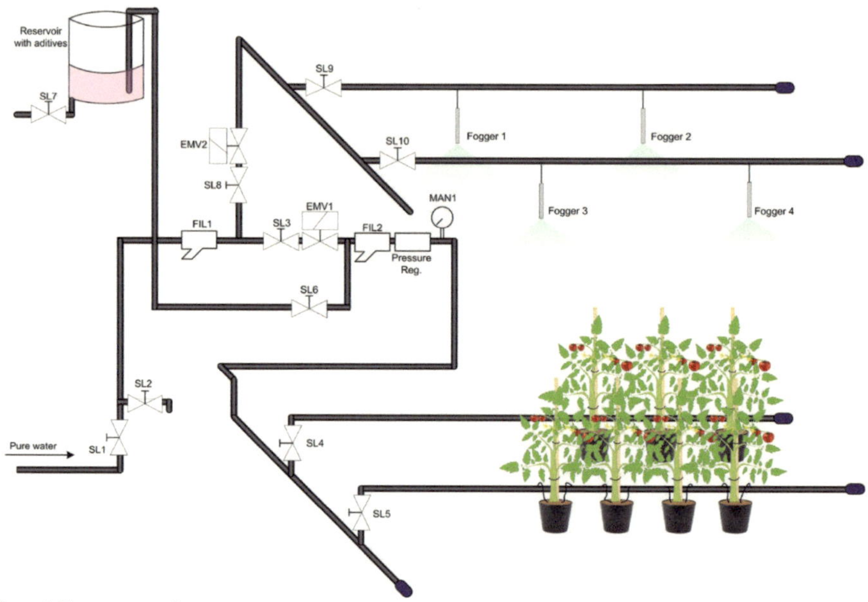

Fig. 9. Humidity control

Climate monitoring is vitally important to the operation in greenhouses and the quality of the collected information has a great influence on the precision and accuracy of control results. Currently, the agro-alimentary market field incorporates diverse data acquisition techniques (S. Scaglia, 2008). Normally, the type of acquisition system is chosen to be optimal for the control algorithm to be used. For traditional climate monitoring and control systems, all sensors are distributed through the greenhouse and connected to the device performing the control tasks. These equipments use time-based data sampling techniques as a consequence of using time-based controllers. Figure 10, presents the temperature controller.

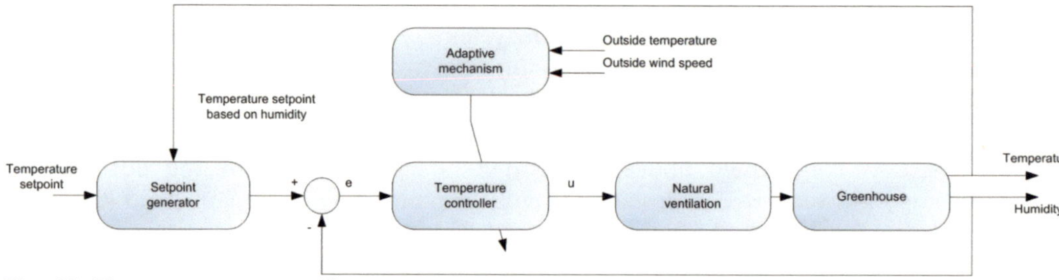

Fig. 10. Temperature controller

Nowadays, commercial systems present more flexibility in the implementation of control algorithms and sampling techniques, especially WSN, where each node of the network can be programmed with a different sampling algorithm or local control algorithm with the main goal of optimizing the overall performance.

4.1 Description of the Sun SPOT WSN module

One of the most poplar technologies in the WSN area is Sun SPOT (Small Programmable Object Technology). It contains 32-bit ARM9 CPU, 512K memory, 2 Mb flash storage and wireless networking is based on ChipCon CC2420 following the 802.15.4 standard with integrated antenna and operates in the 2.4GHz to 2.4835GHz ISM unlicensed bands. The IC contains a 2.4GHz RF transmitter/receiver with digital direct sequence spread spectrum (DSSS) baseband modem with MAC support.

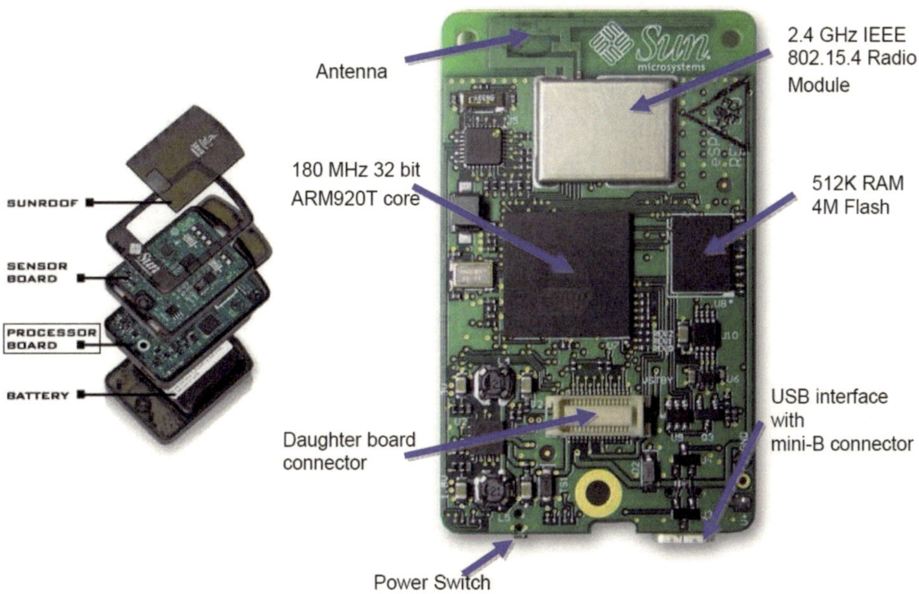

Fig. 11. Sun SPOT processor board

The sensor board integrates multiple sensors, monitoring LED and interactive switches into one board. All the facilities of this board are programmable in Java. The Sun SPOT SDK comes with two important tools for managing the software on your SPOTs: SPOTManager and SPOTWorld. The SPOTManager is a tool for managing the Sun SPOT SDK software. You can use it to download from the Internet both new and old versions of the Sun SPOT SDK. You can use it to make one or another SDK the active SDK on your host workstation, and you can use it to download system software to your Sun SPOTs (Sun Microsystems, 2005).

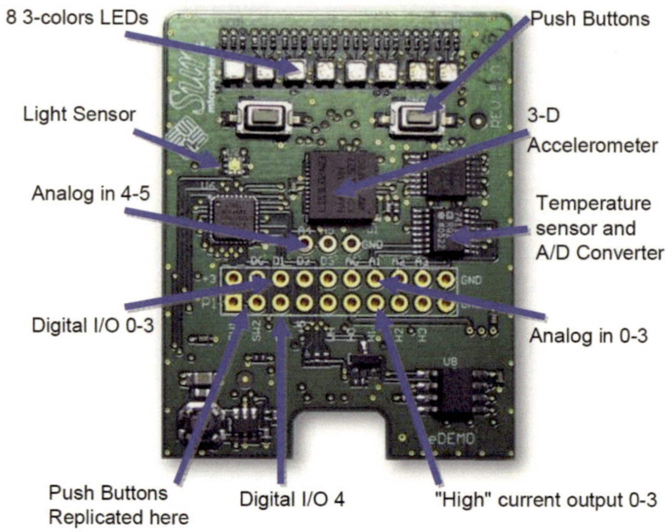

Fig. 12. Sun SPOT sensor board

The facilities of the sensor board are:

- One 2G/6G 3-axis accelerometer
- One temperature sensor
- One light sensor
- Two 8-bit tri-color LEDs
- 6 analog inputs
- Two momentary switches
- 5 general purpose I/O pins

The internal battery is a 3.7V rechargeable lithium-ion prismatic cell. The battery has internal protection circuit to guard against over discharge, under voltage and overcharge conditions. The battery can be charged from either the USB type mini-B device connector or from an external source with a 5V power supply.

4.2 Wireless radio

The wireless network communications uses an integrated radio transceiver, the TI CC2420 (formerly ChipCon). The CC2420 is IEEE 802.15.4 compliant device and operates in the 2.4GHz to 2.4835GHz ISM unlicensed bands. Regulations for these bands are covered by FCC CFR47 part 15 (USA), ETSI EN 300 328 and EN 300 440 class 2 device (Europe) and ARIB STD-T66 (Japan).

The IC contains a 2.4GHz RF transmitter/receiver with digital direct sequence spread spectrum (DSSS) baseband modem with MAC support. Other features include separate TX and RX 128 byte FIFOs, AES encryption (currently not supported), received signal strength indication (RSSI) with 100dB sensitivity and transmit output power setting from -24dBm to 0dBm. Effective bit rate is 250kbps and chip rate is 2000kChips/s. Receive sensitivity is -

90dBm. The digital control and data communications with the CC2420 use PIO port bits and the SPI channel. The CC2420 is a slave SPI bidirectional device addressed when RF_CS (PCS2) is asserted active low. PIO ports reset the CC2420 (RF_RST), power it down (RF_PWDOWN), or check the status of the receive FIFO (FIFO and FIFOP), clear channel assessment (CCA) and start of frame (SFD). There are 33 configuration and status registers, 15 command registers and two 8-bit registers for the separate transmit and receive FIFOs. The first byte sent to the CC2420 is the address made up of 6-bit address, RAM/Register select (Bit 7) and Read/Write select (Bit 6). Following bytes are data read from or written to the CC2420.

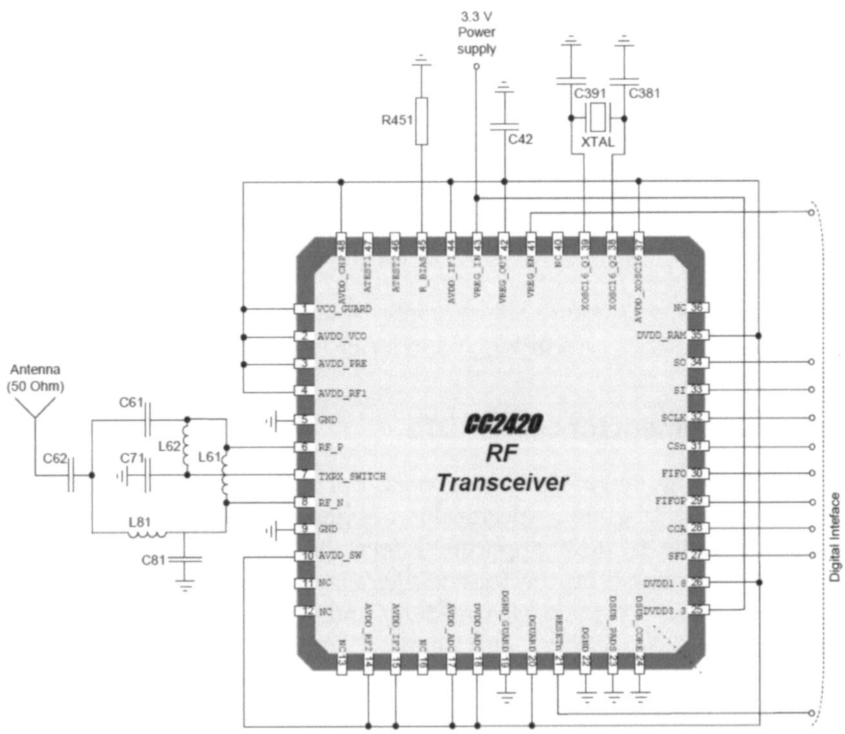

Fig. 13. Typical application circuit for CC2420

The CC2420 is housed in a 48pin quad leadless package (QLP or QFN) that is 7mm square. It is powered with +3.3V Vcc supply. The CC2420 has an internal 1.8V low drop out regulator for powering the internal RF and analog circuitry. It consumes 20mA during receive operation and 18mA for 0dBm transmit. The frequency generation uses an accurate 16MHz crystal with ±10ppm accuracy, ±10ppm stability and ±1ppm aging. The entire RF section is enclosed in an upper and lower RF shield and has modular FCC approval.

4.3 I/O pin Manipulation of the SunSPOT module

The SunSPOT's sensor board has an Atmega88 and operates the 8 tricolor LED's, the accelerometer configuration, and the following pins on the I/O header: I/O pins D0 through D3 can be set as either an output or input.

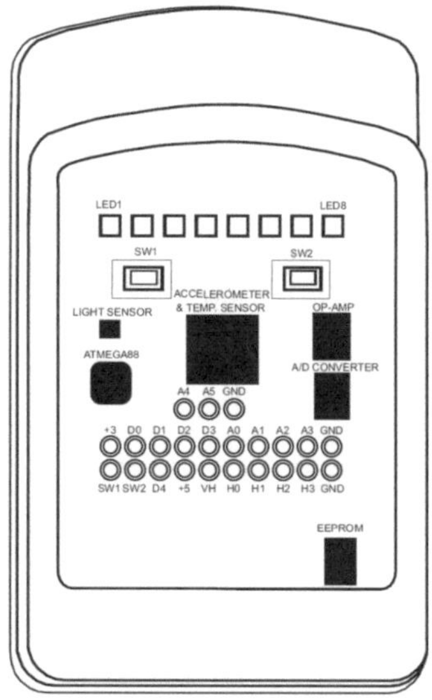

Fig. 14. The SunSPOT's sensor board component location

The high current driver pins, H0 to H3, can only be used as an output. If configured as an output, the pin may be set hi, low, or toggled to its opposite state.

Developing environment for Sun SPOTs

The Java programming language is a general-purpose concurrent class-based object-oriented programming language, specifically designed to have as few implementation dependencies as possible. It allows application developers to write a program once and then be able to run it everywhere on the Internet. Java is a programming language originally developed by James Gosling at Sun Microsystems and released in 1995 as a core component of Sun Microsystems' Java platform. The language derives much of its syntax from C and C++ but has a simpler object model and fewer low-level facilities. Java applications are typically compiled to bytecode that can run on any Java virtual machine (JVM) regardless of computer architecture. The original and reference implementation Java compilers, virtual machines, and class libraries were developed by Sun from 1995. As of May 2007, in compliance with the specifications of the Java Community Process, Sun made available most of their Java technologies as free software under the GNU General Public License. One characteristic of Java is portability, which means that computer programs written in the Java language must run similarly on any supported hardware/operating-system platform. One should be able to write a program once, compile it once, and run it anywhere. This is achieved by compiling the Java language code, not to machine code but to Java bytecode – instructions analogous to machine code but intended to be interpreted by a virtual machine (VM) written specifically for the host hardware. End-users commonly use a Java Runtime Environment (JRE) installed on their own machine for standalone Java applications, or in a Web browser for Java applets. Standardized libraries provide a generic way to access host specific features such as graphics, threading and networking. In some JVM versions, bytecode can be compiled to native code, either before or during program execution, resulting in faster execution (J. Gosling, 2005). The most popular developing environment for Java is Netbeans IDE.

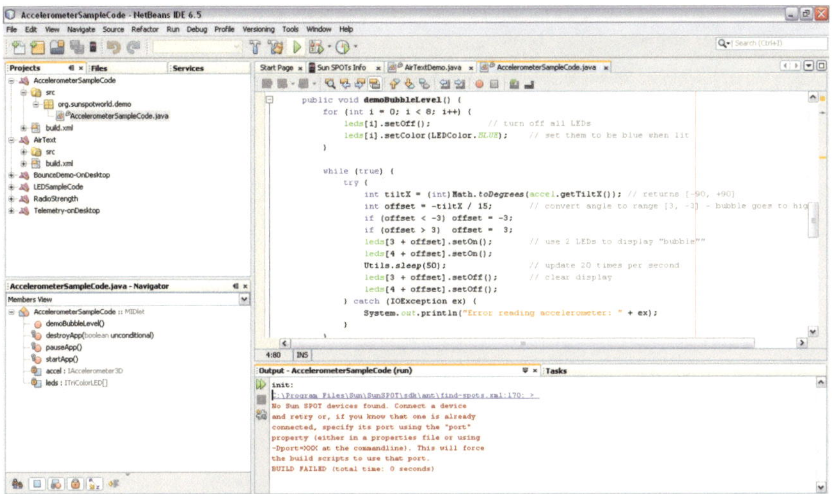

Fig. 15. NetBeans IDE 6.5 – Sun SPOT SDK

A major benefit of using bytecode is porting. However, the overhead of interpretation means that interpreted programs almost always run more slowly than programs compiled to native executables would, and Java suffered a reputation for poor performance. This gap has been narrowed by a number of optimization techniques introduced in the more recent JVM implementations. One such technique, known as just-in-time (JIT) compilation, translates Java bytecode into native code the first time that code is executed, then caches it. This results in a program that starts and executes faster than pure interpreted code can, at the cost of introducing occasional compilation overhead during execution. More sophisticated VMs also use dynamic recompilation, in which the VM analyzes the behavior of the running program and selectively recompiles and optimizes parts of the program. Dynamic recompilation can achieve optimizations superior to static compilation because the dynamic compiler can base optimizations on knowledge about the runtime environment and the set of loaded classes, and can identify hot spots - parts of the program, often inner loops, that take up the most execution time. JIT compilation and dynamic recompilation allow Java programs to approach the speed of native code without losing portability.

4.4 The Sun SPOT emulator

Solarium includes an emulator capable of running a Sun SPOT application on your desktop computer. This allows for testing a program before deploying it to a real SPOT, or if a real SPOT is not available. Instead of a physical sensor board, Solarium displays a virtual SPOT with a control panel where you can set any of the potential sensor inputs (e.g. light level, temperature, digital pin inputs, analog input voltages, and accelerometer values). Your application can control the LEDs' color that is displayed in the virtual SPOT image, just like it would a real SPOT. You can click with the mouse on the push button switches in the virtual SPOT image to press and release the switches. Receiving and sending via the radio is also supported. Each virtual SPOT is assigned its own address and can broadcast or unicast to the other virtual SPOTs. If a shared base station is available a virtual SPOT can also interact over the radio with real SPOTs.

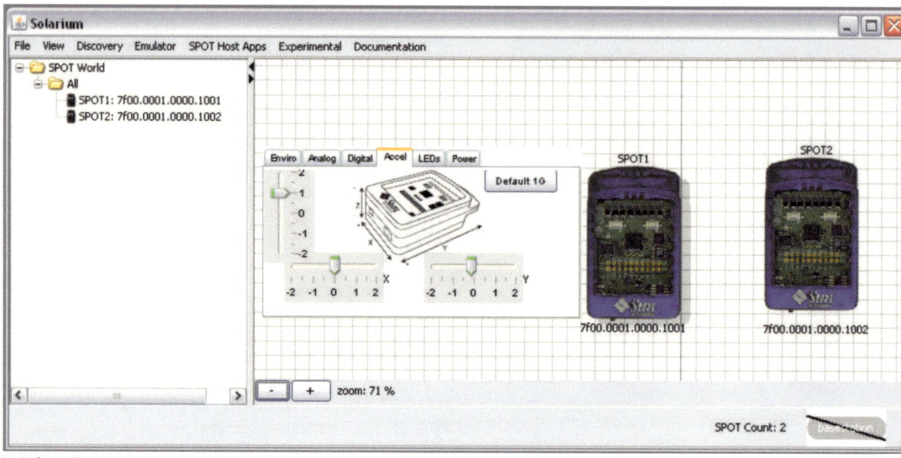

Fig. 16. The Sun SPOT Emulator

Virtual SPOTs can communicate with each other by opening radio connections, both broadcast and point-to-point. Instead of using an actual radio these connections take place over regular and multicast sockets. When a base station SPOT is connected to the host computer and a shared base station is running, virtual SPOTs can also use it to communicate with real SPOTs using the base station's radio. The advantage of using a shared base station is that multiple host applications can then all access the radio. One disadvantage is that communication from a host application to a target SPOT takes two radio hops, in contrast to the one hop needed with a dedicated base station. Another disadvantage is that run-time manipulation of the base station SPOT's radio channel, pan id or output power is not currently possible. Each virtual SPOT has its own Squawk VM running in a separate process on the host computer. Each Squawk VM contains a complete host-side radio stack as part of the SPOT library, which allows the SPOT application to communicate with other SPOT applications running on the host computer, such as other virtual SPOTs, using sockets or real SPOTs via radio if a shared base station is running. The current Solarium implementation is primarily an emulator since it actually runs a SPOT application in a Squawk VM, just like the VM on a real SPOT. Likewise radio interaction between virtual SPOTs is emulated with data sent via packets and streams from one (virtual) SPOT to another. Only the SPOT's interaction with the environment is simulated using a simple model where the user needs to explicitly set the current sensor values. Future versions may incorporate more simulation of SPOT properties like battery level or radio range.

5. Solution

Building and programming a robot is a combination of mechanics, electronics, and problem solving. What you're about to learn while doing the activities and projects in this text will be relevant to "real world" applications that use robotic control, the only difference being the size and sophistication. Robotics has come a long way, especially for mobile robots. In the past, mobile robots were controlled by heavy, large, and expensive computer systems that could not be carried and had to be linked via cable or wireless devices. As shown in Figure 8, the mobile measuring station is navigating inside the greenhouse. Today, however, we can build small mobile robots with numerous actuators and sensors that are controlled by inexpensive, small, and light embedded computer systems that are carried on-board the robot.

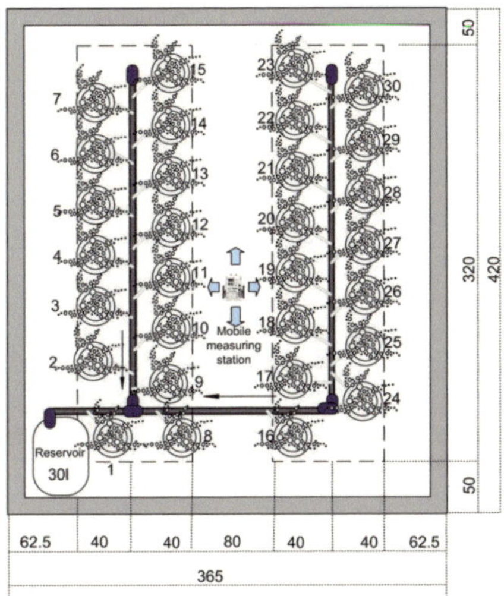

Fig. 17. Greenhouse top view with the mobile measuring station

The mechanical principles, example program listings, and circuits you will use are very similar to, and sometimes the same as, industrial applications developed by engineers. In this project we have used SunSPOT-s to achieve remote control over a Boe-Bot. For this project we have used 2 SunSPOT-s from the kit (free range and base station module) as depicted on Figure 18. SunSPOT's wireless protocol is Zigbee based protocol (I. Matijevics, 2008).

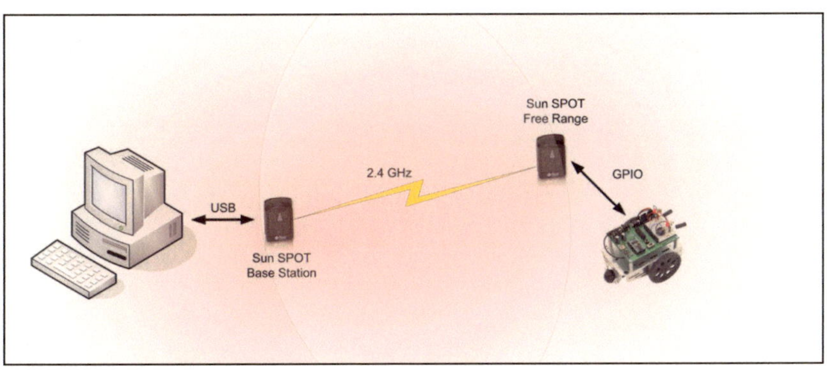

Fig. 18. Connection of the system

The Hardware basically centers around Sun SPOT and DC Motors controlled by Basic Stamp. The Sun SPOT base station will send data to Sun SPOT on the mobile measuring station which will drive the Basic Stamp controller to DC IO pins (J. Simon, 2009). The microcontroller will drive the Motors which will run the measuring station. Figure 19, shows the testing phase of the mobile measuring station.

Fig. 19. Boe-bot with SunSPOT mounted

6. Experimental results

The applications for WSNs are many and varied. They are used in commercial and industrial applications to monitor data that would be difficult or expensive to monitor using wired sensors. They could be deployed in wilderness areas, where they would remain for many years (monitoring some environmental variable) without the need to recharge/replace their power supplies. They could form a perimeter about a property and monitor the progression of intruders (passing information from one node to the next). There are a many uses for WSNs (I. Matijevics, 2009).

Fig. 20. Crops in greenhouse

Typical applications of WSNs include monitoring, tracking, and controlling. Some of the specific applications are habitat monitoring, object tracking, nuclear reactor controlling, fire detection, traffic monitoring, etc. In a typical application, a WSN is scattered in a region where it is meant to collect data through its sensor node. Figure 23, shows the complete control system of the greenhouse. The WSN-based controller has allowed a considerable decrease in the number of changes in the control action and made possible a study of the compromise between quantity of transmission and control performance.

Fig. 21. Capsicum

Motion control of mobile robots is a very important research field today, because mobile robots are a very interesting subject both in scientific research and practical applications. In this paper the object of the remote control is the Boe-Bot. The vehicle has two driving wheels and the angular velocities of the two wheels are independently controlled (A. Pawlowski, 2009). When the vehicle is moving towards the target and the sensors detect an obstacle, an avoiding strategy is necessary. The host system connects to the mobile robot with the SunSPOT module. A remote control program has been implemented as shown on Figure 22.

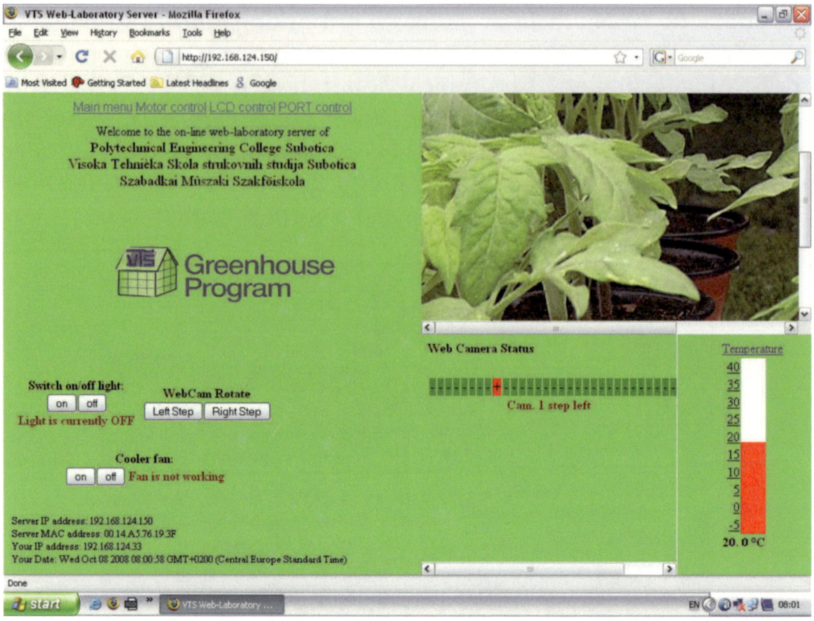

Fig. 22. Control surface of the system

The limit of the level crossing sampling has presented a great influence on the event based control performance where, for the greenhouse climate control problem, the system has provided promising results.

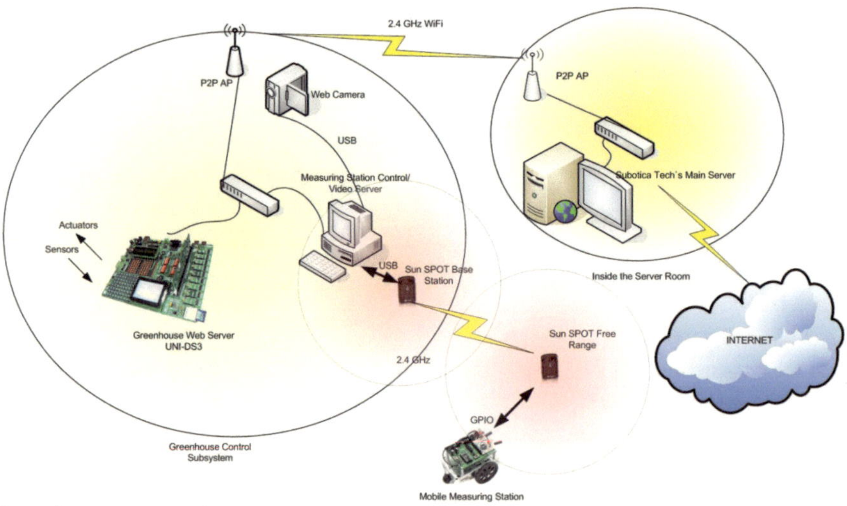

Fig. 23. Greenhouse control system

The code snippet below gives an example for testing the communication devices in broadcast mode as we can see on Figure 24. It is written in Java and runs on SunSPOT modules. Each SPOT is assigned its own address and can broadcast or unicast to the other SPOTs. This code is implemented for testing purposes only.

```
    protected void startApp() throws MIDletStateChangeException {
        System.out.println("Broadcast Counter MIDlet");
        //showColor(color);
        //switches[0].addISwitchListener(this);
        //switches[1].addISwitchListener(this);
        try {
            tx                                                              =
(RadiogramConnection)Connector.open("radiogram://broadcast:123");
            xdg   =   (Radiogram)tx.newDatagram(20);   //transmitting   the
radiogram
            RadiogramConnection                      rx                     =
(RadiogramConnection)Connector.open("radiogram://:123");
            Radiogram rdg = (Radiogram)rx.newDatagram(20);
            //outs[0].setHigh();
            while (true) {
                try {
                    rx.receive(rdg);
                    int cmd = rdg.readInt();
                    //int newCount = rdg.readInt();
                  //int newColor = rdg.readInt();
                    /*if (cmd == CHANGE_COLOR) {
                        System.out.println("Received   packet   from   "   +
rdg.getAddress());
                        //showColor(newColor);
                    } else {
                        //showCount(newCount, newColor);
                    }*/
                    switch (cmd){
                        case           0:            outs[demo.H0].setLow();
outs[demo.H1].setLow();   leds[0].setRGB(200,   0,   0);   leds[0].setOn();
leds[1].setOff();leds[2].setOff();leds[3].setOff(); break;
                        case           4:           outs[demo.H0].setHigh();
outs[demo.H1].setLow();   leds[1].setRGB(200,   0,   0);   leds[1].setOn();
leds[0].setOff();leds[2].setOff();leds[3].setOff(); break;
                        case           3:            outs[demo.H0].setLow();
outs[demo.H1].setHigh();   leds[2].setRGB(200,   0,   0);   leds[2].setOn();
leds[1].setOff();leds[0].setOff();leds[3].setOff(); break;
                        case           1:           outs[demo.H0].setHigh();
outs[demo.H1].setHigh();   leds[3].setRGB(200,   0,   0);   leds[3].setOn();
leds[1].setOff();leds[2].setOff();leds[0].setOff(); break;
//setting up the diagnostic leds
                        default:      leds[4].setRGB(200,      0,      0);
leds[4].setOn(); break;
                    }
                } catch (IOException ex) {
                    System.out.println("Error receiving packet: " + ex);
                    ex.printStackTrace(); // Error detection
                }
            }
        } catch (IOException ex) {
            System.out.println("Error opening connections: " + ex);
            ex.printStackTrace(); // Error detection
        }
    }
```

Fig. 24. Sending broadcast packets via wsn from base station

The Sun SPOT is a Java programmable embedded device designed for flexibility. The basic unit includes accelerometer, temperature and light sensors, radio transmitter, eight multicolored LEDs, 2 push-button control switches, 5 digital I/O pins, 6 analog inputs, 4 digital outputs, and a rechargeable battery. Java implementation and programming the Sun SPOT is surprisingly easy. Experimental testing has demonstrated the validity of our approach.

7. Comparison of the fruit production

Tomatoes are a warm season vegetable crop. They grow best under conditions of high light and warm temperatures. Low light in a fall or winter greenhouse, when it is less than 15% of summer light levels, greatly reduces fruit yield when heating costs are highest. For this reason, it is difficult to recommend that a greenhouse operator should grow and harvest fruit from December 15 to February 15. Based on few years of experience, tomato production is most successful in the spring. Excellent light, moderate heating costs and good prices annually demonstrate this is the best time for greenhouse tomato production. Tomato plants grow best when the night temperature is maintained at 16 - 18 °C. Temperatures below 16 °C will prevent normal pollination and fruit development. In warm or hot outdoor conditions, tomato greenhouses must be ventilated to keep temperatures below 35 °C. High temperatures not only effect the leaves and fruit, but increased soil temperatures also reduce root growth. Table 1, gives an overview of effectiveness of the control system.

Tested plants	Average weight of fruit (with WSN control)	Average weight of fruit (without WSN control)	Average number of fruit per plant (with WSN control)	Average number of fruit per plant (without WSN control)
Tomato	210 g	180 g	17	11
Capsicum	135 g	110 g	15	12
Cucumber	70 g	60 g	13	10

Table 1. The average total weight and number of fruit harvested

Success in greenhouse plants depends completely on fruit yield. Yields of 20 – 25 % gain per plant are very good for annual costs.

8. Conclusion

The system and its implementation have been successful, however there are still possibilities for further development. The first cycle of plant development has just passed, and it has provided numerous valuable data. For the next cycle will better conditions will be provided, with a more experienced staff. With further developments the application of professional industrial electronics will also have to be taken into consideration, which would significantly decrease possible problems. One of the research areas inside the Greenhouse program is wirelessly controlled mobile measuring station. Traditionally research into autonomous robotics has been performed on robotics platforms that cost tens of thousands of dollars. As an alternative, several research groups have developed low-cost robots that are controlled by a SunSPOT node running Java VM. One of the goals of this project is to develop algorithms for coordination and navigation inside the greenhouse.

9. Acknowledgement

This research was partially supported by the TÁMOP 4.2.1/B-09/1/KONV-2010-0005 program of the Hungarian National Development Agency.

10. References

A. Pawlowski, J. Luis Guzman, F. Rodríguez, M. Berenguel, J. Sánchez and S. Dormido, (2009) "Simulation of Greenhouse Climate Monitoring and Control with Wireless Sensor Network and Event-Based Control" Proceedings of the Conference

Gy. Mester, (2009) „Wireless Sensor-based Control of Mobile Robot Motion", Proceeding of the IEEE SISY 2009, pp 81-84, Subotica, Serbia

Gy. Mester, (2009) „Intelligent Wheeled Mobile Robot Navigation", Proceedings of the Conference Európai Kihívások V, pp. 1-5, SZTE, Szeged, Hungary

I. Matijevics, J. Simon, (2008) „Advantages of Remote Greenhouse Laboratory for Distant Monitoring", Proceedings of the Conference ICoSTAF 2008, pp 1-5, Szeged, Hungary

I. Matijevics, J. Simon, (2009) „Comparison of various wireless sensor networks and their implementation", Proceedings of the Conference SIP 2009, pp 1-3, Pécs, Hungary

J. Gosling, (2005) „ The Java™ Language Specification Third Edition"

J. Simon, G. Martinović, (2009) "Web Based Distant Monitoring and Control for Greenhouse Systems Using the Sun SPOT Modules", Proceedings of the Conference SISY 2009, pp. 1-5, Subotica, Serbia

J. Simon, I. Matijevics, (2009) „Distant Monitoring And Control For Greenhouse Systems Via Internet", Zbornik radova konferencije Yuinfo 2009, pp. 1-3, Kopaonik, Srbija

J. Vasu, L. Shahram, (2008) "Comprehensive Study of Routing Management in Wireless Sensor Networks- Part-1"

L. Gonda, , C. Cugnasca, (2006) "A proposal of greenhouse control using wireless sensor networks" In Proceedings of 4thWorld Congress Conference on Computers in Agriculture and Natural Resources, Orlando, Florida, USA

M. J. Matarić, (2007) The Robotics Primer. The MIT Press, 1st edition.

M. A. Goodrich, (2002) Potential fields tutorial.

O. Khatib, (1986) "Real-Time Obstacle Avoidance for Manipulators and Mobile Robots", Int. J. of Robotic Research, Vol.5, No.1, p.60

O. Khatib, (1985) "The Potential Field Approach and Operational Space Formulation in Robot Control" Proc. Fourth Yale Workshop on Applications of Adaptive Systems Theory, Yale University, New Haven, Connecticut, pp. 208-214.

P. Kucsera, (2006) „Sensors For Mobile Robot Systems" , Academic and Applied Research in Military Science, Volume 5, Issue 4, p.645-658,Hungary

P. Kucsera, (2006) "Industrial Component-based Sample Mobile Robot System" , Acta Polytechnica Hungarica, Volume 4 Issue Number 4, Hungary

Roland Siegwart and Illah R., (2004) "Introduction to Autonomous Mobile Robots", Nourbakhsh

Sun Microsystems Inc. , (2007) „Sun™ Small Programmable Object Technology (Sun SPOT)" Owner's Manual Release 3.0
Sun Microsystems Inc. , (2005) „Sun Spot Developer's Guide"
Sun Microsystems Inc. , (2005) „Demo Sensor Board Library"
S. Scaglia, (2008) „The Embedded Internet"
X. Feng, T. Yu-Chu, S. Yanjun, S. Youxian, (2007) "Wireless Sensor/Actuator Network Design for Mobile Control Applications. Sensors" Proceedings of the Conference

7

Model Based WSN System Implementations Using PN-WSNA for Aquarium Environment Control in a House

Ting-Shuo Chen and Chung-Hsien Kuo
Department of Electrical Engineering
National Taiwan University of Science and Technology
Taiwan

1. Introduction

Ubiquitous computing architectures are implemented for cognitive sensor networks. Wireless sensor networks cooperating with cognitive science and artificial intelligence are used to develop cognitive sensor networks (Shenai et al., 2008). Therefore, a cognitive sensor network is generally represented as a closed loop control system (Ruiz et al., 2008), where the feedback data is collected from remote sensor nodes. At the same time, the control approaches are desired to deal with regulations of desired operation scenarios and sensor feedbacks.

Wireless sensor networks (WSN) (Romer et al., 2004; Akyildiz et al., 2008) are developed using autonomous sensor nodes (Dalola et al., 2009) to collect remote sensor data for decision systems with low power consumptions and failure tolerable mechanisms. In general, the WSN system can be applied to the factory automation (Zhuang et al., 2008), intelligent diagnosis (Zhuang et al., 2008), intelligent monitoring and control systems (Sridhar et al., 2007), smart home (Suh et al., 2008), etc. Practically, the challenging issues for developing WSN systems are large amounts of coding and program maintenance efforts for various sensor oriented applications as well as interdisciplinary integrations of domain engineers and WSN engineers.

For the first issue, diverse control and decision scenarios in a WSN system are developed for different sensor nodes. The coding and maintenance for large scale WSN systems would be huge challenges. The second issue is the problems of interdisciplinary integrations. Coding for sensor nodes is a challenge for domain engineers who are not familiar with programming. According to aforementioned challenging issues, model based system implementation approaches are proposed to eliminate the efforts for programming native codes with cross compilers.

In order to perform model based implementation approaches, several discrete event dynamic system (DEDS) modeling approaches are surveyed, such as finite state machine (FSM) (Avnur et al., 1990), unified modeling language (UML) (Manasseh et al., 2010) and Petri net (PN) (Murata et al., 1989; Kuo et al., 2009). The Petri net (Murata et al., 1989) was proposed by C.A. Petri. A PN model may model the system using events and conditions. Events are represented as transitions; conditions are represented as places. Arcs are used to describe pre- and post-conditions between places and transitions. In general, an autonomous sensor node can be also described as conditions, events, and their relationships. Sensor events are generated in terms of the changes of sensor conditions.

Although the PN is suitable for modeling a WSN, the ordinary PN is not applicable due to the lack of interfaces and intercommunications. Therefore, the Petri net based wireless sensor node architecture (PN-WSNA) (Kuo et al., 2009) is selected in this book chapter to model an aquarium environment control in a house. Interface functions are desired for collecting sensor data and controlling actuators. Intercommunication functions provide wireless data exchanges among different autonomous sensor nodes. In our approach, the PN-WSNA system is composed of a PN-WSNA kernel program and a PN-WSNA management program. The PN-WSNA kernel program is developed as an inference engine which is implemented inside the sensor node. The PN-WSNA kernel program is responsible of receiving and interpreting PN-WSNA models, collecting sensor data from analog and digital channels, intercommunication between sensor nodes, PN model inference, decision making, and controlling actuators.

In this book chapter, the Petri net based wireless sensor node architecture (PN-WSNA) (Kuo et al., 2009) is used to construct an aquarium environment control system in a house. This aquarium environment control system demonstrates the modelling and implementation procedures for two PN-WSNA sensor node systems, where one sensor node is deployed for aquarium environment control and the other one is desired for entrance counting system. The entrance counting system counts the people in a house. The aquarium environment control system acquires the data from temperature sensor and dissolved oxygen sensor as well as the people number collected from the entrance counting system. Meanwhile, the light, heater and pump are also activated using the sensor node. As a consequence, the aquarium environment control system is capable of autonomously controlling the temperature and dissolved oxygen concentration in a desired condition. In addition, the light can also be controlled in terms of the presence of people in a house.

2. PN-WSNA Definitions

In this book chapter, the PN-WSNA is developed by inheriting the definitions of the ordinary PN. In order to deal with real-time sensor data acquisitions, intercommunications and actuator controls, additional interface and intercommunication places are defined based upon the ordinary PN for practical cognitive sensor network applications. In addition to the interface and communication places, periodic executions of the system are also defined using timed transitions. The sensor data is further categorized as high enable and low enable situations. Therefore, the PN-WSNA structure is a eleven-tuple, PN-WSNA = (P, P_s, P_{ci}, P_{co}, P_a, T_0, T_t, T_H, T_L, I, O) structure; where P is a finite set of normal places; P_s is a finite set of sensor places; P_{ci} is a finite set of receiver places; P_{co} is a finite set of transmitter places; P_a is a finite set of actuation places; T_0 is a finite set of immediate transitions; T_t is a finite set of timed transitions; T_H is a finite set of high-enable transitions; T_L is a finite set of low-enable transitions; I is the input function; and O is the output functions. The PN-WSNA graphical definitions are shown in Fig. 1. The PN-WSNA definitions are further elaborated as follows.

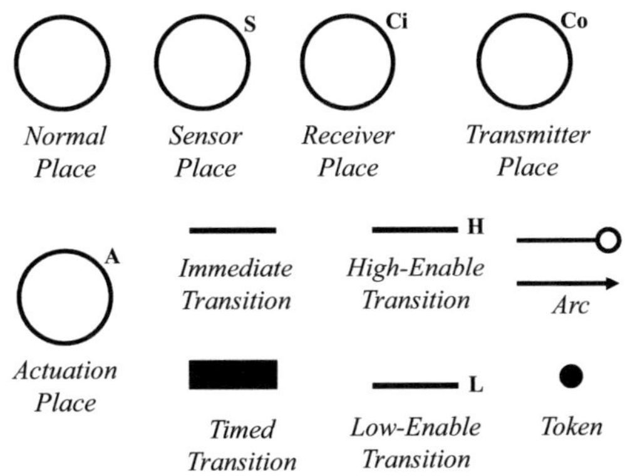

Fig. 1. PN-WSNA graphical definitions.

I. Place: P = {p1,p2,p3,...,pn}: P is a finite set of places, n≧1, and it is denoted a circle. Places of the PN-WSNA are refined as normal places, sensor places, receiver places, transmitter places, and actuation places. Brief introduction is defined as follows. Detailed definitions may refer to (Kuo et al., 2009).

a. Normal places: the definition of a normal place is the same as the place defined in the ordinary PN. Tokens in normal places may represent the corresponding status, condition, command, etc.

b. Sensor places: A sensor place is desired for data collections. For a PN-WSNA, the sensor place may collect sensor signals in terms of analog value (0 – 3 V), binary digits (0 and 1), or serial communication packets (0 – 255) manners, and the sensor data of a place (pi) is denoted as η(pi). A sensor interface is required to be corresponded to an analog-digital-converter (ADC) address, a generalized input-output (GIO) address or the universal asynchronous receiver /transmitter (UART). Because the PN-WSNA does not define the color token (Kuo et al., 2003), the sensor status is eventually represented as "high" or "low" status. Hence, a threshold value is defined for the sensor place to divide the analog value into "high" or "low" status, and the threshold value of a place (pi) is denoted as φ(pi).

c. Receiver and transmitter places: With the PN-WSNA, a receiver place and a transmitter place can be combined as a communication pair, and they appear in different sensor node models. Hence, the transmitter place is a sink place (Kuo et al., 2009); and the receiver place is a source place (Kuo et al., 2009).

d. Actuation places: The actuation place play similar roles to transmitter places; however, the token in an actuation place are converted as actuation signals to control peripheral devices. As a consequence, an actuation place is a sink place, any token in an actuation place may directly control peripheral devices and then the actuation place releases this token.

II. Transitions: T = {t1,t2,t3,...,tm}: T is a finite set of transitions, m≧1, and it is denoted a bar. Transitions of the PN-WSNA are further refined as immediate transitions, timed transitions, high-enable transitions, and low-enable transitions. Detailed definitions are illustrated as below:

a. Immediate transitions: the definition of an immediate transition is the same as the transition defined in the ordinary PN, and it can be used to model events and decisions.
b. Timed transitions: the definition of a timed transition is similar to the transition defined in the ordinary PN; however, tokens in the input places of a timed transition do not deliver to its output places directly. Instead, a fired transition keeps these tokens until a predefined elapsed time is expired. Therefore, an elapsed time factor is further defined for the timed transition.
c. High-enable and low-enable transitions: high-enable and low-enable transitions are defined for sensor places. Basically, high-enable and low-enable transitions serve as output transitions of a sensor place. They must be appeared in a pair configuration; hence conflicts of these transitions are happened. The firing of conflict high-enable and low-enable transitions depends on the sensor data and threshold value defined in the input sensor place. A high-enable transition is fired when the sensor data is greater than or equal to the threshold value defined in the input sensor place; and a low-enable transition is fired when the sensor data is less than the threshold value defined in the input sensor place.

III. In a PN-WSNA model, the places and transitions follow the rules of P∩T = ∅, and P∪T ≠∅.

IV. Token, marking and initial marking: tokens are quantitative representations of bag set in places. The marking is denoted as μ, which represents the token distributions in all places of a PN-WSNA model. μ is a q × 1 column vector, the j-th element of μ indicates the number of tokens in place j. Note that q is a nonnegative integer, and it is equal to the number of places in a PN-WSNA model. The initial marking (μ_0) is defined for the marking of system startup.

VI. Input, output functions, enabling and firing: input and output functions are defined via directed arcs graphically, and they are represented as I(pi,tj)→Ni,j and O(pr,ts) →Nr,s, respectively. Ni,j and Nr,s are nonnegative integers, and they defines the pre- and post-conditions of the PN-WSNA models. In this study, directed and inhibited functions are further defined. A transition (tj) is said to be enabled when (1) satisfies.

$$\prod_{i=1\ Directed}^{k} (I(p_i,t_j)-N_{i,j})>0 \quad and \quad \prod_{i=1\ Inhibited}^{k} (I(p_i,t_j)-N_{i,j})<0 \tag{1}$$

where i = 1 to k, and k equals the number of input places of tj; pi ∈ input places of tj with directed arcs.

At the same time, an enabled transition is not necessarily to be fired because of the conflict situations. The conflict exists when the number of enabled transitions for a place is greater than unity. With a conflict situation, only one of the enabled transitions can be fired. For the PN-WSNA, conflict transitions are resolved in terms of the following approaches.

a. Immediate and timed transitions: Random selections of an enabled and conflict transitions are desired for immediate and timed transitions because of identical token and transition characteristics.
b. High-enable and low-enable transitions: to resolve the conflict situations of a pair of high-enable and low-enable transitions, the sensor data, η(pi), and threshold value, φ(pi), are evaluated for the same input place (pi). A high-enable transition is fired if (2) satisfies.

$$\eta(pi) \geqq \varphi(pi) \tag{2}$$

Meanwhile, a low-enable transition is fired if (3) satisfies.

$$\eta(pi) < \varphi(pi) \tag{3}$$

3. PN-WSNA Based Aquarium Environment Control System

3.1 System Descriptions

To verify the proposed PN-WSNA approaches, an aquarium environment control system is implemented. Fig. 2 shows the facilities used in this experiment. An aquarium is the major environment for this study. A dissolved oxygen meter (with type: DO-5510 from Lutron Co. Ltd.) is used for measuring the dissolved oxygen concentration and the temperature as well in the water. In addition, two infrared human motion detection sensors are used for detecting the entry and exit of visitors. In case of insufficient dissolved oxygen concentration in the water, a pump is activated for increasing the dissolved oxygen. The pump stops when the dissolved oxygen concentration satisfies the setting conditions. On the other hand, in case of low temperature in the water, a heater is also activated for increasing the temperature in the water. Similarly, the heater stops when the water temperature satisfies the setting conditions. It is noted that hysteresis ranges are desired for the activation and termination conditions with respect to their threshold values.

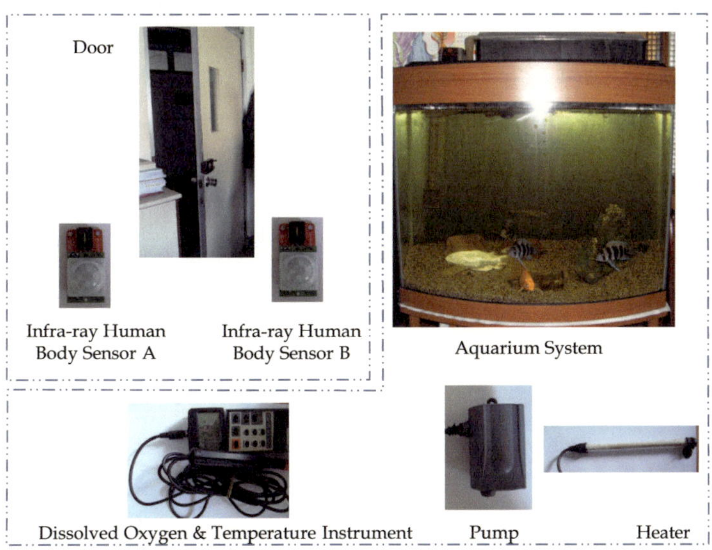

Fig. 2. PN-WSNA model construction architecture.

Two sensor nodes are cooperated to control the proposed aquarium system. PN-WSNA models are implanted inside two sensor nodes to autonomously control aquarium environment system. Fig. 3 shows the architecture this system, respectively. The functions in our system are elaborated as follows.

1. Dissolved oxygen control: The concentration of dissolved oxygen is important for aquarium environment. In general, adequate concentrations of dissolved oxygen depend on the species of fishes. In our system, concentrations of dissolved oxygen are desired for our experiment with 4.5 mg/l. The value of concentration of dissolved oxygen is collected from dissolved oxygen sensor. This sensor can transmit the packets of concentration of dissolved oxygen and temperature via RS-232. In here, an AVR micro-controller is used to collect the data, and then converted it into analog signals (DAC) to meet the interface requirements of the PN-WSNA (ADC). The signal would be transmitted in to Mote-1 via ADC port. If the sensor value is less than 4.5 mg/l, the pump will switch on for increasing concentration of dissolved oxygen. On contrary, if the sensor value is greater than 5.0 mg/l (0.5 mg/l hysteresis range), pump will switch off. The control scenario is shown in Fig. 4.
2. Temperature control: The working process of temperature is similar to concentration of dissolved oxygen. The lower threshold of temperature is 25°C; and the upper threshold of temperature is 27°C. The corresponding action is used to turn on / off the heater. The control scenario is also shown in Fig. 4.
3. Light control: Light control in our system depends on the number of visitors in the house. Counting the number of visitors is realized by comparing the rising edges of two Infra-ray human detection sensors. The control scenario is also shown in Fig. 4.

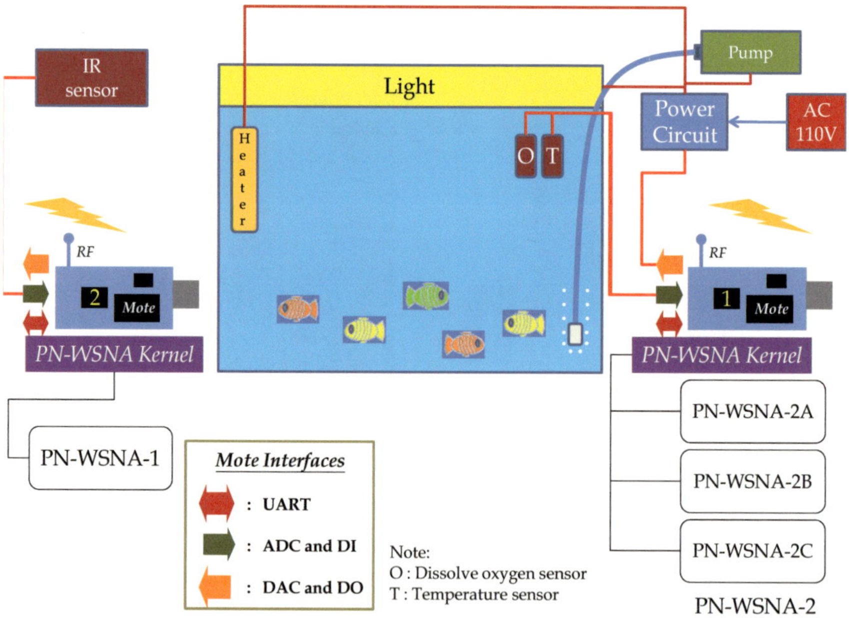

Fig. 3. System architecture of the proposed aquarium system.

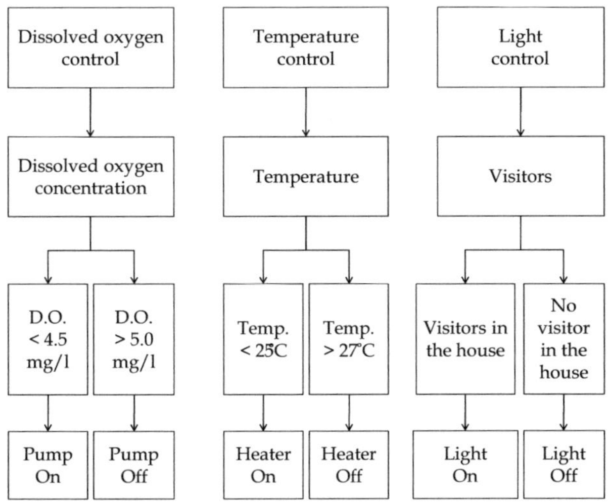

Fig. 4. System operation control scenario.

3.2 PN-WSNA Integrated Development Environment

In order to construct the PN-WSNA models, an integrated development environment (IDE) for constructing the PN-WSNA model is developed. The WSN developer may construct their domain-based PN-WSNA models by using the IDE, and then simulate the PN-WSNA models using the IDE to verify their models before these models are deployed. The PN-WSNA system is composed of the PN-WSNA kernel program and a PN-WSNA management server. Fig. 5 shows the PN-WSNA system implementation architecture. The management server is composed of an IDE which provides a graphical user interface for the domain engineers to construct or modify their PN-WSNA models. At the same time, the databases are also constructed for recording the PN-WSNA models and route tables of sensor nodes. The route tables are used to explore a specific route path for delivering PN-WSNA models to a remote mote in terms of wireless media.

On the other hand, the kernel program is implanted inside a sensor node. In addition, radio frequency (RF) interface with Zigbee protocol (IEEE 802.15.4) as well as physically connected interfaces of UART, ADC, DCA, digital input (DI) and digital output (DO) are also available for cognitive sensing and controls. It is noted that the PN-WSNA IDE is realized using the Microsoft visual C++ and the nesC (Avvenuti et al., 2007) program is used to implement the kernel program.

The PN-WSNA IDE is a plug-and-play model construction environment. Fig. 6 shows the proposed IDE, and the toolbar icons are used for model constructions, editing, revisions, manipulations, run-time simulations, model drawing auxiliaries as well as model deliveries. Detailed descriptions of the tool bars and algorithms for implementing the PN-WSNA inference engine are referred to (Kuo et al., 2009). Finally, the interface functions are coded within the kernel program. Fig. 7 shows the I/O, ADC, DAC and communication interfaces of PN-WSNA motes. The mote is capable of collecting sensor status and actuating the actuators using the sensor interface. At the same time, PN-WSNA motes may also communicate with each other to deliver tokens among different PN-WSNA motes so that distributed decisions can be achieved.

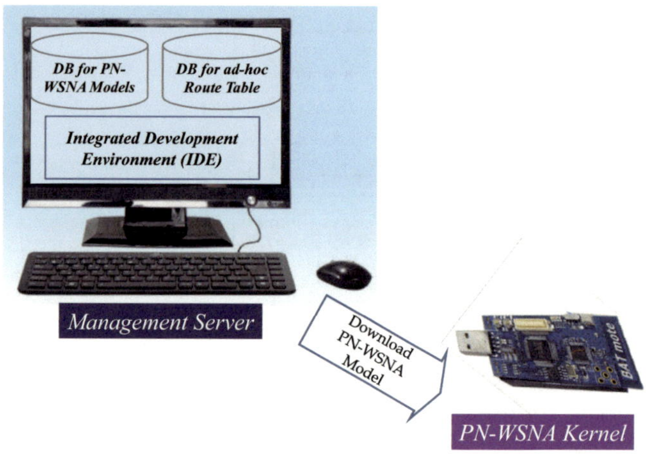

Fig. 5. PN-WSNA model construction architecture.

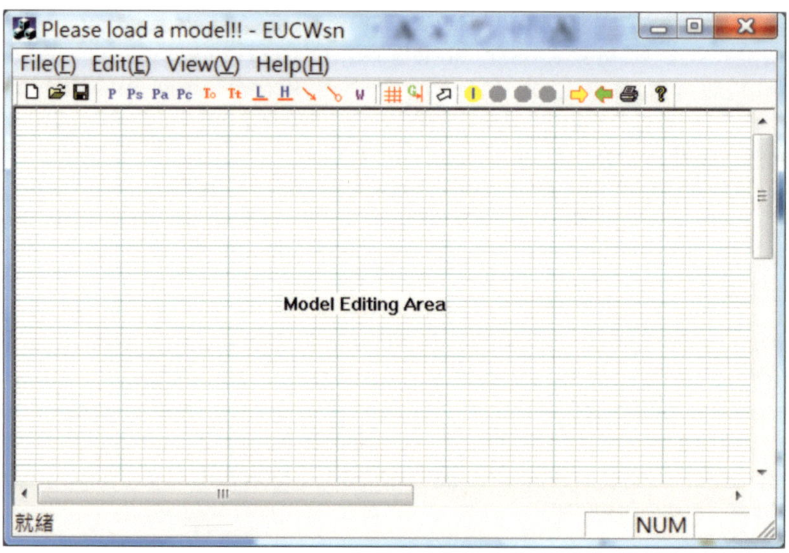

Fig. 6. PN-WSNA IDE workspace.

3.3 PN-WSNA Models

In this subsection, the PN-WSNA models for the proposed aquarium system are presented. These PN-WSNA models are implemented using two PN-WSNA motes. Two motes are communicated via the Zigbee for the delivery of tokens in the corresponding communication places. The proposed overall PN-WSNA architecture was shown in Fig. 3, where PN-WSNA-1 is desired for the entrance counting system using a PN-WSNA mote;

and PN-WSNA-2 is desired for the temperature, dissolved oxygen concentration, and light control system using another PN-WSNA mote.

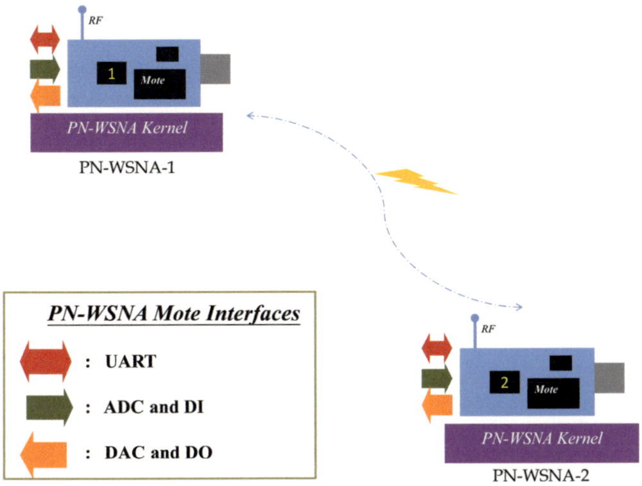

Fig. 7. Interfaces of PN-WSNA motes and their communications.

The first PN-WSNA model (PN-WSNA-1) is an entrance visitor counting and light control system. Fig. 8 shows this model. It is can be classified into three parts, including rising edge detections of two infrared human motion detection sensors, event sequence determinations and light control command generations. For the rising edge detection of an infrared human motion detection sensor, two conditions are considered for pulses generated from each infrared human motion detection sensor including the signals from low-to-high and high-to-low TTL voltage level. The second part is desired for recognizing activated event sequences of two infra-ray sensors (A and B). When infra-ray sensor A is activated first, it means a visitor entering the house. Contrarily, if sensor B is activated first, it means a visitor exiting the house. The last one part is to determine the total number of visitors. If the number of visitors is greater than or equal to one, a command with "turning on the light" is generated; otherwise, a command with "turning off the light" is generated. Because the light is installed a far away mote (PN-WSNA-2), two communication places are desired to transmit the tokens for these light control commands in PN-WSNA1.

For the rising edge detections of two infrared human body sensors model, two similar sub-models are shown first in the left-hand side of the Fig. 8. P001 and P010 indicate the availability of each sensor. T001 and T011 are timed transitions for periodic sampling of the sensors. P003 and P012 indicate the ready signals of sensors A and B, respectively. When the sensor places are ready, the sensor data will be attached within the corresponding places. T002 and T012 are low-enable transitions; T003 and T013 are high-enable transition, and they are used to percept sensor data as high or low status. P004 and P013 indicate the conclusions of low-enable transitions of T002 and T012; P005 and P014 indicate the conclusions of high-enable transitions of T003 and T013. T005, T007, T015 and T017 have the higher priority compared with T004, T006, T014 and T016. It is noted that, the places P006, P007, P015 and P016 are safe (i.e., boundedness with unity token) to keep the current high/

low level status of the sensor. Once the high and low status are both detected (i.e., both P006 and P007 have tokens), the system detects a high/ low level change event. At this moment, the T009, T010, T019 and T020 are used to detect the rising or falling edge event of the sensor in terms of its current level (sensor place P009; the same sensor as P003). For example, if a level change event is detected and its current level is high, then the rising edge from a low-level to high-level voltage would be concluded. It is noted that, only the rising edges (T010 and T020) are used in this project.

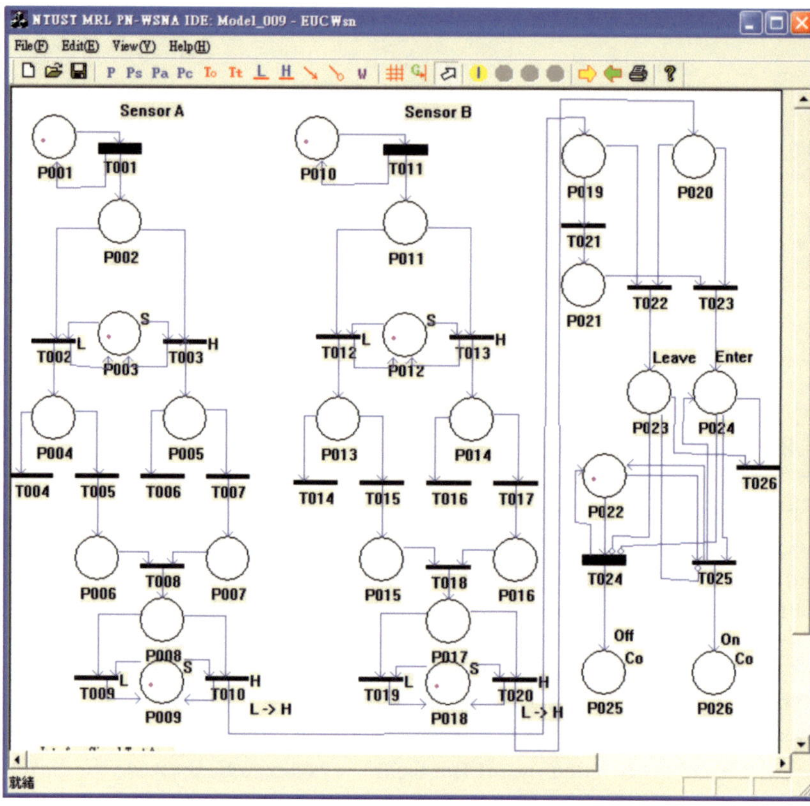

Fig. 8. PN-WSNA model for entrance visitor counting and light control system.

The second part covers the places of P019 - P021 and the transitions of T021 - T023. P019 and P020 represent the rising edges of sensors A and B, respectively. The token arriving sequences determine the entering and exiting events of visitors. If a token arrives at P019 and there is not any token in P020, then the system detects a visor passing through sensor A first. In this situation, the token would enable and fire T021, and then the token enters P021. Once, a token arrives at P020 the T023 will be enabled and fired. As a consequence, a token will be released to P024. On the other hand, a leaving visitors can also be defined similarly. For the case of leaving visitors, token(s) will enter P023.

The remaining part is the light control system. In this sub-model, the tokens in P024 indicate the total number of visitors in a home. The token number would be decreased if a visitor leaves the home (P023). Therefore, the third part of this model realizes such a scenario. If there is no visitor in the home, inhibit arcs from P023 and P024 with respect to T024 would not be inhibited. In this situation, the token in P022 would periodically enable and fire T024 and then the token enters to the transmitter place P025 for delivering tokens to another mote. For another situation, if any visitor is in the home, token(s) would be appeared in P024. Inhibit arc from P024 for T024 would inhibit the activation of T024. For this situation, the token would enable and fire T025, and then the token enters to the transmitter place P026 for delivering tokens to another mote. Other situation is the case of leaving visitors. In this situation, token(s) would be appeared in P023 and P024 as well. Because of using inhibited arcs, the token will enable and fire T026 only.

The second PN-WSNA model (PN-WSNA-2A) is desired for temperature control in our system, as shown in Fig. 9. For conventional temperature control systems, the threshold is setting for turn on and off to modulate the temperature. In order to reduce the switch turn on and off frequency, a hysteresis temperature is desired. In our PN-WSNA, high-enable transition and low enable transition are created. The activation thresholds are defined using specific values. Because of the conflict between high-enable transition and low enable transition, the only one transition would be enabled by comparing the sensor data. In order to create the hysteresis voltage range, two threshold values for upper and lower bounds are desired in this model.

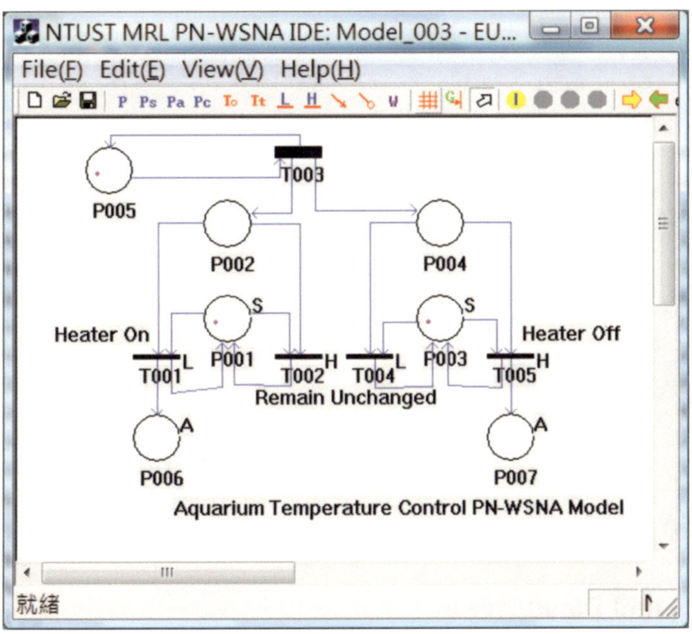

Fig. 9. PN-WSNA model for temperature control.

Two sensory places with the same sensor device and signal as well as two high-enable transitions and low-enable transitions are used in our approach. In this model, P005 indicates the availability of aquarium system. P002 and P004 indicate the ready signals of the temperature sensor. P006 and P007 is the actuation place for turning on and off of the heater. Three tokens are initially assigned to these places for the initial marking. A timed transition (T003) is desired for the periodically sampling and control of this model. If the firing time of T003 is expired, a token in P005 will initiate a decision process. T001 and T004 are low-enable transition; T002 and T005 are high-enable transition, and they are used to percept sensor data as high and low status. Specially, the two different thresholds are defined in the sensory place P001 and P003 for hysteresis ranges. The threshold in sensory place P001 is the lower one. Hence, if data in sensory place P001 is below the threshold, the transition T001 is enabled and then the corresponding actuation place will turn off the heater. The activation of then sensory place P003 is similar with sensory place P001. Therefore, the two thresholds for control the heater is done by aforementioned process. On the other hand, the PN-WSNA model for dissolved oxygen (PN-WSNA-2B) is similar to the temperature model, and it just needs to adjust the threshold of sensory places as specific values. Then, this model can turn on and off of the pump in terms of actuation places.

The last PN-WSNA model (PN-WSNA-2C) is desired for communication between two motes so that the lighting device at a remote mote can be controlled. P001 in PN-WSNA-2C is corresponded P026 in PN-WSNA-1; and P002 in PN-WSNA-2C is corresponded P025 in PN-WSNA-1. Once the receiver place P001 receives a token and then the token would enable fire T001. This token will be released to actuation place P003. The corresponding action of P003 is "turning on the light". The activation process of P004 is similar to P003 and the corresponding action of P004 is "turning off the light". Fig. 10 shows the model of PN-WSNA-2C.

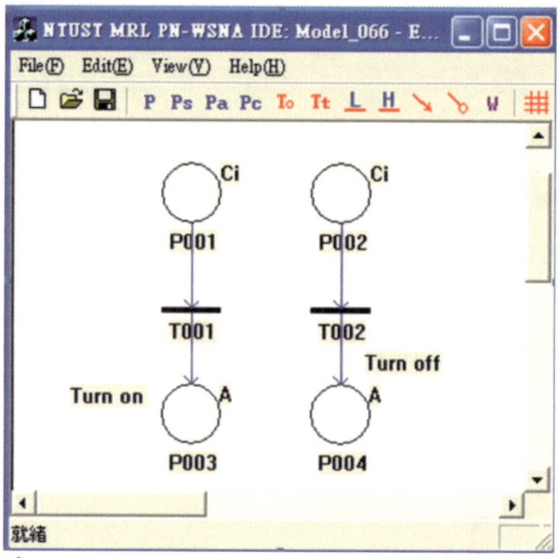

Fig. 10. PN-WSNA model for mote communication and control the light.

4. Experiments and Discussions

In this section, the experiment is proposed and discussed. At first, the initial marking of PN-WSNA-1 was shown in Fig. 8. Tokens are initially paaeared in the sensory places (P003, P009, P011 and P018) and the normal places (P001, P010 and P022). In here, an experimental example is proposed to describe the control procedures of light control system. A time chart of sensor signals with respect to two infrared human motion detection sensors of this experiment is shown in Fig. 11.

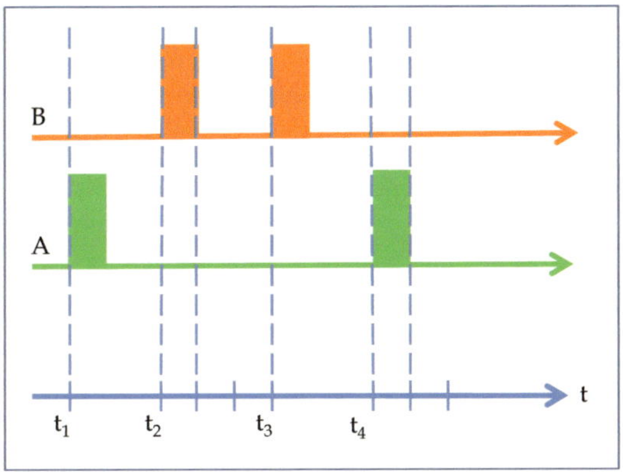

Fig. 11. Time chart of sensor signal

Initial signals of sensor A and B are both in low states before the time t_1. For the sensor A, when time transition T001 is fired, the token enters P002. After token is appeared in P002, the sensory place P003 would enable and fire low-enable transition T002, and then the token enters P004. After that, T005 will fire and then the token transmits to P006. For the setting of safeness, the number of token in P006 is one maximally. The procedures of sensor B are the same as sensor A. The inference of PN-WSNA model for this part is shown in Fig. 12 (a) and Fig. 12 (b).

At the moment of t_1, a high-level signal is generated from sensor A, and then high-enable transition T003 would be fired. The token would consequently enter P007, as shown in Fig. 12 (c). At that time, T008 is fired, and then the token enters P008, as shown in Fig. 12 (d). Because the signal is in a high state, signal from sensor A would enable and fire high-enable transition T010, and then the token enters P019, as shown in Fig. 12 (e). As a consequence, the token in P019 represents a low-to-high rising edge signal is detected, as shown in Fig. 12 (f). After an edge signal is detected from sensor A, token would enter to P021 for waiting edge signal which comes from sensor B, as shown in Fig. 13.

At the moment of t_2, a high-level signal is generated from sensor B. Similar procedures will be executed, and consequently deliver a token to P020. In this situation, the sequence of sensory A and B are that A comes before B. Therefore, T023 is fired, and then the token enters P024 for presenting a visitor entering in the house. After that time, T025 is enabled and fired, and then the token enters the transmitter place P026 to send the command with "turning on the light" to mote-1. The inferences of PN-WSAN model is shown in Fig. 14.

W At the moment of t_1, a visitor leaves the house, and then the corresponding signal of sensory B is activated first. In this situation, T022 is fired and then generates a token for presenting a visitor leaving the house. Then the token in P023 and P024 would be released accordingly. The inferences of PN-WSAN model is shown in Fig. 15.

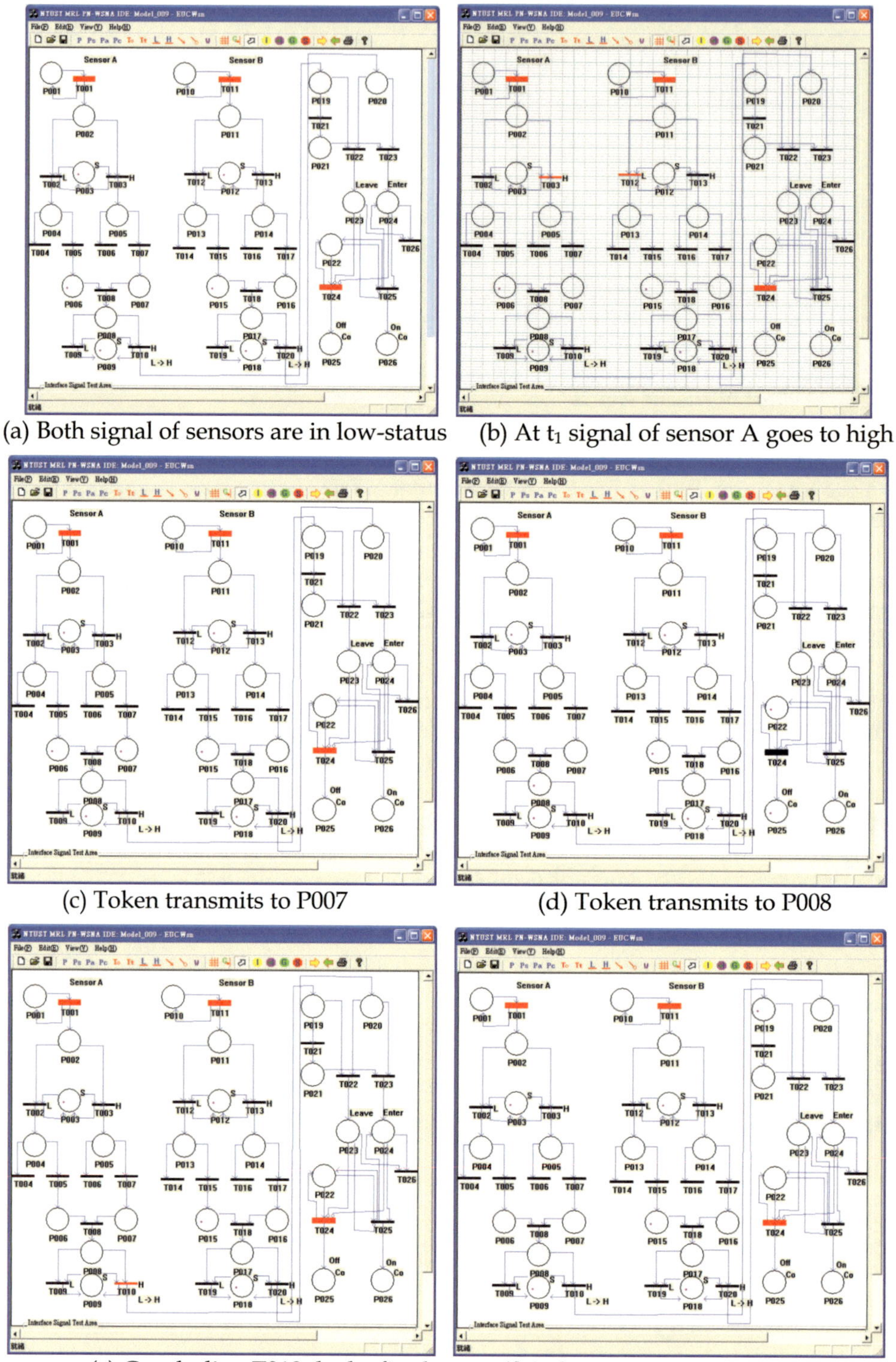

(a) Both signal of sensors are in low-status (b) At t_1 signal of sensor A goes to high

(c) Token transmits to P007 (d) Token transmits to P008

(e) Concluding T010 the be fired (f) Token transmits to P0019

Fig. 12. Decision procedures of sensor A's signal (from low-to-high).

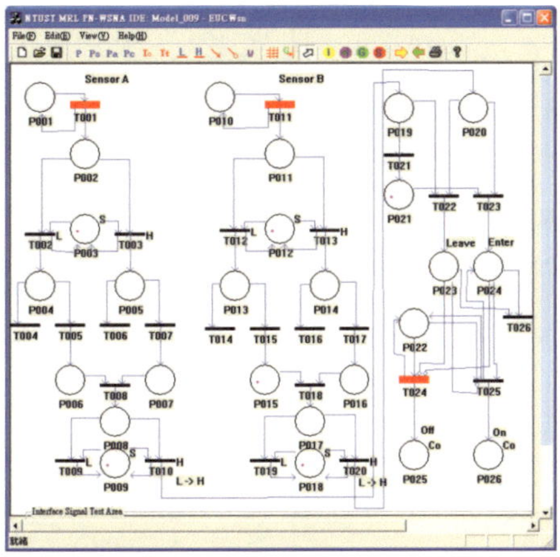

Fig. 13. Concluding rising edge status of sensor A.

5. Conclusions and Future Works

In this book chapter, the PN-WSNA is used to construct an aquarium environment control system in a house. The major advantages of using PN-WSNA are to use a model based WSN realization approach so that the coding efforts from domain engineers can be significantly reduced. In addition, the control scenarios can be verified in terms of the PN-WSNA simulations before the sensor algorithm are deployed. This book chapter use an aquarium environment control system to demonstrate the modelling and implementation procedures for two PN-WSNA sensor node systems, where one sensor node is deployed for aquarium environment control and the other one is desired for entrance counting system. Two PN-WSAN motes are communicated using the communication places of the PN-WSNA. The aquarium environment control system acquires the data from temperature sensor and dissolved oxygen sensor as well as the people number collected from the entrance counting system. Meanwhile, the light, heater and pump are also activated using the sensor node. In the future, the PN-WSNA will be used to construct more complicated WSN system to demonstrate the powerful modelling and control capability of the PN-WSNA.

6. Acknowledgement

This work was supported by the National Science Council, Taiwan, R.O.C., under Grants NSC 98-2218-E-011-017.

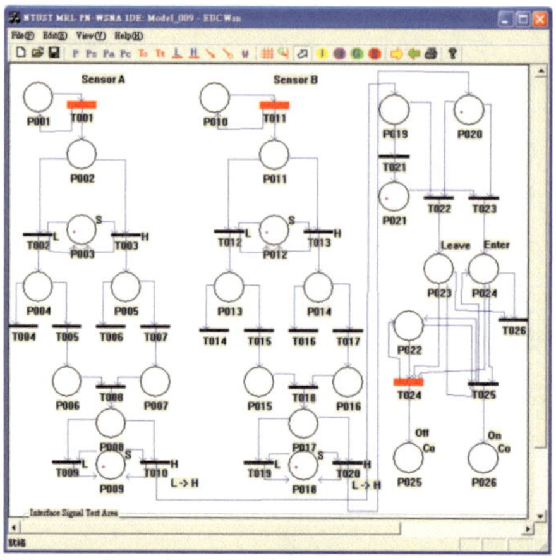

(a) Rising edge detected from sensor B (P020)

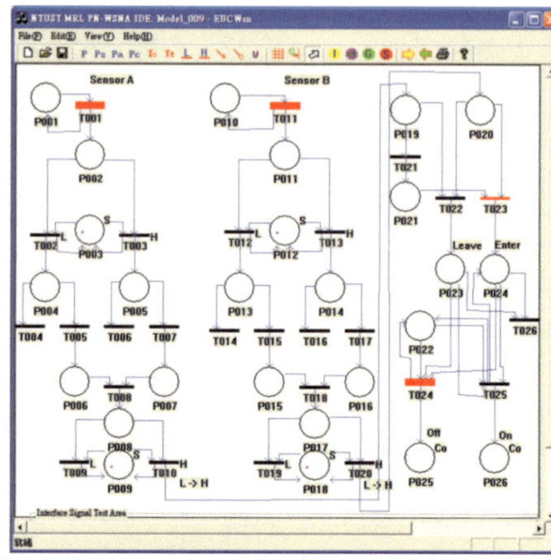

(b) A->B sequence determined

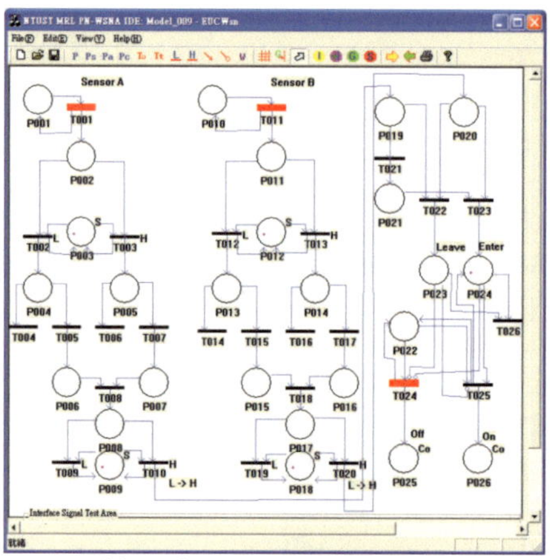

(c) Token enters to P024

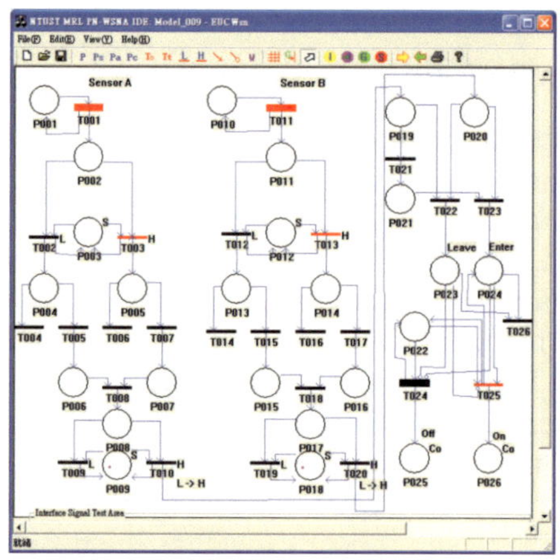

(d) T025 enable fire

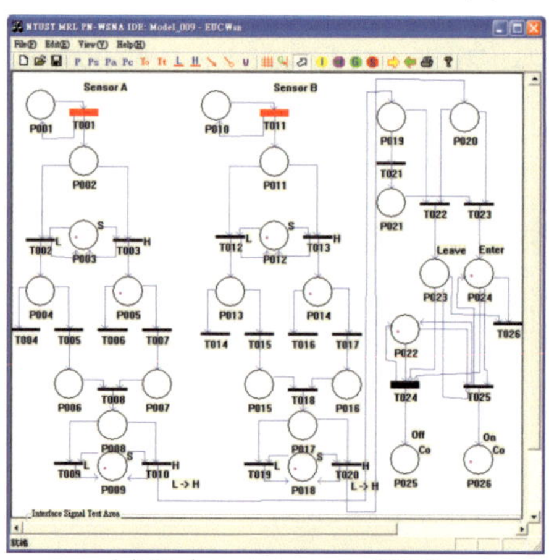

(e) Token enters to P026

Fig. 14. Concluding status of A->B sequence determinations.

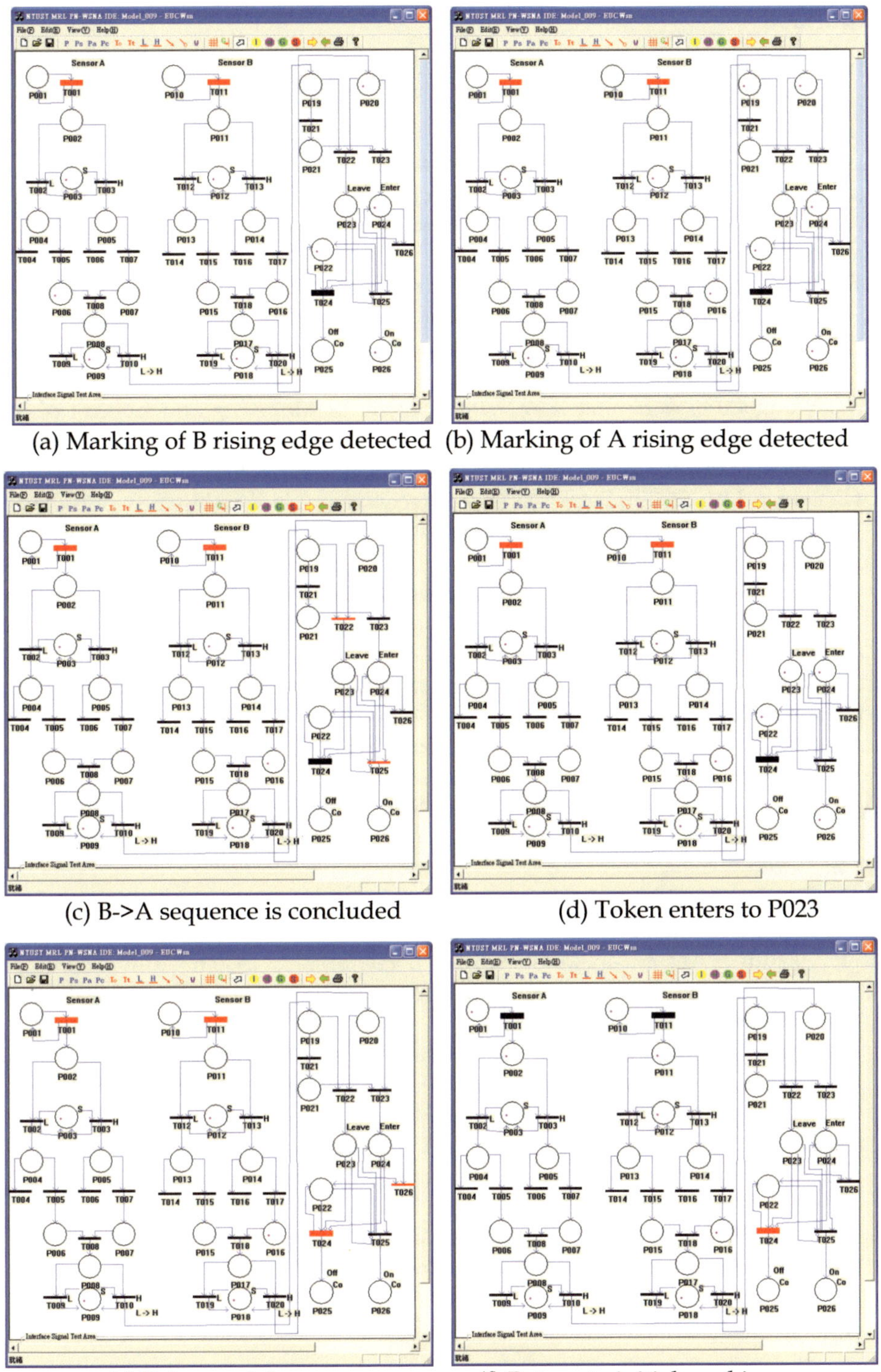

(a) Marking of B rising edge detected (b) Marking of A rising edge detected

(c) B->A sequence is concluded (d) Token enters to P023

(e) Release a token (f) Return to initial marking

Fig. 15. Concluding status of B->A sequence determinations.

7. References

Akyildiz I.F.; Melodia T.; & Chowdhury K.R. (2008). Wireless Multimedia Sensor Networks: Applications and Testbeds, *Proceedings of the IEEE*, Vol. 96, No. 10, pp. 1588 - 1605.

Avnur A. (1990). Finite State Machines for Real-time Software Engineering, *Computing & Control Engineering Journal*, Vol. 1, No. 6, pp. 275 - 278.

Avvenuti M.; Corsini P.; Masci P.; & Vecchio A. (2007). An Application Adaptation Layer for Wireless Sensor Networks, *Pervasive and Mobile Computing*, Vol. 3, No. 4, pp. 413 - 438.

Dalola S.; Ferrari V.; Guizzetti M.; Marioli D.; Sardini E.; Serpelloni M.; & Taroni A. (2009). Autonomous Sensor System with Power Harvesting for Telemetric Temperature Measurements of Pipes, *IEEE Transactions on Instrumentation and Measurement*, Vol. 58, No. 5, pp. 1471 - 1478.

Kuo C.H.; Wang C.H.; & Huang K.W. (2003). Behavior Modeling and Control of 300 mm Fab iIntrabays Using Distributed Agent Oriented Petri Net, *IEEE Transaction on Systems, Man and Cybernetics*, Part A, Vol. 33, No. 5, pp. 641 - 648.

Kuo C.H.; & Siao J.W. (2009). Petri Net Based Reconfigurable Wireless Sensor Networks for Intelligent Monitoring Systems, *International Conference on Computational Science and Engineering*, Vol. 2, pp. 897 - 902.

Manasseh C.; & Sengupta R. (2010). Middleware to Enhance Mobile Communications for Road Safety and Traffic Mobility Applications, *IET Intelligent Transport Systems*, Vol. 4, No. 1, pp. 24 - 36.

Murata T. (1989). Petri Nets: Properties, Analysis and Applications, *Proceedings of the IEEE*, Vol. 77, No. 4, pp. 541 - 580.

Romer K.; & Mattern F. (2004). The Design Space of Wireless Sensor Networks, *IEEE Wireless Communications*, Vol. 11, No. 6, pp. 54 - 61.

Ruiz A.F.; Rocon E.; Raya R.; & Pons J.L. (2008). Coupled Control of Human-exoskeleton Systems: an Adaptative Process, *IEEE Conference on Human System Interactions*, pp. 242 - 246.

Shenai K.; & Mukhopadhyay S. (2008). Cognitive Sensor Networks, *International Conference on Microelectronics*, pp. 315 - 320.

Sridhar P.; Madni A.M.; & Jamshidi M. (2007). Hierarchical Aggregation and Intelligent Monitoring and Control in Fault-tolerant Wireless Sensor Networks, *IEEE Systems Journal*, Vol. 1, No. 1, pp. 38 - 54.

Suh C; & Ko Y.B. (2008). Design and Implementation of Intelligent Home Hontrol Systems Based on Active Sensor Networks, *IEEE Transactions on Consumer Electronics*, Vol. 54, No. 3, pp. 1177 - 1184.

Zhuang L.Q.; Liu W.; Zhang J.B.; Zhang D.H.; & Kamajaya I. (2008). Distributed Asset Tracking Using Wireless Sensor Network, *IEEE International Conference on Emerging Technologies and Factory Automation*, pp. 1165 - 1168.

Zhuang X.; Yang Y.; & Ding W. (2008). The Wireless Sensor Network Node Design for Electrical Equipment On-line Monitoring, *IEEE International Conference on Industrial Technology*, pp. 1 - 4.

8

Wireless Sensor Network for Ambient Assisted Living

Juan Zapata and Francisco J. Fernández-Luque and Ramón Ruiz
Universidad Politécnica de Cartagena
Spain

1. Introduction

There is a great demand, in both the public and the private sector, to take the actions needed to expand uses for electronic devices, assistive and monitoring software, and home health communication technologies to provide assisted living and health care to those in need. By means of using Information and Communication Technologies (ICT) to assist and monitor elderly, disabled, and chronically ill individuals in the home can improve quality of life, improve health outcomes, and help control health care.

Assisted living technologies are for people needing assistance with Activities of Daily Living (ADLs) but wishing to live for as long as possible independently. Assisted living exists to bridge the gap between independent living and nursing homes. Residents in assisted living centers are not able to live by themselves but do not require constant care either. Assisted living facilities offer help with ADLs such as eating, bathing, dressing, laundry, housekeeping, and assistance with medications. Many facilities also have centers for medical care; however, the care offered may not be as intensive or available to residents as the care offered at a nursing home. Assisted living is not an alternative to a nursing home, but an intermediate level of long-term care appropriate for many seniors. Increasing health care costs and an aging population are placing significant strains upon the health care system. Small pilot studies have shown that meeting seniors needs for independence and autonomy, coupled with expanded use of home health technologies, mitigate against the circumstances above, and provide improved health outcomes. Difficulty with reimbursement policies, governmental approval processes, and absence of efficient deployment strategies has hampered adopting such technologies.

Most efforts to-date of applying ICT to health care in the home have focused on telemedicine, in which ICT is used to connect the patient and his biometric information to a health care provider either in real-time or into storage for analysis at a later time (Biemer & Hampe, 2005; Lubrin et al., 2006; Ross, 2004; Rowan & Mynatt, 2005). While such functionality will be included in Information Technology for Assisted Living at Home (ITALH), the main objective in tele-assistence is to provide smart monitors and sensors that will alert the user and/or their care provider (be that a doctor, nurse, family member, neighbor, friend, etc.) of events such as accidents or acute illness and to diagnostic events that could indicate a deterioration in his health condition. This differs from efforts (Sixsmith & Johnson, 2004a), in which IR sensors array are used to detect such events as accidental falls, in that the sensors themselves detect

the event, not a central system. This reduces wireless bandwidth and greatly improves the privacy of the system by not streaming data constantly.
In particular, the chapter is focused in the introduction of an ubiquitous wireless network infrastructure to support an assisted living at home system, called DIA (Dispositivo Inteligente de Alerta, in spanish) which is being developed by Universidad Politécnica de Cartagena, Universidad de Murcia and Ambiental Intelligence & Interaction S.L.L. (Ami2) company. Specifically, the system is constructed based on a wireless communication network in order to transfer data and events of elderly. A typical scenario consists of a private home which is instrumented based on WSN. In this context, the concept of a individual assisted by monitoring via radio-frequency is evident. The wireless infrastructure is a heterogeneous and ubiquitous, being present everywhere at once, wireless network that connects sensor devices within the home to a central Home Health System Gateway and/or a mobile Gateway. The sensor nodes themselves have embedded processing capability and are required to transmit only occasional information about their own status and messages notifying the central system when they detect a significant event. The central system, in a smart sense, connect this network to the outside world via secure Internet and telephone service so that intelligent alerts can be sent out, and authorized caregivers can have access to the system to check up on the user. Privacy and security are fundamental concerns in these systems. The chapter is organized as follows. Section 2 provides an overview of the state of the art in terms of sensor network technology. Section 3 explains the application scenario and where and how the node sensors was deployed and Section 4 discusses data processing issues. The chapter concludes with a brief summary and some final remarks.

2. Technology Overview

2.1 Sensor Network Technology Overview

A sensor network is an infrastructure comprised of sensing (measuring), computing, and communication elements that gives an administrator the ability to instrument, observe, and react to events and phenomena in a specified environment (Sohraby et al., 2007). Typical applications include, but are not limited to, data collection, monitoring, surveillance, and medical telemetry. In addition to sensing, one is often also interested in control and activation.
There are four basic components in a sensor network: (1) a set of distributed or localized sensors; (2) a communication network (usually, but not always, wireless-based); (3) a central point of information clustering (usually called base station or sink); and (4) a set of computing resources at the central point (or beyond, e.g., personal computer board or other device like PDA) to handle data correlation, event trending, status querying, and data mining. In this context, the sensing and computation nodes are considered part of the sensor network; in fact, some of the basic computation may be done in the network itself. The computation and communication infrastructure associated with sensor networks is often specific to this environment and rooted in the device and application-based nature of these networks. Figure 1 shows a generic protocol stack model that can be utilized to describe the WSN. Issues here relate to the following: (1) Physical layer treats about connectivity and coverage. (2) Data Link Layer is the protocol layer which transfers data between adjacent network nodes in a wide area network or between nodes on the same local area network segment. (3) Network Layer is responsible for end-to-end (source to destination) packet delivery including routing through intermediate hosts, whereas the data link layer is responsible for node-to-node (hop-to-hop) frame delivery on the same link. (4) Transport Layer is a group of methods and protocols

within a layered architecture of network components within which it is responsible for encapsulating application data blocks into data units. (5) Upper (application, presentation, session) layer treats about processing application, in what could be an environment with highly correlated and time-dependent arrivals.
Management planes are needed, so that sensor nodes can work together in a power efficient way, route data in a wireless (mobile or not) sensor network, and share resources between sensor nodes. Without them, each sensor node will just work individually. From the whole sensor network point of view, it is more efficient if sensor nodes can collaborate with each other, so the lifetime of the sensor networks can be prolonged (Kahn et al., 1999).

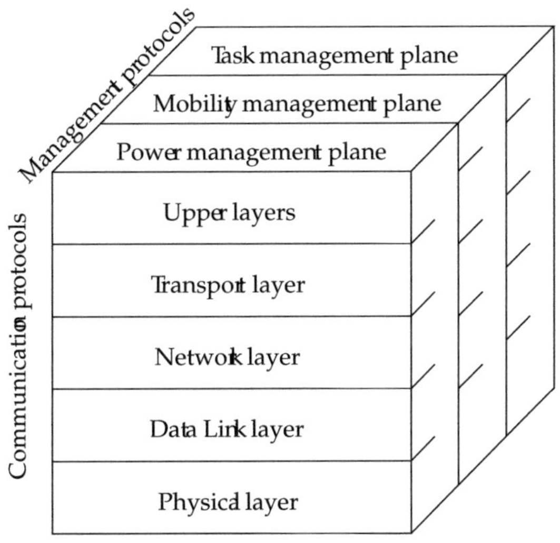

Fig. 1. Generic protocol stack for sensor networks

Sensors in a WSN have a variety of purposes, functions, and capabilities. Sensor networking is a multidisciplinary area that involves, among others, radio and networking, signal processing, artificial intelligence, database management, systems architectures for operator-friendly infrastructure administration, resource optimization, power management algorithms, and platform technology (hardware and software, such as operating systems)
The technology for sensing and control includes electric and magnetic field sensors; radio-wave frequency sensors; optical-, electrooptic-, and infrared sensors; radars; lasers; location/navigation sensors; seismic and pressure-wave sensors; environmental parameter sensors (e.g., wind, humidity, heat); and biochemical national security oriented sensors. Todays sensors can be described as smart inexpensive devices equipped with multiple onboard sensing elements; they are low-cost low-power untethered multifunctional nodes that are logically homed to a central sink node. Sensor devices, or wireless nodes (WNs), are also (sometimes) called motes. Therefore, a WSN consists of densely distributed nodes that support sensing, signal processing, embedded computing, and connectivity; sensors are logically linked by self-organizing means. WNs typically transmit information to collecting (monitoring) stations that aggregate some or all of the information. WSNs have unique characteristics, such as, but not limited to, power constraints and limited battery life for the WNs, redundant data acquisition, low duty cycle, and, many-to-one flows. Power efficiency in WSNs is generally accomplished in three way: Low-duty-cycle operation, Local/in-network processing to reduce data volume

(and hence transmission time), and multihop. Multihop networking reduces the requirement for long-range transmission since signal path loss is an inverse exponent with range or distance. Each node in the sensor network can act as a repeater, thereby reducing the link range coverage required and, in turn, the transmission power.
For a number of years, vendors have made use of proprietary technology for collecting performance data from devices. In the early 2000s, sensor device suppliers were researching ways of introducing standardization, first designers ruled out Wi-Fi (wireless fidelity, IEEE 802.11b) standards for sensors as being too complex and supporting more bandwidth than is actually needed for typical sensors. Infrared systems require line of sight, which is not always achievable; Bluetooth (IEEE 802.15.1) technology was at first considered a possibility, but it was soon deemed too complex and expensive. This opened the door for a new standard IEEE 802.15.4 along with ZigBee (more specifically, ZigBee comprises the software layers above the newly adopted IEEE 802.15.4 standard and supports a plethora of applications). IEEE 802.15.4 operates in the 2.4 GHz industrial, scientific, and medical (ISM) radio band and supports data transmission at rates up to 250 kbit s^{-1} at ranges from 10 to 70 m. ZigBee/IEEE 802.15.4 is designed to complement wireless technologies such as Bluetooth, Wi-Fi, and ultra-wideband (UWB).

2.1.1 Requeriments for Wireless Sensor Network in Ambient Assisted Living Environments

Wireless sensor network integration in Ambient Assisted Living frameworks is usually not described in the literature, and normally is treated as a black box. Following, a list of the functional requirements for WSN in Ambient Assisted Living environments which need to be addressed in order to have a reliable system based on WSN (Martin et al., 2009) is enumerated.

1. *Ambient Assisted Living Systems: Type of Networks.* The design objectives are related to the development of an Ambient Assisted Living System capable of offering its services at home and nursing houses. The AAL systems could be composed by:

    - A *Body Sensor Network (BSN)*, which will include all the devices that a person must wear (accelerometers, gyroscopes, spirometers, oxymeters, etc) or use to allow the services to work. Depending on the elderly profile and the services to be configured, the BSN may include continuous monitoring sensors and other health sensors. To configure the BSN, it is always important to bear in mind the usability restrictions imposed by the users acceptance of personal devices in home environments. Basically, BSN is a mobile subset of the Wireless Sensor Network.
    - A *Wireless Sensor Network*, which will include home infrastructure sensors (ambient, presence, pressure, home automation sensors, etc.), actuators and appliances capable of notifying their status. The Wireless Sensor Network station base will be able to communicate with the BSN by using ad hoc networking capabilities. It will include local intelligent features to dispatch events and orders depending on the situation. These processing capabilities will be part of a home gateway which will connect the home ambient via station base with the Core Care Network.
    - A *Core Care Network*, serving as a bridge of communication between the home sensorial infrastructure and third parties and service providers (caregivers). Services may be enabled through the Core Care Network. It can also authorize the connection of external service providers, centralize system monitoring and guarantee the security of personal data.

2. *Ambient Assisted Living Systems: General design requirements for WSN*. The previous scenario imposes some functional requirements to the Wireless Sensor Network finally composed by infrastructure and personal sensing nodes. Next there is a brief list of the most important features to consider.

   - *Interoperability*. Wireless Sensor Networks in real deployments need to be ready to manage heterogeneous sensors, which need to share a common communication scheme.
   - *Network self-configuration and maintenance*. It is desirable that the WSN demands as little attention from a human operator as possible.
   - *Easy and robust deployment*. When designing WSN functionalities, it is important to consider the deployment requirements to make the network fully operational.
   - *Multihop routing*. A WSN for AAL usually consists of several sensor nodes that send their measurements to a sink node, which collects all the information and typically sends it to a PC, where all the data are stored and elaborated. In a home, the sink node may not provide coverage over the whole area. As a consequence, it is necessary to implement routing algorithms that transmit the information towards the sink through other nodes.
   - *Positioning service*. This kind of service is required for many operational and service purposes. For instance, to process the information related to the place where the user is, avoiding the storage and computation of information that is not relevant in a specific moment.
   - *Energy saving strategies*. As the devices that are integrated in the network have limited computational and radio communication capabilities, collaborative algorithms with energy-aware communication are required to achieve multi-modal collaboration and energy conservation.
   - *Scalability of sensors and actuators*. AALs services may impose different type of sensing and actuation requirements. For example, in a scenario considering services for COPD patients, devices sensing the quality of air may be needed. For that reason, WSN for ACSs need to be ready to include new sensors and actuators, which may be connected to existent network nodes or configure as nodes themselves. Methodologies and software architectures making easier to scale the network sensing capabilities are needed.
   - *Security*. As wireless networks are based on a standard and data are sent over a broadcast channel, it is possible to make packet sniffing and data spoofing attacks. IEEE 802.15.4 MAC layer offers some facilities which can be used by upper layers to achieve a good level of security.

2.2 Sensor Node Technology Overview

Figure 2 shows the general architecture soft and hardware of a sensor node. The terms sensor node, wireless node (WN), Smart Dust, mote, and COTS (commercial off-the-shelf) mote are used somewhat interchangeably in the industry; the most general terms used here are sensor node and WN. WSNs that combine physical sensing of parameters such as temperature, light, or others events with computation and networking capabilities are expected to become ubiquitous in the next future.

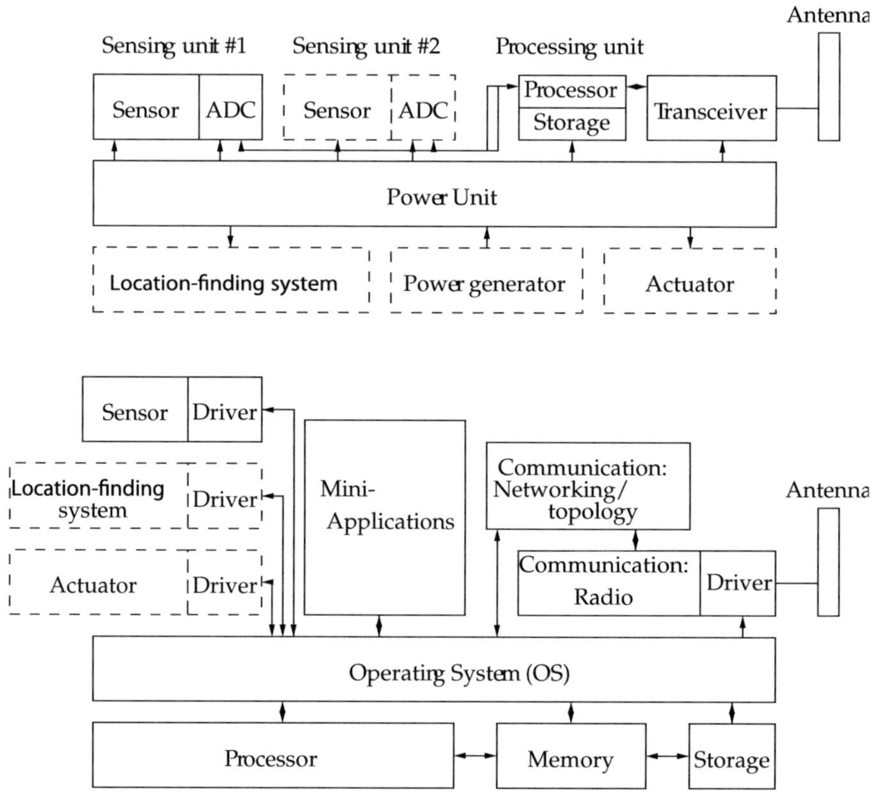

Fig. 2. Hardware and software components of WNs

Many of these examples share some basic characteristics. In most of them, there is a clear difference between sources of data the actual nodes that sense data and sinks nodes where the data should be delivered to. The interaction patterns between sources and sinks show some typical patterns. The most relevant ones are:

- *Event detection.* Sensor nodes should report to the sink(s) once they have detected the occurrence of a specified event. The simplest events can be detected locally by a single sensor node in isolation (e.g., a temperature threshold is exceeded); more complicated types of events require the collaboration of nearby or even remote sensors to decide whether a (composite) event has occurred (e.g., a temperature gradient becomes too steep). If several different events can occur, event classification might be an additional issue.
- *Periodic measurements.* Sensors can be tasked with periodically reporting measured values. Often, these reports can be triggered by a detected event; the reporting period is application dependent.
- *Function approximation and edge detection.* The way a physical value like temperature changes from one place to another can be regarded as a function of location. A WSN can be used to approximate this unknown function (to extract its spatial characteristics), using a limited number of samples taken at each individual sensor node. This approximate mapping should be made available at the sink. How and when to update this

mapping depends on the applications needs, as do the approximation accuracy and the inherent trade-off against energy consumption. Similarly, a relevant problem can be to find areas or points of the same given value. An example is to find the isothermal points in a forest fire application to detect the border of the actual fire. This can be generalized to finding edges in such functions or to sending messages along the boundaries of patterns in both space and/or time.

- *Tracking.* The source of an event can be mobile (e.g., an intruder in surveillance scenarios). The WSN can be used to report updates on the event sources position to the sink(s), potentially with estimates about speed and direction as well. To do so, typically sensor nodes have to cooperate before updates can be reported to the sink.

Embedded sensing refers to the synergistic incorporation of sensors in structures or environments; embedded sensing enables spatially and temporally dense monitoring of the system under consideration (e.g., a home). In biological systems, the sensors themselves must not affect the system or organism adversely. The technology for sensing and control includes electric and magnetic field sensors; radio-wave frequency sensors; optical-, electrooptic-, and infrared sensors; radars; lasers; location and navigation sensors; seismic and pressure-wave sensors; environmental parameter sensors (e.g., wind, humidity, heat); and biochemical national security oriented sensors.

Small, low-cost, robust, reliable, and sensitive sensors are needed to enable the realization of practical and economical sensor networks. Although a large number measurements are of interest for WSN applications, commercially available sensors exist for many of these measurements. Sensor nodes come in a variety of hardware configurations: from nodes connected to a LAN and attached to permanent power sources, to nodes communicating via wireless multi-hop RF radio powered by small batteries. The trend is toward very large scale integration (VLSI), integrated optoelectronics, and nanotechnology; in particular, work is under way in earnest in the biochemical arena.

2.2.1 Hardware and software architecture of WNs

Normally, the hardware components of a WN include the sensing and actuation unit (single element or array), the processing unit, the communication unit, the power unit, and other application-dependent units. Sensors, particularly Smart Dust and COTS motes, have four basic hardware subsystems:

1. *Sensor transducer(s).* The interface between the environment and the WN is the sensor. Basic environmental sensors include, but are not limited to, acceleration, humidity, light, magnetic flux, temperature, pressure, and sound.
2. *Computational logic and storage.* These are used to handle onboard data processing and manipulation, transient and short-term storage, encryption, digital modulation, and digital transmission.
3. *Communication.* WNs must have the ability to communicate either in C1-WSN arrangements (mesh-based systems with multi-hop radio connectivity among or between WNs, utilizing dynamic routing in both the wireless and wireline portions of the network), and/or in C2-WSN arrangements (point-to-point or multipoint-to-point systems generally with single-hop radio connectivity to WNs, utilizing static routing over the wireless network with only one route from the WNs to the companion terrestrial or wireline forwarding node).

4. *Power supply*. An appropriate energy infrastructure or supply is necessary to support operation from a few hours to months or years (depending on the application).

Sensors typically have five basic software subsystems:

1. *Operating system (OS) microcode (also called middleware)*. This is the board common microcode that is used by all high-level node-resident software modules to support various functions. As is generally the case, the purpose of an operating system is to shield the software from the machine-level functionality of the microprocessor. It is desirable to have open-source operating systems designed specifically for WSNs; these OSs typically utilize an architecture that enables rapid implementation while minimizing code size. TinyOS is one such example of a commonly used OS.
2. *Sensor drivers*. These are the software modules that manage basic functions of the sensor transceivers; sensors may possibly be of the modular/plug-in type, and depending on the type and sophistication, the appropriate configuration and settings must be uploaded into the sensor (drivers shield the application software from the machine-level functionality of the sensor or other peripheral).
3. *Communication processors*. This code manages the communication functions, including routing, packet buffering and forwarding, topology maintenance, medium access control (e.g., contention mechanisms, direct-sequence spread-spectrum mechanisms), encryption, and FEC, to list a few.
4. *Communication drivers* (encoding and the physical layer). These software modules manage the minutia of the radio channel transmission link, including clocking and synchronization, signal encoding, bit recovery, bit counting, signal levels, and modulation.
5. *Data processing mini-apps*. These are numerical, data-processing, signal value storage and manipulations, or other basic applications that are supported at the node level for in-network processing.

2.3 A Survey of Sensor Nodes for Wireless Sensor Networks

A significant and prime research of sensor network is a project named *Smart-Dust* developed by University of California at Berkeley, USA (Pister, 2008) in the late 90s. The main objective of the project was to develop a compact size node that includes sensor, capability to compute the sensor data onboard, low cost, minimal power consumption and having bidirectional wireless communication capability. It was later sold commercially by Crossbow Inc. and Moteiv. The first mote was the WeC, which appeared in 1998 and was followed in the next year by the René mote and the next year by the René2 mote and Dot mote. Based on the field trials of these platforms, a second-generation platform called MICA was developed; it appeared in 2001. A third generation of motes were named MICA2 which was appeared in 2002. MICAz was appeared in 2002 and it was the first generation with a 2.4 GHz frequency of radio. Iris Mote is built upon the IEEE 802.15.4 standard. It is regarded as the successor of MICAz, one of the most commonly used mote systems in the world. The most recent significant step in commercial mote development was the Telos in 2004 y Telosb in 2005.

Another interesting research testbed was the *Spec platform* (JLHLabs, 2008), which integrated the functionality of Mica onto a single 5 mm^2 chip. Spec was built with a micro-radio, an analog-to-digital converter, and a temperature sensor on a single chip, which lead to a 30-fold reduction in total power consumption. This single-chip integration also opened the path to low cost sensor nodes. The integrated RAM and cache memory architecture greatly simplified the design of the mote family. However, the tiny footprint also requires a specialized operating

system, which was developed by UC Berkeley, called TinyOS (TinyOS, 2009). TinyOS features component based architecture and event driven model that are suitable for programming with small embedded devices, such as motes. The combination of Motes and TinyOS is gradually becoming a popular experimental platform for many research efforts in the field of WSNs.
The *Medusa MK-2* (UCLA, 2009) sensor node was carried out by the Center for Embedded Networked Sensing (CENS) at UCLA in 2002 to target both high and low-end processing applications. It integrates two microcontrollers, the first one; ATmega128 was dedicated to less computationally demanding tasks, including radio base band processing and sensor sampling. The second one, AT91FR4081, was a more powerful microcontroller (40 MHz, 1 MB flash, 136 kB RAM) that was designed to handle more sophisticated, but less frequent signal processing tasks (e.g., the Kalman filter). The combination of these two microcontrollers provided more flexibility in WSN development and deployment, especially for applications that require both high computation capabilities and long lifetime.
In 2002 the Berkeley Wireless Research Center (BWRC) developed System on Chip (SoC) based sensor node, named *PicoNode II* (Cho et al., 2005). It was built using two ASIC chips that implemented the entire node functionality. In the following year, the same team developed a first radio transmitter (that used power less than 400 μW), *PicoBeacon* was purely powered by solar and vibrational energy sources.
Another ASIC based approach was taken by the *μAMPS* group from MIT. Following their first testbed, *μAMPS-I* (MIT, 2008), the team then tried to build a highly integrated sensor node comprised of a digital and an analog/RF ASIC, *μAMPS-II*. The interesting feature of μAMPS-II was the nodes capability to operate in several modes. It can operate either as low-end stand-alone guarding node, a fully functional node for middle-end sensor networks or as a companion component in a more powerful high-end sensor system. Thus, it favored a network with heterogeneous sensor nodes for a more efficient utilization of resources.
The *Free2move* wireless sensor node (Bilstrup & Wiberg, 2004) is based on a transceiver operating in the 2.4 GHz ISM band. The node was initially thought of as an active RFID tag for monitoring temperature in goods. However, it has been shown that it is also possible to use it as a wireless sensor network node. The node is equipped with an extremely low power microcontroller (Microchip PIC16F87), for executing communication protocols and sensor functionality. The memory and processing resources are very limited to keep the price and energy consumption as low as possible. The node is also equipped with a temperature sensor.

3. Application Scenario

A first prototype scenario has been developed in which a user will have a home assistence system that is able to monitor his or her activity in order to detect incidents and uncommon activities (Fernández-Luque et al., 2009) and (Botía-Blaya et al., 2009). The prototype house or scenario has a bedroom, a hall, a corridor, a toilet, a kitchen, and a living room. Movement infrared sensors are installed in each location. Moreover, in the bedroom there is a pressure sensor in bed; in the hall, a magnetic sensor to detect the opening and closening of the entrance door, and in the sofa of living room another pressure sensor. All sensor boards have a complementary temperature sensor. The data is gathered from sensors mounted in the home. The sensor events are transmitted by the wireless sensor network to the base station by means ZigBee technology. A gateway is also included in the system to allow continuous monitoring. The gateway receives the events from the sensors through base station and decides what the appropriate action to take will be. Options could include querying the user to check on their status, storing (or forwarding) data on the event for future analysis by a assistential care

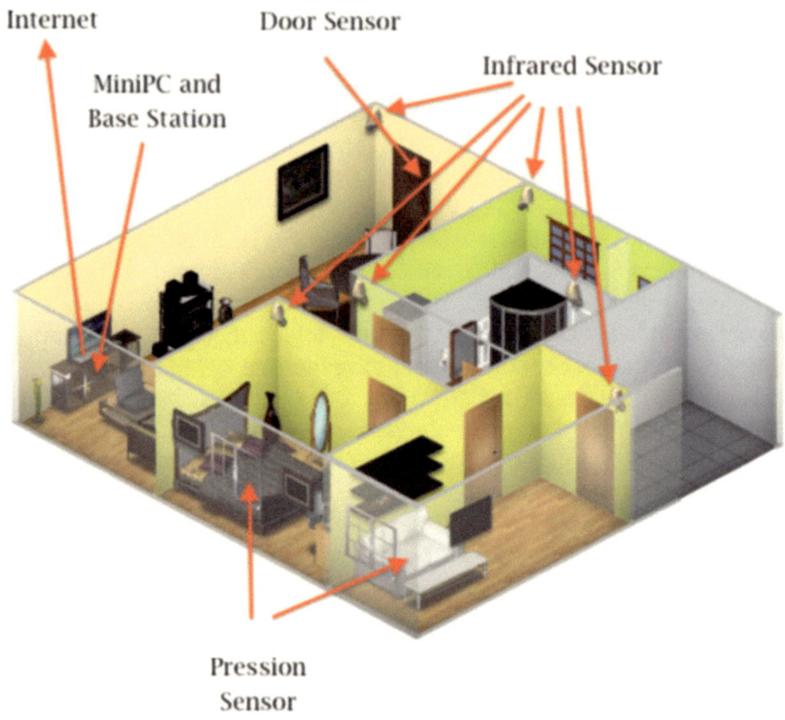

Fig. 3. Schematic overview of the system installed at prototype home

provider, placing a telephone call to a care provider, relative or health care service, or other options. Figure 3 shows a schematic overview of the system.
The main idea consists in monitoring the person living alone in his home without interacting with him. To start, it is needed to know if he is at home in order to activate the ubiquitous custodial care system. It is easy to know by the context if a resident is at home knowing that the entrance door was opened and movement in the hall was detected. By means of distributed sensors installed in each room at home we can know the activities and the elderly location. The sensor boards developed by us allow to distinguish if movement event is induced by a pet or person thanks to a dual passive infrared sensor. On the other hand, as the pressure sensors are located in the bed and the favorite sofa in the living room, we can know more of where he is even if he is not in movement. All this sensorial assembly will be ruled by an artificial intelligent software which will allow to learn of elderly diary activities. If the system detects a suspicious event, i.e., movement in any room at 12 a.m and pressure in the bed, then the system give an alert to the caregiver.

3.1 Assembly of Distributed Sensors

The terms sensor node, wireless node (WN), Smart Dust, mote, and COTS (commercial off-the-shelf) mote are used somewhat interchangeably in the industry; the most general terms used here are sensor node and WN. WSNs that combine physical sensing of parameters such

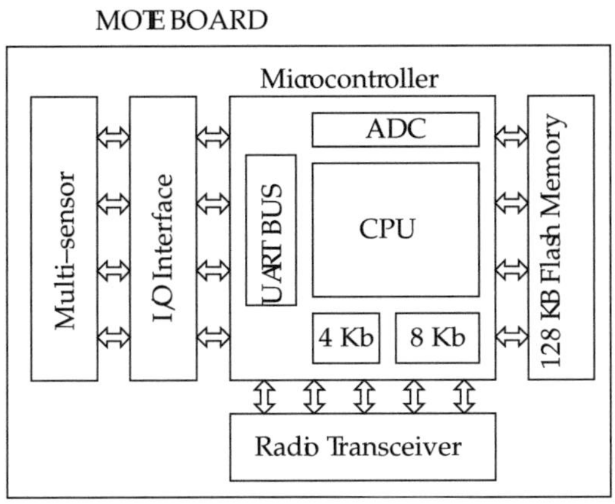

Fig. 4. Sensor node scheme

as temperature, light, or others events with computation and networking capabilities are expected to become ubiquitous in the next future. The basic functionality of a WN generally depends on the application and type of sensor device. Sensors are either passive or active devices. Passive sensors in single-element form include, among others, seismic-, acoustic-, strain-, humidity-, and temperature-measuring devices. Passive sensors in array form include optical- (visible, infrared 1 mm, infrared 10 mm) and biochemical-measuring devices. Arrays are geometrically regular clusters of WNs (i.e., following some topographical grid arrangement). Passive sensors tend to be low-energy devices. Active sensors include radar and sonar; these tend to be high-energy systems.

Activity monitoring can be beneficial for elderly people who live alone at home. By means of using electronic technologies to assist and monitor elderly, disabled, and chronically ill individuals in the home can improve quality of life, improve health outcomes, and help control asistential care. This is done with mote devices developed ad hoc for this purpose which are based on Iris mote from Crossbow (Horton & Suh, 2005). The mote board developed uses a single channel 2.4 GHz radio to provide bi-directional communications at 40 kbit s^{-1}, and an Atmel Atmega 1281 microcontroller running at 8 MHz controls the signal sampling and data transmission. The wireless sensor node is powered by a pair conventional AA batteries and a DC boost converter provides a stable voltage source. Figure 4 and 5 shows a schematic overview of sensor node architecture and station base architecture.

This mote board was designed in order to provide basic environmental sensing, and expansion for other sensing functionality. Actually, wearable sensors are also being included which could measure and analyze the users health as biomedicals signals (ECG, heart rate, etc) and activity such falls. Among other things because we have implemented an integrated antenna on the same board. The assembly of distribuited sensors are integrated in a mesh network. A mesh network is a generic name for a class of networked embedded systems that share several characteristics including: Multi-Hop– the capability of sending messages peer-to-peer to a base station, thereby enabling scalable range extension; Self-Configuring– capable of network formation without human intervention; Self-Healing– capable of adding and removing network nodes automatically without having to reset the network; and Dynamic Routing– capable of adaptively determining the route based on dynamic network conditions (i.e., link quality, hop-count, gradient, or other metric). Multihop protocol is a full featured multi-hop,

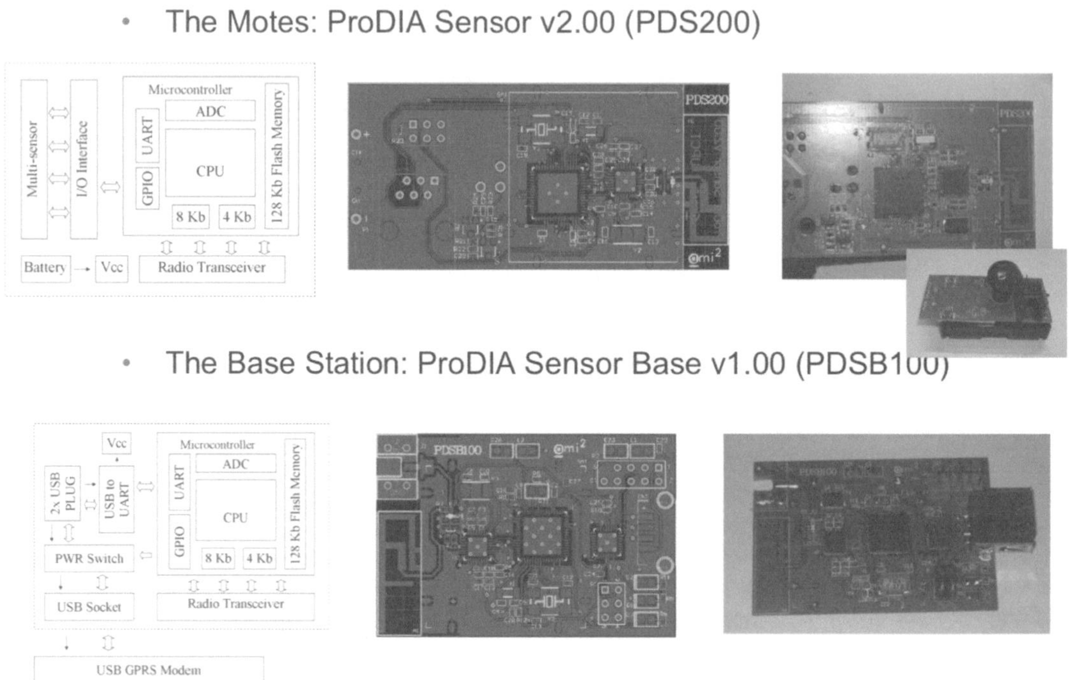

Fig. 5. Motes for PRO(totype)DIA project

ad-hoc, mesh networking protocol driven for events (Al-Karaki & Kamal, 2004; Li et al., 2008; Sagduyu & Ephremides, 2004). This protocol is a modified protocol based on Xmesh developed by Crossbow for wireless networks. A multihop network protocol consists of WN (Motes) that wirelessly communicate to each other and are capable of hopping radio messages to a base station where they are passed to a PC or other client. The hopping effectively extends radio communication range and reduces the power required to transmit messages. By hopping data in this way, our multihop protocol can provide two critical benefits: improved radio coverage and improved reliability. Two nodes do not need to be within direct radio range of each other to communicate. A message can be delivered to one or more nodes in-between which will route the data. Likewise, if there is a bad radio link between two nodes, that obstacle can be overcome by rerouting around the area of bad service. Typically the nodes run in a low power mode, spending most of their time in a sleep state, in order to achieve multi-year battery life. On the other hand, the node is woke up when a event happened by means of an interruption which is activated by sensor board when an event is detected. Also, the mesh network protocol provides a networking service that is both self-organizing and self-healing. It can route data from nodes to a base station (upstream) or downstream to individual nodes. It can also broadcast within a single area of coverage or arbitrarily between any two nodes in a cluster. QOS (Quality of Service) is provided by either a best effort (link level acknowledgement) and guaranteed delivery (end-to-end acknowledgement). Also, XMesh can be configured into various power modes including HP (high power), LP (low power), and ELP (extended low power).

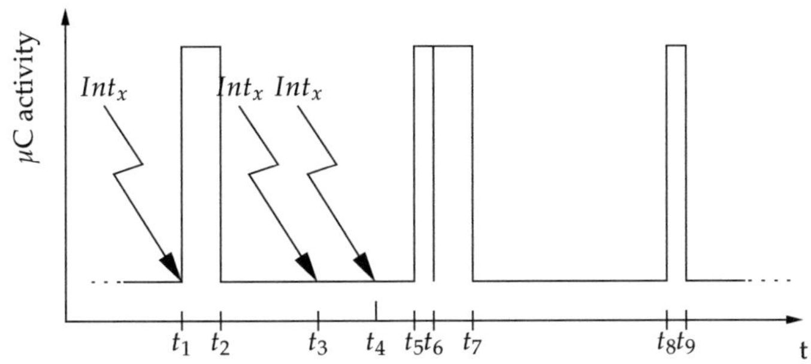

Fig. 6. Composite interruption chronogram

3.2 Sensor Data Monitoring

Inside the sensor node, the microcontroller and the radio transceiver work in power save mode most of the time. When a state change happens in the sensors (an event has happened), an external interrupt wakes the microcontroller and the sensing process starts. The sensing is made following the next sequence: first, the external interrupt which has fired the exception is disabled for a 5 seconds interval; to save energy by preventing the same sensor firing continuously without relevant information. This is achieved by starting a 5 seconds timer which we call the interrupt timer, when this timer is fired the external interrupt is rearmed. For it, there is a fist of taking the data, the global interrupt bit is disabled until the data has been captured and the message has been sent. Third, the digital input is read using the TinyOS GPIO management features. Fourth, battery level and temperature are read. The battery level and temperature readings are made using routines based on TinyOS ADC library. At last, a message is sent using the similar TinyOS routines. In this way, the message is sent to the sensor parent in the mesh. The external led of the multisensor board is powered on when the sending routine is started; and powered off when the sending process is finished. This external led can be disabled via software in order to save battery power.

As an example, an events chronogram driven for interruption is shown in Figure 6, where next thresholds was established: $t_2 - t_1 < 125$ ms, $t_3 - t_1 < 5$ s, $t_4 - t_1 < 5$ s, $t_5 - t_1 = 5$ s, $t_6 - t_5 < 1$ ms, $t_7 - t_6 < 125$ ms, $t_8 - t_6 = 5$ s and $t_9 - t_8 < 1$ ms. Figure 6 can be descripted as follows: at t_1 an external interrupt Int_x has occurred due to a change in a sensor. The external interrupt Int_x is disabled and the interrupt timer started. The sensor data is taken. The message is sent and the external led of our multisensor board is powered on. At t_2 the send process is finished. The external led is powered off. At t_3, an external interrupt Int_x has occurred. The exception routine is not executed because the external interrupt Int_x is disabled. The interrupt flag for Int_x is raised. At t_4, another interruption has occurred but the interruption flag is already raised. At t_5, the interrupt timer is fired. The external interrupt Int_x is enabled. At t_6, the exception routine is executed because the interrupt flag is raised. The external interrupt Int_x is disabled and the interrupt timer started. The sensor data is taken. The message is sent and the external led powered on. At t_7: The send process has finished. The external led is powered off. At t_8, the interrupt timer is fired. The external interrupt Int_x is enabled.At t_9, there are not more pending tasks.

3.3 Base Station

The event notifications are sent from the sensors to the base station. Also commands are sent from the gateway to the sensors. In short, the base station fuses the information and

therefore is a central and special mote node in the network. This USB-based central node was developed by us also. This provides different services to the wireless network. First, the base station is the seed mote that forms the multihop network. It outputs route messages that inform all nearby motes that it is the base station and has zero cost to forward any message. Second, for downstream communication the base station automatically routes messages down the same path as the upstream communication from a mote. Third, it is compiled with a large number of message buffers to handle more children than other motes in the network. These messages are provided for TinyOS, a open-source low-power operative system. Fourth, the base station forwards all messages upstream and downstream from the gateway using a standard serial framer protocol. Five, the station base can periodically send a heartbeat message to the client. If it does not get a response from the client within a predefined time it will assume the communication link has been lost and reset itself.
This base station is connected via USB to a gateway (miniPC) which is responsible of determining an appropriate response by means of an intelligent software in development now, i.e. passive infra-red movement sensor might send an event at which point and moment towards the gateway via base station for its processing. The application can monitor the events to determine if a strange situation has occurred. Also, the application can ask to the sensors node if the event has finished or was a malfunction of sensor. If normal behavior is detected by the latter devices, then the event might just be recorded as an incident of interest, or the user might be prompted to ask if they are alright. If, on the other hand, no normal behavior is detected then the gateway might immediately query the user and send an emergency signal if there is no response within a certain (short) period of time. With the emergency signal, access would be granted to the remote care provider who could log in and via phone call.

3.4 Gateway

Our system has been designed considering the presence of a local gateway used to process event patterns in situ and take decisions. This home gateway is provided with a java-based intelligent software which is able to take decision about different events. In short, it has java application for monitoring the elderly and ZigBee wireless connectivity provided by a USB mote-based base station for our prototype. This layer stack form a global software architecture. The lowest layer is a hardware layer. In the context awareness layer, the software obtains contextual information provided by sensors. The middle level software layer, model of user behavior, obtains the actual state of attendee, detecting if the resident is in an emergency situation which must be solved. The deep reasoning layer is being developed to solve inconsistencies reached in the middle layer.
The gateway is based on a miniPC draws only 3-5 watts when running Linux (Ubuntu 7.10 (Gutsy) preloaded) consuming as little power as a standard PC does in stand-by mode. Ultra small and ultra quiet, the gateway is about the size of a paperback book, is noiseless thanks to a fanless design and gets barely warm. Gateway disposes a x86 architecture and integrated hard disk. Fit-PC has dual 100 Mbps Ethernet making it a capable network computer. A normal personal computer is too bulky, noisy and power hungry.
The motherboard of miniPC is a rugged embedded board having all components– including memory and CPU– soldered on-board. The gateway is enclosed in an all-aluminum anodized case that is splash and dust resistant. The case itself is used for heat removal- eliminating the need for a fan and venting holes. Fit-PC has no moving parts other than the hard-disk. The CPU is an AMD Geode LX800 500 MHz, the memory has 256 MB DDR 333 MHz soldered on-board and the hard disk has 2.5" IDE 60 GB. To connect with base station, the gateway

Fig. 7. Gateway based on miniPC, Mote board and base station

disposes of 2 × USB 2.0 HiSpeed 480 Mbps, also it has 2 × RJ45 Ethernet ports 100 Mbps to connect with Internet. Figure 7 shows the gateway ports base station and our mote board.

4. Results and Discussions

Figure 7 shows the hardware of the built wireless sensor node provides for mote board. In this prototype, a variable and heterogeneous number of wireless sensor nodes are attached to multisensor boards in order to detect the activities of our elderly in the surrounding environment, and they send their measurements to a base station when an event (change of state) is produced or when the gateway requires information in order to avoid inconsistencies. The base station can transmit or receive data to or from the gateway by means of USB interface. It can be seen that the sensor nodes of the prototype house detect the elderly activity. The infrared passive, magnetic and pressure sensors have a high quality and sensitivity. Also, the low-power multihop protocol works correctly. Therefore, the system can determine the location and activity patterns of elderly, and in the close future when the intelligent software will learn of elderly activities, the system will can take decisions about strange actions of elderly if they are not stored in his history of activities. By now, the system knows some habitual patterns of behavior and therefore it must be tuning in each particular case. Additionally, connectivity between the gateway exists to the remote caregiver station via a local ethernet network. The gateway currently receives streamed sensor data so that it can be used for analysis and algorithm development for the intelligent software and the gateway is able potentially to send data via ethernet to the caregiver station.

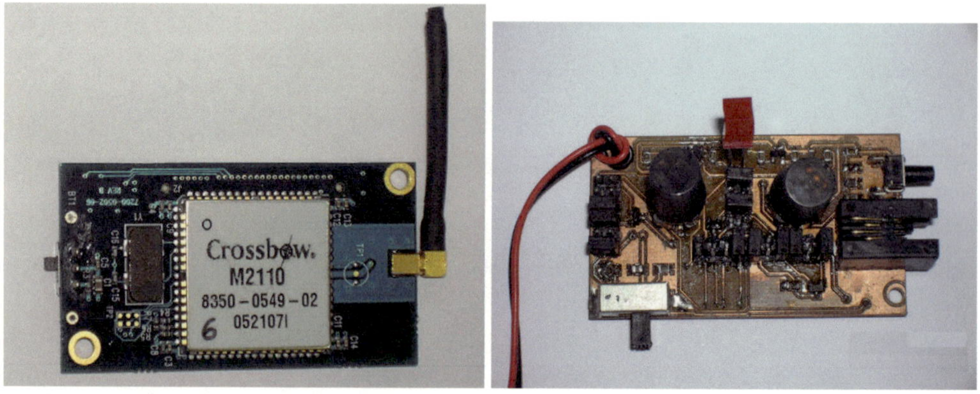

Fig. 8. Iris mote board and our first Multisensor board prototype (2007)

As the transmission is digital, there is no noise in the signals. It represents an important feature because noise effects commonly hardly affect telemedicine and assistence systems. The baud rate allows the transmission of vital and activity signals without problems. The discrete signals (movement, pressure and temperature, for example) are quickly transmitted. Nevertheless, spending 5 s to transmit an signal sample or event does not represent a big problem. Moreover, the system can interact with other applications based on information technologies. Using standards represents an important step for integrating assisted living at home systems. The system was implemented as previously we have described. As mentioned, the system uses Java programming language in order to describe the activity of the elderly and take a decision. The system guaranteed the transmission of a packet per less to 1 seconds, e.g. the baud rate is 57 600 bit s^{-1}. Other signals, such as temperature, need the same time. Furthermore, lost packets are tracked, once it is using a cyclic redundancy code (CRC). There are a lot of sensors which can measure activities and environmental parameters unobtrusively. Among them, just a few sensors are used in our prototype home. In the future, other useful sensors will be used in experiments. For fall measurement (Sixsmith & Johnson, 2004b), a method can be used applied using infrared vision. In addition, microphone/speaker sensors can be used for tracking and ultrasound sensors also can be used for movement. Other sensors can be easily incorporated into our system because we have already developed a small-size multisensor board.

In this sense, we have decided design an accelerometer mote that is small and lightweight that can be worn comfortably without obstructing normal activities. The wearable mote board has mounted a 3-axis accelerometer with high resolution (13-bit) measurement at up to ± 16 g (Analog Devices ADXL345). Digital output data is formatted as 16-bit twos complement and is accessible through either a SPI (3- or 4-wire) (or I2C digital interface). The wearable mote measures the static acceleration of gravity in tilt-sensing applications, as well as dynamic acceleration resulting from motion or shock. High resolution provided by ADXL345 (4 mg/LSB) enables measurement of inclination changes less than 1.0º. Several special sensing functions are provided. Activity and inactivity sensing detect the presence or lack of motion and if the acceleration on any axis exceeds a user-set level. Tap sensing detects single and double taps. Free-fall sensing detects if the device is falling. These functions can be mapped to one of two interrupt output pins. An integrated, patent pending 32-level first in, first out (FIFO) buffer can be used to store data to minimize host processor intervention. Low power modes

Fig. 9. Actor with accelerometer in his waist, log of data and accelometer sensor node prototype

enable intelligent motion-based power management with threshold sensing and active acceleration measurement at extremely low power dissipation. The mote fits inside a plastic box measuring 4×4×1 cm, where the button battery is enclosed in the same package. Clearly, the placement of the device on the body is of primary concern. Some of the criteria are that it should be comfortable and that the device itself should not pose a threat to the wearer in the event of a fall. For our experiments, we attached the mote to a belt worn around the waist. We have not done sufficient experiments on elderly people. In this work, the experiments should be considered preliminary and more data is needed. Figure 9 shows some pictures of accelerometer sensor node and our proofs.

In the literature there is an absence of research data on a persons movement in his or her own house that is not biased by self-report or by third party observation. We are in the process of several threads of analysis that would provide more sophisticated capabilities for future versions of the intelligent software. The assisted living system is a heterogenous wireless network using and ZigBee radios to connect a diverse set of embedded sensor devices. These devices and the wireless network can monitor the elderly activity in a secure and private manner and issue alerts to the user, care givers or emergency services as necessary to provide additional safety and security to the user. This system is being developed to provide this safety and security so that elder citizens who might have to leave their own homes for a group care facility will be able to extend their ability to remain at home longer. This will in most cases provide them with better quality of life and better health in a cost effective manner.

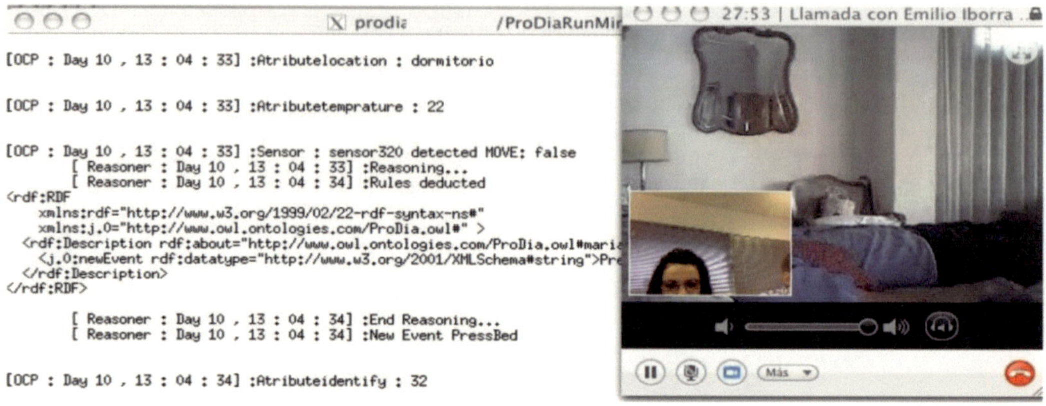

Fig. 10. Monitoring proofs with ssh communication at a patient residence

Also think that this assisted living system can be used in diagnostic because the activity data can show indicators of illness. We think that changes in daily activity patterns can suggest serious conditions and reveal abnormalities of the elderly resident. In summary, we think that our Custodial Care system could be quite well-received by the elderly residents. We think that the infrastructure will need to, i) deal robustly with a wide range of different homes and scenarios, ii) be very reliable in diverse operating conditions, iii) communicate securely with well-authenticated parties who are granted proper access to the information, iv) respect the privacy of its users, and v) provide QoS even in the presence of wireless interference and other environmental effects. We are continuing working on these issues. Figure 10 shows a real scenario where we can see the log in the left when a resident is lying in the bed.

5. Summary

Assistence living at home care represents a growing field in the social services. It reduces costs and increases the quality of life of assisted citizen. As the modern life becomes more stressful and acute diseases appear, prolonged assistence become more necessary. The same occurs for the handicapped patients. Home care offers the possibility of assistence in the patients house, with the assistance of the family. It reduces the need of transporting patients between house and hospital. The assistence living at home routines can be switched by telemedicine applications. Actually, this switch is also called telehomecare, which can be defined as the use of information and communication technologies to enable effective delivery and management of health services at a patients residence.

Summing up, we have reviewed the state of the art of technologies that allow the use of wireless sensor networks in AAL. More specifically, technology based on the sensor nodes (WNs) that conform it. We have proposed a wireless sensor network infrastructure for assisted living at home using WSNs technology. These technologies can reduce or eliminate the need for personal services in the home and can also improve treatment in residences for the elderly and caregiver facilities. We have introduced its system architecture, power management, self-configuration of network and routing. In this chapter, a multihop low-power network protocol has been presented for network configuration and routing since it can be considered as a natural and appropriate choice for ZigBee networks. This network protocol is modified of original protocol of Crossbow because our protocol is based in events and is not based

in timers. Moreover, it can give many advantages from the viewpoint of power network and medium access. Also, we have developed multisensors board for the nodes which can directly drive events towards an USB base station with the help of our ZigBee multihop low-power protocol. In this way, and by means of distributed sensors (motes) installed in each of rooms in the home we can know the activities and the elderly location. A base station (a special mote developed by us too) is connected to a gateway (miniPC) by means an USB connector which is responsible of determining an appropriate response using an intelligent software, i.e. passive infra-red movement sensor might send an event at which point and moment towards the gateway via base station for its processing. This software is in development in this moment therefore is partially operative.
DIA project intends to be developed with participatory design between the users, care providers and developers. With the WSN infrastructure in place, sensor devices will be identified for development and implemented as the system is expanded in a modular manner to include a wide selection of devices. In conclusion, the non-invasive monitoring technologies presented here could provide effective care coordination tools that, in our opinion, could be accepted by elderly residents, and could have a positive impact on their quality of life. The first prototype home in which this is being tested is located in the Region de Murcia, Spain. Follow these tests, the system will be shared with our partners for further evaluation in group care facilities, hospitals and homes in our region.

6. Acknowledgments

The authors gratefully acknowledge the contribution of Spanish Ministry of Ciencia e Innovación (MICINN) and reviewers' comments. This work was supported by the Spanish Ministry of Ciencia e Innovación (MICINN) under grant TIN2009-14372-C03-02.

7. References

Al-Karaki, J. & Kamal, A. (2004). Routing techniques in wireless sensor networks: a survey, **11**(6): 6–28.

Biemer, M. & Hampe, J. F. (2005). A mobile medical monitoring system: Concept, design and deployment, *ICMB '05: Proceedings of the International Conference on Mobile Business*, IEEE Computer Society, Washington, DC, USA, pp. 464–471.

Bilstrup, U. & Wiberg, P.-A. (2004). An architecture comparison between a wireless sensor network and an active rfid system, *Local Computer Networks, 2004. 29th Annual IEEE International Conference on*, pp. 583–584.

Botía-Blaya, J., Palma, J., Villa, A., Pérez, D. & Iborra, E. (2009). Ontology based approach to the detection of domestic problems for independent senior people, *IWINAC09*, International Work-Conference on the Interpalay Between Natural and Artificial Computation, IWINAC, pp. 55–64.

Cho, N., Song, S.-J., Kim, S., Kim, S. & Yoo, H.-J. (2005). A 5.1-µw uhf rfid tag chip integrated with sensors for wireless environmental monitoring, *Solid-State Circuits Conference, 2005. ESSCIRC 2005. Proceedings of the 31st European*, pp. 279–282.

Fernández-Luque, F., Zapata, J., Ruiz, R. & Iborra, E. (2009). A wireless sensor network for assisted living at home of elderly people, *IWINAC '09: Proceedings of the 3rd International Work-Conference on The Interplay Between Natural and Artificial Computation*, Springer-Verlag, Berlin, Heidelberg, pp. 65–74.

Horton, M. & Suh, J. (2005). A vision for wireless sensor networks, *Proc. IEEE MTT-S International Microwave Symposium Digest*, p. 4pp.

JLHLabs (2008). WSN Lab.
URL: `http://www.jlhlabs.com/`

Kahn, J. M., Katz, R. H. & Pister, K. S. J. (1999). Next century challenges: mobile networking for "smart dust", *MobiCom '99: Proceedings of the 5th annual ACM/IEEE international conference on Mobile computing and networking*, ACM Press, New York, NY, USA, pp. 271–278.
URL: `http://dx.doi.org/10.1145/313451.313558`

Li, Y., Thai, M. T. & Wu, W. (2008). *Wireless Sensor Networks And Applications*, Springer.

Lubrin, E., Lawrence, E. & Navarro, K. F. (2006). Motecare: an adaptive smart ban health monitoring system, *BioMed'06: Proceedings of the 24th IASTED international conference on Biomedical engineering*, ACTA Press, Anaheim, CA, USA, pp. 60–67.

Martin, H., Bernardos, A., Bergesio, L. & Tarrio, P. (2009). Analysis of key aspects to manage wireless sensor networks in ambient assisted living environments, *Applied Sciences in Biomedical and Communication Technologies, 2009. ISABEL 2009. 2nd International Symposium on*, pp. 1 –8.

MIT (2008). MIT WSN Research Group.
URL :http://mtlweb.mit.edu/researchgroups/icsystems/gallery.html.

Pister, K. (2008). Smart dust, autonomous sensing and communication in a cubic millimeter.
URL: `http://robotics.eecs.berkeley.edu/~pister/SmartDust/`

Ross, P. (2004). Managing care through the air [remote health monitoring], *Spectrum, IEEE* **41**(12): 26–31.

Rowan, J. & Mynatt, E. D. (2005). Digital family portrait field trial: Support for aging in place, *CHI '05: Proceedings of the SIGCHI conference on Human factors in computing systems*, ACM, New York, NY, USA, pp. 521–530.

Sagduyu, Y. & Ephremides, A. (2004). The problem of medium access control in wireless sensor networks, **11**(6): 44–53.

Sixsmith, A. & Johnson, N. (2004a). A smart sensor to detect the falls of the elderly, *Pervasive Computing, IEEE* **3**(2): 42–47.

Sixsmith, A. & Johnson, N. (2004b). A smart sensor to detect the falls of the elderly, **3**(2): 42–47.

Sohraby, K., Minoli, D. & Znati, T. (2007). *Wireless Sensor Networks: Technology, Protocols, and Applications*, John Wiley and Sons.
URL : http://www.wiley.com/WileyCDA/WileyTitle/productCd-0471743003.html.

TinyOS (2009). Tinyos website.
URL: `http://www.tinyos.net/`

UCLA (2009). UCLA WSN Projects.
URL: `http://nesl.ee.ucla.edu/projects/ahlos/mk2`

9

Monitoring of human movements for fall detection and activities recognition in elderly care using wireless sensor network: a survey

Stefano Abbate
IMT Institute for Advanced Studies Lucca
Italy

Marco Avvenuti, Paolo Corsini and Alessio Vecchio
University of Pisa
Italy

Janet Light
University of New Brunswick
Canada

1. Introduction

The problem with accidental falls among elderly people has massive social and economic impacts. Falls in elderly people are the main cause of admission and extended period of stay in a hospital. It is the sixth cause of death for people over the age of 65, the second for people between 65 and 75, and the first for people over 75. Among people affected by Alzheimer's Disease, the probability of a fall increases by a factor of three.

Elderly care can be improved by using sensors that monitor the vital signs and activities of patients, and remotely communicate this information to their doctors and caregivers. For example, sensors installed in homes can alert caregivers when a patient falls. Research teams in universities and industries are developing monitoring technologies for in-home elderly care. They make use of a network of sensors including pressure sensors on chairs, cameras, and RFID tags embedded throughout the home of the elderly people as well as in furniture and clothing, which communicate with tag readers in floor mats, shelves, and walls.

A fall can occur not only when a person is standing, but also while sitting on a chair or lying on a bed during sleep. The consequences of a fall can vary from scrapes to fractures and in some cases lead to death. Even if there are no immediate consequences, the long-wait on the floor for help increases the probability of death from the accident. This underlines the importance of real-time monitoring and detection of a fall to enable first-aid by relatives, paramedics or caregivers as soon as possible.

Monitoring the activities of daily living (ADL) is often related to the fall problem and requires a non-intrusive technology such as a wireless sensor network. An elderly with risk of fall can be instrumented with (preferably) one wireless sensing device to capture and analyze the

body movements continuously, and the system triggers an alarm when a fall is detected. The small size and the light weight make the sensor network an ideal candidate to handle the fall problem.
The development of new techniques and technologies demonstrates that a major effort has been taken during the past 30 years to address this issue. However, the researchers took many different approaches to solve the problem without following any standard testing guidelines. In some studies, they proposed their own guidelines.
In this Chapter, a contribution is made towards such a standardization by collecting the most relevant parameters, data filtering techniques and testing approaches from the studies done so far. State-of-the-art fall detection techniques were surveyed, highlighting the differences in their effectiveness at fall detection. A standard database structure was created for fall study that emphasizes the most important elements of a fall detection system that must be considered for designing a robust system, as well as addressing the constraints and challenges.

1.1 Definitions

A *fall* can be defined in different ways based on the aspects studied. The focus in this study is on the kinematic analysis of the human movements. A a suitable definition of a fall is "Unintentionally coming to the ground or some lower level and other than as a consequence of sustaining a violent blow, loss of consciousness, sudden onset of paralysis as in stroke or an epileptic seizure." (Gibson et al., 1987). It is always possible to easily re-adapt this definition to address the specific goals a researcher wants to pursue.
In terms of human anatomy, a fall usually occurs along one of two planes, called *sagittal* and *coronal* planes. Figure 1(a) shows the sagittal plane, that is an X-Z imaginary plane that travels vertically from the top to the bottom of the body, dividing it into left and right portions. In this case a fall along the sagittal plane can occur forward or backward. Figure 1(b) shows the coronal Y-Z plane, which divides the body into dorsal and ventral (back and front) portions. The coronal plane is orthogonal to the sagittal plane and is therefore considered for lateral falls (right or left). Note that if the person is standing without moving, that is, he or she is in a *static* position, the fall occurs following in the down direction. The sense of x, y and z are usually chosen in order to have positive z-values of the acceleration component when the body is falling.

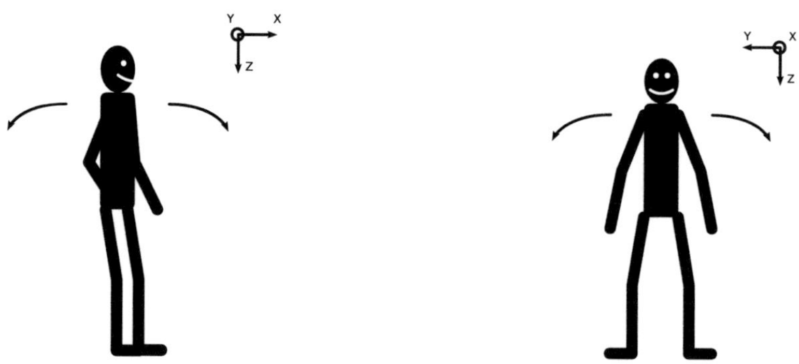

(a) Along sagittal plane (b) Along coronal plane

Fig. 1. Fall directions

Toppling simply refers to a loss in balance. Figure 2(a) shows the body from a kinematic point of view. When the vertical line through the center of gravity lies outside the base of support the body starts *toppling*. If there is no reaction to this loss of balance, the body falls on the ground (Chapman, 2008).
Let us now consider the fall of a body from a stationary position at height $h = H$. Initially the body has a potential energy mgh which is transformed into kinetic energy during the fall with the highest value just before the impact on the floor ($h = 0$). During the impact the energy is totally absorbed by the body and, after the impact, both potential and kinetic energy are equal to zero. If the person is conscious the energy can be absorbed by the his muscles, for example, using the arms (see Figure 2(b)), whereas if the person is unconscious it can lead to sever injuries (see Figure 2(c)).

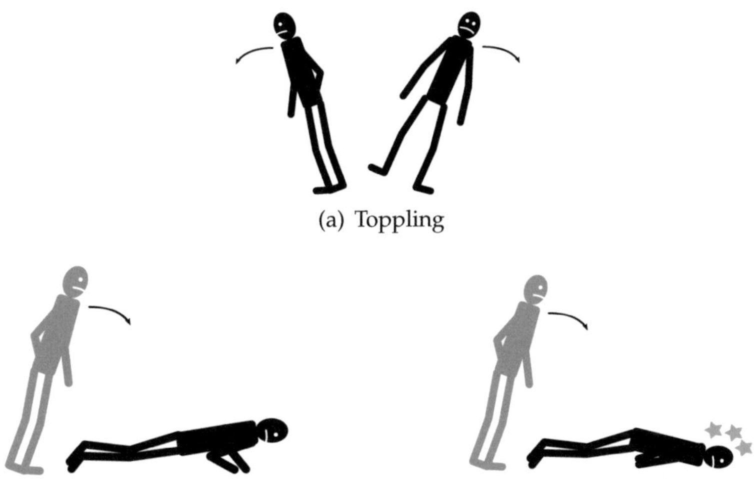

(a) Toppling

(b) Conscious fall

(c) Unconscious fall

Fig. 2. Kinematic analysis of a fall

Strictly related to a fall is the *posture*, a configuration of the human body that is assumed intentionally or habitually. Some examples are standing, sitting, bending and lying. A posture can be determined by monitoring the tilt transition of the trunk and legs, the angular coordinates of which are shown in Figure 3(a) and Figure 3(b) (Li et al., 2009; Yang & Hsu, 2007). The ability to detect a posture helps to determine if there has been a fall.

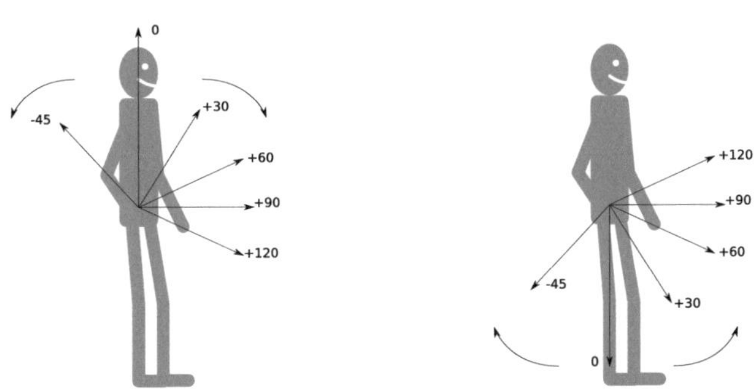

(a) Trunk

(b) Legs

Fig. 3. Angular coordinates

1.2 Related Surveys of Research on Patient Monitoring Technologies

So far, a few surveys on fall detection systems have been written and extended. Some of them propose their own standards and this is useful for people already working on the problem of fall detection. This survey provides a comprehensive, if not exhaustive, guide from the first-hand approach of the problem, highlighting the best practices to merge valid but heterogeneous procedures.

The first survey on fall detection by Noury et al. (2007) describes the systems, algorithms and sensors used in the detection of a fall in elderly people. After an overview of the state-of-the-art techniques, they discovered the lack of a common framework and hence proposed some performance evaluation parameters in order to compare the different systems. These parameters had to be evaluated for a set of falling scenarios that included real falls and actions related to falls.

Yu (2008) focused on a classification of the approaches and principles of existing fall detection methods. He also provided a classification of falls and a general framework of fall detection, alert device and system schema.

The authors of Noury et al. (2007) described the in-depth sequence of falling (Noury et al., 2008). They stated that it was difficult to compare academic studies because the conditions of assessment are not always reported. This led to the evaluation of not only the above described parameters and scenarios, but also of other objective criteria such as detection method, usability and lifespan of a device.

In applications involving accelerometers, Kangas et al. (2007) used accelerometry-based parameters to determine thresholds for fall detection. The posture information was used to distinguish between falls and activities of daily living. Their experiments showed the most suitable placement for the sensor to be waist and the head, whereas placing the sensor on the wrist gave rise to additional problems.

2. Fall risk factors

A person can be more or less prone to fall, depending on a number of risk factors and hence a classification based on only age as a parameter is not enough. In fact, medical studies have determined a set of so called *risk factors*:

- Intrinsic:
  - Age (over 65)
  - Low mobility and bone fragility
  - Poor balance
  - Chronic disease
  - Cognitive and dementia problems
  - Parkinson disease
  - Sight problems
  - Use of drugs that affect the mind
  - Incorrect lifestyle (inactivity, use of alcohol, obesity)
  - Previous falls
- Extrinsic:
  - Individual (incorrect use of shoes and clothes)
  - Drugs cocktail

- Internal Environment:
  Slipping floors
  Stairs
  Need to reach high objects
- External Environment:
  Damaged roads
  Crowded places
  Dangerous steps
  Poor lighting

There is a clear correlation between the above list and the probability of fall. The number of people that fall are as follows (Tinetti et al., 1988):

- 8% of people without any of risk factors
- 27% of people with only one risk factor
- 78% of people with four or more risk factors

The history of the falls is also important since people who have already fallen two times are more at risk to fall again. This can be due to psychological (fear, shame, loss of self-esteem), and/or physical (injuries, lack of exercise) reasons.

3. How, where and why people fall

Among elderly people that live at home, almost half of the falls take place near or inside the house (Campbell et al., 1990; Lipsitz et al., 1991). Usually women fall in the kitchen whereas men fall in the garden (Lord et al., 1993).
The rate of falls increases significantly among elderly people living in nursing homes: at least 40% of the patients fell twice or more within 6 months. This rate is five times more with respect to the rate of fall when people live at home. This may be due to people having to aquaint themselves with the new living enviroment and its obstacles.

3.1 Physical causes

The factors that lead to most of the falls in people over 65 are to stumble on obstacles or steps and to slip on a smooth surface. The fall is usually caused by loss of balance due to dizziness. Approximately 14% of people do not know why they fall and a smaller number of people state that the fall is due to the fragility of the lower limbs (Lord et al., 1993).
Further researchers determined that traditional fall prevention measures such as bed rails can make the fall worse (Masud & Morris, 2001).

3.2 Activities

Most of the falls happen during the activities of daily living (ADL) that involve a small loss of balance such as standing or walking. Fewer falls happen during daily activities that involve a more significant movement such as sitting on a chair or climbing the stairs. Conversely, activities usually defined "dangerous", such as jogging or physical exercises are less likely to increase the probability of a fall (Tinetti et al., 1988). There are more falls during the day than during the night (Campbell et al., 1990).

3.3 Consequences

Accidental falls are the main cause of admission in a hospital and the sixth cause of death for people over 65. For people aged between 65 and 75 accidental falls are the second cause of death and the first cause in those over 75 (Bradley et al., 2009).

3.3.1 Physical damage

Scratches and bruises are the soft injures due to a fall (Bradley et al., 2009). In the worst cases the injuries are concentrated on the lower part of the body, mainly on the hip. On the upper part of the body the head and the trunk injuries are the most frequent. About 66% of admissions to an hospital are due to at least one fracture. The fracture of elbow and forearm are more frequent but hip fracture is the most difficult to recover from. Such a fracture in fact requires a long recovery period and involves the loss of independence and mobility.
Sometimes, when a person falls and is not able to stand up by himself, he lies down on the floor for long time. This leads to additional health problems such as hypothermia, confusion, complications and in extreme cases can cause death (Lord et al., 2001).

3.3.2 Psychological damage

A fall also involves hidden damages that affect the self-confidence of a person (Lord et al., 2001). Common consequences are fear, loss of independence, limited capabilities, low self-esteem and generally, a lower quality of life.

3.3.3 Economic damage

The direct costs associated with falls are due to the medical examinations, hospital recoveries, rehabilitation treatments, tools of aid (such as wheelchairs, canes etc.) and caregivers service cost (Englander & Hodson, 1996).
Indirect costs concern the death of patients and their consequences. Recent studies have determined that in the year 2000 alone fall-related expenses was above 19 billion dollars and it is estimated to reach 54.9 billion in 2020. This shows that year by year, health costs due to the falls are increasing dramatically (Massachusetts Department of Public Health, 2008).

3.4 Anatomy of a fall

A fall is generally the consequence of a normal activity of daily living and is triggered by a hard-predictable event such as tripping over, slipping or loss of balance. Once the fall and thus the impact on the floor occur, the subject usually lies down for some seconds or even hours and then tries to recover by himself or with the help of someone else. Just before the impact, the body of the subject is in a free-fall, its acceleration is the same as the gravitational acceleration. Thus, it is possible to distinguish five phases as depicted in Figure 4:

1. Activity of Daily Living
2. Hard-predictable event
3. Free-fall
4. Impact
5. Recovery (optional)

(a) Activity of Daily Living (b) Hard predictable event (c) Free-fall (d) Impact (e) Recovery

Fig. 4. Anatomy of a fall

Note that there are activities of daily living that can be wrongly detected as falls, e.g. "falling" on a chair.

4. Typical fall scenarios

The most important scenarios of falls are described by Yu (2008) in detail:

- **Fall from standing**
  1. It lasts from 1 to 2 seconds.
  2. In the beginning the person is standing. At the end the head is stuck on the floor for a certain amount of time.
  3. A person falls along one direction and the head and the center of mass move along a plane.
  4. The height of the head varies from the height while standing and the height of the floor.
  5. During the fall the head is in free-fall.
  6. After the fall the head lays in a virtual circle that is centered in the position of the feet before the fall and has radius the height of the person.
- **Fall from chair**
  1. It lasts from 1 to 3 seconds.
  2. In the beginning the height of the head varies from the height of the chair to the height of the floor.
  3. During the fall the head is in free-fall.
  4. After the fall the body is near the chair.
- **Fall from bed**
  1. It lasts from 1 to 3 seconds.
  2. In the beginning the person is lying.
  3. The height of the body varies from the height of the bed to the height of the floor.
  4. During the fall the head is in free-fall.
  5. After the fall the body is near the bed.

With the description of the main falls it is possible to simplify the complexity of a fall. This enables in turn to focus on the resolution of the detection fall problem, rather than on the reconstruction of a detailed scenario. The simplified and theoretical description often reflects the practical sequence of a fall.

5. Risk assessment tools

A risk assessment tool determines which people are at risk of falls that invoke specific countermeasures, to avoid or at least reduce any injuries (Perell et al., 2001; Vassallo et al., 2008). There are three fundamental types:

1. Medical exams performed by a geriatrician or other qualified people.
2. Risk factors evaluation performed in a hospital.
3. Evaluation of movement ability performed by a physiotherapist.

Medical exams take into account many parameters including the history of falls, drug therapy, strength, balance, diet and chronic diseases. However, they are only "descriptive" tools and hence do not provide numerical indexes. The risk factors evaluation is performed once a patient is admitted to a hospital and is based on specific methods and indexes. The evaluation is then periodically updated and is therefore more useful than the single assessment in the previous category. The analysis of a person at home performed by a physiotherapist can be more detailed but also more intrusive. Nevertheless, many researchers do not agree on the validity of such tools. Oliver (2008) suggests the characteristics essential to an effective risk assessment tool:

- Short-time period to be completed
- Parameters to address:
    1. High-risk faller
    2. Low-risk faller
    3. Falls prediction probability
    4. Non-falls prediction probability
    5. Prediction accuracy

An integrated and on-line monitoring service would provide updated data about the condition of a patient, a condition that can vary frequently especially in elder people. A step further from the monitoring of human movements is the monitoring of physiological parameters.

6. Technological approaches to fall detection

There are three main categories of devices based on the technology used:

- Vision-based
- Environmental
- Wearable

A *Vision-based* approach uses fixed cameras that continuously record the movement of the patients. The acquired data is submitted to specific image algorithms that are able to recognize the pattern of a fall to trigger an alarm. Vision-based approaches can be classified as:

1. *Inactivity detection*, based on the idea that after a fall, the patient lies on the floor without moving.
2. *Body shape change analysis*, based on the change of posture after the fall.
3. *3D head motion analysis*, based on the monitoring the position and velocity of the head.

The main limits of this approach are the time and cost of installation, the limited space of application (only where there are the cameras) and privacy violation.
The use of *Environmental* devices is an approach based on the installation of sensors in the places to be monitored. When people interact with the environment, infrared or pressure sensors on the floor are able to detect a fall. The problem here is the presence of false-negatives, for example, a fall that occurs on a table is not detected.
Both Visual-based and Environmental device approaches require a pre-built infrastructure, and this enables their use in hospitals and houses, but it is hard to use them outdoor.
In the *Wearable* approach, one or more wearable devices are worn by the patient. They are usually equipped with movement sensors such as accelerometers and gyroscopes, whose values are transmitted via radio and analyzed. This solution offers advantages such as low installation cost (indoor and outdoor), small size and offers the possibility to also acquire physiological data (blood pressure, ECG, EEG etc.).

7. Wireless sensor networks and general system architecture

A wireless sensor network is a set of spatially distributed sensing devices, also called nodes, that are able to communicate with each other in a wireless ad-hoc network paradigm (Akyildiz et al., 2002). Each device is usually battery-powered and can be instrumented with one or more sensors which enable acquisition of physical data such as temperature, body acceleration and so on. The nodes are able to organize themselves in order to create an ad-hoc routing tree, whose root is represented by a *sink* node. The sink node is usually connected to a personal computer, also called the *base station*, that will receive all the data sent by nodes (see Figure 5). Besides the sensing and wireless communication capabilities, the nodes feature a processing unit that enables local data treatment and filtering. This is important in order to reduce the use of the radio communication which is the most energy expensive task performed by a node with respect to sensing and processing.

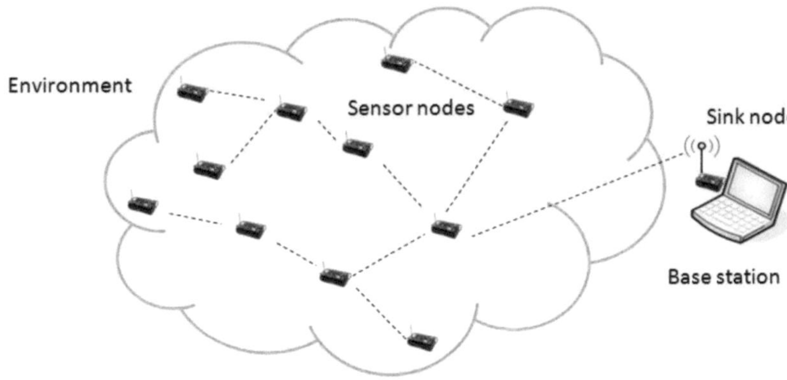

Fig. 5. Wireless Sensor Network topology

The light-weight characteristics of a wireless sensor network perfectly fit the needs of a fall detection system based on the wearable approach. The size, shape and weight of the nodes enable them to be worn easily by a person. Moreover, many general purpose nodes are commercially available at low-cost. According to the specific need of the study it is possible to obtain customized hardware with reduced form factor still maintaining the same functional characteristics. Figure 6(a) shows Tmote-Sky, a general purpose node that is able to sense temperature, humidity and light (Polastre et al., 2005), whereas Figure 6(b) shows SHIMMER, a smaller size version of the Tmote Sky which is more suitable to be worn by a person (Realtime Technologies LTD, 2008). The SHIMMER is equipped with a tri-axial accelerometer for movement monitoring and a Secure Digital (SD) slot to locally log a large amount of data. These platforms enable addition of other sensors such as gyroscopes, in the same board.

(a) Tmote-Sky

(b) SHIMMER

Fig. 6. Examples of nodes

Figure 7 shows the general architecture for a human movement monitoring system based on a wireless sensor network. One or more sensing nodes are used to collect raw data. Analysis of the data can be performed on the node or on the base station by a more powerful device such as a smartphone or a laptop. The wireless connectivity standard between the nodes (e.g. ZigBee) can be different from the one that connects the sink node with the base station (e.g. Bluetooth). The base station in turn acts as a gateway to communicate with the caregivers through wireless and/or wired data connection (e.g. Internet or other mobile phones).

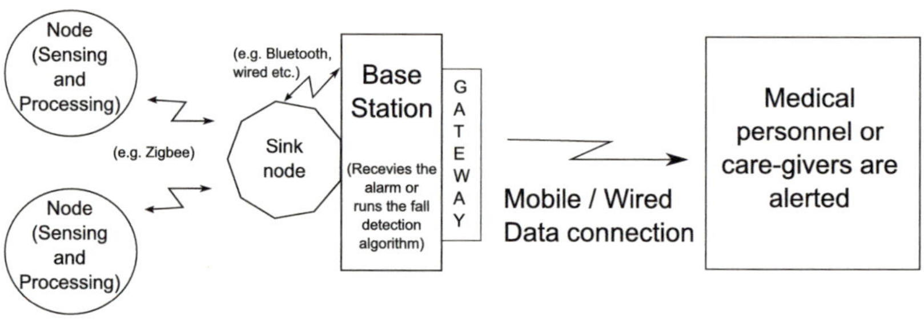

Fig. 7. Traditional system architecture

7.1 Node sensors and position

A node for kinematic monitoring is typically instrumented with the following sensors:

- Accelerometer, to measure the acceleration.
- Gyroscope, to measure the angular velocity.

In particular, the gyroscope requires more energy than the accelerometer. If we connect the acceleration of the movements with the position of the node worn by the patient, it would be possible to detect the posture of a person.
The placement of one or more nodes on the body is the key to differentiate the influences of various fall detection algorithms. It is not possible to neglect the usability aspect, since it strongly affects the effectiveness of the system. A node placed on the head gives an excellent impact detection capability, but more hardware efforts are required to ensure its usability for wearing the node continuously. The wrist is not recommended to be a good position, since it is subject to many high acceleration movements that would increase the number of false positives. The placement at the waist is more acceptable from the user point of the view, since this option fits well in a belt and it is closer to the center of gravity of the body. There are many other node locations selected by researchers, such as the armpit, the thigh or the trunk, quoting their own advantages and disadvantages as explained later. Sometimes the nodes are inserted in clothes, for example jackets, or in accessories such as watches or necklaces.

8. Performance evaluation parameters and scenarios

8.1 Indexes

A real working fall detection system requires to be sufficiently accurate in order to be effective and alleviate the work of the caregivers. The quality of the system is given by three indexes that have been proposed based on the four possible situations shown in Table 1:

	A fall occurs	A fall does not occur
A fall is detected	*True Positive* (TP)	*False Positive* (FP)
A fall is not detected	*False Negative* (FN)	*True Negative* (TN)

Table 1. Possible outputs of a Fall Detection system

- *Sensitivity* is the capacity to detect a fall. It is given by the ratio between the number of detected falls and the total falls that occurred:

$$\text{Sensitivity} = \frac{\text{TP}}{\text{TP} + \text{FN}} \tag{1}$$

- *Specificity* is the capacity to avoid false positives. Intuitively it is the capacity to detect a fall only if it really occurs:

$$\text{Specificity} = \frac{\text{TN}}{\text{TN} + \text{FP}} \tag{2}$$

- *Accuracy* is the ability to distinguish and detect both fall (TP) and non-fall movement (TN):

$$\text{Accuracy} = \frac{\text{TP} + \text{TN}}{\text{P} + \text{N}} \tag{3}$$

Where P and N are, respectively, the number of falls performed and the number of non-falls performed.

Accuracy (Equation 3) is a global index whereas sensitivity and specificity (Equations 1 and 2) enable a better understanding of the some limits of a system.
A fall exhibits high acceleration or angular velocity which are not normally achievable during the ADL. If we use a fixed low threshold to detect a fall, the sensitivity is 100% but the specificity is low because there are fall-like movements like sitting quickly on a chair, a bed or a sofa which might involve accelerations above that threshold.

8.2 Amplitude parameters

The logged data is sometimes pre-processed by applying some filters: a low-pass filter is used to perform posture analysis and a high-pass filter is applied to execute motion analysis. However, this processing is not mandatory and it strongly depends on the fall detection algorithm. The calibration of the sensors is sometimes neglected or not mentioned in research studies, but it is an important element that ensures a stable behavior of the system over time.
Amplitude parameters are useful during specific phases of the fall (Dai et al., 2010; Kangas et al., 2007; 2009).
The *Total Sum Vector* given in Equation 4 is used to establish the start of a fall:

$$\mathrm{SV}_{\mathrm{TOT}}(t) = \sqrt{(A_x)^2 + (A_y)^2 + (A_z)^2} \tag{4}$$

where A_x, A_y, A_z are the gravitational accelerations along the x, y, z-axis.
The *Dynamic Sum Vector* is obtained using the Total Sum Vector formula applied to accelerations that are filtered with a high-pass filter taking into account fast movements.
The *MaxMin Sum Vector* given in Equation 5 is used to detect fast changes in the acceleration signal, which are the differences between the maximum and minimum acceleration values in a fixed-time ($\Delta t = t_1 - t_0$) sliding window for each axis.

$$\mathrm{SV}_{\mathrm{MaxMin}}(\Delta t) = \max_{t_0 \leq i \leq t_1} \mathrm{SV}_{\mathrm{TOT}}(i) - \min_{t_0 \leq j \leq t_1} \mathrm{SV}_{\mathrm{TOT}}(j) \tag{5}$$

Vertical acceleration given in Equation 6 is calculated considering the sum vectors $\mathrm{SV}_{\mathrm{TOT}}(t)$ and $\mathrm{SV}_{\mathrm{D}}(t)$ and the gravitational acceleration G.

$$Z_2 = \frac{\mathrm{SV}^2_{\mathrm{TOT}}(t) - \mathrm{SV}^2_{\mathrm{D}}(t) - G^2}{2G} \tag{6}$$

8.3 Fall Index

Fall Index in Equation 7 is proposed by (Yoshida et al., 2005). For any sample *i* in a fixed time window, the Fall Index can be calculated as:

$$FI_i = \sqrt{\sum_{i-19}^{i} ((A_x)_i - (A_x)_{i-1})^2 + \sum_{i-19}^{i} ((A_y)_i - (A_y)_{i-1})^2 + \sum_{i-19}^{i} ((A_z)_i - (A_z)_{i-1})^2} \tag{7}$$

Since the Fall Index (FI) requires high sampling frequency and fast acceleration changes, it will miss falls that happen slowly. Hence, FI is not used unless researchers want to compare the performances of their systems with previous studies that have used it.

8.4 Standard trial scenarios and characteristics

Researcher should agree on a common set of trials in order to test and compare different fall detection systems. In Table 2 we propose a set of actions for which a fall detection system should always detect a fall. In Table 3 we propose a set of fall-like activities of daily living that can lead the system to output false positives. In addition to performing tests on all the listed 36 actions, each research group can combine them in sequential protocols, called *circuits* (e.g. sitting, standing, walking, falling).

#	Name	Symbol	Direction	Description
1	Front-lying	FLY	Forward	From vertical going forward to the floor
2	Front-protecting-lying	FPLY	Forward	From vertical going forward to the floor with arm protection
3	Front-knees	FKN	Forward	From vertical going down on the knees
4	Front-knees-lying	FKLY	Forward	From vertical going down on the knees and then lying on the floor
5	Front-right	FR	Forward	From vertical going down on the floor, ending in right lateral position
6	Front-left	FL	Forward	From vertical going down on the floor, ending in left lateral position
7	Front-quick-recovery	FQR	Forward	From vertical going on the floor and quick recovery
8	Front-slow-recovery	FSR	Forward	From vertical going on the floor and slow recovery
9	Back-sitting	BS	Backward	From vertical going on the floor, ending sitting
10	Back-lying	BLY	Backward	From vertical going on the floor, ending lying
11	Back-right	BR	Backward	From vertical going on the floor, ending lying in right lateral position
12	Back-left	BL	Backward	From vertical going on the floor, ending lying in left lateral position
13	Right-sideway	RS	Right	From vertical going on the floor, ending lying
14	Right-recovery	RR	Right	From vertical going on the floor with subsequent recovery
15	Left-sideway	LS	Left	From vertical going on the floor, ending lying
16	Left-recovery	LR	Left	From vertical going on the floor with subsequent recovery
17	Syncope	SYD	Down	From standing going on the floor following a vertical trajectory
18	Syncope-wall	SYW	Down	From standing going down slowly slipping on a wall
19	Podium	POD	Down	From vertical standing on a podium going on the floor
20	Rolling-out-bed	ROBE	Lateral	From lying, rolling out of bed and going on the floor

Table 2. Actions to be detected as falls

#	Name	Symbol	Direction	Description
21	Lying-bed	LYBE	Lateral	From vertical lying on the bed
22	Rising-bed	RIBE	Lateral	From lying to sitting
23	Sit-bed	SIBE	Backward	From vertical sitting with a certain acceleration on a bed (soft surface)
24	Sit-chair	SCH	Backward	From vertical sitting with a certain acceleration on a chair (hard surface)
25	Sit-sofa	SSO	Backward	From vertical sitting with a certain acceleration on a sofa (soft surface)
26	Sit-air	SAI	Backward	From vertical sitting in the air exploiting the muscles of legs
27	Walking	WAF	Forward	Walking
28	Jogging	JOF	Forward	Running
29	Walking	WAB	Backward	Walking
30	Bending	BEX	Forward	Bending of about X degrees (0-90)
31	Bending-pick-up	BEP	Forward	Bending to pick up an object on the floor
32	Stumble	STU	Forward	Stumbling with recovery
33	Limp	LIM	Forward	Walking with a limp
34	Squatting-down	SQD	Down	Going down, then up
35	Trip-over	TRO	Forward	Bending while walking and than continue walking
36	Coughing-sneezing	COSN	-	-

Table 3. Activities that must not be detected as falls

8.4.1 Participant characteristics

Different people have different physical characteristics and therefore it is extremely important to specify, for each trial, the following five parameters:

- Gender
- Age
- Weight
- Height
- Body Mass Index [1]

8.4.2 Hardware characteristics

Variation among the technology of the nodes depends on their level of the development and manufacturing cost. It is therefore important to define some basic characteristics for the hardware used in trials:

- Model
- Sampling frequency
- Update rate
- Movement detection delay time
- Range of measurement
- Size

[1] Body mass index (BMI) is a measure of body fat based on height and weight that applies to adult men and women.

- Weight
- Wired/wireless communication protocol

9. Falls study database

Data acquisition is probably the most difficult and time-consuming portion in a fall-detection study. In the best case, log files of fall trials contain raw accelerations measured during the simulation of an action (fall or ADL). If other researchers want to access and use such raw accelerations, it is necessary to provide an accurate description of the trials. Moreover, previous studies generally describe the tests performed and the results obtained, but the acceleration data is usually not publicly made available. This points out the need for a database with a standard structure to store all the logs. Such a database is intended to be available to the scientific community and has two main advantages: on one hand the possibility of storing and sharing data coming from sensors following a standard format; on the other hand, the availability of raw sensed data before, during, and after a fall or an activity of daily living that enables the researchers to test and validate fall detection algorithms using the same test-beds. A trial or *experiment* is described in terms of the *action* performed, the *configuration* used for the wearable device and the *user's profile*. Human actions under study are all characterized by the following aspects: *i*) *posture*: users have a particular body orientation before and after the action is performed; *ii*) *surface*: user's body is supported by a particular kind of surface before and after the action is performed. A configuration establishes a particular way to sense kinematic data, and it can be described in terms of the following: *i*) *position*: the device is worn at some body position; *ii*) *device used*: the type of sensor node adopted for the collection of data. The Entity-Relationship model depicted in Figure 8 is derived from the previous considerations.

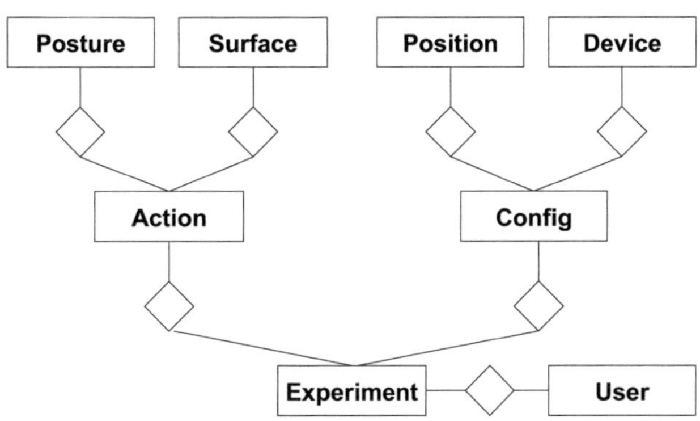

Fig. 8. Database Entity-Relationship diagram

A possible structure of the table is the following:

```
Postures (ID, posture)

Surfaces (ID, surface)

Action (ID, starting_posture, starting_surface, ending_posture,
        ending_surface, description)
```

```
Position (ID, position)

Device (ID, manufacturer, model, description, characteristics)

Configuration (ID, record_content, Mote, scale_G, sample_frequency,
               Body_position, x_direction, y_direction, z_direction)

Users (ID, age, gender, height_cm, weight_kg, body_mass_index)

Experiments (ID, Configuration, Action, User, content)
```

Note that we decided to collect, represent, and store extra information, such as the posture of the user before and after a potential fall, the separate acceleration values and acceleration magnitude as-well. This has been done to foster the reuse of the collected data and to enable the evaluation of future techniques on the same sets of data.

10. Overview of fall detection algorithms

From what has been explained so far, many different approaches have been taken to solve the fall detection problem using accelerometers. The basic and trivial system uses a threshold to establish if a person falls, which is subject to many false positives. Some researchers have tried to introduce computationally-hard type of intensive algorithms but the goal has been always to find a trade-off between the system accuracy and the cost.

Depeursinge et al. (2001) used a two-level neural network algorithm to analyze the accelerations given by two sensors placed in distinct parts of the body. Such accelerations are translated into spatial coordinates and fed into the algorithm. The output of the system represents the probability that a fall is happening: if the probability is low, the system continues monitoring whereas if the probability is medium or high, the system generates an alarm unless the person presses a button.

Clifford et al. (2007) developed a system composed of a series of accelerometers, a processor and a wireless transceiver. The acquired acceleration data is constantly compared with some standard values. If there is a fall event, the processor sends an alarm signal to a remote receiver. A similar approach is given by Lee et al. (2007) using a sensor module and an algorithm to detect posture, activity and fall. For long range communication with the base station, there are intermediate nodes that act as repeaters. The sensitivity was 93.2%.

Lindemann et al. (2005) used an acoustic device on the rear side of the ear, to measure velocity and acceleration. Also Wang et al. (2008) used a sensor on the head of the patient since it increases the accuracy of the detection.

The Inescapable Smart Impact detection System ISIS (Prado-Velasco et al., 2008) used a sensor with an accelerometer and a smartphone as base station. Moving the processing to the smartphone extended the lifetime of the batteries and the usability of the sensor. They achieved 100% sensitivity with reduction in specificity.

Other methods are based on the body posture and use more than one sensor. Some researchers divided the human activities into two parts: static position and dynamic transition (Li et al., 2009). They used two sensors both with an accelerometer and a gyroscope, one placed on the chest and the other on the thigh. The gyroscope helped to decrease the false positives.

Noury et al. (2003) used a sensor with two accelerometers, one orthogonal to the other and placed under the armpit. The fall is detected on the basis of the inclination of the chest and its velocity. The alarm is not raised if the patient presses a button on time, avoiding thus false alarms. An experimental evaluation showed levels of sensitivity and specificity equal to 81%.

In a similar study researchers used a device with three different sensors for body posture detection, vibration detection and to measure vertical acceleration (Noury et al., 2000). Data was processed by the base station. The sensitivity and specificity here were 85%.
Other researchers developed a real-time algorithm for automatic recognition of physical activities and their intensities (Tapia et al., 2007). They used five accelerometers placed on the wrist, the ankle, the upper arm, the upper thigh and the hip. In addition, they used a heart rate monitor placed on the chest. Trials have been conducted on 21 people for 30 different physical activities such as lying down, standing, walking, cycling, running and using the stairs. Data analyzed both in time and frequency domain were classified using the Naive Bayes classifier. Results showed an accuracy of 94.6% for a person using the training set of that person, whereas the accuracy was 56.3% using the training sets of all the other people.
Another research work exploited an accelerometer placed on the waist (Mathie et al., 2001). The device was so small that it fitted in a belt. The authors analyzed the duration, velocity, angle of a movement and its energy consumption to distinguish between activity and rest. The processing of the information was conducted by a base station. The authors used a threshold of 2.5G to detect a fall under the assumption that the subjects are not in good health and therefore unable to perform actions with acceleration above that threshold. This means that, to avoid false positives, they had to reduce the activity recognition capability of the system.
Hwang et al. (2004) used a node placed on the chest featuring an accelerometer, a gyroscope, a tilt sensor, a processing unit and a Bluetooth transmitter. The accelerometer measured the kinetic force whereas the tilt sensor and the gyroscope estimated the body posture. The goal was to detect some activities of daily living and falls. The authors experimented on three people, aged over 26 years, studying the four activities: forward fall, backward fall, lateral fall and sit-stand. In this study, the system could distinguish between fall and daily activities. The accuracy of fall detection was 96.7%.
Recently, smartphones with embedded accelerometers have been used to act both as fall detector and as gateway to alert the caregivers (Dai et al., 2010; Sposaro & Tyson, 2009). The problems associated with this approach are related to the device placement (in a fixed position or not) and to the short battery lifetime. Usually in these applications there is a trivial fall detection algorithm and to avoid false positives, the user should press a button to dismiss the alarm when there is no real fall.

11. Issues and challenges for designing a robust system

The review of the above proposed solutions shows some pitfalls for a real implementation. The system found more promising is the one that takes into account postures given by the accelerometers and gyroscopes to reduce false positives (Li et al., 2009). But the authors used two nodes and did not detect activities of daily living such a "falling" on a chair or a bed. The reported sensitivity is 92% and specificity 91%.
Hence the first challenge is to improve the performance of systems, to assist the patient only when there is a real fall. If we imagine to deploy the system in a hospital, it would be very annoying to run frequently to a patient because of false alarms.
The next challenge is to take into account the usability. The ideal system should be based on only one wearable sensor with small form factor, possibly placed in a comfortable place such as a belt. This may complicate the posture detection. Moreover the energy consumption must be low to extend the battery lifetime. This requires careful management of radio communications (the activity with the highest consumption of energy), flash storage and data sampling

and processing. To support clinical requirements battery lifetime is a major concern: the minimum battery lifetime should be at least one day, in order to avoid stressing the caregivers with the tasks of recharging and replacing the devices, considering that longer the battery life better the continuity and the effectiveness of the system.

12. Conclusion

The development of a fall detection system requires a non-negligible warm-up time to fully understand the problem of falls. In this survey the basics of the fall-problem together with the most relevant approaches have been described. The aim is to provide guidelines to speed-up the design process of a new fall detection system by compiling the merits of efforts taken during the past 30 years in developing a fall detection system. The researchers took many different approaches to solve the problem of falls among elderly with the lack of any standard testing guidelines. They proposed their own guidelines but they did not cover the problem from the beginning. The review shows the different approaches and presents a standard procedure by collecting the most relevant parameters, data filtering and testing protocols. This study also provided a standard structure for a database considering the issues and challenges of a fall detection system.
A step further from the *detection* is the *prediction* of non-accidental falls. Some papers left prediction as a future work, suggesting consideration of the physiological state of elderly. The first problem to face is the selection of the physiological measurements that are relevant to a fall and the ways to measure them. It is presumable that even if the complexity of such a predictive system increases, the advantages are much more.

13. References

Akyildiz, I. F., Su, W., Sankarasubramaniam, Y. & Cayirci, E. (2002). Wireless sensor networks: a survey, *Computer Networks* **38**: 393–422.

Bradley, C., Pointer, S., of Health, A. I., Welfare. & of South Australia., F. U. (2009). *Hospitalisations due to falls by older people, Australia, 2005-06 / Clare Bradley, Sophie Pointer*, Australian Institute of Health and Welfare, Canberra :.

Campbell, A. J., Borrie, M. J., Spears, G. F., Jackson, S. L., Brown, J. S. & Fitzgerald, J. L. (1990). Circumstances and Consequences of Falls Experienced by a Community Population 70 Years and over during a Prospective Study, *Age Ageing* **19**(2): 136–141.

Chapman, A. (2008). *Biomechanical Analysis of Fundamental Human Movements*, Human Kinetics, United States.

Clifford, M. A., Borras, R. L. & Gomez, L. (2007). System and method for human body fall detection.
URL: *http://www.freepatentsonline.com/7248172.html*

Dai, J., Bai, X., Yang, Z., Shen, Z. & Xuan, D. (2010). Perfalld: A pervasive fall detection system using mobile phones, *Pervasive Computing and Communications Workshops (PERCOM Workshops), 2010 8th IEEE International Conference on*, pp. 292 –297.

Depeursinge, Y., Krauss, J. & El-khoury, M. (2001). Device for monitoring the activity of a person and/or detecting a fall, in particular with a view to providing help in the event of an incident hazardous to life or limb.
URL: *http://www.freepatentsonline.com/6201476.html*

Englander, F.and Terregrossa, R. & Hodson, T. (1996). Economic dimensions of slip and fall injuries, *Journal of Forensic Sciences (JOFS)* **41**(05): 733–746.

Gibson, M., Andres, R., Isaacs, B., Radebaugh, T. & Worm-Petersen, J. (1987). The prevention of falls in later life. a report of the kellogg international work group on the prevention of falls by the elderly., *Danish Medical Bulletin* **34**(4): 1–24.

Hwang, J., Kang, J., Jang, Y. & Kim, H. (2004). Development of novel algorithm and real-time monitoring ambulatory system using bluetooth module for fall detection in the elderly, *Engineering in Medicine and Biology Society, 2004. IEMBS '04. 26th Annual International Conference of the IEEE*, Vol. 1, pp. 2204 –2207.

Kangas, M., Konttila, A., Winblad, I. & Jamsa, T. (2007). Determination of simple thresholds for accelerometry-based parameters for fall detection, *Engineering in Medicine and Biology Society, 2007. EMBS 2007. 29th Annual International Conference of the IEEE*, pp. 1367 – 1370.

Kangas, M., Vikman, I., Wiklander, J., Lindgren, P., Nyberg, L. & Jämsä, T. (2009). Sensitivity and specificity of fall detection in people aged 40 years and over., *Gait Posture* .

Lee, Y., Kim, J., Son, M. & Lee, M. (2007). Implementation of accelerometer sensor module and fall detection monitoring system based on wireless sensor network, *Engineering in Medicine and Biology Society, 2007. EMBS 2007. 29th Annual International Conference of the IEEE*, pp. 2315–2318.

Li, Q., Stankovic, J. A., Hanson, M. A., Barth, A. T., Lach, J. & Zhou, G. (2009). Accurate, fast fall detection using gyroscopes and accelerometer-derived posture information, *BSN '09: Proceedings of the 2009 Sixth International Workshop on Wearable and Implantable Body Sensor Networks*, IEEE Computer Society, Washington, DC, USA, pp. 138–143.

Lindemann, U., Hock, A., Stuber, M. & Keck, W. (2005). Evaluation of a fall detector based on accelerometers: A pilot study, *Medical and Biological Engineering and Computing* **43**(5): 548–551.

Lipsitz, L. A., Jonsson, P. V., Kelley, M. M. & Koestner, J. S. (1991). Causes and Correlates of Recurrent Falls in Ambulatory Frail Elderly, *Journal of Gerontology* **46**(4): M114–M122.

Lord, S., C., S. & Menz, H. (2001). Falls in older people: Risk factors and strategies for prevention, *Ageing and Society* **21**(05): 667–675.

Lord, S., Ward, J., Williams, P. & Anstey, K. (1993). An epidemiological study of falls in older community-dwelling women: the Randwick falls and fractures study, *Australian Journal of Public Health* **17**(3): 240–245.

Massachusetts Department of Public Health (2008). *Unintentional Fall-Related Injuries among Massachusetts Older Adults*, Commonwealth of Massachusetts, Department of Public Health.

Masud, T. & Morris, R. (2001). Epidemiology of falls, *Age Ageing* **30**(4): 3–7.

Mathie, M., Basilakis, J. & Celler, B. (2001). A system for monitoring posture and physical activity using accelerometers, *Engineering in Medicine and Biology Society, 2001. Proceedings of the 23rd Annual International Conference of the IEEE*, Vol. 4, pp. 3654 – 3657 vol.4.

Noury, N., Barralon, P., Virone, G., Boissy, P., Hamel, M. & Rumeau, P. (2003). A smart sensor based on rules and its evaluation in daily routines, *Engineering in Medicine and Biology Society, 2003. Proceedings of the 25th Annual International Conference of the IEEE*, Vol. 4, pp. 3286–3289 Vol.4.

Noury, N., Fleury, A., Rumeau, P., Bourke, A., Laighin, G., Rialle, V. & Lundy, J. (2007). Fall detection - principles and methods, *Engineering in Medicine and Biology Society, 2007. EMBS 2007. 29th Annual International Conference of the IEEE*, pp. 1663–1666.

Noury, N., Herve, T., Rialle, V., Virone, G., Mercier, E., Morey, G., Moro, A. & Porcheron, T. (2000). Monitoring behavior in home using a smart fall sensor and position sensors, *Microtechnologies in Medicine and Biology, 1st Annual International, Conference On. 2000*, pp. 607–610.

Noury, N., Rumeau, P., Bourke, A., ÓLaighin, G. & Lundy, J. (2008). A proposal for the classification and evaluation of fall detectors, *IRBM* **29**(6): 340 – 349.

Oliver, D. (2008). Falls riskprediction tools for hospital inpatients. time to put them to bed?, *Age Ageing* **37**(3): 248–250.

Perell, K., Nelson, A., R.L., G., Luther, S., Prieto-Lewis, N. & Rubenstein, L. (2001). Fall risk assessment measures: an analytic review, *Journal of Gerontology* **56A**(12): M761–M766.

Polastre, J., Szewczyk, R. & Culler, D. (2005). Telos: enabling ultra-low power wireless research, *IPSN '05: Proceedings of the 4th international symposium on Information processing in sensor networks*, IEEE Press, Piscataway, NJ, USA.

Prado-Velasco, M., del Rio-Cidoncha, M. G. & Ortiz-Marin, R. (2008). The inescapable smart impact detection system (isis): An ubiquitous and personalized fall detector based on a distributed ?divide and conquer strategy?, *Engineering in Medicine and Biology Society, 2008. EMBS 2008. 30th Annual International Conference of the IEEE*, pp. 3332–3335.

Realtime Technologies LTD (2008). Shimmer: Sensing health with intelligence modularity, mobility and experimental reusability.

Sposaro, F. & Tyson, G. (2009). ifall: An android application for fall monitoring and response, *Engineering in Medicine and Biology Society, 2009. EMBC 2009. Annual International Conference of the IEEE*, pp. 6119 –6122.

Tapia, E. M., Intille, S. S., Haskell, W., Larson, K., Wright, J., King, A. & Friedman, R. (2007). Real-time recognition of physical activities and their intensities using wireless accelerometers and a heart rate monitor, *ISWC '07: Proceedings of the 2007 11th IEEE International Symposium on Wearable Computers*, IEEE Computer Society, Washington, DC, USA, pp. 1–4.

Tinetti, M. E., Speechley, M. & Ginter, S. F. (1988). Risk factors for falls among elderly persons living in the community, *New England Journal of Medicine* **319**(26): 1701–1707.

Vassallo, M., Poynter, L., Sharma, J. C., Kwan, J. & Allen, S. C. (2008). Fall riskassessment tools compared with clinical judgment: an evaluation in a rehabilitation ward, *Age Ageing* **37**(3): 277–281.

Wang, C.-C., Chiang, C.-Y., Lin, P.-Y., Chou, Y.-C., Kuo, I.-T., Huang, C.-N. & Chan, C.-T. (2008). Development of a fall detecting system for the elderly residents, *Bioinformatics and Biomedical Engineering, 2008. ICBBE 2008. The 2nd International Conference on*, pp. 1359–1362.

Yang, C.-C. & Hsu, Y.-L. (2007). Algorithm design for real-time physical activity identification with accelerometry measurement, *Industrial Electronics Society, 2007. IECON 2007. 33rd Annual Conference of the IEEE*, pp. 2996 –3000.

Yoshida, T., Mizuno, F., Hayasaka, T., Tsubota, K., Wada, S. & Yamaguchi, T. (2005). A wearable computer system for a detection and prevention of elderly users from falling, *The 12th International Conference on Biomedical Engineering* .

Yu, X. (2008). Approaches and principles of fall detection for elderly and patient, *e-health Networking, Applications and Services, 2008. HealthCom 2008. 10th International Conference on*, pp. 42–47.

10

Odor Recognition and Localization Using Sensor Networks

Rabie A. Ramadan
Cairo University
Cairo, Egypt

1. Introduction

Odor usually quantified by five parameters which are 1) intensity, 2) degree of offensiveness, 3) character, 4) frequency, and 5) duration. It has different forms including Gas, Chemical, Radiation, Organic Compounds, and Water odors including different water contaminations. For such odors, there are many of the traditional methods that have been used for a number of years. However, these methods suffer from different problems including the detection cost, the long time taken for analysis and detection, and exposing human to danger. On the other hand, the advances in sensing technology lead to the usage of sensor networks in many applications. For instance, sensors have been used to monitor animals in habitat areas and monitor patients' health. In addition, sensor networks have been used to monitor critical infrastructures such as gas, transportation, energy, and water pipelines as well as important buildings. Sensors are tiny devices that can be included in small areas. At the same time, they are capable of capturing different phenomena from the environment, analyze the collected data, and take decisions. In addition, sensors are able to form unattended wireless ad hoc network that can survive for long time. Such features enable wireless sensor networks (WSN) to play an essential role in odor detection. In fact, odor detection became one of the important applications due to the terrorist attack that started by the one occurred at Tokyo Subway in 1995. Since this time, odor detection and localization is considered as one of the important applications. Researchers believe that sensors and sensor networks will play an important role in odor detection and localization. In this chapter, we generalize the term odor to include the radiation detection and localization since the radiation in most of the recent work is considered as an odor.

Odor detection in human depends on the smell sense; therefore, it is considered as other senses such as vision and hearing that have a theory behind it. The theory behind the smell is explained in (Al-Bastaki, 2009) where olfactory systems simulate the neurobiological information processing systems (biological neural networks) as shown in Figure 1. The collected olfactory information is processed in both the olfactory bulb and in the olfactory cortex. The function of the cortex, then, is to perform the pattern classification and recognition of the odors. Once the odor is identified, its information is transmitted to hippocampus, limbic system and the cerebral cortex. At this moment, the conscious perception of the odor and how to act on it takes place.

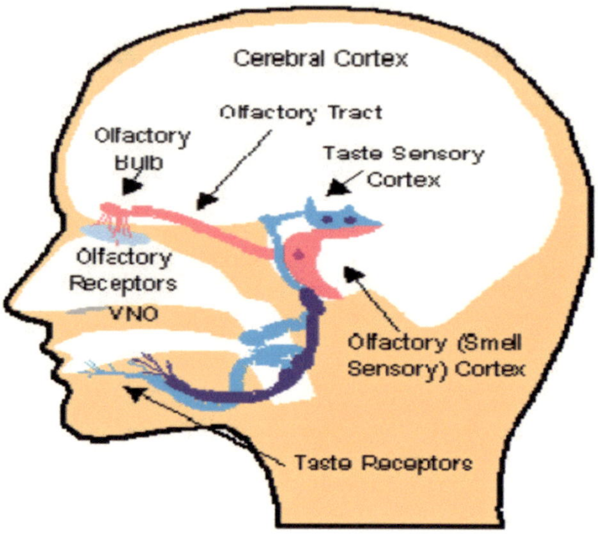

Fig. 1. The major processes of the olfactory system (Al-Bastaki, 2009)

To simulate such process, electronic noses have been developed. As can be understood, the main components of such noses are the sensing and the pattern recognition components. The first part consists of many of the sensors including gas, chemical, and many other sensors. The term chemical sensors refer to a set of sensors that respond to a particular analyte in a selective way through a chemical reaction. The second part, pattern recognition, is the science of discovering regular and irregular patterns out of a given materials. Many Artificial Intelligence (AI) algorithms and techniques are utilized in this part. Some of these techniques will be explained later in this chapter. To simplify the idea of the electronic noses, Figure 2 shows the basic components of an electronic nose. The figure shows that an electronic nose must contain a processor and a memory for analyzing the received digital data. At the same time, it has to have the appropriate set of sensors that identifies the smell print of an odor.

Once the odor is detected, its source has to be localized and contaminated if it is dangerous such as chemicals or radiations. There are different localization methods including the one that use mobile robots as well as different AI algorithms. Therefore, for odor manipulation, we have three phases as shown in figure 3 which are odor sensing, recognition, and localization. In each phase different techniques and algorithms are used. In the following sections we explore some of the detection and localization methods. Then, we propose a hybrid odor localization method that is based on Genetic Algorithms (GA), Fuzzy Logic Controller (FLC), and Swarm Intelligence. The initial results showed some significant results in localizing the odor sources.

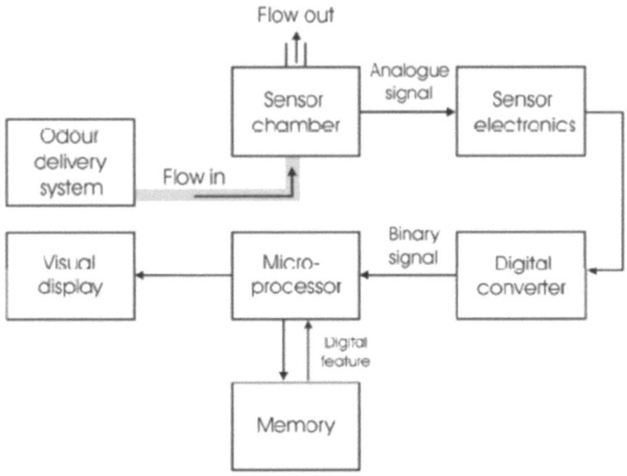

Fig. 2. Basic components of an electronic nose (Gardner et al., 1999)

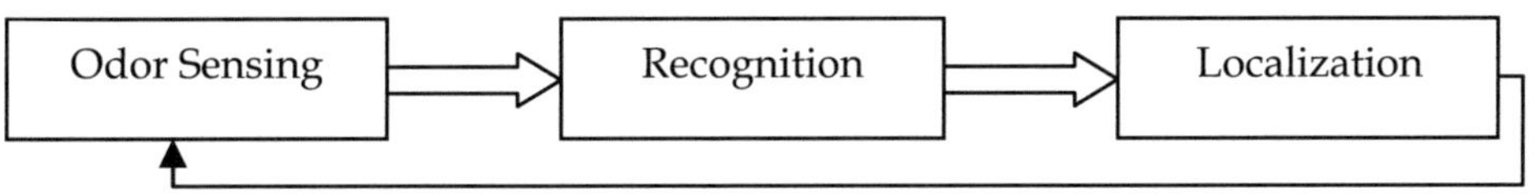

Fig. 3. Odor manipulation phases

2. Odor Sensing

As mentioned, there are many types of odors including the odor in environment pollution. In fact, the need for detecting odors that causes pollution and the need for clean environment are leading the research in this field. Reliable real time detection techniques are urgently required especially with the increasing of new diseases that are caused by these odors. Sampling and analytical procedures are no longer on call due to the availability of other techniques that can produce the results on demand.

Different sensors have been used to detect odors including Conductivity, Piezoelectric, Metal-oxide-silicon field-effect-transistor (MOSFET), Optical Fiber, and Spectrometry-Based Sensors. The idea behind the operation of these sensors are explained in (Korotkaya, 2010) and summarized as follows:
The conductivity sensors exhibit a change in resistance when exposed to volatile organic compounds. Such sensors respond to water vapor, humidity difference, but not too sensitive for specific odorants. The Piezoelectric sensors are used to measure temperature, mass changes, pressure, force, and acceleration. The main idea behind these sensors is that the gas sample is adsorbed at the surface of the polymer, increasing the mass of the disk-polymer device and thereby reducing the resonance frequency. The reduction is inversely proportional to odorant mass adsorbed by the polymer. The MOSFET odor-sensing devices are based on the principle that volatile odor components in contact with a catalytic metal can produce a reaction in the metal. The reaction's products can diffuse through the gate of a MOSFET to change the electrical properties of the device. Optical-fiber sensors utilize glass fibers with a thin chemically active material coating on their slides or ends. A light source at

a single frequency (or at a narrow band of frequencies) is used to interrogate the active material, which in turn responds with a change in color to the presence of the odorant to be detected and measured. Finally, Spectrometry-Based sensors use the principle that each molecular has a distinct infrared spectrum

The state-of-the-art method for detecting odor emissions is the classical olfactometry. In this method, the odor detection is based on a group of people (panelists) with 95% probability of average odor sensitive. The results depend on the smelling capabilities of the panel members. This method is expensive and it is not feasible where continuous and online monitoring is required. Devices like electronic noses that utilize different odor sensors, stated previously, are more effective and much cheaper. The following are some examples on the usage of electronic noses in odors' detection.

In (Staples & Viswanathan, 2005), the authors used an electronic nose named zNose that simulates an almost unlimited number of specific virtual chemical sensors, and produces olfactory images based upon aroma chemistry. Figure 3 shows the zNose device and its componenets. Using zNose odor detection is done in seconds. The nose uses a handheld air sampler that consists of a battery operated sample pump and a tenax® filled probe. This enables organic compounds associated with odors to be remotely collected.

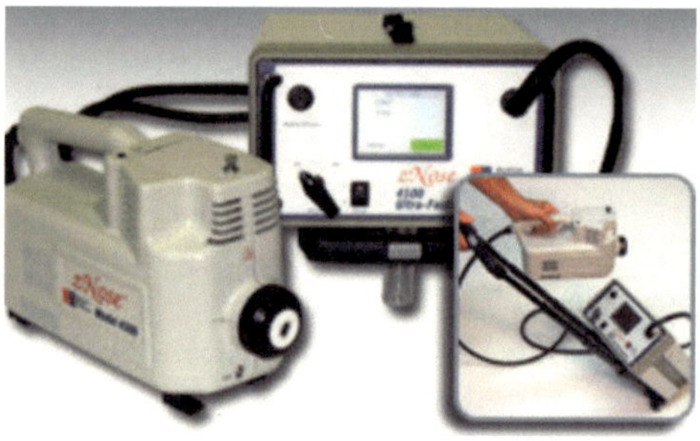

Fig. 4. zNose device (zNose, 2010)

Another example of electronic noses is in detecting the human body odor. One may think that it is not that important to use an electronic device for human body odor detection. However, the argument is if we succeeded in detecting the human body odor, the same device and experience will be applicable for other important applications such as healthcare monitoring, biometrics and cosmetics. Further to these applications, human odor could be used to uniquely identify a person which might be important in other fields such as in security. In addition, the electronic noses could be used to diagnose the urine odor of the patients with kidney disorders (Natale et al., 1999). Moreover, the human odor detection, armpit odor, is studied in (Chatchawal et al., 2009) using an electronic nose. The authors reported two problems they face when trying to detect the human body odors; the first problem is that the human may have different sweat at different time and environment; therefore, the humidity might affect the odor detection procedure in which it affects the gas sensors quality. The second problem people may use deodorants to reduce unpleasant body odor. The authors controlled the humidity using two methods including hardware-based and software-based methods through adding some of the sweat thresholds.

Electronic noses also used to detect explosives to help fighting international terrorism. The main idea behind a nose that detects explosives is to emulate the dog nose capabilities avoiding some of the dogs' drawbacks such as rigorous training, testing, and validation exercises in various operational scenarios with different types of explosives. Various types of noses have been developed; some of them are based on sensor arrays to detect different combinations of explosions. An example on this type of nose is the one developed by Walt and his group where they developed an expensive senor array of fiber optics cable. Some others are based on vapour sensors to detect different vapors such as DNT and TNT emanating from landmines. A good source on reviewing about the electronic noses that are used for explosive detection could be found in (Jehuda, 2003).

A network of the previous electronic noses might be efficient in detecting the odors or explosives in a specific area. However, the difficulties of building a network of different noses involve the wireless media problems, the data analysis coming from all of these nodes, and the deployment of these nodes. One prototype of wireless sensor (e-noses) networks is mentioned in (Jianfeng el at., 2009) where the authors developed a network to monitor odorant gases and accurately estimating odor strength in and around livestock farms. Each nose consists of four metal-oxide semiconductor (MOS) gas sensors. Figure 5 shows the e-nose with gas chamber, pump and sensor array. For communication and networking purposes, this node (e-nose) is mounted on a MICAZ board (MICAZ, 2010). The sink node uses a modified Kalman filter for the data filtering.

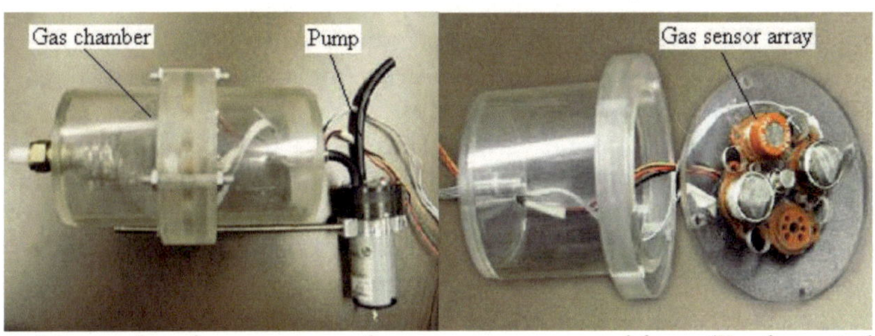

Fig. 5. E-nose with gas chamber, pump and sensor array exerted from (Jianfeng et al. , 2009)

3. Odor Recognition

Once the odors are sensed, they have to be analyzed for decision making. However, such decision is not easy since there are many factors that affect for instance (Roppel and Wilson, 2000) (i) sensors have some overlapping specificities, (ii) not all the sensors will have the same performance in sensing the odors, and (iii) no general agreement has been reached on what constitutes the fundamental components of odor space. Therefore, recognizing the odor and taking the appropriate decision still a problem to many of the applications. In the following paragraph, we will go through some of the artificial intelligence techniques that try to solve such problem including neural networks and fuzzy logic.

In (Linder et al., 2005), the authors use a standard feedforward network for odor intensity recognition for multiclass classification problem. As shown in Figure 6, the network consists of the following inputs: (1) peaks D1 (acetone), (2) D2 (ethanol), (3) C1+D3 (first peak of matrix, i. e. lemon oil, and isopropanol) , (4) C2+D4 (second peak of matrix and isoamyl acetate), and (5) C3 (third matrix peak). The output of the network is three classes which are weak, distinct, strong odour. The network has been trained based on some of the medical data prepared by the authors. The authors claimed that such network was very successful in detecting different odors' intensity.

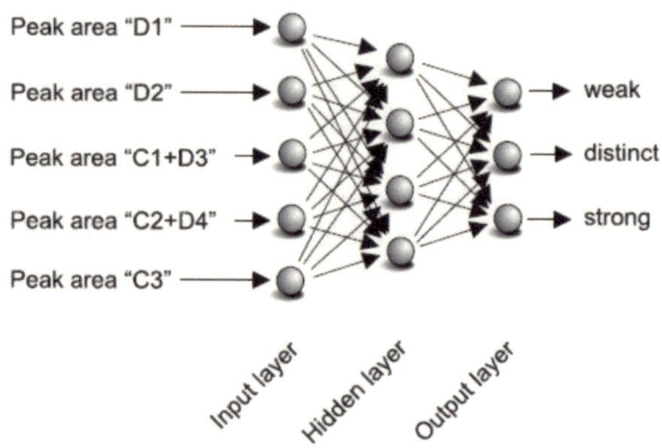

Fig. 6. Structure of a standard feedforward network that could serve for the assignment of different classes of odour intensity (Linder et al., 2005).

In (Bekiret al., 2007) , the authors combined the fuzzy logic and neural networks for better odor recognition. The sensing is done by a handheld odor meter, OMX-GR sensor which is a commercial product. Figure 7 shows the used system diagram where a gas sensor array is used to differentiate between 11 types of gases. Once the sensors collect the odors data, the data is transferred to the unsupervised feature extraction block where a fuzzy c-mean algorithm is applied for data clustering. As it is known, fuzzy c-mean divides the data into fuzzy partitions, which overlap with each other. Therefore, the containment of each data to each cluster is defined by a membership grade in (0, 1). This clustering reduces the number of inputs given to the neural networks given in the following block. The function of the neural network is to classify the sensor array output into odor categories. The authors trained the network using 16 types of perfumes with 20 samples of each. The accuracy of the developed algorithm claimed by the authors was 93.75%.

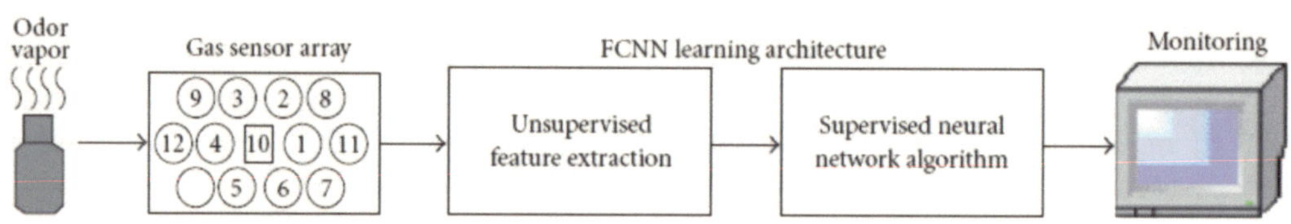

Fig. 7. Fuzzy Neural system for odor recognition (Bekiret et al., 2007)

Another AI technique is used for odor recognition is proposed in (İhsan et al. , 2009). The authors use Cerebellar Model Articulation Controller (CMAC) based neural networks. The CMAC concept is not new; The CMAC was firstly proposed during the 1970s by James Albus, whose idea was based on a model of the cerebellum which is a part of the brain responsible for learning process (Cui et al., 2004). The CMAC can generally be described as a transformation device that transforms given input vectors into associated output vectors (Wisnu et al., 2008). It plays an important role in nonlinear function approximation and system modeling. CMAC is extremely fast learning technique compared to multiple layer perceptron (MLP) neural networks. As shown in Figure 8, the CMAC basically consists of three layers which are the normalized input space, basis functions, and weight vector. The output of the network could be considered as associative memory that holds the odor detection decision.

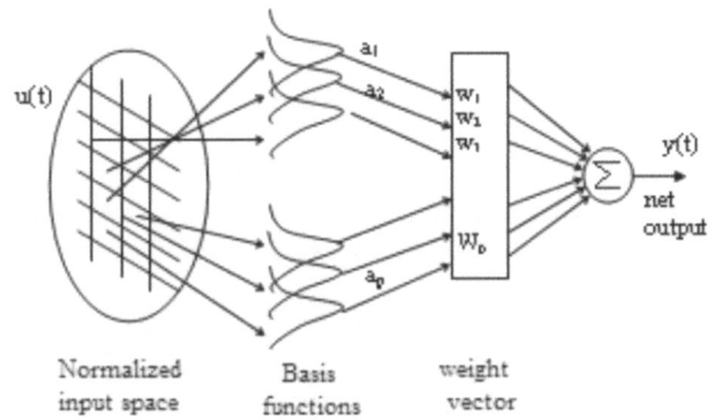

Fig. 8. CMAC layers

4. Odor Localization

Odor localization has gained a lot of attention after some of the terrorist attacks such as the one occurred at Tokyo Subway in 1995. Odor localization problem falls into one of two main approaches which are Forward problems and Inverse problems. In the forward problems , the state of the odor is estimated in advance while in the inverse problems the prior state of the odor is estimated based on the its current state. Odor localization usually done through using robots which is an emulation to the usage of dogs to find bombs, mines, and drugs. However, the dog has the capability to sense, recognize, and analyze the data; then naturally takes the decision. For robots, there is a need for localization algorithms. In the following paragraphs, we review the main ideas behind the work done in this area. Then, in the next section, we introduce our approach for odor localization.

Odor localization using robots has attracted many of the researchers and most of the work done in this area is trying to imitate the animals, birds, and swarms behaviors. For instance, insects fly in a zigzag pattern towards their matting partners. Ants use the same strategy to follow the pheromones trails. Therefore, robots could also do the same to locate the odor sources (Wei et al. , 2001). On the other hand, animals may follow spiral surge algorithm to locate the odor sources (Hayes et al. , 2002) where animals move in spiral path until it perceives a certain concentration and then moves straight upwind. As soon as it loses the scent, it starts spiraling again. A simple example on the zigzag and spiral procedures is shown in Figure 9. A comparison between the spiral and the zigzag algorithms are presented in (Lochmatter el at., 2008). The results show that the spiral is more efficient than the zigzag algorithm.

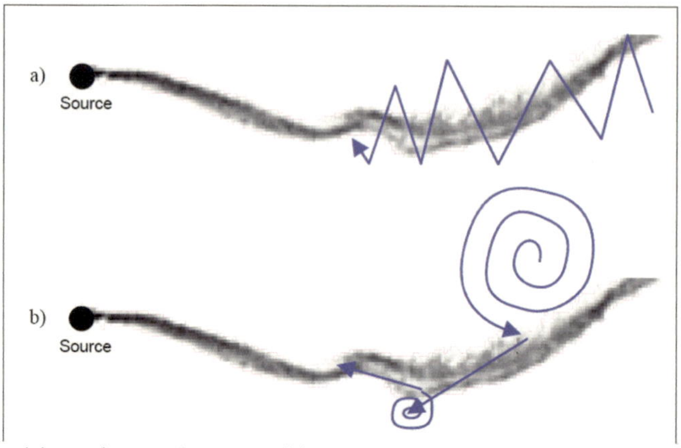

Fig. 9. Zigzagging (a) and spiral surge (b) are two bio-inspired odor source localization algorithms for single-robot systems (Lochmatter et al. , 2007).

Another method is reported in (Duckett et al., 2001) where the mobile robot turns around for 360◦ to point to the direction of the odor. Then it moves in the odor source direction. A similar technique has been used in (Loutfi and Coradeschi, 2002) where two gas sensors are mounted on the front and back of a Koala mobile robot given in Figure 10. The robot has programmed to use a wall following algorithm to avoid obstacles. The odor tracking algorithm used in (Loutfi and Coradeschi, 2002) is very simple. The default move direction for the robot is to move forward. If the front senor has a strong reaction to the odor, the robot turns 180◦ and move forward. This turn is used to avoid sensors saturation. If the rear sensor has stronger reaction to the odor than the front sensor, that means the robot has passed the odor source and it has to turn around. If both sensors have the same reaction, the robot is adjusted to move forward.

Fig. 10. Koala mobile robot

In (Wisnu et al. , 2008) , the authors try to localize the odor sources using swarm intelligence by a group of robots. The main challenge in the swarm intelligence in the authors' design is the interaction. The interaction process is divided into three phases which are encoding, synchronization, and comparison. The output of the overall interaction process is one of two outputs according to the robot that has the interaction box and the one is communicating with it. The encoding phase deals with formatting the received signals in a suitable format to the robot processing component. The synchronization phase defines which robot the holder of the interaction box can communicate with. In the third phase, comparison, the robot compares its sensed data to the received ones. Figure 11 shows the interaction state diagram for a robot. As stated in (Wisnu et al. , 2008), a robot is wandering in the arena is in the state of wandering. If it perceives some robots who arc in beaconing state, it selects one of them randomly to make a contact. The robot compares the odor concentrations received by the two robots. If the concentration received by its mate is higher, the robot approaches its mate; otherwise it keeps wandering. To get out of the wandering state, either the robot loses its mate signal or its concentration indicates inverse result. If no way the robot changes its state to attraction state, after certain period of time, the robot changes its beaconing state. After another period of time, the robot goes back to wandering state. If two robots move close to each other, they should avoid collision which is the obstacle avoidance state. The overall designed swarm behavior follows spatio-temporal model based on Eulerian framework.

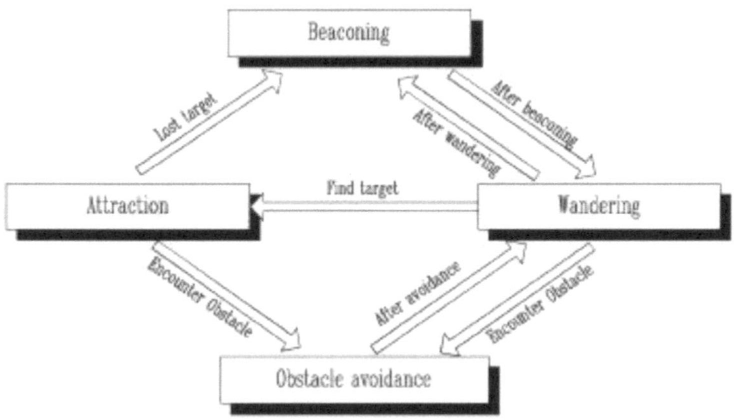

Fig. 11. Interaction state diagram for a robot (Wisnu et al. , 2008).

A swarm based fuzzy logic controller approach is also proposed (Cui el at. , 2004) for odor localization. The swarms are assumed to be the mobile robots deployed in the odor field. Robots form a wireless network to exchange their findings using table-driven routing protocol (Royer and TohToh, 1999). The monitored odor field is divided into a grid of cells that are marked differently by the robots as given in Figure 12. To avoid collision and wasting of time and energy, each robot avoids exploration of a cell occupied by other robot. To keep the robots connected, the swarm cohesion property is utilized for this purpose. The expansion cell in the Figure is the cell in the grid map that is unexplored and unoccupied. The fuzzy logic controller is used to avoid the uncertainties in the collected information. The rule-based fuzzy logic controller is used as shown in Figure 13 where the robot movement direction is decided by the controller based on the collected information from other robots. The directions are identified by 8 linguistic variables shown in Figure 14. As shown in the Figure, the direction limited to only 45° only.

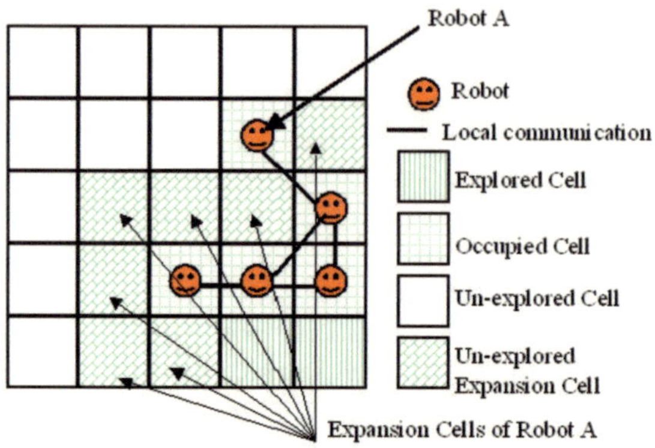

Fig. 12. Monitored field cells and their marking (Cui el at. , 2004)

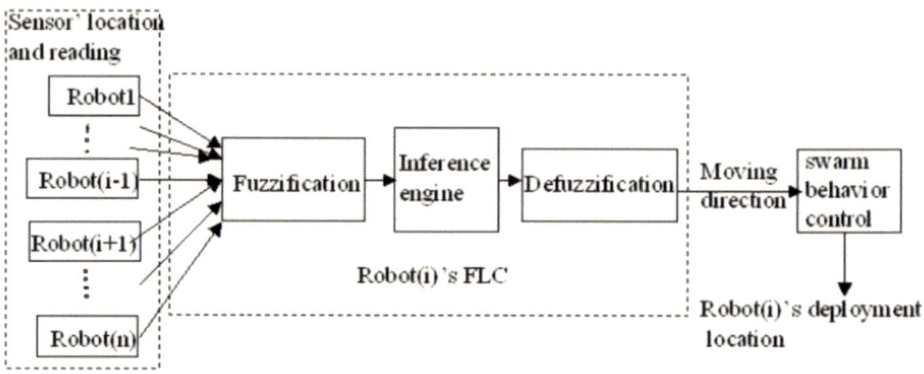

Fig. 13. Robot's Fuzzy Logic Controller (Cui el at. , 2004).

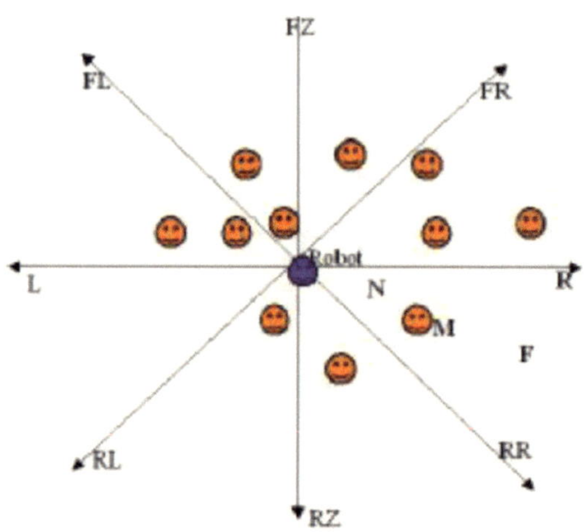

Fig. 14. Fuzzy Logic used linguistics (Cui el at. , 2004).

A combination of a static Wireless Sensor Network (WSN) and mobile robots are also used for muli-odor sources localization (Zhen el at. , 2010). The WSN is assumed to be deployed in the monitored area to collect information about the odor and the wind speed and direction. The WSN works as backbone to the robots where nodes may sense the environment odor, transmit the data to other nodes, sense and transmit at the same time, and may be nodes could be in non-active state. for multi-hop routing , the authors uses GPSR-based protocol. Robots on the other side are assumed powerful machines where they are mounted with odor and vision sensors, Global Positioning System (GPS), and may be some of the tools to prevent odor from spreading. Robots could communicate to each other or to the WSN nodes using their wireless devices. However, robots are assumed to form an ad hoc network with one as a manager and others as workers. The manager collects the sensed information from the robots and the sensors; then it takes the decision to move the workers towards the correct direction. Again, the monitored field is divided into small zones and based on the historical data, the manager can assign different number of robots.

5. Localization and Hybrid Approach

In the following subsections, we introduce a new hybrid solution for odor localization. The solution uses Genetic algorithms for local search, fuzzy logic to select the suitable direction and the swarm intelligence to keep robots connected. This work is ongoing work and it is not finalized yet; however, the initial results seem promising and outperform the previous solutions.

5.1. The Problem

The problem that we tackle in this section is to find a single source of odor in a certain space R. The space is described as a 2D environment with known dimensions. R is assumed to contain a dangerous odor source(s) in unknown location x_s and y_s ,where s refers to the hazardous source. Here, we investigate single odor source localization only; multiple odor sources localizations problem is considered as part of our future work. The space R could be divided into a grid of cells with known dimensions. In order to save the human rescue lives as well as to speed up the search in the contaminated space, a number of robots with appropriate sensors are used to locate the odor source.

Robots are assumed to have a wireless communication that enables them to form an ad hoc wireless network. The communication range does not have to be large enough to cover all of the search space. In addition, robots are assumed having the capability of running different algorithms such as Genetic Algorithms (GA) , Fuzzy Logic Controller (FLC), and Swarm Intelligence. Robots are also assumed to be capable of identifying their location from a priori given reference point. Moreover, we assume that the robots will have enough energy sources to complete their task. Robots are initially deployed close to each other to guarantee their connectivity. However, the deployment points could be anywhere in the search space.

Another fact that needs to be considered during the localization process is the wind speed which might cause sensors' readings uncertainties. The wind speed has its most effect at outdoor environments. However, indoor environments might have another source of readings noise. Therefore, an appropriate method is required to handle the readings uncertainties. We propose a fuzzy logic approach to solve this problem.

5.2 Hybrid Computational Intelligence Algorithm (H-CIA) for odor Source Localization

In this section, we explain the details of our algorithm for odor source localization. The algorithm consists of three phases; the first phase allows the robot to locally search its local area and come up with the best location to start from. This phase utilizes the GA for identifying the best location (x, y) to start from. A space search is proposed in the second phase where robots benefits from the neighbors robots to identify the moving direction. Since the neighbors readings might not be accurate and involve uncertain information, a fuzzy logic approach is proposed to identify the moving direction. In the third phase, to keep the sensors together and get the best of their readings, swarm intelligence is exploited. Through the next subsections, we explain the details of each phase.

5.2.1 Phase One: Local Search

As mentioned in the problem statement, the search space is divided into a grid of cells. These cells are assumed equal in terms of their area. Also, robots are initially loaded with the space map and its cells dimensions. Therefore, a robot *i* is assumed capable of localizing itself and identifying its location in the cell. A certain point $Pr(x_r, y_1)$ is selected to be the robots' localization reference. Once robots are deployed in the search space, they start to localize themselves and determine their deployed cells. Our proposal in this phase is to allow the sensor to locally search their cell for the odor source. Our methodology in this search is the usage of GA. GA is a well know optimization algorithm and proven to be efficient in many of the search problems. Therefore, we exploit the GA in the local search phase and we propose our own GA representation to the problem. This representation showed good results from among other representations.

First the robot's chromosome is its position in terms of x and y coordinates regarding the given reference point in the search space. A certain set of chromosomes *C* are initially generated to represent the GA initial population. This set of chromosomes is limited to the robot's current cell dimensions to limit the GA search space. The crossover in done between two chromosomes in two steps: 1) randomly select x or y from the chromosome to exchange and 2) exchange the selected coordinate. The mutation is done based on a probability *Pm* where $0< Pm >1$. If *Pm* is greater than 0.5 , a random chromosome is selected from the initial population and its x coordinate replaces x coordinate in one the current used chromosomes. On the next iterations, if the evaluation function enhanced, we keep replacing the x_s' coordinates; otherwise we exchange y_s' coordinates instead. The evaluation function is represented by the measured odor concentration value at each new location which in our case are two locations based on the current generated chromosomes. The odor distribution is assumed to follow the Gaussian distribution. The GA algorithm runs for certain number of iterations and terminates. During these iterations, the best chromosome is stored and it will be the final robot's location in this phase.

5.2.2 Phase Two: Space Search

In this phase of the solution, we try to move the robot towards the odor source and avoid the GA local minima. At the same time, we make the most out of robots readings so far. As stated previously, robots readings are uncertain due to the wind speed in outdoor environments or other noise in indoor environments. In this case Fuzzy logic comes to to handle the robots' readings uncertainties. The objective of fuzzy logic is to identify the correct direction of the robot to move to based on the current sensors hazardous concentration readings as well as neighbors robots' positions.

Therefore, for the fuzzy logic to work, there are two inputs with two membership functions. The first input is the robot's neighbor's readings and the second is their positions in terms of x and y coordinates. These coordinates are converted to a certain linguistics that relates the robots positions to the reference point. Here, we assume the reference point is located at the center of the space and each robot could identify its location to the reference points in terms of the following 8 directions: 1) Left (L) , 2) Right (R), 3) Top (T), 4) Down (D), 5) Right Top (RT), 6) Left Top (LT), 7) Left Down (LD), and 8) Right Down (RD). The membership of the readings input is chosen to be dynamic based on the readings values. Five linguistics are chosen for this input which are Very-Low, Low, Medium, High, and Very-High. Each one falls in almost 20% of the input's range. For the defuzzification, it seems that the center of gravity (COG) of fuzzy sets is an essential feature that concurrently reflects the location and shape of the fuzzy sets concerned. Therefore, we use COG as our defuzzification process as shown in equation (1).

$$COG = \frac{\sum \mu_A(a) * a}{\sum \mu_A(a)} \tag{1}$$

Where, $\mu_A(a)$ is the membership function of set A.

The output membership function produces the same eight directions as presented in the input. Once the robot decides on the direction, it moves according to the next phase restrictions. When it reaches a new zone, it starts again the GA.

5.2.3 Phase Three: Swarm Movement Control

Once the direction is identified, if the robots move freely, another problem might occur in which the robots might get disconnected from each other. A similar Swarm intelligence model presented in (Fei el at., 2008) is used to keep robots together. Probability Particle Swarm Optimization (P-PSO) algorithm uses probability to express the local and global fitness functions. More detailed description about the P-PSO algorithm can be found in (Fei el at., 2008). The coherence characteristic of the P-PSO moves only the robot that does not affect the robots connectivity.

It is worth mentioning that the initial results based on simulation setup shows that 85% of the time the robots reach the odor source. However, the cell size has to be small to prevent the robot in moving in a zigzag form and taken long time to converge. Currently we prepare a real robot experiments to check the performance of the hybrid approach in odor localization. Several parameters might not easy to handle in real environments such as the effect of obstacles, robots speed, robots batteries, etc.

6. Conclusion

In this chapter, we explored some of techniques and algorithms used for odor recognition and localization. We started by introducing different types of sensors that are currently used for odor sensing. We focused on one of the famous devices which is the electronic nose. Then, we reviewed some of the concepts used for odor recognition including neural networks and swarm intelligence. This phase is followed by odor localization using single and multi-robots. Finally, we proposed a new hybrid method based on Genetic Algorithms, Fuzzy Logic, and Swarm Intelligence. The Genetic Algorithm is used for local search and the fuzzy logic identifies the direction of the robot's movement. Then, the Swarm Intelligence is used for robots' cohesion and connectivity.

7. References

Al-Bastaki, Y. (2009). An Artificial Neural Networks-Based on-Line Monitoring Odor

Bekir, K. & Kemal , Y. (2007) Fuzzy Clustering Neural Networks for Real-Time Odor Recognition System. Journal of Automated Methods and Management in Chemistry, Vol. 2007.

Chatchawal, W. ; Mario, L.; & Teerakiat, K. (2009). Detection and Classification of Human Body Odor Using an Electronic Nose. Sensors 2009, Vol. 9, pp.7234-7249.

Cui , X.; Hardin , T. ; Ragade, R.; Elmaghraby, A. (2004). A Swarm-based Fuzzy Logic Control Mobile Sensor Network for Hazardous Contaminants Localization. In Proceedings of the IEEE International Conference on Mobile Ad-hoc and Sensor Systems (MASS'04).

Duckett, T.; Axelsson, M.; Saffiotti, A. (2001) Learning to locate an odour source with a mobile robot. In Proc. ICRA-2001, IEEE International Conference on Robotics and Automation, pp. 21–26.

Fei, L.; Qing-Hao, M.; Shuang, B.; Ji-Gong, L.; Dorin, P. (2008). Probability-PSO Algorithm for Multi-robot Based Odor Source Localization in Ventilated Indoor. Intelligent Robotics and Applications, Lecture Notes in Computer Science Springer, Vol. 5314,pp. 1206-1215.

Gardner, J. & Bartlett, P. (1999). Electronic Noses Principles and Applications. Oxford University Press. Oxford, UK.

Hayes, A.; Martinoli, A.; Goodman, R. (2002) Distributed Odor Source Localization. *IEEE Sensors Journal*, Vol. 2, No. 3, pp. 260-271.

http://www.it.lut.fi/kurssit/03-04/010970000/seminars/Korotkaya.pdf as visited on 06/09/2010

İhsan, Ö. & Bekir, K. (2009) . Hazardous Odor Recognition by CMAC Based Neural Networks. *Sensor 2009*, Vol. 9, pp. 7308-7319;

Jehuda, Y. (2003). Detection of Explosives by Electronic Noses. *Analytical Chemistry* , Vol. *75, No.* 5, pp. 98 A-105 A.

Jianfeng, Q.; Yi, C.; & Simon, X. (2009). A Real-Time De-Noising Algorithm for E-Noses in a Wireless Sensor Network. *Sensor 2009.* Vol. 9, pp.895-908;

Korotkaya, Z. "Biometric Person Authentication: Odor", Pages: 1 - 6,

Linder, R. ; Zamelczyk, M. ; Pöppl, R. ; Kośmider, J. (2005). *Polish Journal of Environmental Studies.* Vol. 14, No. 4, pp. 477-481

Lochmatter, L.; Raemy,X.; Matthey, L.; Indra, S.; Martinoli, A. (2008). A Comparison of Casting and Spiraling Algorithms for Odor Source Localization in Laminar Flow. In Proceedings of the 2008 IEEE International Conference on Robotics and Automation (ICRA 2008), pp. 1138-1143.

Lochmatter, T.; Raemy, X.; Martinoli, A. (2007). Odor Source Localization with Mobile Robots. Bulletin of the Swiss Society for Automatic Control, Vol. 46: pp.11-14.

Loutfi, A. & Coradeschi , S. (2002). Relying on an electronic nose for odor localization," IEEE International Symposium on Virtual and Intelligent Measurement Systems. VIMS '02.

MICAZ, http://www.openautomation.net/uploadsproductos/micaz_datasheet.pdf as visited on 06/09/2010

Natale, C.D.; Mantini, A.; Macagnano, A.; Antuzzi, D.; Paolesse, R.; D'Amico, A. (1999). Electronic Nose Analysis of Urine Samples Containing Blood. Phys. Meas., pp. 377-384.

Roppel, T. & Wilson, D. (2000). Biologically-inspired pattern recognition for odor detection. Pattern Recogn. Lett. 21, No.3, pp. 213-219.

Royer, E. & TohToh, C. (1999). A review of current routing protocols for ad hoc mobile, IEEE Personal Communication, Vol. 6, No.2, pp. 46—55.
Sensing System, *Journal of Computer Science* , Vol. 5, No. 11, pp. 878-882,

Staples, E. & Viswanathan,S. (2005). Odor Detection and Analysis using GC/SAW zNose. Electronic Sensor Technology, School of Engineering and Technology, National University.

Wei, L.; Jay, A.; Ring, T. (2001). Tracking of Fluid-Advected Odor Plumes: Strategies Inspired by Insect Orientation to Pheromone. *Journal of Adaptive Behavior*, Vol. 9, No. 3-4, pp. 143-170.

Wisnu, J. ; Petrus, M. ; Benyamin, K. ; Kosuke, S. ; Toshio, F. (2008) Modified PSO Algorithm Based on Flow of Wind for Odor Source Localization Problems in Dynamic Environments. In Wseas transactions on Systems, Vol. 7.

Zhen, F.; Zhan, Z. ; Xunxue, C. ; Daoqu, G. ; Yundong, X. ; LiDong, D. ; ShaoHua, W. (2010). Multi-odor Sources Localization and Tracking with Wireless Sensor Network and Mobile Robots. 1st IET international Conference on Wireless Sensor Networks. IET-WSN-2010.

zNose device available at http://www.estcal.com/ as visited on 06/09/2010.

Part 2

Communication and Networking Technologies

11

Modelling Underwater Wireless Sensor Networks

Jesús Llor and Manuel P. Malumbres
Universidad Miguel Hernández de Elche
Spain

1. Introduction

The study of Underwater Wireless Networks as a research field has grown significantly in recent years offering a multitude of proposal to resolve the communication between the nodes and protocols for information exchange networks. Acoustics has been used by nature for years to communicate in the underwater environment using it as a language, dolphins and whales for instance are able to use it to send information between their groups. The first reference to the underwater sound propagation can be found in what Leonardo Da Vinci wrote in 1490: "If you cause your ship to stop and place the head of a long tube in the water and place the outer extremity to your ear, you will hear ships at a great distance from you".

Years later in 1826 the first scientific studies were done by picking real data measures (Colladon, 1893). The physicist Jean-Daniel Colladon, and his partner Charles-Francois Sturm a mathematic, made the first recorded attempt at Lake Geneva, Switzerland, to find out the speed of sound in water. After experimenting with an underwater bell with ignition of gunpowder on a first boat, the sound of the bell and flash from the gunpowder were observed 10 miles away on a second boat. With this collection of data of the time between the gunpowder flash and the reception of the sound reaching to the second boat they were able to establish a pretty accurate value for the speed of the sound in water, tested with this empirical method.

In the early XX century in 1906, the first sonar type was developed for military purpose by Lewis Nixon; there was a great interest in this technology during World War I so as to be able to detect submarines. It was in 1915, when the "echo location to detect submarines" was released by physicist Paul Langévin and the engineer Constantine Chilowski, device capable for detecting submarines using the piezoelectric properties of the quartz. It was not useful for the war as it arrived too late, but it established the roots of the upcoming design for sonar devices.

The first targets in which the developing of underwater sound technology was involved were to determine the distance to the shore or to other ships. After experimenting it was quickly discovered by researchers that pointing the sound device down towards the seafloor, the depth could also be collected with enough precision. Then, by picking a lot of values it was used for new purposes like measuring the relief of the ocean (bathymetry), seafloor shape registering, search for geological resources (i.e. oil, gas, etc.), detecting and tracking fish banks, submarine archaeology, etc.

These were the main underwater acoustic application mainly use for the exploration of seafloor and fishery with sonar devices. In the 90's the researchers became aware of a new feature applicable to underwater communications, multipoint connections could be capable of translating the networked communication technology to the underwater environment. One of the former deployments was the Autonomous Oceanographic Surveillance Network (AOSN), supported by the US Office of Naval Research (ONR) (Curtin et al, 1993). It calls for a system of moorings, surface buoys, underwater sensor nodes and Autonomous Underwater Vehicles (AUVs) to coordinate their sampling via an acoustic telemetry network.

Wireless terrestrial networking technologies have experienced a considerably development in the last fifteen years, not only in the standardization areas but also in the market deployment of a bunch of devices, services and applications. Among all these wireless products, wireless sensor networks are exhibiting an incredible boom, being one of the technological areas with greater scientific and industrial development step (Akyildiz et al, 2002).

The interest and opportunity in working on wireless sensor network technologies is endorsed by (a) technological indicators like the ones published by MIT (Massachusetts Institute of Technology) in 2003 (Werff, 2003) where wireless sensor network technology was defined as one of the 10 technologies that will change the world, and (b) economic and market forecasts published by different economic magazines like (Rosenbush et al, 2004), where investment in Wireless Sensor Network (WSN) ZigBee technology was estimated over 3.500 Million dollars during 2007.

Recently, wireless sensor networks have been proposed for their deployment in underwater environments where many of applications such us aquiculture, pollution monitoring, offshore exploration, etc. would benefit from this technology (Cui et al, 2006). Despite having a very similar functionality, Underwater Wireless Sensor Networks (UWSNs) exhibit several architectural differences with respect to the terrestrial ones, which are mainly due to the transmission medium characteristics (sea water) and the signal employed to transmit data (acoustic ultrasound signals) (Akyildiz et al, 2006).

Then, the design of appropriate network architecture for UWSNs is seriously hardened by the conditions of the communication system and, as a consequence, what is valid for terrestrial WSNs is perhaps not valid for UWSNs. So, a general review of the overall network architecture is required in order to supply an appropriate network service for the demanding applications in such an unfriendly submarine communication environment.

Major challenges in the design of underwater acoustic networks (Llor & Malumbres 2009) are:

- Battery power is limited and usually batteries cannot be recharged because solar energy cannot be exploited;
- The available bandwidth is severely limited;
- The channel suffers from long and variable propagation delays, multi-path and fading problems;
- Bit error rates are typically very high;
- Underwater sensors are prone to frequent failures because of fouling, corrosion, etc.

This chapter will give an overview of underwater wireless networks going-through all the layers with emphasis on the physical layer and how it behaves in different and changing environment conditions. Besides a brief outline of the most outstanding MAC layer protocols as the ones of the routing layer algorithms will be presented. Also the main application are presented and finally the conclusions.

In the next section, we briefly describe the main issues in the design of efficient underwater wireless sensor networks. Following a bottom-to-top approach, we will review the network architecture, highlighting some critical design parameters at each of the different network layers, and overcoming the limitations and problems introduced by UWSN environments.

2. Topology

In (Partan et al, 2006), taxonomy of UWSN regimes is outlined. They classify different UWSNs in terms of both spatial coverage and node density. For every kind of network topology, different architectural approaches have to be considered in order to improve the network performance (throughput, delay, power consumption, packet loss, etc.). So, it is important to design the network architecture taking into account the intended network topology.

3. Physical Layer: Acoustic Link

The most common way to send data in underwater environments is by means of acoustic signals, dolphins and whales use it to communicate. Radio frequency signals have serious problems to propagate in sea water, being operative for radio-frequency only at very short ranges (up to 10 meters) and with low-bandwidth modems (terms of Kbps). When using optical signals the light is strongly scattered and absorbed underwater, so only in very clear water conditions (often very deep) does the range go up to 100 meters with high bandwidth modems (several Mbps).

The theory of the sound propagation is according to the description by Urick (Urick & Robert, 1983), a regular molecular movement in an elastic substance that propagates to adjacent particles. A sound wave can be considered as the mechanical energy that is transmitted by the source from particle to particle, being propagated through the ocean at the sound speed. The propagation of such waves will refract upwards or downwards in agreement with the changes in salinity, temperature and the pressure that have a great impact on the sound speed, ranging from 1450 to 1540 m/s.

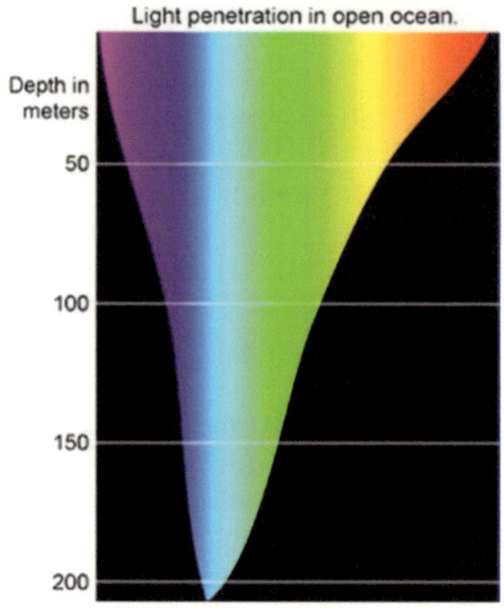

Fig. 1. This diagram offers a basic illustration of the depth at which different colors of light penetrate ocean waters. Water absorbs warm colors like reds and oranges and scatters the cooler colors

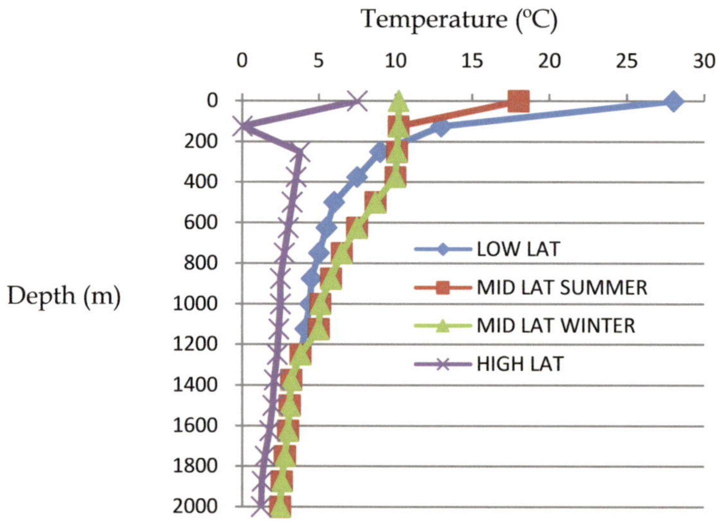

Fig. 2. Temperature variation depending on latitude and season

Depth (m)	Salinity (ppm)
0	37.45
50	36.02
100	35.34
500	35.11
1000	34.90
1500	34.05

Table 1. Salinity depending on the depth

The transmission loss (TL) is defined as the decrease of the sound intensity through the path from the sender to the receiver. There have been developed diverse empirical expressions to measure the transmission loss. Thorp formula (Urick & Robert, 1983) defines the signal transmission loss as:

$$\propto = \frac{0.11\,f^2}{1+f^2} + \frac{44\,f^2}{4100+f^2}\,[db/Km]$$
$$SS = 20\log r \qquad (1)$$
$$TL = SS + \propto x\ 10^{-3}$$

where f is frequency in kHz, r is the range in meters; SS is the spherical spreading factor and α is the attenuation factor. Then a more accurate expression for the attenuation factor was presented, the one proposed in the Thorp formula in (Berkhovskikh & Laysanov, 1982):

$$\propto = \frac{0.11\,f^2}{1+f^2} + \frac{44\,f^2}{4100+f^2} + 2.75\ x\ 10^{-4} f^2 + 0.003 \qquad (2)$$

Since acoustic signals are mainly used in UWSNs, it is necessary to take into account the main aspects involved in the propagation of acoustic signals in underwater environments, including: (1) the propagation speed of sound underwater is around 1500 m/s (5 orders of magnitude slower than the speed of light), and so the communication links will suffer from large and variable propagation delays and relatively large motion-induced Doppler effects; (2) phase and magnitude fluctuations lead to higher bit error rates compared with radio channels' behaviour, this makes necessary the use of forward error correction codes (FEC); (3) as frequency increases, the attenuation observed in the acoustic channel also increases, which is a serious bandwidth constraint; (4) multipath interference in underwater acoustic communications is severe due mainly to the surface waves or vessel activity, that are an important issue to attain good bandwidth efficiency.

Several works in the literature propose models for an acoustic underwater link, taking into account environment parameters as salinity degree, temperature, depth, environmental interference, etc. Other physical aspects of the ocean as noise in the medium (Coates, 1989), the wind, thermal noise, the turbulence and the ship noise are included by these formulas, depending on the frequency and this factors:

$$\begin{aligned} 10\log N_t(f) &= 17 - 30\log f \\ 10\log N_s(f) &= 40 + 20\,(s-0.5) + 26\log f \\ 10\log N_w(f) &= 50 + 7.5\,w^{\frac{1}{2}} + 20\log f - 40\log(f+0.4) \\ 10\log N_{th}(f) &= -15 + 20\log f \end{aligned} \qquad (3)$$

where N_t is the noise due to turbulence, N_s is the noise due to shipping, N_w is the noise due to wind, and N_{th} represents the thermal noise. The overall noise power spectral density for a given frequency f is then:

$$N\ (f) = N_t(f) + N_s(f) + N_w(f) + N_{th}(f) \tag{4}$$

In (Xie & Gibson, 2006) the authors present the Monterey-Miami Parabolic Equation to describe the behavior of the propagation of the sound. In (Porter & Liu 2010) Bellhop, a ray tracing tool shows how the physical environment conditions and terrain shapes have a great impact in the sound attenuation.

3.1 MMPE

Monterey-Miami Parabolic Equation (MMPE) model is used to predict underwater acoustic propagation using a parabolic equation which is closer to the Helmholtz equation (wave equation), this equation is based on Fourier analysis. The sound pressure is calculated in small increments changes in range and depth, forming a grid. If we increase the step size, we can obtain better performance. The propagation loss formula based on the MMPE model:

$$PL(t) = m(f, s, d_A, d_B) + w(t) + e() \tag{5}$$

where:
PL(t): propagation loss while transmitting from node A to node B.
m(): propagation loss without random and periodic components; obtained from regression using MMPE data.
f: frequency of transmitted acoustic signals (in kHz).
d_A: sender's depth (in meters).
d_B: receiver's depth (in meters).
r: horizontal distance between A and B nodes, called range in MMPE model (in meters).
s: Euclidean distance between A and B nodes (in meters).
w(t): periodic function to approximate signal loss due to wave movement.
e(): signal loss due to random noise or error.

The *m()* function represents the propagation loss provided by the MMPE model. According to the logarithmic nature of the data, a nonlinear regression is the best option to provide an approach to the model based on the coefficients supplied by the preliminary model. The proposed expression to calculate this function is the following one:

$$m(f,s,d_A,d_B) = \log\left(\left|\frac{\left(\frac{s}{0.914}\right)^{A0} (d_A)^{A9} s^{A7} ((d_A - d_b)^2)^{A10}}{(s * d_B)^{10\,A5}}\right|\right) + \left(f^2\left(\frac{A1}{1+f^2} + \frac{40}{4100+f^2} + 0.00275\right) + 0.003\right) * \left(\frac{s}{914}\right) + A6 * d_B + A8 * s \tag{6}$$

The *w()* function considers the movement of a particle that will oscillate around its location in a sinusoidal way. That movement is represented as circular oscillations that reduce their radius as the depth of the particle increases. The length of that radius is dependent of the energy of the wave and is related to the height of the wave. The common waves have hundreds of meters of wavelength and have an effect up to 50 meters of depth

For the calculation of the effects of the wave we will consider:

$$w(t) = h(l_w, d_B t, h_w, T_w)\ E(t, T_w) \tag{7}$$

where
w(t): periodic function to approximate the lost signal by the wave movement.
h(): scale factor function.
l_W: ocean wave length (meters).
d_B: depth of the receiver node.
h_W: wave height (meters).
T_W: wave period (seconds).
E(): function of wave effects in nodes.

This function contains the elements that are resembled the node movement, first calculating the scale factor h() and then the wave effect in a particular phase of the movement. The calculation of the scale factor is as follows:

$$h(l_w, d_B, t, h_w, T_w) = \frac{\left(h_w\left(1 - \left(\frac{2d_B}{l_w}\right)\right)\right)}{0.5} * \left|\sin\left(\frac{2\pi(\text{mod } T_w)}{T_w}\right)\right| \tag{8}$$

The *e()* function represents a random term to explain background noise. As the number of sound sources is large and undetermined, this random noise follows a Gaussian distribution and is modeled to have a maximum of 20dB at the furthest distance. This function is calculated by the following equation:

$$e() = 20\left(\frac{s}{s_{max}}\right) R_N \tag{9}$$

where:

e(): random noise function
s: distance between the sender and receiver (in meters).
s_{max}: maximum distance (transmission range)
h_w: height of the wave (in meters).
R_N: random number, Gaussian distribution centered in 0 and with variance 1.

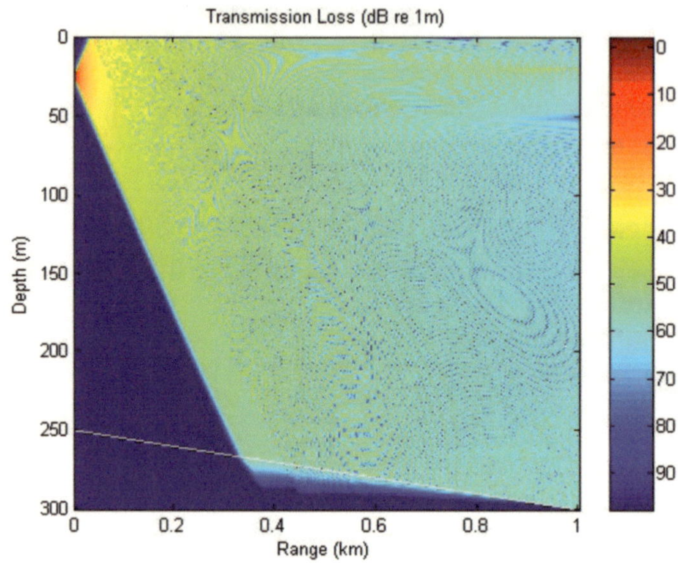

Fig. 3. MMPE Transmission Loss (db)

3.2 Bellhop

Ray tracing requires the solution of the ray equations to determine the ray coordinates. Amplitude and acoustic pressure requires the solution of the dynamic ray equations. For a system with cylindrical symmetry the ray equations can be written:

$$\frac{dr}{ds} = c\xi(s) \quad , \quad \frac{d\xi}{ds} = -\frac{1}{c^2}\frac{\vartheta c}{\vartheta r} \tag{10}$$

where *r(s)* and *z(s)* represent the ray coordinates in cylindrical coordinates and s is the arclenght along the ray; the pair *c(s) [ξ (s),ζ(s)]* represents the tangent versor along the ray. Initial conditions for and *r(s)*, and *z(s)* , *ξ(s)* and *ζ(s)* are

$$r(0) = r_s \quad , \quad z(0) = z_s \quad , \quad \xi(0) = \frac{\cos\theta_s}{c_s} \quad , \quad \zeta(0) = \frac{\sin\theta_s}{c_s} \tag{11}$$

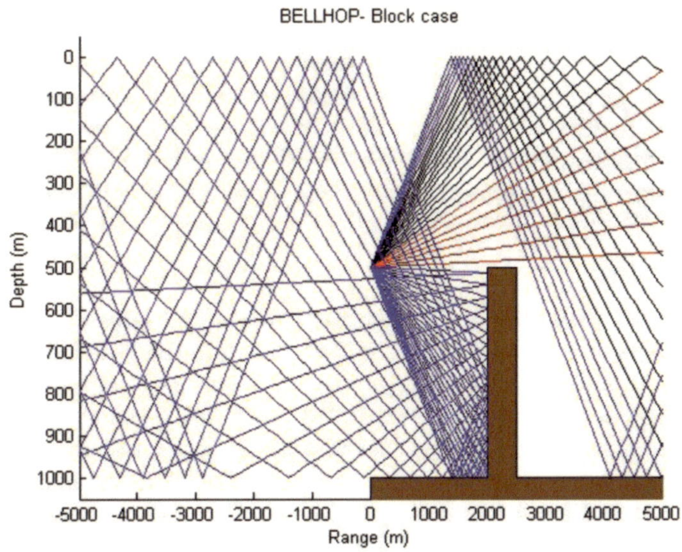

Fig. 4. Bellhop ray trace

where θ_s represents the launching angle, (r_s , z_s) is the source position, and c_s is the sound speed at the source position. The coordinates are sufficient to obtain the ray travel time:

$$\tau = \int_{\Gamma} \frac{ds}{c(s)} \tag{12}$$

which is calculated along the curve, $[r , z_s]$.

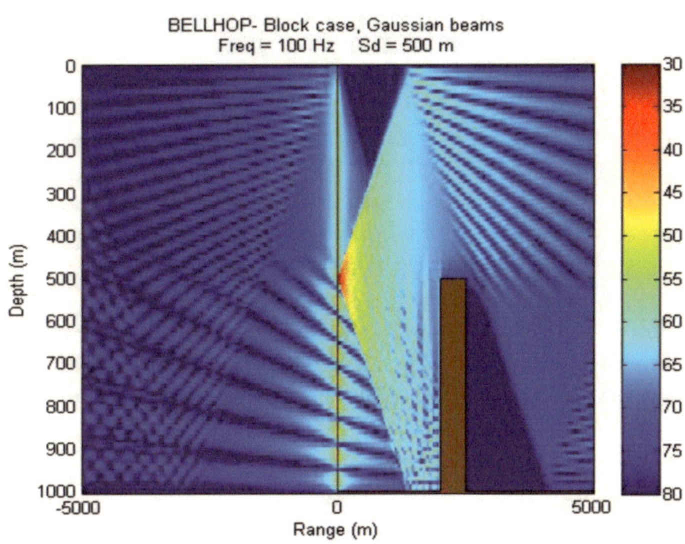

Fig. 5. Bellhop pressure

4. Mac Layer

The main task of MAC protocols is to provide efficient and reliable access to the shared physical medium in terms of throughput, delay, error rates and energy consumption. However, due to the different nature of the underwater environment, there are several drawbacks with respect to the suitability of the existing terrestrial MAC solutions for the underwater environment. In fact, channel access control in UWSNs poses additional challenges due to the aforementioned peculiarities of underwater channels (Molins and Stojanovic, 2006).

In the MAC Layer we have different types of protocols and ones of the latest examples presented in each one of the categories. The first ones are the non handshaking protocol such as Aloha-like, an on the opposite side we have the handshaking protocols like DACAP (Distance Aware Collision Avoidance Protocol) a collision avoidance protocol based on virtual Medium Access Control carrier sensing. A number of adaptations have been proposed to adopt MACA (Multiple Access with Collision Avoidanc) (Karn, 1990), MACAW (Media Access Protocol for Wireless LAN's) (Bharghavan et al, 1994), and FAMA (Floor Acquisition Multiple Access) (Fullmer & Luna-Acebes, 1995) for underwater networks in (Molins & Stojanovic, 2006). But also new protocols such as T-Lohi (between handshake and non-handshake) are becoming more popular as they have also a great efficiency in terms of the battery use.

4.1 Aloha-like

In Aloha like mode the source node sends its data frames as soon as it receives a packet from the upper-layer protocol. It does not check the medium to see if it is busy and so it does not perform any back-off, it. The node that receives the data will answer with and acknowledge data frame, if there was no problem at the reception such as a collision or packet lost during the transmission (i.e. when there is overlapping of the receiving periods of two or more frames at the destination location, or the receiver was transmitting).

If the source, does not receive an ACK because either the frame was not correctly delivered or the ACK was lost, the sender will timeout, wait a random period (back-off) and retransmit the frame. This protocol follows the stop-and-wait paradigm. That is, the source must receive an acknowledgement for each data frame before the next frame can be sent. In addition, after a successful frame transmission, the sender will perform a back off, even if it has additional frames to send from the same packet or from a new packet.

There different version called Aloha-based protocols, a couple of proposals modification for underwater networks of the protocol can be found in (Chirdchoo, 2007) that presents Aloha with Collision Avoidance (Aloha-CA) and Aloha with Advance Notification (Aloha AN).

- Aloha-CA: Pays close attention to every packet it overhears, picking the information of who are the sender and receiver. With this information it can easily calculate the busy duration due to the packet at every one of the nodes. Each node will store in a database table the information of the monitored packets with the busy durations of every node of the neighbourhood.

- Aloha-AN: This protocol has all the features of Aloha-CA and adds the sending of a small advanced notification packet with the necessary information to let the other nodes build the databases tables. The sender will wait after this packet for a lag time before sending the actual data packet. Whenever a node has a packet to transmit, it will check the database table to ensure that the packet does not result in a collision at any other reachable nodes.

4.2 DACAP

DACAP is a handshaking protocol Distance Aware Collision Avoidance Protocol (Peleato & Stojanovic 2007) for Ad Hoc Underwater Acoustic Sensor Networks. This protocol is a non-synchronized data access that uses control messages to decide when to start the communication Request-To-Send (RTS) / Clear-To-Send (CTS) handshake. If a collision happens during the handshake, the transmitter will not receive a CTS that enables him to send the data packet.

The protocol is describe in Figure 6 and Figure 7 and explained in the following steps:

- RTS reception: After this event, the node sends a CTS to the sender, and waits for a data packet. If another RTS is received the node sends a warning short packet to give and advice to the sender of the last one that the medium is in use.

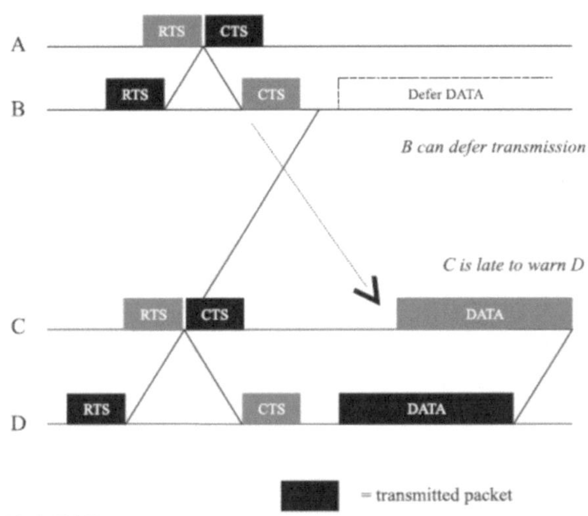

Fig. 6. Transmission in DACAP

- CTS reception: When the transmitter receives this message it waits for those nodes whose transmissions are still happening to avoid any collisions. If it happens to receive another CTS or a warning packet reception the current data packet to send will be deferred for a random back-off time, if the waiting time expires the transmission proceeds normally.

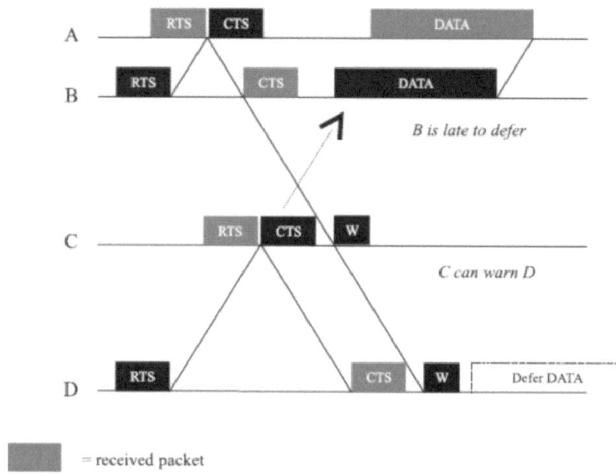

Fig. 7. Reception in DACAP

DACAP is a collision avoidance protocol with an easy scalable adaptation to big networks involving more nodes and a greater area. The protocol is aware of power consumption by avoiding collisions at the same time that maximizes the throughput. It minimizes the handshake time using the tolerance to interference of the receiver node, when this one is close to the limit of the range reception range. It works with a half-duplex communication link, the nodes do not need to be synchronized and it supports mobile nodes.

The throughput with this protocol is several times higher than the one achieved with Slotted FAMA, while offering similar protection to collisions, i.e. savings in energy. Although CS-ALOHA offers higher throughput in most cases, it wastes too much power on collisions.

4.3 T-Lohi

Tone-Lohi is a contention-based Mac protocol that uses short packet as wake up tone to reserve the medium. It is a full distributed reservation process and one of its main features is the power consumption, the nodes will be in an idle mode with low energy requirement until it receives the wake up tone.

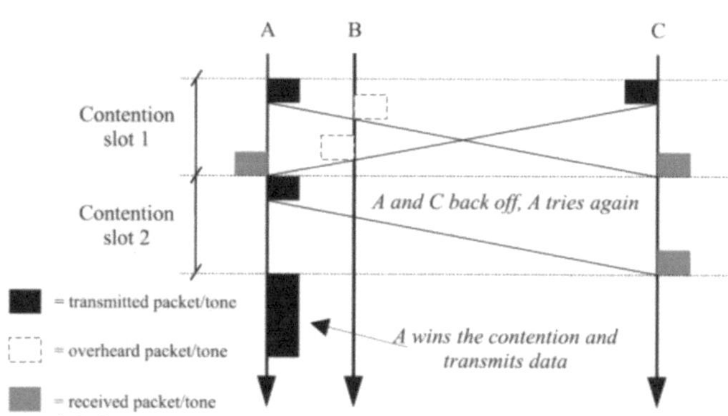

Fig. 8. T-Lohi Protocol Scheme

The main goals of T-Lohi are to make an efficient use of channel utilization, achieve a stable throughput, and save as much energy as possible without having an impact in the performance. This energy conservation is approached in two ways: (1) the reservation to prevent data packet collisions or at least reduce them, (2) and the usage of wake-up tones for the receivers to keep them in low power while in listening mode.

5. Network Layer

This layer is mainly responsible of routing packets to the proper destinations. So, a routing protocol is required when a packet must go through several hops to reach its destination. It is responsible for finding a route for the packet and making sure it is forwarded through the appropriate path. The way paths are selected for every source destination pair will have a direct impact on the overall network performance.

Most of the routing proposals for UWSN are based on the ones developed for terrestrial ad-hoc and wireless sensor networks. Some of the protocols designed exclusively for underwater wireless networks are:

5.1 DBR

Depth-Based Routing (Xie et al, 2008). It can handle network dynamics efficiently without the assistance of a localization service, it needs only local depth information. It is a greedy algorithm that tries to deliver a packet from a source node to sinks.

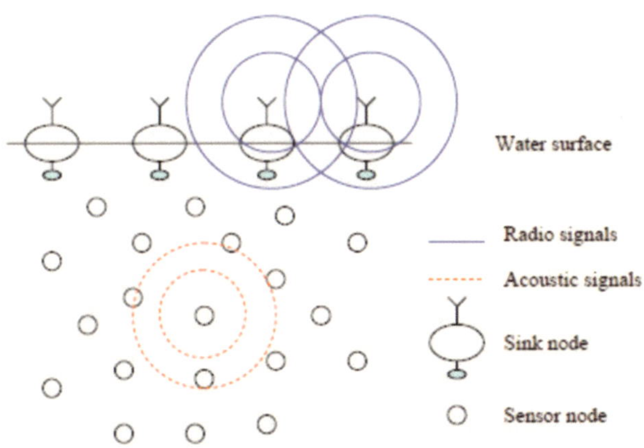

Fig. 9. Multiple sink underwater sensor network architecture

The source nodes send the data packets seeking for the sinks, in this process as the packet hops from one node to another the depth decreases as it gets closer to the final sink receiver. Finally the packet can achieve to reach to the surface. The decision that is taken in each one of the nodes during the transmission is based on its own depth and the depth of the previous sender.

When a node receives a packet it extracts the information of the depth of the previous node and compares against its own depth. After comparing the node will have two behaviours:

(1) The node is closer to the surface $d_c < d_p$ so it will forward the packet. (2) If the current node depth is greater $d_c > d_p$, it will discard the packet as it comes from a node with a better position.

Probably specially at the beginning of the transmission in the first hops, a lot of receivers of the packet will decide to forward it. To avoid the collisions that these retransmissions would bring and the high power requirements needed in the network, the number of forwarded messages must be controlled.

Also it can happen that as it is been using a multiple omnidirectional path algorithm to route the packets, that a node receive several times the same packet and in the same way forward it the same number of times. In order to save energy a node will know which packets have already sent so as not to send a packet more than one time.

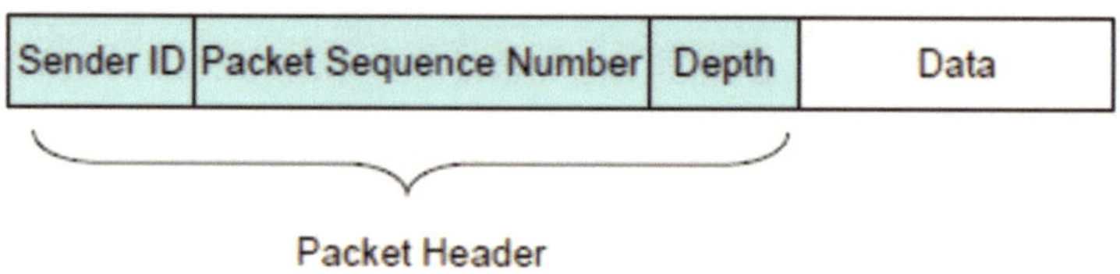

Fig. 10. DBR Packet Format

The Packet Format will be divided in:

- Sender ID: is the identifier of the source node.
- Packet Sequence Number: is a unique sequence number assigned by the source node to the packet. Together with Sender ID, Packet Sequence Number is used to differentiate packets in later data forwarding.
- Depth: is the depth information of the recent forwarder, which is updated hop-by-hop when the packet is forwarded.

As mention before there is need to reduce power consumption forwarding only the necessary packets. To achieve this protocol uses the Redundant Packet Suppression, which consist of two features for avoiding redundant packet. One is that multiple paths are naturally used to forward packets. The other is that a node may send a packet many times. Although multiple paths in DBR cannot be completely eliminated, a priority queue is created to reduce the number of forwarding nodes, and thus control the number of forwarding paths. To solve the second problem, a packet sent buffer is used in DBR to ensures that a node forwards the same packet only once in a certain time interval.

5.2 VBF

Vector-Based Forwarding protocol (Xie et al, 2006) is an algorithm that allows the nodes to weigh the benefit to forward packets and reduce energy consumption by discarding the low benefit packets. One of the main factors in underwater wireless networks is to safe power so as not to let nodes run out of batteries, not being able to recharge them during long period of times. This protocol tries to focus its features in this direction. To aim this target each packet will include the information of the location of the sender the final receiver and the one of next hop of the packet.

To be able to run this protocol it is assumed that every node have the capacity of measuring the distance and angle of arrival (AOA) of the signal. The route of the packet is compute in the sender and included in the packet. When a node receives a packet it calculates its relative position towards the target. This works recursively in all the nodes during the transmission. If the node knows that is close enough to the routing vector (it will be under the threshold value established for this purpose) it will include its position and forward the packet, in other case it drops the packet. In this way, all the packet forwarders in the sensor network form a "routing pipe": the sensor nodes in this pipe are eligible for packet forwarding, and those which are out do not forward.

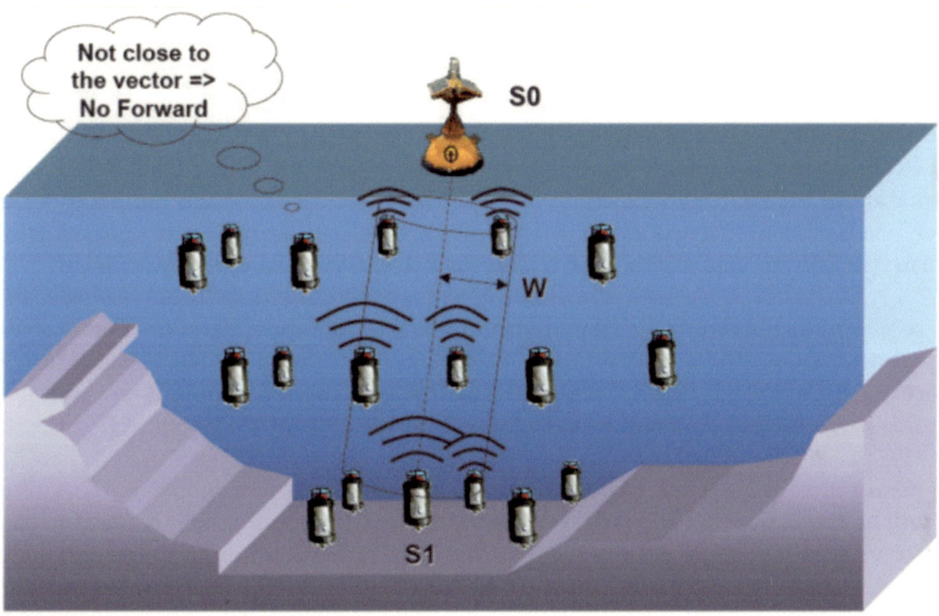

Fig. 11. High Level view of VBR for UWSN.

The figure represents the nodes that are within the routing pipe that forward the packet, "w" the threshold used to measure the width of the pipe. And the nodes that are out of this path discard the packets. This protocol is scalable to the size of the network. This kind of forwarding path (specified by the routing vector) involving nodes for packet routing has as result the energy of the network.

5.3 FBR

Focus Beam Routing, a routing protocol (Jornet & Stojanovic 2008) based on location is presented as a way to find the path between two nodes in a random deployed network. The figure 8 shows a simple two-dimensional network to explain the protocol, although it works in the same way in three-dimensional scenarios.

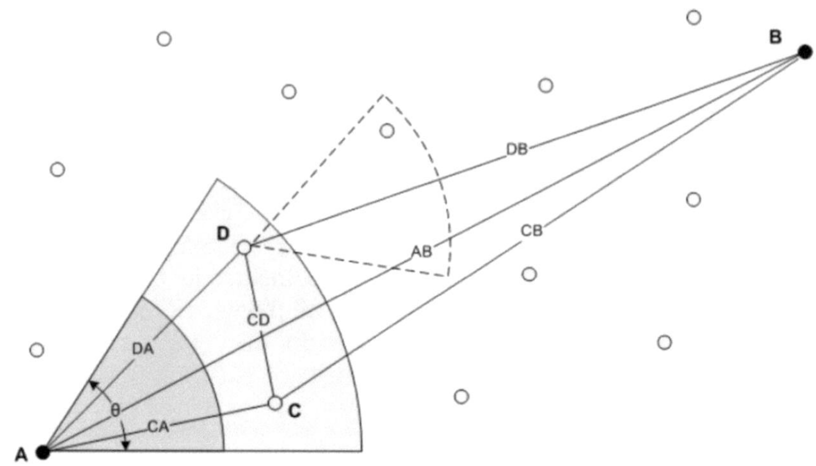

Fig. 12. Nodes within the transmitter's cone θ are candidate relays.

Assuming a communication between node A and node B, node A will send and Request to Send (RTS) multicast message to all the reachable neighbours. This packet will include the information of the source (node A) and final receiver (node B). As the protocol works with power levels, the first try is done at the lowest level and it increases if there is a need because it receives no answers within a wait time established for each power level.

This request is a short control packet that contains the location of the source node (A) and of the final destination (B). Note that this is in fact a multicast request. The initial transaction is performed at the lowest power level and the power is increased only if necessary. Power control is performed as an integral part of routing and medium access control.

Each power level will have a radius and though it will reach to a certain number of nodes. This will be the nodes that receive the RTS and its information that will be used to calculate the relative position to the AB line. This is done to know if the node is a candidate to be a relay node. Candidate nodes are inside a cone of $\pm\theta/2$ from the line AB. Every candidate node will answer to with a Clear to Send (CTS) to the transmitter; the nodes out of the candidate zone will stay in silence.

After sending the RTS the transmitter will wait to receive CTS messages from other nodes, three possible things can occur: (1) The transmitter receive no answers, the RTS has not reach any neighbour, therefore the transmitter increases the power level an tries again as it is shown in the example. (2) The transmitter receives one CTS, the sender of this message is selected as a relay for the next hop, sending him the DATA message. (3) The transmitter receives more than one CTS message, looking at the location information of the candidates included in CTS message the node that is closer to the final destination is selected as relay receiving the DATA message. After sending data the transmitter will wait for an acknowledgement message. This process will continue until we reach the final destination.

Packet collisions can happen but always will involve short packet as the link is safe for data packets which have no risk of collisions. Although the chances of collision are small, if the source node detects a collision, it will detect signal but it will not decode the information of the data, it will resend the RTS once again, without increasing the power level.

6. Applications

As established in the introduction, underwater acoustic wireless sensor networks can be used in a wide range of different applications, as it is done by radio frequency air networks. Ones of the main places we can use UWSN are:

- Environmental Monitoring. Pollution is nowadays one of the greatest problems, oil spills from ships or broken tubes can make a lot of harm to the marine biological activity, the industry and tourist places. Monitoring ecosystems can help understanding and predicting the human and climate or weather effect in underwater environment.
- Prevention of natural disaster. By measuring the seismic activity from different remote location the sensors could alert to the coast places by detecting tsunami or submarine earthquakes alarms.
- Underwater Navigation. The sensor can be placed to make routing, identifying hazards on the seafloor, rocks or shoals in shallow water,
- Assisted Navigation. Sensors can be used to identify hazards on the seabed, locate dangerous rocks or shoals in shallow waters, mooring positions and drawing the bathymetry profile of the area.
- Underwater Discovery. Underwater wireless sensor networks can be used to find oilfields or reservoirs, locate routes for placing connections for intercontinental submarine cables. Also they could seek for shipwrecks or archaeology or lost sink cities.
- Underwater Autonomous Vehicles (UAVs). Distributed sensor in movement can help monitoring area for surveillance, recognition and intrusion detection.

7. Future Trends

A lot of advantages can be achieved by using underwater sensor networks, but a lot of research must be done in the next years. The developing of this technology will have a great impact in the industry.

It is necessary to improve the physical layer performance in terms of efficiency, building low power acoustic modem that are able to make a make best use of the bandwidth, reducing the error rate with forward error correcting coders.

Although there many proposal in the MAC layer, it seems that the collision avoidance protocols in the top for been chosen for underwater networks. In any case this decision can vary according to the application or topology type. These protocols should be also aware of power consumption, making one of their main objectives.

Currently there is a lot of works related to MAC layer proposals since this is one of the more sensible parts of the UWSN architecture. It seems that distributed CDMA-based schemes are the candidates for underwater environments, but it depends of many factors such as the application and network topology. Also, MAC protocols should be designed taking energy consumption into account as a main design parameter.

According to the routing layer protocols, there is a need to be able to adapt to the changing conditions, to include mobility patterns and also be capable of saving energy. Most of the routing protocols need to know the location of all the nodes, geographically-based algorithms may be appropriate for underwater networks. They have to include methods to avoid errors, deal with shadow zones or disconnections or failures and mobility. Cross layer communication between the layers should be required to share the information and adjust the parameters depending on the environment condition variation.

8. Conclusions

Underwater Acoustic Wireless Sensor Networks is still growing and following the path of Radio Frequency in Terrestrial Networks, although having a very different environment with a lot of challenges to achieve in changing conditions. There are many potential research fields in which it can be applied, and needs to solve some open issues so as to be able to provide a reliable and efficient way to communicate in the network.

The development of new modem and the incorporation of companies and research to the UWSN technology will come with new commercial products and solutions to take advantage of the possibilities that underwater bring us, building and industry around submarine technologies.

9. References

Akyildiz, F., Su, W., Sankarasubramaniam, Y., and Cayirci, E. (2002). Wireless sensor networks: A survey. *Computer Networks*, 38(4):393–422.

Akyildiz, F., Pompili, D., and Melodia, T. (2006). State of the Art in Protocol Research for Underwater Acoustic Sensor Network. *Proceedings of the ACM International Workshop on UnderWater Networks (WUWNet)*, Los Angeles CA, September 25th, pp. 7-16.

Berkhovskikh. L., Laysanov, Y. (1982). Fundamentals of Ocean Acoustics, N.Y: Springer

Bharghavan, V., Demers, A., Shenker, S., Zhang, L. (1994). MACAW: A Media Access Protocol for Wireless LAN's. *Proceedings of the ACM SIGCOMM Conference*, London, UK, August 31- September 2, pp. 212-225.

Chirdchoo, N., Soh, W., Chua, K. C. (2007). *The IEEE Conference on Computer Communications (infocom)* , Anchorage, Alaska, USA, May

Coates, R. (1989), Underwater Acoustic Systems, Wiley, New York.

Colladon, J. D. (1893). Souvenirs et Memoires. Geneva, CH: Albert-Schuchardt.

Cui, J-H., Kong, J., Gerla, M. and Zhou, S. (2006). Challenges: Building scalable mobile underwater wireless sensor networks for aquatic applications. *IEEE Network* 3:12-18.

Curtin, T. B., Bellingham, J. G., Catipovic, J., and Webb, D. (1993). Autonomous oceanographic sampling networks. *Oceanography* 6:86–94.

Freitag, L., Stojanovic, M., Kifoyle, D., and Preisig, J. (2004). High-Rate Phase-Coherent Acoustic communication: A Review of a Decade of Research and a Perspective on Future Challenges. *Proceedings of the 7th European Conf. on Underwater Acoustics*, Delft, HL, 5-8 July.

Fullmer, C.L., García-Luna-Acebes, J.J. (1995). Floor acquisition multiple access (FAMA) for packet-radio networks. *Computer Communication Review* 25(4):262-273.

Jornet, J.M., Stojanovic, M. (2008). Focused beam routing protocol for underwater acoustic networks. *Proceedings of the third ACM international workshop on Underwater Networks*, San Francisco, California, USA, September 15.

Karn, P. (1990). MACA - A new channel access method for packet radio. *Proceedings of the ARRL 9th Computer Networking Conference*, London, Ontario (CA), September 22, pp. 134-140.

Molins, M., Stojanovic, M. (2006). Slotted FAMA: A MAC Protocol for Underwater Acoustic Networks. *Proceedings of the IEEE OCEANS*, Singapore, 16-19 May, pp. 1-7.

Llor, J., Malumbres M.P. (2009). Analyzing the Behaviour of Acoustic Link Models in Underwater Wireless Sensor Networks. *PM2HW2N*

Partan, J., Kurose, J., Neil Levine, B. (2006). A Survey of Practical Issues in Underwater Networks. Proceedings of the ACM International Workshop on UnderWater Networks (WUWNet), Los Angeles CA, September 25th, pp. 17-24.

Peleato, B., Stojanovic, M. (2007). Distance Aware Collision Avoidance Protocol for Ad Hoc Underwater Acoustic Sensor Networks, *IEEE Comm. Letters*

Porter, M., Liu, Yong-Chun Bellhop (2010). [Online]. Code and manual available at: http://oalib.hlsresearch.com/Rays/index.html

Rosenbush, S., Crockett, R. O., Yang, C. (2004). Sin cables, sin normas y a bajo precio. Dinero: Inteligencia empresarial 932:70-73.

Sozer, Ethem M., Stojanovic, Milica and Proakis, John g. (1999), Design and Simulation of an Underwater Acoustic Local Area Network, Northeastern University, *Communications and Digital Signal Processing Center*, 409 Dana Research Building, Boston, Massachusetts.

Urick, Robert J. (1983). Principles of Underwater Sound. NewYork, USA: McGraw-Hill (3rd edition).

van der Werff, T.J. (2003). 10 Emerging Technologies That Will Change the World. Technology Review (MIT). [Online]. http://www.technologyreview.com/read_article.aspx?id=13060&ch=infotech

Xie, G, Gibson, J, Diaz-Gonzalez, L, (2006). Incorporating Realistic Acoustic Propagation Models in Simulation of Underwater Acoustic Networks: A Statistical Approach, *Proceedings of MTS/IEEE Oceans Conference, Boston, September 2006.*

Xie, P., Cui, J.H., Lao L. (2006). VBF: Vector-Based Forwarding Protocol for Underwater Sensor Networks. *Proceedings of IFIP Networking*, May 2006.

Yan, H., Shi, Z., Cui, J.H., (2008). DBR: Depth-Based Routing for Underwater Sensor Networks. *Proceedings of Networking'08*, Singapore. , May 5-9

12

Prospects and Problems of Optical Diffuse Wireless Communication for Underwater Wireless Sensor Networks (UWSNs)

Davide Anguita, Davide Brizzolara, Giancarlo Parodi
University of Genoa
Italy

1. Introduction

The aim of this paper is to highlight the prospects and problems of optical wireless communication for applications in the field of Underwater Wireless Sensor Networks (UWSNs). The necessity of wireless underwater connections, especially in UWSN, has dramatically increased in the last few years for a wide range of applications, from environmental monitoring to surveillance. The problem of wireless underwater communication is a challenging field of investigation since common terrestrial devices, equipped with radio wireless links, cannot be employed underwater and current underwater available technologies, mainly based on acoustical solutions, still pose problems due to low propagation speed of sound in water and low data-rate.

In the last years, the interest towards optical wireless communication has increased both for terrestrial (Akella et al., 2005) than for underwater applications (Doniec et al., 2010) since it allows to overcome some of the problems of acoustical communication (Akyildiz et al., 2005). In fact low data rate and low propagation speed require specific solutions and make it difficult the integration of underwater technologies with current available terrestrial modules. Transferring the paradigm of Smart Dust (V. Hsu & Pister, 1998) to UWSNs, largely unaffected by WSN revolution, is a challenge that can open great opportunities since current underwater solutions are still of large dimension, expensive and power-consuming.

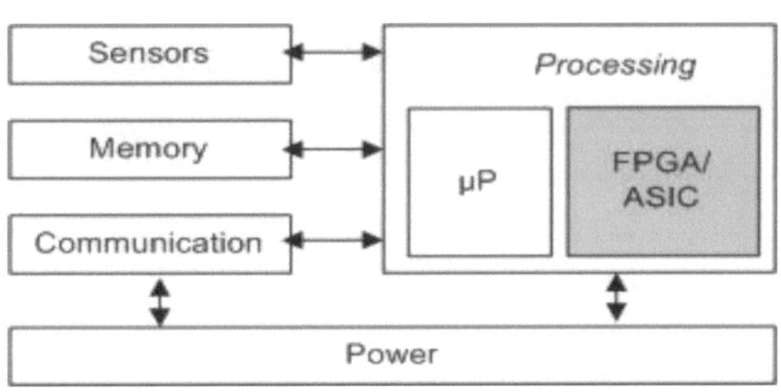

Fig. 1. Wireless Node Structure equipped with FPGA

This paper highlight the relatively less explored and innovative possibility of optical underwater communication as an interesting and feasible alternative solution for underwater networking targeting small, low-cost and low-power devices. In particular it focuses both the problems related to specific characteristics of the underwater channel targeting the interface with current terrestrial technologies available for Wireless Sensor Network (WSN) such as the one developed at DIBE WiseLab (Akyildiz et al., 2005) depicted in Fig.1 .
The goal of our work is to build a prototype of an UWSN based on optical communication among nodes. The optical PHY and MAC Layers have been developed considering the characteristics of IEEE 802.11 Infrared (IR) PHY layer and the compatibility with the current IEEE 802.15.4 protocol for terrestrial Wireless Sensor Networks (WSNs). This can be considered as a first step which, of course, should be followed by modifications and adaptation of the upper layers.
The network architecture can be organized as a three or bi-dimensional underwater network in which each node floats at different depths trying to maintain a fixed position with a good approximation, for instance by an anchorage to the sea bed (as detailed in Chapter 11). The nodes should be designed of slight dimensions (between 15 and 20 cm), they should be densely deployed, maintaining a distance between 10 and 30 meters from each other. As detailed in the next paragraph, each node should be able to communicate, by using only a wireless optical communication link, to the neighbors and, by using a multi hop path, to a base station placed on the water surface.
The interesting aspects which are reported in this work are related to:

- the use of optical communication (the adaptation of current available protocols, the choice of modulation, the adaptation to water channel variability, etc ...);
- the interface with current technology for WSN;
- the design and implementation of circuits for diffuse generation and reception of light impulses targeting low-cost and low-power components;

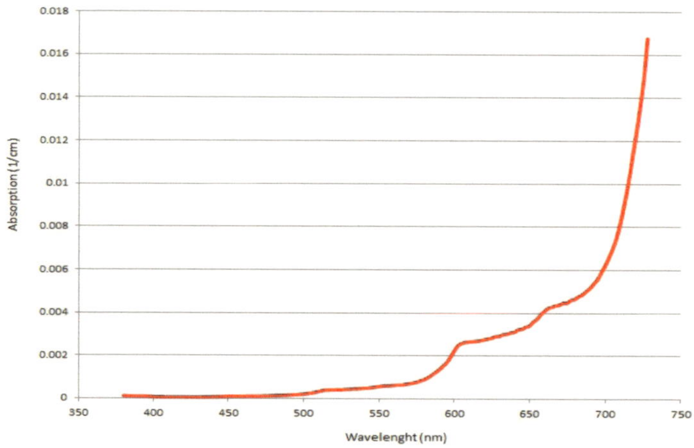

Fig. 2. Underwater Attenuation of Light from experimental data

	Acoustic	Electromagnetic	Optical
Nominal speed (m/s)	$\approx 1,500$	$\approx light - speed$	$\approx light - speed$
Power Loss	$0.1dB/m/Hz$	$\approx 28dB/1km/100MHz$	$\sim turbidity$
Bandwidth	$\approx kHz$	$\approx MHz$	$10 - 150MHz$
Frequency band	$\approx KHz$	$\approx MHz$	$10^{14} - 10^{15}Hz$
Antenna size	$\approx 0.1m$	$\approx 0.5m$	$\approx 0.1m$
Effective range	$\approx km$	$\approx 10m$	$\approx 10 - 100m$

Table 1. Comparison of Different Technologies for Underwater Wireless Communication

2. Underwater Wireless Communication: physical aspects

Due to the impossibility of using Radio Frequencies (RF), traditionally wireless underwater communication employs acoustic waves because sound propagates well in water and its range can be very long (~km). However, it has several disadvantages such as narrow bandwidth and latency in communication due to the slow speed of acoustic wave in water. For instance, at ranges of less than 100 m the data transmission rates of these systems in shallow littoral waters are ~10 kb/s.

Experimental tests have shown that an alternative feasible solution is optical communication especially in blue/green light wavelengths, even if limited to short distances (up to 100 m)(Lanbo et al., 2008). Compared to acoustic communication it offers a practical choice for high-bandwidth communication and it propagates faster in the water (2.255 x 10^8).

Nevertheless it is affected by different factors to take into account for an efficient design. The attenuation of a light beam between two points can be described as in 1 where d_1 and d_2 are the positions of the points.

$$A = e^{-k(d_1 - d_2)} \left(\frac{d_1}{d_2}\right)^2 \tag{1}$$

In the first term, k is defined as $k = a(\lambda) + b(\lambda)$ and it is dependent by the wavelength: a is the term related to the absorption of water while b models the scattering which depends both on light wavelength and turbidity. The second term, instead, models the quadratic attenuation. A comparison between different wireless underwater technologies for underwater communication is illustrated in Table 1.

3. Wireless Underwater Communication Systems: analysis of the state of art

In this chapter will be briefly investigated the state of art related to underwater wireless communication.

Currently the use of wireless communication is very common in a wide range of terrestrial devices. In particular, one of the most innovative application is related to Wireless Sensor Networks (WSNs), as well detailed in (Akyldiz et al., 2001) (Puccinelli & Haenggi, 2005), where a large number of small nodes communicate and work by using an optical wireless link.

3.1 Underwater Acoustical Communication

As discussed in the introduction, although there are many recently developed solutions for WSNs, the unique characteristics of the underwater acoustic communication channel, such as limited bandwidth capacity and variable delays, require very efficient and reliable new data communication protocols and UWSNs nodes are still an open field of research (Akyldiz et al., 2001) (Akyildiz et al., 2005) (Lanbo et al., 2008). The main limitations due to acoustic communication, which will be described can be summarized as follows:

- The available bandwidth is severely limited;
- The underwater channel is severely impaired, especially due to multi-path and fading;
- Propagation delay in underwater is five orders of magnitude higher than in radio frequency (RF) terrestrial channels, and extremely variable;
- High bit error rates and temporary losses of connectivity (shadow zones) can be experienced, due to the extreme characteristics of the underwater channel.

Considering low-power and low-cost devices and targeting short-medium distances, the most important available products are described in (Yan et al., 2007) (Singh et al., 2006) and they can reach data-rate in the order of Kb/s.

3.2 Underwater Electromagnetic Communication

As regards Electromagnetic Communication, extremely low frequency radio signals have been used in military applications: Germans pioneered electromagnetic communication in radio frequency for submarines during World War II, where the antenna was capable of outputting up to 1 to 2 Mega-Watt (MW) of power. An extremely low frequency (ELF) signal, typically around 80 Hz at much lower power, has been used to communicate with naval submarines globally today. This is possible mainly because most of the transmission paths are through the atmosphere (Shelley, 2005).
In (Al-Shamma'a et al., 2004) a theoretical analysis and experiments show that radio waves within a frequency range 1 to 20MHz is able to propagate over distances up to 100 m by using dipole radiation with transmission powers in the order of 100 W. This will yield high data rates beyond 1 Mbps which allows video images to be propagated at standard camera frame rates (25Hz) (Lucas et al., 2004). The antenna design in such case is very different from that of the antennas used for conventional service in the atmosphere: in fact, instead of having direct contact with seawater, the metal transmitting and receiving aerials are surrounded by waterproof electrically insulating materials. This way, an EM signal can be launched from the transmitter into a body of seawater and picked up by a distant receiver.
Recently, in September 2006, the first commercial underwater radio-frequency (RF) modem in the world, model S1510, was released by Wireless Fibre Systems (Fibre, 2008). Its data rate is 100 bps, and communication range is about several tens of meters. In January 2007, a broadband underwater RF modem, model S5510, came into birth. It supports 1-10 Mbps within 1 meter range (Fibre, 2008). Due to the propagation property of EM waves, EMCOMM is an appealing choice only for very short range applications. One example is the communication between autonomous underwater vehicles (AUVs) and base stations, where the AUVs can move within the communication range of a base station to offload data and receive further instructions (Shelley, 2005).

3.3 Optical Underwater Wireless Communication

Underwater optical communication has been investigated both through theoretical studies (Jaruwatanadilok, 2008) (Lanbo et al., 2008) (Kedar, 2007) and in experimental tests (Feng Lu, 2009)(Hanson & Radic, 2008), mainly developed in USA, Canada and Australia.
Currently, there are not many research activities on underwater optical communication, and few commercial optical modems are available specifically for underwater communication. As well detailed in the next paragraphs, recent interests in underwater sensor networks and

sea floor observatories have greatly stimulated the interest in short-range high-rate optical underwater communication.

3.3.1 Point-to-point Communication

In (Tivey et al., 2004) a low power and low cost underwater optical communication system is proposed by using inespensive components. It is based on IrDa protocol and the adaptation to underwater channel is performed by replacing Infrared Communication with light generated by LEDs. Also in (Shill et al., 2004) an underwater communication system has been implemented for a swarm of submersibles. It combines the IrDA physical layer with 3 Watt high power light emitting diodes, emitting light in the green and blue part of the visible spectrum. As in the previous example, the approach is to use the IrDA physical layer modulation replacing the infrared light emitting diodes (LEDs) with high power green or blue LEDs, and also the photodiode with a type which is sensitive in the visible part of the spectrum. The prototype transceiver costs approximately $45 per unit and, on this low hardware level, no link management or higher level error correction is done. The wide angular coverage, the uniform emission footprint and very high light intensity allow for either omnidirectional coverage up to 2 metre radius with only five transmitters, when using simultaneously transmitting expensive LEDs that consume 2W or, with additional lenses, long range directional links having a collimated beam (up to 5 meters).
In (Feng Lu, 2009) and (Lee et al., 2008) a low-cost (in the order of $10)medium-range optical underwater modem is proposed. A sophisticated detection algorithm is exploited (based on spread spectrum) to maximize the communication range. Tests have been performed up to 10 meters and a data-rate of 310bps has been achieved.
In (Vasilescu et al., 2007) and (Vasilescu et al., 2005) the underwater wireless sensor network AquaNodes is described. In this application the use of optical communication is shown as an efficient solution for data muling in the network proposed by CSIRO ICT Centre (Australia) and MIT CSIAL (USA). This research proposes a network where data exchange is performed by connecting, both optically and acoustically, previously deployed static nodes and mobile (AUVs) vehicles, so that the characteristics of the two communication strategies can be exploited. Each device is provided with an optical modem, which can perform a transmission up to 300 kb/s in a range below 8 meters. This approach clearly shows that the use of an optical communication system allows for a considerable reduction in terms of energy consumption with respect to a multi-hop acoustic transmission and a resulting increase of operational life of the system. Moreover, considering this work, the use of optical modems instead of acoustic ones for short range communications leads to a remarkable reduction of costs, since the cost of an optical modem is of the order of $50 node against approximately $3000 node of an acoustic modem.
In (Ito et al., 2008a) is illustrated a careful analysis by modeling the underwater channel based on underwater optics. Through this analysis, it is showed that a single color LED is very weak in a wavelength dependent underwater channel and, to overcome this problem, a multi-wavelength adaptive scheme combined with rate adaptive transmission is proposed taking inspiration from already developed algorithms. The proposed system can adapt to the channel by considering the change in power for each wavelength band, and controlling the data rate.
In (Hanson & Radic, 2008) the work is motivated by the need to demonstrate error-free underwater communication at qualitatively higher data rates than previously reported with either optical or acoustic methods. In this case a Laser source is used and an error-free underwater

optical transmission measurements at 1 Gbit/s over a 2 m path in a laboratory water pipe.
In (Chancey, 2007) a short range underwater optical communication link is established by using an open source free space modem proposed by the project RONJA. Unfortunately, the used equipment is very expensive and difficult to use in real applications; as reported, the experimental results achieved 10 Mb/s at a distance up to 5 meters.
A commercial product targeting optical underwater communication, from Ambalux (Ambalux, 2008), includes high-bandwidth transceivers, which allow a point-to-point transmission at a data rate of 10Mbs up to 40m, depending on environmental conditions. A set of tests based on this commercial products has been described in (Baiden & Bissiri, 2007), where a solution for omni-directional communication is also proposed in order to improve the efficacy of the connection with an underwater untethered vehicle. This approach, however, is not suitable for a dense UWSN or small and low power devices, where nodes are deployed with low accuracy and concerns on system size and energy consumption are addressed.
Recently an underwater optical wireless modem (AquaOptical)(Doniec et al., 2009) has been implemented. It is designed for integration in robotic application and allows to support a wireless communication of 2 Mb/s up to a range of 30m.

3.4 Diffusive underwater optical communication

In these paragraph are reported some researches related to the use of diffusive underwater optical communication for different applications.
In (Baiden et al., 2009) the point-to-point optical communication system developed by Ambalux (Ambalux, 2008) is used to perform a wide number of tests with the aim of paving the way for a future underwater omni-directional wireless optical communication systems. The LEDs used in the test emitted light in the green and blue light spectrum and were tested in a pool and in a tank filled with lake water. The primary objective of these tests was to get profiles of the behaviors of such communication systems with respect to water characteristics such as turbidity levels, prior to building an omni-directional optical communication. The results of the tests indicated that turbidity level, viewing angle and separation distance plays a significant role in the behavior of blue light in water. In particular, the aim was to define a threshold viewing angle (TVA), the minimum viewing angle at which communication is lost when one of the communication devices is rotated with respect to a virtual axis that contains the segment represented by the center of gravities of the two devices when they are aligned. On the basis of the previous studies, two geometric forms were modelled: icosahedron and spherical hexagon. The icosahedron was retained because of its simplicity in geometry and its ability to provide complete free space coverage using the selected LED.
In (Baiden & Bissiri, 2007), the previous illustrated configuration has been tested to implement a high bandwidth optical networking for underwater untethered tele-robotic operation. A rate of 4 Mb/s was achieved in hemispherical configuration up to 5 meters. Experimentation in the field achieved initially 115 Kb/s over a distance of 15 metres. A second experiment, after the modification of the transmitter and the receiver software, was attempted: in this case the transmission was increased to 1.5 Mb/s and the first wireless underwater video pictures were transmitted. Despite the originally anticipated different kinds of turbidity (fish, plankton, rocks or other objects) in the water, the experiment still received video feedback of the anchor of a floating laboratory at around 15 metres deep in the bottom of the lake.
In (Fair et al., 2006) an optical modem technology for seafloor observatories is illustrated and problems related to design and implementation of a system for underwater optical wireless communication are focused. The idea is to implement an optical modem system which should

provide sufficient bandwidth to allow transmission of compressed high resolution video (i.e., 2-10 Mbit/s for studio quality video), allowing nearly unrestricted motion on the part of the mobile sensor (UUV), and working at ranges below 100 m. In this case, the design of the system have been developed considering the following aspects: (1) selection of a transmission light source (wavelength, power, beamwidth), (2) selection of a detector (field of view, quantum efficiency, gain), and (3) selection of an aiming and tracking strategy. As regards transmission light source, LEDs are suggested to be used for the transmitters as they are switchable at high rates, can be reasonably collimated, and are easily arrayed to increase transmit power. High intensity blue LEDs such as those fabricated from Gallium Indium Nitride (InGaN) on a silicon carbide (SiC) substrate can provide on the order of 10 mW each and they can be configured in arrays to increase the total radiant flux (optical power). The light from one or more LEDs can be collimated with a lens to focus its beam. While their bandwidth is low by optical standards (30 ns rise time), a 100 ns bit duration meets the design goal of 10 Mbit/s. LED transmitters provide significant flexibility as well: both OOK and PPM are possible with the same hardware. As regards the receiver, in this work a photomultiplier tube (PMT) is chosen for the detector because it provides higher sensitivity and less noise than photodiodes (including avalanche or PIN types) although the tube size can be a limiting factor in certain conditions. Considering the aiming and tracking strategies, three possible configurations for optical communications between a fixed node and a free-swimming vehicle are suggested in this paper:

1. Pointed transmission (Tx) and reception (Rx) with acoustic or optical aiming is the most efficient scenario from an optical transmission standpoint. The transmitter consists of several collimated LEDs or a laser diode, and the receiver is a PMT. This method requires a search and acquisition mode by both the Rx and Tx which consumes time and energy (regardless of whether the aiming is acoustic or optical);
2. Directional-Tx to omni-directional-Rx scenario can be accomplished with a hemispherical PMT detector and either a laser diode or a hemispherical array of LEDs. In this case, the aiming problem is one-sided and could be accomplished via acoustics or with optics;
3. Omni-directional Tx and Rx which is the mechanically simplest solution. This can be accomplished using the large area PMT detector with a moderate output-power blue laser diode or LED and diffusing optics. The diffusing optics can be done either with discrete reflective and refractive elements or with a high transmission scattering medium.

Taking into account the previous proposed theoretical studies, some optical systems have been implemented and tested, as reported in (Farr et al., 2005). The omni-directional light source used was blue-green (470nm) to take advantage of the low attenuation in seawater at those wavelengths, and produced a uniform light field over a 2π steradian hemisphere. Six commercially available 470nm (blue) light emitting diodes (LED) were arranged in a hemispherical geometry and encapsulated in a weekly scattering potting material to provide some diffusion of the light field over the full hemisphere of operation. The receiver consisted of a large-aperture, hemispherical photomultiplier tube (PMT) chosen for its high sensitivity, low noise and high speed. Apertures and mirrors have been used to obtain a 91 m path length in a 15 m deep pool and to prevent reflections from the surrounding walls from contaminating the results. A 100 meter bench test has been performed to verify the range and geometry calculation using neutral density filters to simulate the attenuation of water.
In the same work (Farr et al., 2005) different tests are performed to validate the concept of

omni-directional free water optical communication in the range of 10 meters. In particular, the power spectrum of the received signal during the transmission of a pulse train at different repetition rates is reported: 1.25 MHz, 5 MHz and 10 MHz.
This work focuses in particular the communication between Underwater Mobile Vehicles and a fixed node targeting application in sea oor observatories. The achieved distances and data-rate are interesting but it could be difficult to apply these results to low-cost and small devices where transmitters and receivers should be placed on the same small surface.
In (Liu & Ge, 2006) the concept of underwater laser sensor network is illustrated as a new approach for broadband communication in the underwater environment based on the blue-green laser has been proposed. In this paper the applications of the underwater sensor network in the undersea exploration are discussed with the difficulties in the traditional the underwater acoustic sensor network. A basic of prototype of underwater laser sensor network is described: it includes the architecture of laser sensor node and protocol stack for underwater laser sensor network, but it does not shown implementation or interesting practical results.
The here described devices show that the use of optical communication can be a feasible solutions for underwater wireless communication. In the comparison with the here proposed works, the aim of this paper is to illustrate the problems and possible solutions for the application of optical wireless communication in order to support a UWSN of dense deployed nodes as detailed in Chapter 1 and 11.

4. Application of underwater optical wireless communication

Optical underwater communication is an effective alternative to current underwater technology especially in some particular environments such as, for instance, shallow, coastal and fresh inland waters where the use of this approach is useful to overcome all the shortcomings related to the use of acoustic communication and to allow a wide adoption of underwater monitoring systems. In particular the possibility of transferring high amount of data in a limited amount of time reducing power consumption can support the transmission of short video and pictures for a reliable monitoring and surveillance. Small dimensions and low-cost components allow to establish a dense deployed networks performing an effective fine grained sampling in the area of interest. It could be possible, for instance, to perform pollution monitoring and frequent data collection (water temperature, specific conductivity, pH, turbidity, and possibly oxygen concentration) and, by using a high-data rate optical link, periodically deliver data reducing the time devoted to transmission and network congestion.

5. Design consideration for a diffuse optical underwater communication

The design of underwater optical systems for underwater networking should takes into account different aspects which are illustrated in the next paragraphs:

- the Physical(PHY) Layer to manage the trasmission and reception of data by using devoted circuits;
- the Medium Access Control(MAC) Layer to manage the possibility of more than one devices to communicate each others;
- the aspects related to the circuits design and implementation (such as choice of the Led and photodiodes and their collocation on a surface).

In the next paragraph some design consideration for each of the previous aspect will be presented and our design and implementation results will be briefly illustrated. The here described system is an based on a previous implementation illustrated in (Anguita et al., 2010) (Anguita et al., 2008) (Anguita et al., 2009) and it implements PHY and MAC Layer modules which have been described by using an Hardware Description Language (HDL), in particular VHDL (Very High Speed Hardware Description Language), as suggested in (Pang et al., 2007) (Dasalukunte & Owall, 2008) and well detailed in Chapter 6.3.

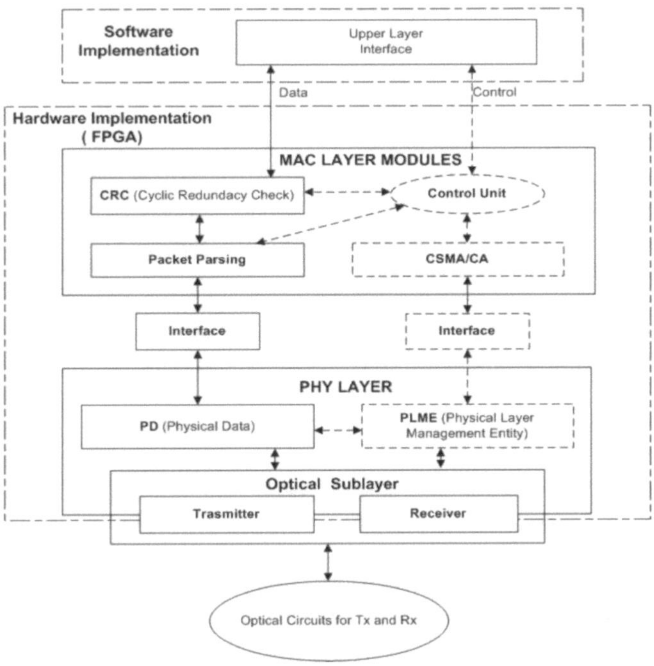

Fig. 3. General block diagram of the Optical Transmission System

6. The Physical Layer

The Physical Layer has to support:

1. the transmission and reception of data by using optical communication;
2. the interface with a MAC Layer of the communication stack by providind the services required by the upper layers.

In particular, the design should take into account the peculiarity of underwater channel and focus on the following elements:

1. efficient wireless optical modulation techniques;
2. adaptation to underwater channel variability;
3. adaptation to data-rate communication requirements;
4. interface with current standard used for WSN (i.e. IEEE 802.15.4).

Modulation	Implementation Complexity	Sensitivity to multi-path delay	Required transmitted power	Bandwidth
OOK	simple	general	P_{R-OOK}	$2R_B$
FSK	most complex	most anti-sensitive	$\frac{1}{2}P_{R-OOK}$	$\|f_1 - f_0\| + 2R_B$ (2 FSK)
DPSK	more complex	more sensitive	$\frac{1}{8\sqrt{\ln 2}}P_{R-OOK}$	$2R_B$ (2 DPSK)
PPM	simple	general	$\frac{2}{P_{R-OOK}}$	$\frac{L}{log_2 L}R_B$ (L-PPM)

Table 2. Comparison of Various Modulations Techniques

6.1 Wireless optical modulation techniques

As well detailed in literature (Ghassemlooy et al., 2007) (Park & Barry, 1995) (Kahn & Barry, 1997) basic modulation techniques mainly contain three formats, such as Amplitude Modulation (ASK), Frequency Modulation (FSK) and Phase Modulation (PSK).
Taking into account previous reported studies (Lee et al., 2008), Pulse-Position Modulation (PPM) can be considered in order to further improve the anti-disturbance capacity of information transmission and it proposed in different systems both for terrestrial (IEEE 802.11 IR) than for underwater systems (Lee et al., 2008).
Pulse Position modulation is a well-known orthogonal modulation technique, in which M message bits are encoded by transmitting a single pulse in one of 2^M possible time-shift. PPM scheme, at each time interval is T_s and $L = 2^M$ time-shifts constitute a PPM frame. At the transmitter end, the signal will be launched by the light pulse to form a specific time slot, and in the receiver end, the photoelectric diode detects the light pulse and then to judge its time slot to evaluate his position and resume the signal. In L-PPM system, a block of input bits is mapped on to one of distinct waveforms, each including one $"on"$ chip and M-1 $"off"$ chips. A pulse is transmitted during the $"on"$ chip, it can be defined as (Meihong, Xinsheng & Zhangguo, 2009):

$$P_m = \begin{cases} i_s & t \in [m-1]T_f/L, mT_f/L \\ 0 & else \end{cases} \tag{2}$$

Considering the characteristics of the different modulation the Table 2 can be proposed for a comparison between different techniques (R_B is the base-band signal bandwidth).
Since the characteristics of PPM appeared to be interesting for a wide range of applications some alternative and modified PPM modulations can be considered such as Differential Pulse Position Modulation (DPPM), Pulse Interval Modulation (PIM) and Dual Header Pulse Interval Modulation (DH-PIM).

6.1.1 BER vs. SNR

Considering the previous described modulations it is interesting to evaluate the Bit Error Rate versus SNR. The following results are mainly based on (Meihong, Xinsheng & Zhangguo, 2009) (Meihong, Xinsheng & Fengli, 2009). For OOK demodulation format a threshold is used to compare the received voltage to decide "1" or "0". In AWGN channel model, the received voltage is:

$$p(t) = \begin{cases} i_s + n_{c(t)}, & "1" \\ n_{c(t)}, & 0" \end{cases} \tag{3}$$

where $n_{c(t)}$ is the Gaussian process. For the data "1", the probability density of $x(t)$ is:

$$p_1(x) = \frac{1}{\sqrt{2\pi}\sigma} exp(\frac{-(x - i_s)^2}{2\sigma^2}) \tag{4}$$

for the data bit "0" the probability density of x(t) is:

$$p_0(x) = \frac{1}{\sqrt{2\pi}\sigma} exp(\frac{-x^2}{2\sigma^2}) \tag{5}$$

By fixing the the judgment threshold to $\frac{1}{2}i$, the bit error rate is defined as:

$$p_e(ook) = \frac{1}{2} erfc \frac{i_s}{2\sqrt{2}\sigma} = \frac{1}{2} erfc \frac{\sqrt{S}}{2\sqrt{2}} \tag{6}$$

where erfc is the complementary error function. According to (Lee & Kahn, 1999), in additive Gaussian noise channel, the bit error rate of 2FSK coherent modulation and 2DPSK coherent modulation are given by the following equations:

$$p_e(FSK) = \frac{1}{2} erfc \sqrt{\frac{S}{4}} \tag{7}$$

$$p_e(DFSK) = erfc \sqrt{\frac{S}{2}} (1 - \frac{1}{2} \frac{S}{2}) \tag{8}$$

For the L-PPM modulation in the Gaussian white noise channel, there are many performance evaluation methods for the BER as suggested in (C.X.Fan et al., 2001) (Ma, 2003) (Malik et al., 1996). The following formula is used in (Meihong, Xinsheng & Fengli, 2009):

$$p_e(L - PPM) = \frac{1}{L} [\frac{1}{2} erfc(\frac{1-k}{2\sqrt{2}\sqrt{LS}}) + \frac{L-1}{2} erfc(\frac{k}{2\sqrt{2}} \sqrt{LS})] \tag{9}$$

in order to show the Bit Error Rate in different SNR conditions. The 8-PPM is reported to be a good choice to reduce the BER.

6.1.2 Power Consumption

Also problems related to power consumption has been investigated in literature (Tivey et al., 2004) (Meihong, Xinsheng & Zhangguo, 2009) since they are crucial in underwater devices. Small and light systems are desirable in order to improve underwater system performance and considerations about power consumption are important in the choice of modulation. The following considerations can be carried out for the different modulations:

1. for the Frequency Shift Key (FSK) modulation a specific frequency carrier wave for digital "1", and a different frequency carrier wave for digital "0" are generated. In this case the optical transmit power required as the transmitting duration is always on and it appears to be inefficient for applications on power-constrained devices.

	OOK	FSK	DPSK	4-PPM	8-PPM
Maximum rate	$\frac{1}{2P}$	$\frac{1}{2P}$	$\frac{1}{2P}$	$\frac{1}{2P}$	$\frac{3}{8P}$
Transmit power	Middle	Higher	Highest	Low	Lowest
Complexity of modulation	Low	Higher	Highest	Lower	Lower

Table 3. Transfer rate versus implementation complexity from (Meihong, Xinsheng & Fengli, 2009)

2. the Phase Shift Key (PSK) modulator generates an in phase signal for digital "1" and an out of phase signal for a digital "0". In this case the PSK demodulator are complex since the different coherent demodulation are needed to compare the current phase with the previous phase and it can be inefficient for power-constrained devices.
3. the On-Off Keying (OOK) and Pulse Position Modulation (PPM) do not use completely the frequency or phase information, but the design of the receiver and the transmitter is simple. The data throughput of PPM modulation is smaller than OOK modulation, but the required receive power is just:

$$\frac{1}{\sqrt{\frac{L}{2} log_2 L}}$$

   of OOK modulation at the same error rate performance. It means that PPM could transmit longer distance than OOK at the same transmitting power condition.

If P represent the smallest pulse width, the comparison of different modulation techniques, taking into account the Maximum rate, the Transmit power and the Complexity of modulation is shown in Table 3.

6.2 Adaptation to underwater channel

As well detailed in theoretical studies (Jaruwatanadilok, 2008) (Giles & Bankman, 2005) (Smart, 2005) optical underwater transmission is deeply influenced by turbidity and the received power is highly dependent by the wavelenght of transmitted signal and the dimensions of the particles and substances dissolved in water.
To overcome this problem the use of different wavelenghts for transmission can improve SNR. By using visible optical communication, this results in having different colours which can be used to generate the transmitted signal. This solution has been successfully explored in (Ito et al., 2008a) where it is shown, by comparing theoretical and experimental results, that the better wavelength lies around 420 nm (blue) and increase in presence of turbidity (Chancey, 2007).
The design of the PHY Layer should take into account the possibility of switching from one colour to another considering underwater conditions; the design can be based on different parameter such as the evaluation of decrease or improvement of SNR or comparison between different transmission lines. By using a system equipped with different wavelength LEDs for transmission it could be possible to evaluate in parallel the performance of each transmission line (Diana & Kahn, 1999). The modules described in the following paragraphs, as for the channels in radio based transmission, allow the management of different transmission lines by using an ad-hoc parameter. Currently the evaluation of an automatic system is under evaluation taking into account the previous reported works.

The transmission system has to consider also the adaptation of the transmission date-rate which can be motivated both by the evaluation of underwater channel conditions and by the information which has to be transmitted. If a PPM modulation is used for the transmission the impulse duration has also a strong influence on the power comsuption since power consumption increase when LEDs for transmission are activated.
An approximate evaluation of power consumption (Energy/bit) can be based on the following formula:

$$E = P \times t_{pulse} \times pulses/bit \tag{10}$$

with P the input power to the LED, t_{pulse} the time for each pulse and taking into account the number of pulses per bits.

6.3 PHY Layer HDL modules implementation

The PHY Layer has been completely described by using an Hardware Description Language (HDL), while a small subset of functions of the MAC Layer for testing purposes have been implemented by a software interface. The description of each module has been carried out by using an Hardware Description Language because:

1. it could easily integrated in WSN nodes 1 developed for terrestrial application equipped by a Field Programmable Gate Array (FPGA), such as the one developed by the WISI Laboratory at DIBE (WiseLab, 2010);
2. it could be used for the implementation of an ASIC/ASIP specifically devoted to manage the transmission and reception of data by using the optical. In this case the implementation of a prototype on FPGA is to be considered as a passage for simulation and optimization whose aim is then to build a specific CHIP for the management of the optical communication.

The used HDL Language is VHDL (Very High Speed Hardware Description Language). It is defined in IEEE as a tool of creation of electronics system because it supports the development, verification, synthesis and testing of hardware design, the communication of hardware design data and the maintenance, modification and procurement of hardware.
The here proposed work has the aim to implement and design modules for the management of optical communication in a Underwater Wireless Sensor Network (UWSN), targeting the interface with current terrestrial technologies (in particular those based on IEEE 802.15.4).
The functions of the **Transmitter** are in synthesis:

1. the generation of a synchronization signal which is used by the transmitter in order to synchronize the impulse duration which can defined by the user;
2. a transmission signal based on PPM modulation (4 or 16 PPM)in which bits are encoded by the position of the light pulse in time slots.

The synchronization is performed by alternating 32 presences and absences of a pulse in consecutive time slots, allowing the receiver to calculate the pulse duration chosen for the following transmission of data.
The choice of PPM is based on the evaluation of the performance of different modulation schemes for underwater optical wireless communication illustrated in the previous chapter. Although most reported underwater communication systems use OOK modulation technique because of its simplicity for implementation, the proposed results argue that OOK has the disadvantages in power efficiency and control capacity of the error rate for underwater optical

channel. DPSK has good error control capability and high bandwidth, but it consumes large power and it is more complex to implement in embedded devices.
The frame format is organized as described in 4. As usual it is composed by different fields which have the following functions:

1. a Preamble composed by:
    - a field **SYNC** which is used to perform the synchronization, composed by 64 slots (32 alternating impulse of a prefixed duration), and SFR (Start Frame Delimiter) to indicate the end of the synchronization.
    - a field **DR** which is used to identify which modulation is chosen (in this case 4-PPM or 16-PPM).
2. a **Frame Length** used to communicate the number of byte in the PSDU (Physical layer Standard Data Unit) transmitted by using PPM modulation;
3. a Physical layer Standard Data Unit (**PSDU**) which is composed by the data transmitted to MAC Layer;

Preamble			Header	
SYNC	SFD	DR	Frame Length	PSDU

64 slot	4 slot	3 slot	7 bit	0-127 byte

Transmitted by PPM

Fig. 4. Frame Format

The **Receiver** has been designed to synchronize automatically with the transmitter baud rate: it determines how many clock cycles lasts a time slot for the transmitter by counting 32 transitions of the input. This value is the output prescaler and it is used to decode the following PPM transmission. The receiver detects the modulation (4 or 16 PPM) and performs also a clock correction in order to maintain synchronization by using two different methods. The first one is based on clock modification considering the truncating error made during mean calculation of the time slot duration. The second one modifies system operations by verifying that the input value is constant for the entire duration of the time slot. These methods are carefully described in the following paragraphs. The possibility of modify the duration of a single impulse can support the adaptation to different external circuits and the automatic variation of impulse duration.
Two different algorithms to maintain the synchronization have been implemented, since receiver and transmitter clocks could not be perfectly syncronised:

1. the automatic correction of time slot duration during the reception of data on the basis of the truncating error in the calculation of the time-slot period.
2. the dynamic correction by evaluating that the sampling should always be in the center of the transmitted impulse and modifying the sampling clock if this condition is not reached.

The **Physical Layer Management Entity (PLME)** provides the layer management service interfaces through which layer management functions may be invoked.
In this optical PHY Layer the PLME can manage the following services:

1. PLME-ED (Energy Detection) which reply by sending informations about the detect energy of the received signal (in this case the information can be received as input or a default value can be set if the detection is not supported);
2. PLME-GET read settings of the PHY Layer;
3. PLME-SET modify the settings of the PHY Layer;
4. PLME-CCA performs the CCA (clear channel assessment) and send the collected information to the MAC Layer;
5. PLME-SETTRX- STATE allows to change the internal operating state of the transceiver;
6. PLME-PPM is a specific service has been added to support the optical communication and the choice of a transmission based on 4 or 16-PPM;

The choice of the primitives, inspired to those of RF PHY Layer, have been adapted to optical communication in particular considering:

1. the management of the CCA (Clear Channel Assessment) which evaluates the current state of reception and the current state of the channel: the evaluation of this parameter is based on the state of the receiver, since, while the system is receiving, a transmission is not allowed, or on the evaluation of a period of time previously defined if;
2. the management of modulation by adding the possibility of choosing between 4-PPM and 16-PPM.
3. the adaptation of some attributes related to the state and function of the PHY Layer.

The **Physical Data (PD)** service enables the transmission and reception of PHY protocol data units (PPDUs) across the physical channel.
It is implemented by using 3 different sub-modules: the PDTR for the transmission, the PDREC for the reception and the PDSAP which allows coordination among internal sub-modules and the rest of the PHY Layer by implementing FIFO buffers. In addition it is interfaced to the MAC Layer. The PD manages four I/O interfaces to communicate through four FIFO buffers: two for upper layer (reading and writing) and two for PDTR and PDREC. This PD communicates to MAC Layer by using the same primitives structure performed in the PLME, with different codes, The operations performed by the PD does not require shared resources since the 3 modules, which have been easily optimized, can work in parallel.
The implemented PHY Layer are interfaced to the MAC modules described in the next paragraph.

7. The MAC Layer

7.1 Mac Layer Design Consideration

The previous described optical PHY Layer can be interfaced with an appropriate Multiple Access Control (MAC) module for providing addressing and channel access control mechanisms that make it possible for several network nodes to communicate within a multi-point network. As detailed in advance the idea is to support the interface with current terrestrial technologies developed for Wireless Sensor Networks based on IEEE802.15.4.

In this paragraph some considerations for the design of an hardware friendly MAC Layer are proposed taking into account the previous implemented optical PHY Layer and the current terrestrial technologies for WSN.
The current IEEE 802.15.4 standard proposes different MAC frame format (Beacon Frame format, Data and Acknowledgment frame format, MAC command frame format). The possibility of choosing the duration of the Mac Protocol Data Unit (MSPU) in the structure of the PHY Layer can easily allows the adaptation to the different format. The creation of the MSPU can be performed by using the hardware modules similar to those described in the previous paragraph (for instance in the description of the PD module).
The MAC Layer hardware implementation can be based on a modular approach similar to that used for PHY Layer implementation. The MAC Layer handles the access to the PHY Layer through the PD-SAP and the PLME-SAP, while the MAC Management Entity (MLME) should manage the different services for the Upper Layers, similarly to the PLME in the PHY Layer.
The use of optical communication allows to overcome the problems related to low propagation speed of acoustical communication which required complex access algorithms.
Considering power consuption, it is unlike to have a continuous communication between nodes and the communication activity is not regular between the node. A good choice appears to be a competition for channel access based on a CSMA/CA protocol in which a device, when it wishes to transmit data, waits for a random number of back-off periods before sensing the channel. If the channel is busy, the device increases the number of attempts by one and checks if the maximum number of attempts has been reached. If the limit is exceeded, the device generates a channel access error and reports this event to upper layers. If the number of attempts is below the limit, the device reiterates this procedures until it either captures the channel successfully or the number of attempts exceeds the limit.
In particular, as concerns the parameters specifically related to the optical communication two main aspects will be considered:

1. the possibility of changing the wavelength of the emitter, by choosing different colors (blue, red or green) if water turbidity varies. It can be implemented, for instance, by using a mechanism such as the one described in (Ito et al., 2008b) by managing the corresponding parameters in the PHY Layer;
2. the possibility of choosing between a 4 or 16 PPM;
3. the support of an higher data rate in comparison with the Radio based technologies for WSN. For instance, the IEEE 802.15.4 can achieve 250 Kb/s and the implementation of the Optical Management modules on a dedicated FPGA or by using an ASIP/ASIC should target the synchronization with the Upper Layers of the node.

If the transmission and reception of data should be encrypted for reasons of security, and encryption module can be easily integrated in the design as suggested in (Dasalukunte & Owall, 2008).

7.2 MAC Layer HDL implemented modules

The MAC Layer hardware design is based on a modular approach similar to that used for PHY Layer implementation.
The MAC Layer handles the access to the PHY Layer through the PD and the PLME, while it manages different services for the Upper Layers, similarly to the PLME in the PHY Layer.
Currently the following functions have been implemented:

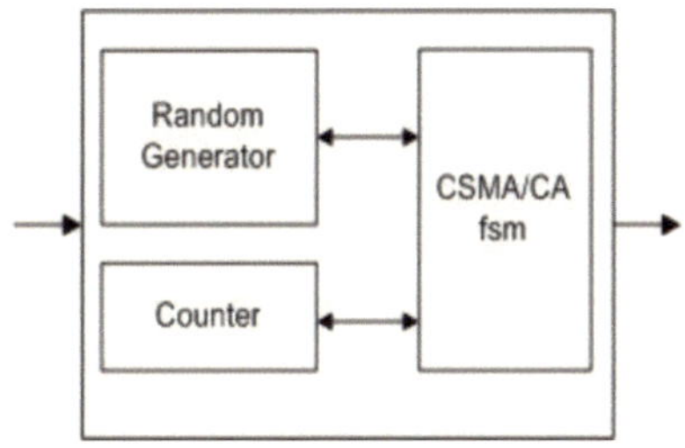

Fig. 5. CSMA-CA implementation

1. Packet parsing and addresses verification;
2. Cyclic Redundancy Check (CRC): to check the integrity of transferred data;
3. a module to manage the transmission and reception of the MAC Payload to the upper layer;
4. the CSMA/CA mechanism for channel access.

The previous described CSMA/CA mechanism has been implemented in hardware by using a dedicated module composed by a random number generator, for the calculation of the random delay, a counter and an other module, based on a FSM, to manage the CSMA/CA algorithm as depicted in Figure 5.

8. Implementation Results for PHY and MAC Layer

As a first step, the previous described modules have been synthesized and implemented on a Xilinx Spartan-3 FPGA in order to use a previously implemented testbed (Anguita et al., 2010). The second step, reported in this paper, has been the implementation an optimization of the previously described modules on an Actel IGLOO low-power FPGA in order to integrate the system in a WSN node (WiseLab, 2010)(Figure 1). A dedicated program in C# has been integrated from the previous version (Anguita et al., 2010) in order to manage the interface and to perform tests of transmission by setting different parameters of the PHY and MAC Layer (Figure 6).

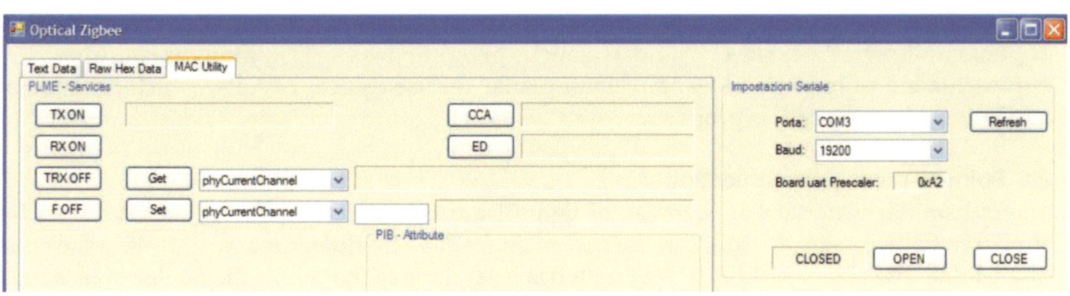

Fig. 6. User interface to optical PHY and MAC Layer

In Table 4 the results of the implementation on Actel IGLOO AGL250 are reported in terms of core cells. The maximum frequency that can be achieved by the system is 20 MHz which can support a transmission in the order of Mb/s.

Module	**Core Cells**	**%**
PHY LAYER	**2405**	**40**
Trasmitter	*347*	*6*
Receiver	*735*	*12*
PLME	*442*	*7*
PD	*881*	*15*
MAC Layer	**962**	**16**
CSMA-CA	*601*	*10*
Packet Parsing	*238*	*4*
Others	*123*	*2*

Table 4. Implementation on IGLOO Devices (AGLN250)

Wavelength of emitted light	Photodiode Sensitivity	Absorption Coefficient
450nm (blue)	0.37 A/W	9.2 x $10^{-5}cm^{-1}$
532nm (green)	0.39 A/W	4.4 x $10^{-4}cm^{-1}$
650nm (red)	0.42 A/W	3.4 x $10^{-3}cm^{-1}$
890nm (infrared)	0.63 A/W	6.0 x $10^{-2}cm^{-1}$

Table 5. Sensitivity for PDB-S5971 High-speed photodiodes and absorption coefficients

9. Wireless optical communication circuits

The implementation of an efficient diffuse underwater communication require the implementation of circuits for transmission and reception of light impulse. Considering the application of the communication system in a USWN the focus should be on using low-cost and low-power components targeting a low/medium distance up to 30 m. They are the basis to implement a diffuse optical communication which could support a dense network of small nodes able to exchange data at high data-rate (in the order of Mb/s).
As regards the transceiver components, a careful evaluation should be performed. For the transmitter the use of LED seems to be more interesting for low cost and low power application, offering small, flexible, cheap devices which are easy to array. For the receiver, the use of PIN-photodiodes could be attractive taking into account their fast response, but also the use of photomultiplier tubes (PMT's) and avalanche photodiodes (APD's) could be consider.
A presentation of both point-to-point than planar diffuse optical circuits, which have been implemented and tested, is proposed.

9.1 Point-to-point communication

The transmitter generates an impulse of light of a fixed duration (250 ns) which allows to support a transmission at 1Mb/s in the case of an 16-PPM modulation or at 2Mb/s in the case of a 4-PPM. The choice of LED wavelength has been done to maximize the power of received signal as illustrated below (Figure 8).
Particular attention has been posed on the circuit for the reception, since the reciprocal distance of the underwater devices cannot be fixed in advance and the receiver has to maintain his functioning in all the coverage area of the transmitter.
The receiver has been designed considering different blocks: a photodiode; a transresistance amplifier, to have a conversion from current to voltage; a bandpass filter to eliminate noise

Fig. 7. Experimental set-up for point-to-point tests (a) - Bidirectional Transceiver (b)

below 10 kHz and above 20 MHz; an Automatic Gain Control (AGC), based on a Linear Technology LT1006, used to amplify the signal received by the first part of the circuit and to automatically increase or decrease the gain according to the signal amplitude; a comparator, to determine the output value by fixing a threshold.

The receiver has been implemented by using the following component: Si PIN photodiode Hamamatsu S5971 - high-speed photodiodes, with $1mm^2$ surface area. To evaluate the better wavelength for the transmitter Table 5 has been compiled considering the absorption coefficients for clear water and the photodiode sensitivity reported in the component datasheet. The output current of the photodiode is proportional to:

$$S \times e^{-k(d)} \times P \quad (11)$$

where S is sensitivity an P is the power in watts, d the distance and k the absorption coefficient. Figure 8 shows how the output current varies according to distance for different wavelengths of light, by using the equation 11.

Due to the severe attenuation in water, the output current for infrared is less than for blue/green light at 10m, even if the sensitivity of the photodiode is higher for infrared. It is possible to note that red light outperforms green light up to 1.5/2 m but blue and green are better beyond. Taking into account the previous results, also considering that the attenuation is strongly influenced by turbidity, the better choice appear to be blue or green light. These considerations are very important because the system performance is determined by the detector when signal attenuation along a wireless link is considered. It is crucial for the receiver to detect low-level optical signals maintaining a Signal-to-Noise Ratio (SNR) sufficiently large to yield an acceptable Bit Error Rate (BER).

The receiver circuit has been tested in air and in water. Tests in clear water allows to receive correctly the transmitted sequence generated by the system (Figure 7) up to 2 meters: a more accurate evaluation of the BER is planned for turbid water. Considering the limitation of the available testbed, tests in air been performed at different distances: a reception of the transmitted sequence is possible starting from few centimeters, since the AGC avoids saturation, up to 10 meters, while below light impulses are not clearly detected. Even if the AGC stage has to be modified to allow the reception in case of higher distances, considering that our

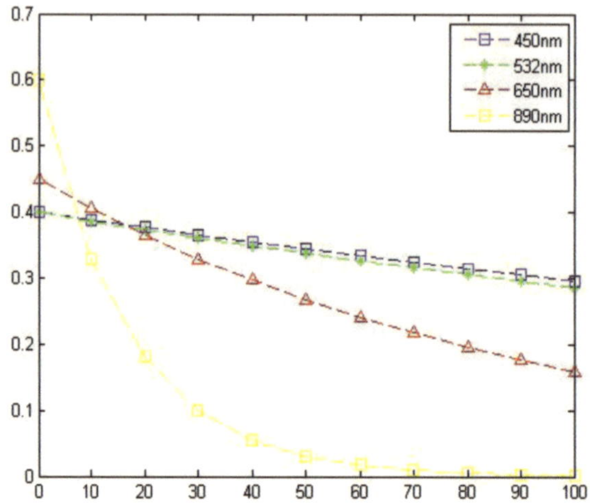

Fig. 8. Comparison between different wavelengths

target is up to 10-15 meters underwater, the tested circuit is a good starting point for further improvements.
The cost of the components is less than 30 euros, very cheap in comparison with some acoustic modem.

10. Design Consideration and Implementation of Planar Optical Circuits

The possibility of targeting the connection between more than two devices leads to the necessity of implementing a directional, up to omni-directional, transceiver able to send and detect optical impulse. Taking into account some previous works (Akella et al., 2005), in this paragraph some considerations regarding the design and implementation of a circular transceiver are reported with a preliminary implementation description.

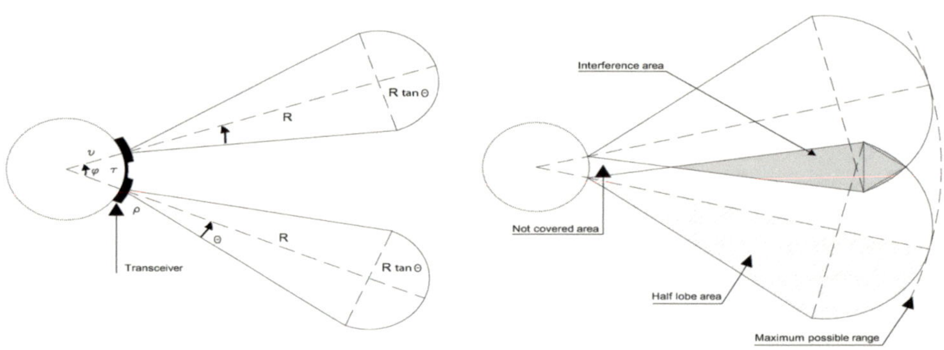

Fig. 9. Transmitter disposed to cover a 2-D circular area with (a) and without overlap (b)

In particular the collocation of components (LED or photodiodes) on a 2D structure is considered. Assuming that n transmitter are placed at equal distance gaps on the circular node (radius r), considering that the diameter of a transmitter is 2ρ:

$$\tau = \frac{2\pi r - 2n\rho}{n} \tag{12}$$

The angular difference between any two neighboring transceiver is given as:

$$\varphi = 360^0 \frac{\tau}{2\pi r} \tag{13}$$

The coverage area L of a single transmitter can be given by (Akella et al., 2005):

$$L = R^2 tan(\vartheta) + 0.5\pi R tan(\vartheta)^2 \tag{14}$$

Two cases can happen for the effective coverage area C of a single transmitter, based on the value of φ, θ, R, and r:

1. if coverage area of the neighbor transmitter do not overlap (Figure 9):

$$R tan(\vartheta) \leq (R + r) tan(0.5\varphi) \tag{15}$$

In this case, the effective coverage area is equivalent to the coverage area, i.e. C=L.

2. Coverage area of the neighbor transmitter overlap (Figure 9):

$$R tan(\vartheta) > (R + r) tan(0.5\varphi) \tag{16}$$

In this case, the effective coverage area is equivalent to the coverage area excluding the area that interferes with the neighbor transceiver. If I is the interference area that overlaps with the neighbor transmitter's coverage, then C = L-I. Considering that the target of our work is to have a directional or at least omni-directional a good design approach should minimize the interfere area.

If the target is to have a transmission and reception of data by using all the elements placed on the circular structure, the overlap is not a problem for the communication. But, since the final idea is to add the possibility of supporting also a directional communication in which a single LED could be activated, the idea is to minimize the interference between adjacent LEDs and receivers.

Taking into account the previous circuits, a 2-d transceiver has been implemented. The LEDs disposition on a 2-d structure has been performed using 12 Ledman LL1503PLBL1-301 blue LED (with 30^0 of FOV) disposed on a circular disk of 10 cm diameter. The reduction of the overlap between the signal generated by each LED has been targeted so to allow the possibility of supporting also a directional transmission, by using only one or a reduced set of LEDs. The same approach has been considered for the placement of the photodiodes for the 2-d receiver. Tests of the transmitter have been carried out by using a single receiver, equipped with a SFH-2013P photodiode. Different measures have been considered at different distances. The profile of the received optical signal, determined by considering the maximum value in a line-of-sight condition and the minimum value which is measured between a LED and another, show that the generated light impulses is uniformly distributed on the surface. Tests on the receiver have shown a performance decrease in comparison with the point-to-point circuit due to the noise generated by the photodiodes which are not directly exposed to the light impulses. Nevertheless a reception up to 4 meters in air can be achieved.

11. Future Research Directions

Starting from the previous described results, the creation of an innovative long survival optical Underwater Wireless Sensor Network (UWSN) could be targeted. It will able to sense, compute, communicate and cooperate in an underwater environment by using long-survival, low-cost and eventually disposable nodes. Innovative approaches could be developed to address:

1. wireless, adaptive and low-power optical underwater communication;
2. underwater energy scavenging and harvesting;
3. autonomous and self-governing behavior of the nodes in the underwater environment, including intelligent energy storing and utilization, to guarantee long-term operation under a wide range of conditions and avoid frequent and costly rescue procedures.

The first innovation will address the communication capabilities. Efficient optical communication could be implemented thank to a low-cost, efficient and omni-directional communication system, based on short wavelength LEDs, and able to exploit the minimum absorption wavelength window, shifting toward longer wavelengths in turbid waters. The optical communication, despite reaching shorter distances respect to acoustic communication, will allow to achieve high data rates with lower energy requirements.

Fig. 10. optical UWSN Concept

Finally, the implementation of adaptive directionality, by activating only some of the transducers will allow to spare energy and optimize the communication efficiency. Online nonlinear modeling and adaptation to the time-varying aquatic channel characteristics of the optical system can be used to advance the state-of-the-art and achieve effective and efficient Free Space Omni-directional Optical (FSOO) underwater communication.
The second innovative field of research could be underwater energy scavenging for self-supply or for increasing the lifetime of both each node and the entire network. While solar

power can be exploited on the water surface and in shallow but clear waters, other techniques must be explored for producing and storing energy in a general underwater environment. Terrestrial energy scavenging for artificial artifacts and autonomous sensors relies mainly on exploiting environmental vibrations, converting mechanical energy in electrical energy. In the static underwater nodes, random movements forced by the water flow, underwater currents and noise, can be exploited to perform underwater energy scavenging (D.Zhu, 2010). The delicate equilibrium between energy harvesting, storage and consumption is to addressed by developing adaptive behavior, to optimize the survivability of both the nodes and the entire network.

The third field of research should address the development of low-power, low-cost miniaturized nodes able to sense, compute, communicate and cooperate in an aquatic environment. The development of new nodes, with volumes that are orders of magnitude smaller than current generation equipments, will allow the development of new applications, where tiny and, eventually, disposable nodes will be able to perform 4D monitoring of aquatic environments. The increased density of the network, which will be possible thank to the miniaturization and low-cost of the nodes, will allow to compensate for the relatively short range of the underwater optical communication channel, which is, as shown, well below 100m. Furthermore, in settings where environmental issues are of limited concern, the use of low-cost disposable nodes will avoid frequent and costly rescue procedures, by simply adding new nodes to areas not covered by the network or to overcome the malfunctioning of old nodes.

The development of the previous described fields of research could lead to the implementation of low-cost optically communicating nodes, able to be deployed with low accuracy on the area of interest, and capable of self-configuring as a sensing network as depicted in Figure 10.

12. References

Akella, J., Liu, C., Partyka, D., Yuksel, M., Kalyanaraman, S. & Dutta, P. (2005). Building blocks for mobile free-space-optical networks, *Second IFIP International Conference on Wireless and Optical Communications Networks, 2005. WOCN 2005.*, pp. 164 – 168.

Akyildiz, I. F., Pompili, D. & Melodia, T. (2005). Underwater acoustic sensor networks: research challenges, *Ad Hoc Networks* **3**(3): 257 – 279.

Akyldiz, I. F., Su, W., Sankarasubramaniam, Y. & Cayirci, E. (2001). A survey on sensor network.

Al-Shamma'a, A., Shaw, A. & Saman, S. (2004). Propagation of electromagnetic waves at mhz frequencies through seawater, *Antennas and Propagation, IEEE Transactions on* **52**(11): 2843–2849.

Ambalux (2008). http://www.ambalux.com.

Anguita, D., Brizzolara, D., Ghio, A. & Parodi, G. (2008). Smart plankton: a nature inspired underwater wireless sensor network, *Natural Computation, 2008. ICNC '08. Fourth International Conference on* **7**: 701 –705.

Anguita, D., Brizzolara, D. & Parodi, G. (2009). Building an underwater wireless sensor network based on optical communication: Research challenges and current results, *Sensor Technologies and Applications, 2009. SENSORCOMM '09. Third International Conference on* pp. 476 –479.

Anguita, D., Brizzolara, D. & Parodi, G. (2010). Design and implementation of hdl modules and circuits for underwater optical wireless communication, *Proceedings of the 9th WSEAS International Conference on TELECOMMUNICATIONS and INFORMATICS* pp. 132 – 137.

Baiden, G. & Bissiri, Y. (2007). High bandwidth optical networking for underwater untethered telerobotic operation, *OCEANS 2007*, pp. 1–9.

Baiden, G., Bissiri, Y. & Masoti, A. (2009). Paving the way for a future underwater omni-directional wireless optical communication systems, *Ocean Engineering* **36**(9-10): 633 – 640.

Chancey, M. A. (2007). Degree of master of science: Short range underwater optical communication links.

C.X.Fan, F.Zhang, Xu, B. & Wu, C. (2001). Communication theory, **Defense Industry Press**.

Dasalukunte, D. & Owall, V. (2008). A generic hardware mac for wireless personal area network platforms, *Proceedings of the 11th International Symposium on Wireless Personal Multimedia Communication (WPMC'08)* .

Diana, L. & Kahn, J. (1999). Rate-adaptive modulation techniques for infrared wireless communication, *Proceedings of ICC '99* **1**: 597–603.

Doniec, M., Detweiler, C., Vasilescu, I. & Rus, D. (2010). Using optical communication for remote underwater robot operation.

Doniec, M., Vasilescu, I., Chitre, M., Detweiler, C., Hoffmann-Kuhnt, M. & Rus, D. (2009). Aquaoptical: A lightweight device for high-rate long-range underwater point-to-point communication, *OCEANS 2009, MTS/IEEE Biloxi - Marine Technology for Our Future: Global and Local Challenges* pp. 1 –6.

D.Zhu, M.J.Tudor, S. (2010). Strategies for increasing the operating frequency range of vibration energy harvesters: a review, *Measurement Science and Technology* .

Fair, N., Chave, A., Freitag, L., Preisig, J., White, S., Yoerger, D. & Sonnichsen, F. (2006). Optical modem technology for seafloor observatories, *OCEANS 2006* pp. 1–6.

Farr, N., Chave, A., Freitag, L., Preisig, J., White, S., Yoerger, D. & Titterton, P. (2005). Optical modem technology for seafloor observatories, *OCEANS, 2005. Proceedings of MTS/IEEE* pp. 928–934 Vol. 1.

Feng Lu, Sammy Lee, J. M. C. S. (2009). Low-cost medium-range optical underwater modem, *WUWNET 2009*, pp. 1–6.

Fibre, W. (2008). http://www.wirelessfibre.co.uk/.

Ghassemlooy, Z., Popoola, W., Rajbhandari, S., Amiri, M. & Hashemi, S. (2007). A synopsis of modulation techniques for wireless infrared communication, *ICTON Mediterranean Winter Conference, 2007. ICTON-MW 2007*, pp. 1–6.

Giles, J. & Bankman, I. (2005). Underwater optical communications systems. part 2: basic design considerations, *IEEE Military Communications Conference. MILCOM 2005.* pp. 1700–1705 Vol. 3.

Hanson, F. & Radic, S. (2008). High bandwidth underwater optical communication, *Applied Optics* **36**: 277–283.

Ito, Y., Haruyama, S. & Nakagawa, M. (2008a). Short-range underwater wireless communication using visible light leds.

Ito, Y., Haruyama, S. & Nakagawa, M. (2008b). Short-range underwater wireless communication using visible light leds.

Jaruwatanadilok, S. (2008). Underwater wireless optical communication channel modeling and performance evaluation using vector radiative transfer theory, *Selected Areas in Communications, IEEE Journal on* **26**(9): 1620–1627.

Kahn, J. & Barry, J. (1997). Wireless infrared communications, *Proceedings of the IEEE* **85**(2): 265–298.

Kedar, D. (2007). Underwater sensor network using optical wireless communication, *SPIE Newsroom - The International Society for Optical Engineering* .

Lanbo, L., Shengli, Z. & Jun-Hong, C. (2008). Prospects and problems of wireless communication for underwater sensor networks, *Wirel. Commun. Mob. Comput.* **8**(8): 977–994.

Lee, D. & Kahn, J. (1999). Coding and equalization for ppm on wireless infrared channels, *IEEE Transaction on Communication* **47**: 255–260.

Lee, S., Mounzer, J., Mirza, D. & Schurgers, C. (2008). Demo abstract: Low cost, medium range optical, communication for underwater test beds, *The Third ACM International Workshop on UnderWater Networks (WUWNet)*.

Liu, Y. & Ge, X. (2006). Underwater laser sensor network: A new approach for broadband communication in the underwater, *Proceedings of the 5th WSEAS International Conference on Telecommunications and Informatics*, pp. 421–425.

Lucas, J., Al-ShammaŠa, A., Seim, J., Loehr, W., Puchbauer, G. & McGregor, D. (2004). Underwater communications using electromagnetic waves (emcomms), Vol. Proceeding of European Conference on Marine Science and Ocean Technology.

Ma, H. (2003). Research on optical modulation and demodulation techniques in mobile atmospheric laser communication, **National University of Defense Technology**.

Malik, D., Joseph, M. & John, R. (1996). Performance of pulse-position on measured non-directed indoor infrared channels, *IEEE Transaction on Communications* **44**: 173–177.

Meihong, S., Xinsheng, Y. & Fengli, Z. (2009). The evaluation of modulation techniques for underwater wireless optical communications, *Communication Software and Networks, International Conference on* **0**: 138–142.

Meihong, S., Xinsheng, Y. & Zhangguo, Z. (2009). The modified ppm modulation for underwater wireless optical communication, *Communication Software and Networks, International Conference on* **0**: 173–177.

Pang, W., Chew, W., Choong, F. & Teoh, E. (2007). Vhdl modeling of the ieee802.11b dcf mac, *Proceedings of the 6th WSEAS International Conference on Instrumentation, Measurement, Circuits and Systems* .

Park, H. & Barry, J. (1995). Modulation analysis for wireless infrared communications, *Communications, 1995. ICC '95 Seattle, 'Gateway to Globalization', 1995 IEEE International Conference on*, Vol. 2, pp. 1182–1186 vol.2.

Puccinelli, D. & Haenggi, M. (2005). Wireless sensor networks: applications and challenges of ubiquitous sensing, *IEEE, Circuits and Systems Magazine*.

Shelley, T. (2005). Radio waves transmit information underwater, *Eureka Magazine*.

Shill, F., Zimmer, U. R. & Trumpi, J. (2004). Visible spectrum optical communication and distance sensing for underwater applications.

Singh, S., Grund, M., Bingham, B., Eustice, R., Singh, H. & Freitag, L. (2006). Underwater acoustic navigation with the whoi micro-modem, *OCEANS 2006*, pp. 1–4.

Smart, J. (2005). Underwater optical communications systems part 1: variability of water optical parameters, *IEEE Military Communications Conference. MILCOM 2005.* pp. 1140–1146 Vol. 2.

Tivey, M., Fucile, P. & Sichel, E. (2004). A low power, low cost, underwater optical communication system, *Ridge 2000 Events* pp. 27–29.

V. Hsu, J. M. K. & Pister, K. S. J. (1998). Wireless communications for smart dust, *Electronics Research Laboratory Technical Memorandum Number M98/2*.

Vasilescu, I., Detweiler, C. & Rus, D. (2007). Aquanodes: an underwater sensor network, *WuWNet '07: Proceedings of the second workshop on Underwater networks*, ACM, New York, NY, USA, pp. 85–88.

Vasilescu, I., Kotay, K., Rus, D., Dunbabin, M. & Corke, P. (2005). Data collection, storage, and retrieval with an underwater sensor network, *SenSys '05: Proceedings of the 3rd international conference on Embedded networked sensor systems*, ACM, New York, NY, USA, pp. 154–165.

WiseLab (2010). http://www.wise-laboratory.it/test/index.htm, *DIBE: Department of Biophysical and Electronic Engineering* .

Yan, H., Zhou, S., Shi, Z. J. & Li, B. (2007). A dsp implementation of ofdm acoustic modem, *WuWNet '07: Proceedings of the second workshop on Underwater networks*, ACM, New York, NY, USA, pp. 89–92.

13

Estimation of Propagation Characteristics along Random Rough Surface for Sensor Networks

Kazunori Uchida and Junichi Honda
Fukuoka Institute of Technology
Japan

1. Introduction

The main focus in the development of wireless communications engineering is providing higher data rates, using lower transmission power, and maintaining quality of services in complicated physical environments, such as an urban area with high-rise buildings, a randomly profiled terrestrial ground and so on. In order to achieve these goals, there has been substantial progress in the development of low-power circuits, digital algorithms for modulation and coding, networking controls, and circuit simulators in recent years [Aryanfar,2007]. However, insufficient improvement has been made in wireless channel modeling which is one of the most basic and significant engineering problems corresponding to the physical layer of the OSI model.

Recently, the sensor network technologies have attracted many researchers' interest especially in the fields of wireless communications engineering as well as in the fields of sensor engineering. The sensor devices are usually located on the terrestrial surfaces such as dessert, hilly terrain, forest, sea surface and so on, of which profiles are considered to be statistically random. In this context, it is very important to investigate the propagation characteristics of electromagnetic waves traveling along random rough surfaces (RRSs) and construct an efficient as well as reliable sensor network over terrestrial grounds with RRS-like profiles [Uchida,2007], [Uchida,2008], [Uchida,2009], [Honda,2010].

In the early years of our investigations, we applied the finite volume time domain (FVTD) method to estimate the electromagnetic propagation characteristics along one-dimensional (1D) RRSs [Honda,2006], [Uchida,2007]. The FVTD method, however, requires too much computer memory and computation time to deal with relatively long RRSs necessary for a sensor network in the realistic situation. To overcome this difficulty, we have introduced the discrete ray tracing method (DRTM) based on the theory of geometrical optics, and we can now deal with considerably long RRSs in comparison with the operating wavelength. The merit of using DRTM is that we can treat very long RRSs compared with the wavelength without much computer memory nor computation time. Thus, the DRTM has become one of the most powerful tools in order to numerically analyze the long-distance propagation characteristics of electromagnetic waves traveling along RRSs [Uchida,2008], [Uchida,2009], [Honda,2010].

In this chapter, we discuss the distance characteristics of electromagnetic waves propagating along homogeneous RRSs which are described statistically in terms of the two parameters, that is, height deviation h and correlation length $c\ell$. The distance characteristics of propagation are estimated by introducing an amplitude weighting factor α for field amplitude, an

order β for an equivalent propagation distance, and a distance correction factor γ. The order yields an equivalent distance indicating the distance to the β-th power. The order was introduced by Hata successfully as an empirical formula for the propagation characteristics in the urban and suburban areas [Hata,1980]. In the present formulations, we determine these parameters numerically by using the least square method. Once these parameters are determined for one type of RRSs, we can easily estimate the radio communication distance between two sensors distributed on RRSs, provided the input power of a source antenna and the minimum detectable electric field intensity of a receiver are specified.
The contents of the present chapter are described as follows. Section 1 is the introduction of this chapter, and the background of this research is denoted. Section 2 discusses the statistical properties of 1D RRSs and the convolution method is introduced for RRS generation. Section 3 discusses DRTM for evaluation of electromagnetic waves propagating along RRSs. It is shown that the DRTM is very effective to the field evaluation especially in a complicated environment, since it discretizes not only the terrain profile but also the procedure for searching rays, resulting in saving much computation time and computer memory. Section 4 discusses a numerical method to estimate propagation or path loss characteristics along 1D RRSs. An estimation formula for the radio communication distance along the 1D RRSs is also introduced in this Section. Section 5 is the conclusion of this chapter, and a few comments on the near future problems are remarked.

2. Generation of 1D random rough surface

As mentioned in the introduction, the sensor network has attracted many researchers' interest recently in different technical fields like signal processing, antennas, wave propagation, low power circuit design and so forth, just as the same as the case of radio frequency identification (RFID) [Heidrich,2010]. The sensor devices are usually located on terrestrial surfaces such as dessert, hilly terrain, forest, sea surface and so on. Since these surfaces are considered to be statistically random, it is important to study statistics of the RRSs as well as the electromagnetic wave propagation along them in order to construct reliable and efficient sensor network systems [Honda,2009].
In this section, we describe the statistical properties of RRSs and we show three types of spectral density functions, that is, Gaussian, n-th order of power-law and exponential spectra. We also discuss the convolution method for RRS generation. The convolution method is flexible and suitable for computer simulations to attack problems related to electromagnetic wave scattering from RRSs and electromagnetic wave propagation along RRSs.

2.1 Spectral density function and auto-correlation function

In this study, we assume that 1D RRSs extend in x-direction and it is uniform in z-direction with its height function as denoted by $y = f(x)$. The spectral density function $W(K)$ for a set of RRSs is defined by using the height function and the spatial angular frequency K as follows:

$$\int_{-\infty}^{\infty} W(K)dK = h^2 \tag{1}$$

where h is the standard deviation of of the height function or height deviation, and

$$W(K) = \lim_{L\to\infty} \frac{1}{2\pi} \left\langle \frac{1}{L} \left| \int_{-L/2}^{L/2} f(x)e^{-jKx}dx \right|^2 \right\rangle \tag{2}$$

where $<>$ indicates the ensemble average of the RRS set. As is well-known, the auto-correlation function is given by the Fourier transform of the spectral density function as follows:

$$\rho(x) = \int_{-\infty}^{\infty} W(K) e^{jKx} dK \,. \tag{3}$$

Now we summarize three types of spectral density functions that are useful for numerical simulations of the propagation characteristics of electromagnetic waves traveling along RRSs.

1. Gaussian Type of Spectrum:
   The spectral density function of this type is defined by

$$W(K) = \left(\frac{clh^2}{2\sqrt{\pi}} \right) e^{-\frac{K^2 cl^2}{4}} \tag{4}$$

   where cl is the correlation length, and the auto-correlation function is given by

$$\rho(x) = h^2 e^{-\frac{x^2}{cl^2}} \,. \tag{5}$$

2. N-th Order Power-Law Spectrum:
   The spectral density function of this type is given by

$$W(K) = \left(\frac{clh^2}{2\sqrt{\pi}} \right) \left\{ 1 + \frac{\Gamma^2(N - \frac{1}{2})}{\Gamma^2(N)} \frac{K^2 cl^2}{4} \right\}^{-N} \tag{6}$$

   where $\Gamma(N)$ is the Gamma function with $N > 1$, and the auto-correlation function is given by

$$\rho(x) = \frac{h^2}{[1 + \frac{x^2}{Ncl^2}]^N} \,. \tag{7}$$

3. Exponential Spectrum:
   The spectral density function of this type is given by

$$W(K) = \left(\frac{clh^2}{\pi} \right) \left\{ 1 + K^2 cl^2 \right\}^{-1} \tag{8}$$

   and the auto-correlation function is given by

$$\rho(x) = h^2 e^{-\frac{|x|}{cl}} \,. \tag{9}$$

2.2 Convolution method for RRS generation

As is well-known, we should not use discrete Fourier transform (DFT) but fast Fourier transform (FFT) for practical applications to save computation time. For simplicity of analyses, however, we use DFT only for theoretical discussions. Now we consider a complex type of 1D array $\mathbf{f}$ corresponding to a discretized form of $f(x)$ and its complex type of spectral array $\mathbf{F}$ defined by

$$\mathbf{f} = (f_0, f_1, f_2, \cdots, f_{N-2}, f_{N-1}) \,, \quad \mathbf{F} = (F_0, F_1, F_2, \cdots, F_{N-2}, F_{N-1}) \,. \tag{10}$$

The complex type of spectral array **F** is the DFT of **f**. And the complex type of DFT is defined as follows:

$$\mathbf{F} = \mathrm{DFT}(\mathbf{f})\,, \quad F_\nu = \sum_{n=0}^{N-1} f_n e^{-j2\pi\frac{n\nu}{N}} \quad (\nu = 0, 1, \cdots, N-1)\,. \tag{11}$$

Moreover, the inverse DFT is defined by

$$\mathbf{f} = \mathrm{DFT}^{-1}(\mathbf{F})\,, \quad f_n = \frac{1}{N}\sum_{\nu=0}^{N-1} F_\nu e^{j2\pi\frac{n\nu}{N}} \quad (n = 0, 1, \cdots, N-1)\,. \tag{12}$$

First, we discretize the spectral density function discussed in the preceding section by introducing the discretized spatial angular frequency K_n as follows:

$$K_n = \frac{2\pi n}{N_1 c\ell} \quad (n = 0, 1, 2, \cdots, N-1) \tag{13}$$

where $N = N_1 N_2$. It is assumed that N_2 is the number of discretized points per one correlation length $c\ell$ and correlation beyond the distance $N_1 c\ell$ is negligibly small. Then we can obtain the real type of 1D array **w** by using the spectral density function $W(K)$ at the discretized spatial angular frequencies as follows:

$$\mathbf{w} = (w_0, w_1, \cdots, w_{N-1}) \tag{14}$$

where the elements w_n of the array are expressed as follows:

$$\begin{aligned} w_n &= 2\pi W(K_{n'})/N_1 c\ell \\ n' &= \begin{cases} n & (0 \le n < N/2) \\ N-n & (N/2 \le n < N)\,. \end{cases} \end{aligned} \tag{15}$$

It should be noted that the DFT of the above 1D array corresponds to the discretized autocorrelation function of $\rho(x)$ as follows:

$$\mathrm{DFT}(\mathbf{w}) \leftrightarrow \rho(x)\,. \tag{16}$$

Thus we can utilize this relation to check the accuracy of the discretized numerical results for the spectral density function of the RRSs we are dealing with.
Second, we introduce another real 1D array $\tilde{\mathbf{w}}$ by taking the square root of the former array as follows:

$$\tilde{\mathbf{w}} = (\sqrt{w_0/N}, \sqrt{w_1/N}, \cdots, \sqrt{w_{N-1}/N})\,. \tag{17}$$

Performing the DFT of the above 1D array leads to a new weighting array defined by

$$\tilde{\mathbf{W}} = (\tilde{W}_0, \tilde{W}_1, \cdots, \tilde{W}_{N-1}) = \mathrm{DFT}(\tilde{\mathbf{w}}) \tag{18}$$

This weighting array includes all the information about the spectral properties of the RRSs, and also it plays an important role as a weighting factor when we generate RRSs by the convolution method.

Third, we consider the random number generator necessary for computer simulations. C programming language provides us the software rand(a) which produces a sequence of random numbers ranging in $[0, a]$ [Johnsonbaugh,1997]. Then we can generate another sequence of random number x_i in the following way:

$$\begin{aligned} u_1 &= \text{rand}(2\pi), \quad u_2 = \text{rand}(1) \\ x_i &= \sqrt{-2\log(u_2)}\cos(u_1) \quad (i = 0, 1, 2, ...). \end{aligned} \tag{19}$$

It can be proved that the random numbers obtained by the above functions belong to the normal distribution as follows:

$$(x_0, x_1, x_2, \cdots) \in N(0, 1)\,. \tag{20}$$

As a result, we can generate a sequence of the discrete random rough surface with arbitrary N' points by performing the discrete convolution between the sequence of the Gaussian random number $x_i \in N(1, 0)$ given by Eq.19 and the weighting array $\tilde{W}_k$ given by Eq.18. The final results are summarized as follows:

$$\begin{aligned} f_n &= \sum_{k=0}^{N-1} \tilde{W}_k x_{n+k} \quad (n = 0, 1, 2, 3, \cdots, N' - 1) \\ \{x_i\} &\in N(1, 0) \quad (i = 0, 1, 2, \cdots, N + N' - 1)\,. \end{aligned} \tag{21}$$

Eq.21 is the essential part of the convolution method, and it provides us any type of RRSs with arbitrary length [Uchida,2007], [Uchida,2008].
It is worth noting that correlation of the generated RRSs is assumed to be negligibly small outside the distance of $N_1 c\ell$ and the minimum discretized distance is $c\ell/N_2$. One of the advantages of the present convolution method is that we can generate continuous RRSs with an arbitrary number of sample points $N' > N$, provided that the weighting array $\tilde{W}_k$ in Eq.18 is computed at the definite number of points $N = N_1 N_2$. The other advantage is that the present method is more flexible and it saves more computation time than the conventional direct DFT method [Thoros,1989], [Thoros,1990], [Phu,1994], [Tsang,1994], [Yoon,2000], [Yoon,2002].

3. Discrete ray tracing method (DRTM)

In this chapter, we apply DRTM to the investigation of propagation characteristics along random rough surfaces whose height deviation h and correlation length $c\ell$ are much longer than the wavelength, that is, $h, c\ell >> \lambda$. In the past, we used the ray tracing method (RTM) to analyze electromagnetic wave propagation along 1D RRSs. The RTM, however, requires lots of computer memory and computation time, since its ray searching algorithm is based only on the imaging method. The present DRTM, however, requires much less computer memory and computation time than the RTM. This is the reason why we employ the DRTM for ray searching and field computing. First, we discretize the rough surface in term of piecewise-linear lines, and second, we determine whether two lines are in the line of sight (LOS) or not (NLOS), depending on whether the two representative points on the two lines can be seen each other or not.
The field analyses of DRTM are based on the well-known edge diffraction problem by a conducting half plane which was rigorously solved by the Wiener-Hopf technique [Noble,1958]. The Wiener-Hopf solution cannot be rigorously applied to the diffraction problem by a plate

of finite width. When the distance between the two edges of the plate is much longer than the wavelength, however, it can be approximately applied to this problem with an excellent accuracy. This is the basic idea of the field analyses based on DRTM. Numerical calculation are carried out for the propagation characteristics of electromagnetic waves traveling along RRSs with Gaussian, n-th order of power-law and exponential types of spectra.

3.1 RRS discretization in terms of piecewise-linear profile

A RRS of arbitrary length can be generated by the convolution method discussed in the preceding section. We treat here three types of spectral density functions for generating RRSs. The first is Gaussian, the second is n-th order of power-law and the third is exponential distribution, where the RRS parameters are correlation length cl and height deviation h. Fig.1 shows four examples of RRSs with Gaussian, first and third order of power-law and exponential spectra, and the parameters are selected as $cl = 10.0$ $[m]$ and $h = 1.0$ $[m]$. It is shown that the Gaussian spectrum exhibits the smoothest roughness.

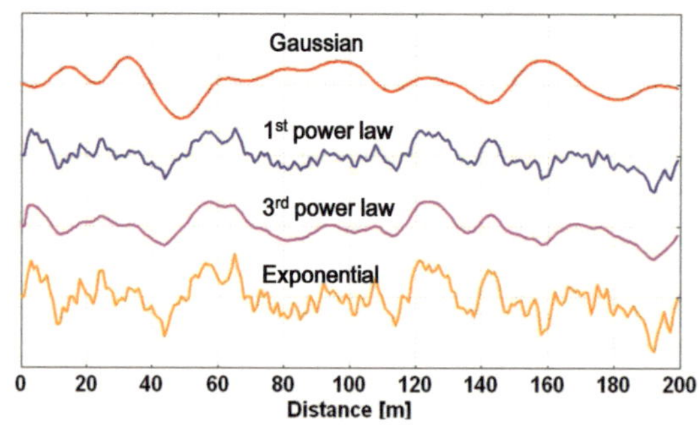

Fig. 1. Examples of random rough surface.

The convolution method introduced in the preceding section provides us the data of position vectors corresponding to the discretized RRS points as follows:

$$\mathbf{r_n} = (n\Delta x, f_n) \quad (n = 0, 1, 2, \cdots, N' - 1) \tag{22}$$

where the minimum discretized distance is given by

$$\Delta x = c\ell / N_2 \, . \tag{23}$$

On the other hand, we can determine the normal vector of each straight line as follows:

$$\mathbf{n_n} = (\mathbf{u_z} \times \mathbf{a_n}) / |\mathbf{u_z} \times \mathbf{a_n}| \quad (n = 0, 1, 2, \cdots, N' - 1) \tag{24}$$

where $\mathbf{u_z}$ is the unit vector in z-direction. Moreover, the vector corresponding to each straight line is given by

$$\mathbf{a_i} = (\mathbf{r_{i+1}} - \mathbf{r_i}). \tag{25}$$

Thus all the informations regarding traced rays can be expressed in terms of the position vectors $\mathbf{r_n}$ in Eq.22 and the normal vectors $\mathbf{n_n}$ in Eq.24, resulting in saving computer memory.

3.2 Algorithm for searching rays based on DRTM

Now we discuss the algorithm to trace discrete rays with respect to a discretized RRS. We propose a procedure to approximately determine whether the two straight lines $\mathbf{a}_i$ and $\mathbf{a}_j$ $(i \neq j)$ are in the line of sight (LOS) or not (NLOS) by checking whether the two representative points on the two lines can be seen from each other or not. The representative point of a line may be its center or one of its two edges, and in the following discussions, we employ the central point as a representative point of a line. Thus the essence of finding rays is reduced to checking whether the representative point of one straight line is in LOS or in NLOS of the representative point of the other line.

One type of ray is determined by constructing the minimum distance between the two representative points which are in NLOS, while the other type of ray is determined by connecting the two representative points which are in LOS. The traced rays obtained in this way are approximate, but the algorithm is simple and thus we can save much computation time. Moreover, we can modify the discrete rays into more accurate ones by applying the principle of the shortest path to the former case and the imaging method to the latter case. The former type is shown in (a) of Fig.2, and the latter type is depicted in (b) of Fig.2. In these figures, S denotes a source point and R indicates a receiver point.

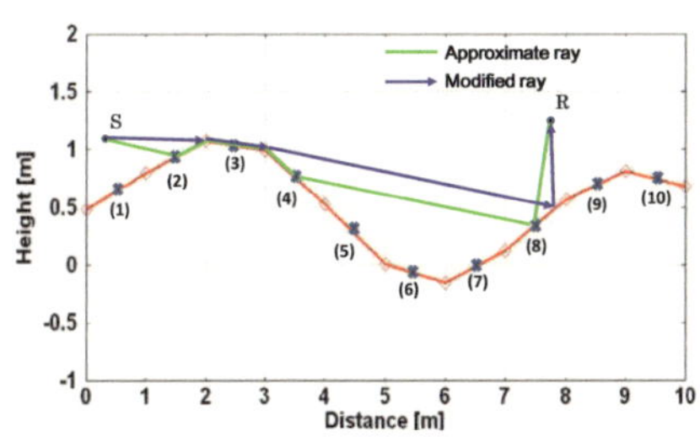

(a) in case of NLOS

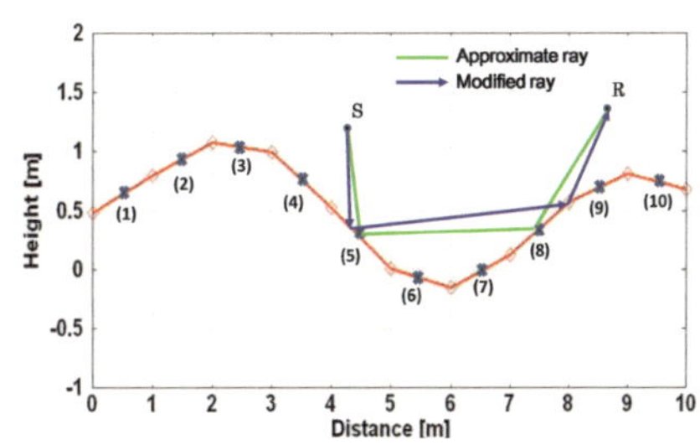

(b) in case of LOS

Fig. 2. Examples of searching rays.

Let us explain the example of searched ray in (a) of Fig.2. First we find a shortest path from (2) to (4) which are in NLOS, and we also find a straight line (4) to (8) which are in LOS. Moreover, we add the straight line from S to (2) which are in LOS as well as the straight line from (8) to R which are also in LOS. Thus we can draw an approximate discrete ray from S to R through (2), (3), (4) and (8). The discrete ray is shown by green lines. In order to construct a more accurate ray, we modify the discrete ray so that the distance from S to (8) may be minimum, and we apply the imaging method to the discrete ray from (4) to R through (8). The final modified ray is plotted in blue lines in (a) of Fig.2. The ray from S to (8) constitutes a diffraction. We call it as a source diffraction, because it is associated with shadowing of the incident wave from source S by the line (3).

Let us explain another example of searched ray shown in (b) of Fig.2. First, we find the straight line from (5) to (8) which are in LOS. Second, we add the lines from S to (5) and from (8) to R, since S and (5) as well as (8) and R are in LOS. Thus we obtain an approximate discrete ray from S to R through (5) and (8) as shown in green lines. In order to obtain more accurate ray, we can modify the discrete ray based on the imaging method. The final ray plotted in blue lines shows that the ray emitted from S is first reflected from the line at (5) and next diffracted at the right edge of the line at (8), and finally it reaches R. We call this type of diffraction as an image diffraction, since it is associated with reflection and the reflection might be described as an emission from the image point with respect to the related line.

3.3 Reflection and diffraction coefficients

The purpose of this investigation is to evaluate the propagation characteristics of electromagnetic waves traveling along RRSs from a source point S to a receiver point R. We assume that the influences of transmitted waves through RRSs on propagation are negligibly small. As a result, the received electromagnetic waves at R are expressed in terms of incident, reflected and diffracted rays in LOS region, and they are denoted in terms of reflected and diffracted rays in NLOS region.

First we consider electromagnetic wave reflection from a flat ground plane composed of a lossy dielectric. The lossy dielectric medium, for example, indicates a soil ground plane. Fig.3 shows a geometry of incidence and reflection with source point S and receiver point R together with the source's image point S_i. In Fig.3, the polarization of the incident wave is assumed such that electric field is parallel to the ground plane (z-axis) or magnetic field is parallel to it. We call the former case as E-wave or horizontal polarization, and we call the latter case as H-wave or vertical polarization [Mushuake,1985], [Collin,1985].

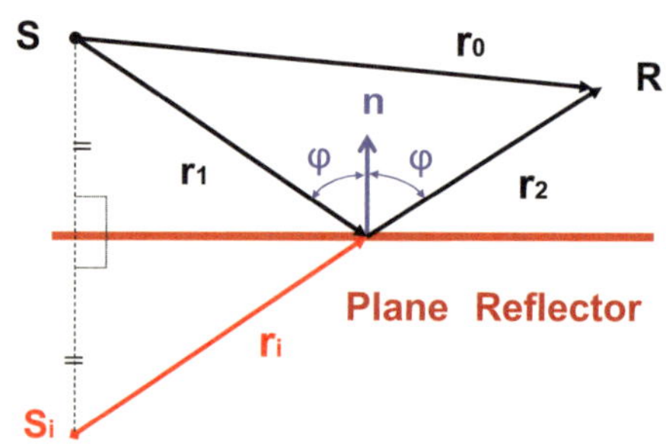

Fig. 3. Incidence and reflection.

The incident wave, which we also call a source field, and the reflected wave, which we also call an image field, are given by the following relations:

$$E_z, H_z = \Psi(r_0) + R^{e,h}(\phi)\Psi(r_1 + r_2) \tag{26}$$

where E_z and H_z indicate E-wave (e) and H-wave (h), respectively. The distances r_0, r_1 and r_2 are depicted in Fig.3, and the complex field function expressing the amplitude and phase of a field is defined in terms of a propagation distance r as follows:

$$\Psi(r) = \frac{e^{-j\kappa r}}{r}\,. \tag{27}$$

In the field expressions, the time dependence $e^{j\omega t}$ is assumed and suppressed through out this chapter. The wavenumber κ in the free space is given by

$$\kappa = \omega\sqrt{\epsilon_0\mu_0} \tag{28}$$

where ϵ_0 and μ_0 denote permittivity and permeability of the free space, respectively. E_z and H_z in Eq.26 correspond to $R^e(\phi)$ and $R^h(\phi)$, respectively. As mentioned earlier, E or H-wave indicates that electric or magnetic field is parallel to z-axis, respectively.
The reflection coefficients are expressed depending on the two different polarizations of the incident wave as follows:

$$\begin{aligned} R^e(\phi) &= \frac{\cos\phi - \sqrt{\epsilon_c - \sin^2\phi}}{\cos\phi + \sqrt{\epsilon_c - \sin^2\phi}} \\ R^h(\phi) &= \frac{\epsilon_c\cos\phi - \sqrt{\epsilon_c - \sin^2\phi}}{\epsilon_c\cos\phi + \sqrt{\epsilon_c - \sin^2\phi}} \end{aligned} \tag{29}$$

where ϕ is the incident angle as shown in Fig.3. Moreover, the complex permittivity of the medium is given by

$$\epsilon_c = \epsilon_r - j\frac{\sigma}{\omega\epsilon_0} \tag{30}$$

where ϵ_r is dielectric constant and σ is conductivity of the medium.

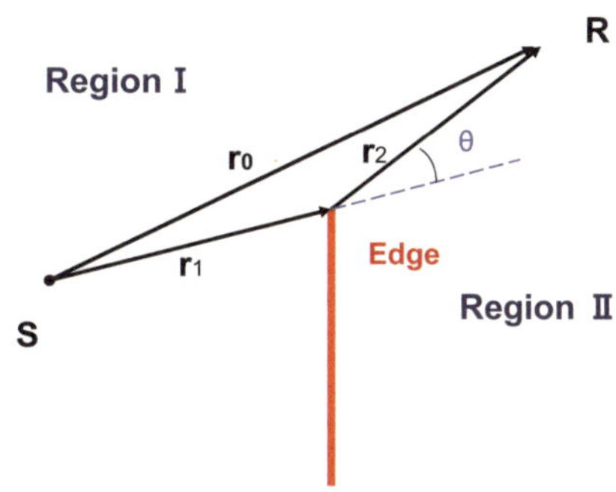

Fig. 4. Source diffraction from the edge of an half plane.

According to the rigorous solution for the plane wave diffraction by a half-plane [Noble,1958], diffraction phenomenon can be classified into two types. One is related to incident wave or field emitted from a source, which we call source field in short, as shown in Fig.4, and we call this type of diffraction as a source diffraction. The other is related to reflected wave or field emitted from an image, which we call image field in short, as shown in (a) of Fig.5, and we call this type of diffraction as an image diffraction. It should be noted that the rigorous solution based no the Wiener-Hopf technique is applicable only to the geometry of a semi-infinite half plane, and its extension to finite plate results in an approximate solution. However, it exhibits an excellent accuracy when the plate width is much longer than the wavelength. This is the starting point of the field analysis based on DRTM.
First we consider the source diffraction shown in Fig.4. In this case, we assume that the diffracted wave is approximated by the Winner-Hopf (WH) solution [Noble,1958]. The total diffracted fields for different two polarizations, that is E and H-wave, are given by

$$E_z, H_z = \begin{cases} D^s\Psi(r_0) & \text{(Region I)} \\ D^s\Psi(r_1 + r_2) & \text{(Region II)} \end{cases} \tag{31}$$

where the distances r_0, r_1 and r_2 are depicted in Fig.4 and the complex field function is defined by Eq.27. The source diffraction coefficient is defined as follows:

$$D^s = \begin{cases} 1 - D(r_1, r_2, \theta)\Psi(r_1 + r_2)/\Psi(r_0) & \text{(Region I)} \\ D(r_1, r_2, \theta) & \text{(Region II)} \end{cases} \tag{32}$$

where the diffraction function used in the above equation is defined by

$$D(r_1, r_2, \theta) = e^{jX^2} F(X) . \tag{33}$$

The complex type of Fresnel function for the diffraction function is defined by [Noble,1958]

$$F(X) = \frac{e^{\frac{\pi}{4}j}}{\sqrt{\pi}} \int_X^{\infty} e^{-ju^2} du \;\; (X > 0) \tag{34}$$

where the argument is defined by

$$X = \sqrt{\kappa(r_1 + r_2 - r_0)} \tag{35}$$

and the distances r_0, r_1 and r_2 are shown in Fig.4.
Finally, we discuss the image diffraction as shown in (a) of Fig.5 where S is a source point, S_i is its image point and R is a receiver point, respectively. Two edges of the line of the dicretized rough surface are given by the position vectors defined by Eq.22 as follows:

$$\mathbf{r_{i1}} = (i\Delta x, f_i), \;\; \mathbf{r_{i2}} = ((i+1)\Delta x, f_{i+1}) . \tag{36}$$

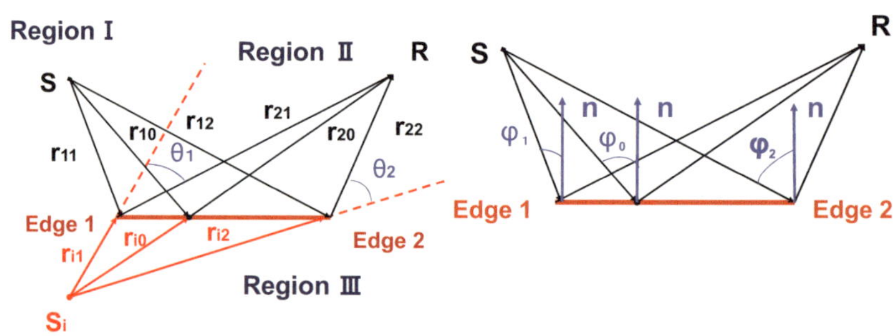

(a) Diffraction angles θ_1 and θ_2 (b) Reflection angles ϕ_1, ϕ_2 and ϕ_0

Fig. 5. Image diffraction.

Intersection between one straight line from $\mathbf{r_{i1}}$ to $\mathbf{r_{i1}}$ and the other straight line from the image point S_i to the receiver point R can be expressed as

$$\mathbf{r_{i0}} = \mathbf{r_{i1}} + \eta(\mathbf{r_{i2}} - \mathbf{r_{i1}}) \tag{37}$$

where η is a constant to be determined. It is worth noting that the constant η can be easily determined in terms of the position vectors, and $\eta < 0$ĄC$0 \le \eta \le 1$ and $1 < \eta$ correspond to regions I, II and III in (a) of Fig.5, respectively.

By use of the image diffraction coefficients as well as the field function in Eq.27, the image diffraction fields corresponding to the three regions are summarized as follows:

$$E_z, H_z = \begin{cases} D^{ie,h}\Psi(r_{i1}+r_{21}) & \text{(Rgion I)} \\ D^{ie,h}\Psi(r_{i0}+r_{20}) & \text{(Region II)} \\ D^{ie,h}\Psi(r_{i2}+r_{22}) & \text{(Region III)}\ . \end{cases} \tag{38}$$

The image diffraction coefficients are expressed in terms of the reflection coefficients in Eq.29, the diffraction function in Eq.33 and the field function in Eq.27 as follows:

$$D^{ie,h} = \begin{cases} -R^{e,h}(\phi_2)D(r_{i2},r_{22},\theta_2)\Psi(r_{i2}+r_{22})/\Psi(r_{i1}+r_{21}) \\ +R^{e,h}(\phi_1)D(r_{i1},r_{21},\theta_1) \quad \text{(RgionI)} \\ -R^{e,h}(\phi_1)D(r_{i1},r_{21},\theta_1)\Psi(r_{i1}+r_{21})/\Psi(r_{i0}+r_{20}) \\ -R^{e,h}(\phi_2)D(r_{i2},r_{22},\theta_2)\Psi(r_{i2}+r_{22})/\Psi(r_{i0}+r_{20}) \\ +R^{e,h}(\phi_0) \quad \text{(RegionII)} \\ -R^{e,h}(\phi_1)D(r_{i1},r_{21},\theta_1)\Psi(r_{i1}+r_{21})/\Psi(r_{i2}+r_{22}) \\ +R^{e,h}(\phi_2)D(r_{i2},r_{22},\theta_2) \quad \text{(RegionIII)} \end{cases} \tag{39}$$

where the two diffraction angles θ_1 and θ_1 are shown in (a) of Fig.5 and the three reflection angles ϕ_0, ϕ_1 and ϕ_1 are depicted in (b) of Fig.5.

The main feature of the image diffraction is described in the following way by using the traced rays shown in (a) of Fig.5. In region I, we have two image diffraction rays ($S \rightarrow \mathbf{r_{i1}} \rightarrow R$) and ($S \rightarrow \mathbf{r_{i2}} \rightarrow R$). In region II we have two image diffraction rays ($S \rightarrow \mathbf{r_{i1}} \rightarrow R$) and ($S \rightarrow \mathbf{r_{i2}} \rightarrow R$) and a reflection ray ($S \rightarrow \mathbf{r_{i0}} \rightarrow R$). In region III, we have two image diffraction rays ($S \rightarrow \mathbf{r_{i1}} \rightarrow R$) and ($S \rightarrow \mathbf{r_{i2}} \rightarrow R$). It should be noted that all these fields are continuous at the two boundaries of the three regions, that is, from I to II and from II to III.

3.4 Field evaluations based on DRTM

In the preceding sections, we have firstly proposed an algorithm to construct a discrete ray starting from a source, repeating source or image diffractions successively, and terminating at a receiver. Secondly, we have discussed the algorithm of DRTM to approximately evaluate the electromagnetic fields in relation to source or image diffractions based on the modified rays derived from the approximate discrete rays. As a result, electromagnetic fields along RRSs can be calculated numerically by repeating the DRTM computations step by step.

We assume that the source antenna is a small dipole antenna with gain $G = 1.5$ [Mushiake,1985] and input power P_i $[W]$. The direction of the source antenna is denoted by a unit vector $\mathbf{p}$, and the E or H-wave corresponds to the antenna direction parallel to z or y-direction, respectively. The fields of the small dipole antenna are classified into three types, that is, static, induced and radiated fields. We call the first two terms as the near fields and the last one as the far field [Mushiuake,1985],[Collin,1985]. Contrary to RFID where the near fields are mainly used [Heidrich,2010], sensor networks mainly use the far fields which are predominant in the region where the distance from the source is much longer than the wavelength ($r >> \lambda$).

Thus neglecting the near fields, the electric field radiating from a small dipole antenna is expressed in the following form [Mushiake,1985], [Collin,1985]

$$\mathbf{E_0} = \sqrt{30GP_i}[(\mathbf{u_r} \times \mathbf{p}) \times \mathbf{u_r}]\Psi(r) \tag{40}$$

where $\mathbf{r}$ is a position vector from the source to a receiver point and $r = |\mathbf{r}|$. The unit vector $\mathbf{u_r} = \mathbf{r}/r$ is in the direction from the source to the receiver point, and $|\mathbf{u_r} \times \mathbf{p}| = \sin\theta$

is the directivity of the small dipole antenna. The electric field radiated from the source antenna propagates along a RRS to a receiver point, decaying due to repeated source and image diffractions as discussed in the preceding section. We have assumed that the propagation model is 2D, which means that the RRSs are uniform in z-direction and the direction of propagation is restricted only to the (x, y)-plane. This assumption indicates that the back and forward diffractions are predominant and the side diffractions are negligibly small, and this assumption might be valid as long as the isotropic 2D RRSs are concerned.
At a source diffraction point, the electric field is subject to both amplitude and phase conversions according to Eq.32, but this source diffraction gives rise to no conversion of polarization. At an image diffraction point, however, not only amplitude and phase conversions but also conversion of polarization occur. The latter conversion is described in such a way that E-wave conversion occurs for electric field component parallel to z-direction and H-wave conversion does for electric field components perpendicular to z-direction as shown in Eq.39. At a source diffraction point, of course, the electric field receives the same conversion both for E-wave and H-wave as expressed in Eq.32.
Thus we can summarize the electric field at a receiver point in the following dyadic form

$$\mathbf{E} = \sum_{n=1}^{N} \left[\prod_{m=1}^{m=M_n^i} (\mathbf{D^i_{nm}}) \cdot \prod_{k=1}^{k=M_n^s} (\mathbf{D^s_{nk}}) \cdot \mathbf{E_0} \right] \Psi(r_n) \tag{41}$$

where $\mathbf{E_0}$ is the electric field vector of the n-th ray at the first source or diffraction point. N is the number of rays included for the field computations, M_n^s is the number of source diffractions of the n-th ray, and M_n^i is the number of its image diffractions. Moreover, $\mathbf{D^s_{nk}}$ is a dyadic source diffraction coefficient at the k-th source diffraction point of the n-th ray, and the coefficient can be computed by using the source diffraction coefficient D^s defined in Eq.32. On the other hand, $\mathbf{D^i_{nm}}$ is a dyadic image diffraction coefficient at the m-th image diffraction point of the n-th ray, and the coefficient can be computed by using the image diffraction coefficients D^{ie}_{nm} and D^{ih}_{nm} defined in Eq.39. It should be noted that the source diffraction coefficient is the same both for E and H-waves, while the image diffraction coefficient is different depending on the polarization of the fields at the image diffraction points. Moreover, the total distance of the n-th ray is given by

$$r_n = \sum_{m=0}^{m=M_n^s+M_n^i} r_{nm} \quad (n = 1, 2, \ldots, N). \tag{42}$$

where M_n^s and M_n^i are the number of source and image diffraction points of the n-th ray, respectively.
It is worth noting that the specular reflection from a plate is included automatically as a special case of the image diffraction in the reflection region II of (a) in Fig.5. In the present DRTM computations, of course, the more the diffraction times increases, the more computation time is required. However, we can neglect the higher order of diffractions because their effects are small. In the present analyses, we include at most the three order of diffractions, resulting in saving much computation time compared to the method of moments (MoM) [Yagbasan,2010].

4. Propagation characteristics of electromagnetic waves along RRSs

In the preceding sections, we have proposed the convolution method for generating RRSs with two parameters, height deviation h and correlation length $c\ell$. We have also proposed the

DRTM to compute electric fields which are first radiated from a small dipole antenna, next propagate along RRSs repeating source and image diffractions, and finally arrive at a receiver. According to the DRTM process, the electromagnetic waves emitted from a source antenna propagate along RRSs, repeating reflection, diffraction and shadowing, and thus resulting in a more attenuation than in the free space.
The field distribution along one pattern of RRS exhibits one pattern of field variation with respect to propagating distance, and that of the other pattern of RRS shows another pattern of field variation. Accordingly every field distribution is different depending on the seed of RRS generation. However, as is evident from the theory of statistics, the ensemble average of the field distributions may show a definite propagation characteristics in a simple analytical form. This situation was empirically confirmed by Hata in case of the propagation characteristics in urban or suburban areas [Hata,1980].

4.1 Distance characteristics of averaged field distribution

Now we show a numerical example to explain the statistical properties of electromagnetic wave propagation along RRSs. In this numerical simulation, the source antenna is placed at x=0 [m] and at 0.5 [m] high above RRSs, and the receiver point is movable along RRSs at 0.5 [m] high above them. The operating frequency is chosen as f=1 [GHz]. The RRS parameters are selected as height deviation h=10 [m] and correlation length cl=50 [m], and the material constants are chosen as dielectric constant ϵ_r=5 and conductivity σ=0.0023 [S/m]. We assume here that the terrestrial ground is composed of a dry soil [Sato,2002].

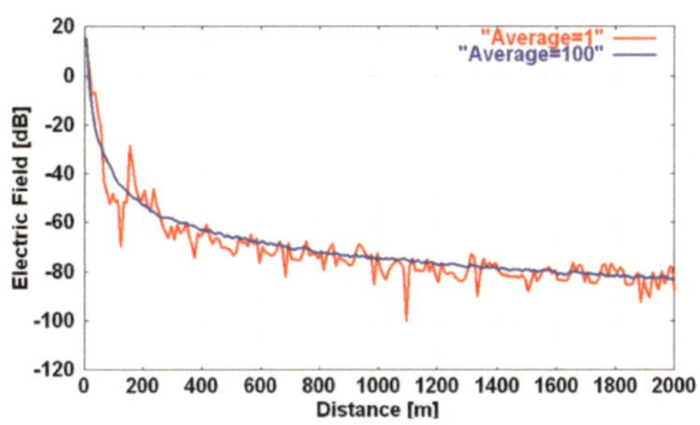

Fig. 6. Field distribution along one generated Gaussian RRS together with the ensemble average of 100 samples.

Fig.6 shows two electric field distributions along RRSs; one curve in red is the field distribution for one generated Gaussian RRS, and the other in blue is the ensemble average of the field distributions for 100 generated RRSs. In Fig.6, it is well demonstrated that one pattern of the field distributions is varying rapidly along the propagating distance, while the ensemble average of them is expressed in terms of a smooth and monotonic curve. As a result, it is concluded that we can approximate the ensemble average of the field distributions in terms of a simple analytic function.
Now we approximate the ensemble average of the electric field intensity by an analytic function with three constants, α, β and γ, as follows:

$$E = \frac{10^{\frac{\alpha}{20}}}{(r+\gamma)^{\beta}} \quad [V/m] \tag{43}$$

where the unit input power or P_i=1 [W] in Eq.40 is assumed, and α is an amplitude weighting factor, β is an order of propagation distance, and γ is a distance correction factor. Rewriting the above relation in dB leads to the following equation [Hata,1980]:

$$E = \alpha - 20\beta \log_{10}(r + \gamma) \ \ [dB]. \tag{44}$$

Next, we determine the three unknown constants α, β and γ based on the method of least mean square (LMS) with the objective function defined by

$$\Phi(\alpha, \beta, \gamma) = \frac{1}{N} \sum_{n=1}^{N} (E_n^d - E_n)^2 \longrightarrow \min \tag{45}$$

where N is the number of data. Moreover, E_n^d is the averaged value of simulated or experimental data at $r = r_n$, and E_n is the data computed by putting $r = r_n$ in Eq.43. Applying the conjugate gradient method (CGM) [Press,1992] to Eq.45, we can numerically determine the constants α, β and γ explicitly.

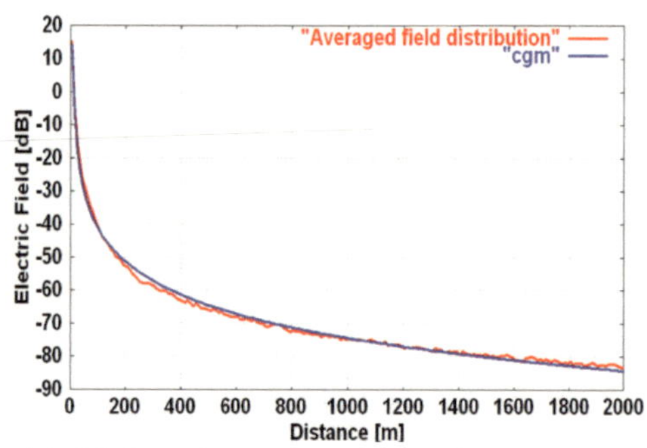

Fig. 7. Ensemble average of field distribution compared with estimated field distribution.

Fig.7 shows a comparison of the ensemble average of the field distribution with the estimated function based on the proposed algorithm. The estimated curve is given by Eq.43 with the constants determined by CGM such as α=22.6 [dB], β=1.62 and γ=-6.85 [m]. It is worth noting that the distance correction factor γ is small in comparison with the radio communication distance r, and thus we can neglect it as $\gamma \simeq 0$. Consequently, when we estimate the propagation characteristics of electromagnetic waves traveling along RRSs, the most important parameters are α and β in Eq.43.

4.2 Procedure for estimation of radio communication distance

In the preceding section, we have demonstrated that the ensemble average of the field distributions of electromagnetic waves traveling along RRSs is well expressed in a simple and analytic function with three parameters α, β and γ. In this section, we propose an algorithm to estimate the radio communication distance when the two parameters for RRSs as well as another two parameters for source and receiver are specified. The first two parameters are the height deviation h and the correlation length cl for RRSs, and the last two parameters are the input power P_i of a source antenna and the sensitivity of a receiver or the minimum detectable electric intensity E_m of the receiver.

Feasibility of the present method comes from the simplified form of the electric field distribution given by Eq.43. Once the three constants α, β and γ are determined based on the simulated or experimental data, we can analytically evaluate the radio communication distance from a source with an arbitrary input power P_i to a receiver with an arbitrary minimum detectable electric intensity E_{min}. In this case we have assumed that both the source and receiver are on the homogeneous RRSs with the same parameters. Considering the dB expression such as [dBm] for the input power P_i and [dBμ V/m] for the minimum detectable electric intensity E_{min}, we can rewrite Eq.44 in the following form

$$E_{min} - 120 = \alpha - 20\beta \log_{10}(r + \gamma) + P_i - 30. \tag{46}$$

Solving for r in the above equation, we can analytically determine the radio communication distance r by the following relation:

$$r = 10^{(\alpha + P_i - E_{min} + 90)/20\beta} - \gamma \simeq 10^{(\alpha + P_i - E_{min} + 90)/20\beta}. \tag{47}$$

Eq.47 is the estimation formula for the radio communication distance between the source antenna and receiver placed on homogeneous RRSs. The parameters of the RRSs are the height deviation h [m] and correlation length $c\ell$ [m]. Moreover, the input power of the source is denoted by P_i [dBm] and the minimum detectable electric intensity of th receiver is described by E_{min} [dBμ V/m]. As mentioned earlier, we can neglect the distance correction factor γ, and we have the more simplified estimation formula in Eq.47.

We show some numerical examples of the radio communication distances along RRSs computed by Eq.47. Fig.8 (a) shows radio communication distance r [m] versus the minimum detectable electric field intensity E_{min} [dBμV/m] of the receiver with the input power P_i [dBm] of the source antenna as a parameter. The height deviation of the RRSs is selected as $h = 10$ [m] and the correlation length is chosen as $cl = 50$ [m]. The spectrum of the RRSs is assumed to be Gaussian, and the material constants of the RRS are chosen such that dielectric constant is ϵ_r=5 and conductivity is $\sigma = 0.023$ [S/m]. The operating frequency is chosen as $f = 1$ [GHz]. The minimum detectable electric intensity E_{min} is varied from 40 [dBμ V/m]to 80 [dBμV/m], where the parameters for input power of source antenna are selected as $P_i = -10, -8, \cdots, 8, 10$ [dBm].

Fig.8 (b) shows radio communication distance r [m] versus minimum detectable electric field intensity E_{min} [dBμV/m] where the RRS height deviation is selected as $h = 10$ [m] and its correlation length is chosen as $c\ell = 100$ [m]. Other parameters are chosen as the same values as in Fig.8 (a). In these two figures, it is demonstrated that the larger the input power of a source antenna becomes, the longer is the radio communication distance. It is also shown that the smaller the minimum detectable electric field intensity of a receiver becomes, the longer is the radio communication distance. Thus, based on the proposed procedure, we can estimate the radio communication distances along RRSs.

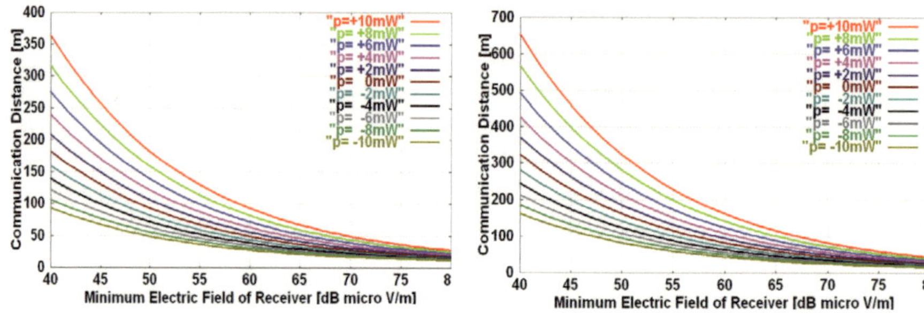

(a) $h = 10\ m$ and $c\ell = 50\ m$ (b) $h = 10\ m$ and $c\ell = 100\ m$

Fig. 8. Radio communication distances along RRSs with Gaussian spectrum.

Fig.9 (a) and Fig.9 (b) show the radio communication distance versus minimum detectable electric intensity of a receiver located above the RRSs with the 1st order of power-law spectrum. The RRS parameters are chosen as the height deviation is $h = 10$ [m] and the correlation length is $c\ell = 50$ [m] in (a), and $h = 10$ [m] and $c\ell = 100$ [m] in (b), respectively. Other parameters are selected as the same as the former two Gaussian cases in Fig.8.

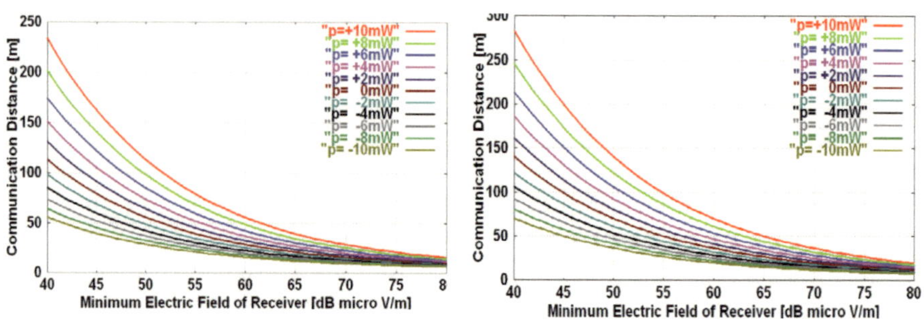

(a) $h = 10\ m$ and $c\ell = 50\ m$ (b) $h = 10\ m$ and $c\ell = 100\ m$

Fig. 9. Radio communication distances along RRSs with power-law spectrum.

Fig.10 (a) and Fig.10 (b) show the radio communication distance versus minimum detectable electric intensity along the RRSs with exponential type of spectrum. The RRS parameters are chosen as $h = 10$ [m] and $c\ell = 50$ [m] in (a), and $h = 10$ [m] and $c\ell = 100$ [m] in (b), respectively. Other parameters are the same as the former examples in Figs.8 and 9.

It is evident from the above numerical examples that the radio communication distances vary depending on the types of the RRS spectra, even though the RRS parameters h and $c\ell$ are the same. It is shown that the radio communication distances along the RRSs with the Gaussian type of spectrum are the longest of the three types of spectra and those with the exponential type of spectrum are the shortest of the three.

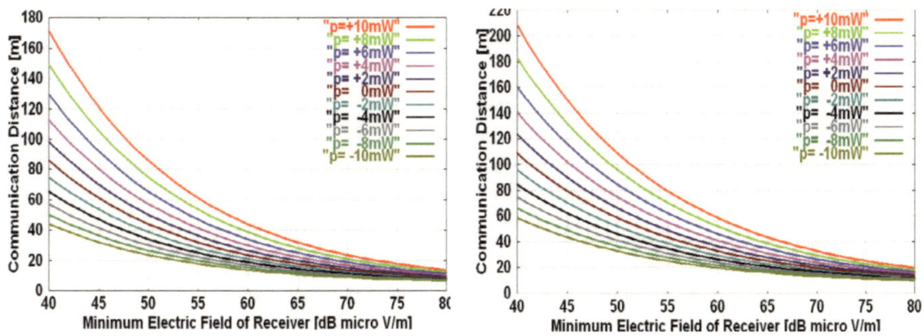

(a) $h = 10\ m$ and $c\ell = 50\ m$ (b) $h = 10\ m$ and $c\ell = 100\ m$

Fig. 10. Radio communication distances along RRSs with exponential spectrum.

5. Conclusion

In this chapter, from a viewpoint of the application of radio communications to sensor networks, we have first discussed the convolution method to generate random rough surfaces (RRSs) to numerically simulate the propagation characteristics of electromagnetic waves traveling along the RRSs. Second, we have introduced the discrete ray tracing method (DRTM) to numerically evaluate the distributions of the electromagnetic waves along the RRSs. The remarkable point of the present method is to discretize not only the RRS's profile but also the ray tracing itself, resulting in saving much computation time. Third, we have proposed an algorithm to estimate the ensemble average of the field distributions in a simple and analytic expression by introducing the amplitude weighting factor α, the order of propagation distance β and the distance correction factor γ. Finally, we have introduced a formula to estimate the radio communication distance along RRSs, provided that the input power of a source antenna and the minimum detectable electric field intensity of a receiver are specified.

Numerical calculations were carried out for the dependence of radio communication distance versus minimum detectable electric field intensity of a receiver with the input power of a source antenna as a parameter. The results of computer simulations have revealed us that the longer the correlation length of the RRSs becomes, the longer are the radio communication distances. It has also been found that the radio communication distances along the RRSs vary depending on the types of spectra of the RRSs, and those of the Gaussian type of spectrum are longer than those of any other types of spectra.

We have treated only the homogeneous 1D RRSs to estimate the radio communication distances. In a more realistic situation, however, it is required to deal with inhomogeneous 2D RRSs [Uchida,2009]. This deserves as a near future investigation.

6. References

Aryanfar, F. & Sarabandi, K. (2007). Validation of Wireless Channel Models Using a Scaled mm-Wave Measurement System, *IEEE Antennas and Propagation Magazine,* vol.49, no.4, pp.124-134.

Collin, R.E. (1985). Antennas and Radio Wave Propagation, *McGraw-Hill Inc., New York*, pp.13-86.

Hata, M. (1980). Empirical formula for propagation loss in land mobile radio services, *IEEE Trans. Veh. Technol.*, VT-29(3), pp.317-325.

Heidrich, J. et al. (2010). The Roots, Rules, and Rise of RFID, *IEEE Microwave Magazine*, vol. 11, no. 3 , pp.78-86.

Honda, J. et al. (2006). Effect of rough surface spectrum on propagation characteristics, *IEEJ Technical Reports*, EMT-06-128, pp.65-70.

Honda, J. & Uchida, K. (2009). Discrete Ray-Tracing Method (DRTM) Analysis of Propagation Characteristics along Random Rough Surface in Relation to Development of Wireless Sensor Network, *Proceedings of RWS 2009*, TU2P-4, pp.248-251.

Honda, J., Uchida, K., Yoon, K.Y. (2010). Estimation of radio communication distance along random rough surface, *IEICE Trans. Electron.*, E93-C(1), pp.39-45.

Johnsonbaugh, R. & Kalin, M. (1997). C for Scientists and Engineers, *Prentice-Hall, Inc., New Jersey*, pp.191-195.

Mushiake, Y. (1985). Antennas and Radio Propagation, *Corona Publishing Co., Ltd.*, p.39.

Noble, B. (1958). Methods based on the Wiener-Hopf technique, *Pergamon Press.*

Phu, P., Ishimaru, A., Kuga Y. (1994). Co-polarized and cross-polarized enhanced backscattering from two-dimensional very rough surfaces at millimeter wave frequencies, *Radio Sci.*, 29(5), pp.1275-1291.

Press, W.H. et al. (1992). Numerical Recipes in Fortran 77: The Art of Scientific Computing, *Cambridge University Press, New York,* pp.387-448.

Sato, M. (2002). Subsurface imaging by ground penetrating radar, *IEICE Trans. Electron.*, vol.J85-C, no.7, pp.520-530.

Thoros, E.I. (1988). The validity of the Kirchhoff approximation for rough surface scattering using a Gaussian roughness spectrum, *J. Acoust. Soc. Am.*, 83(1), pp.78-92.

Thoros, E.I. (1990). Acoustic scattering from a Pierson-Moskowitz sea surface, *J. Acoust. Soc. Am.*, 88(1), pp.335-349.

Tsang, L., Chan, C.H., Pak, K. (1994). Backscattering enhancement of a two-dimensional random rough surface (three-dimensional scattering) based on Monte Carlo simulation, *J. Opt. Soc. Am. A*, 11(2), pp.711-715.

Uchida, K. et al. (2007). FVTD analysis of electromagnetic wave propagation along random rough surface, *IEICE Trans. Commun.*, J90-B(1), pp.48-55.

Uchida, K. et al. (2008).Ą@Analysis of electromagnetic wave propagation along rough surface by using discrete ray tracing method, *Proceedings of ISAP 2008*, pp.939-942.

Uchida, K., Honda, J., Yoon, K.Y. (2009). Distance characteristics of propagation in relation to inhomogeneity of random rough surface, *Proceedings of ISMOT 2009*, pp.1133-1136.

Yagbasan, A. et al. (2010). Characteristic Basis Function Method for solving Electromagnetic scattering Problems Over Rough Terrain Profiles, *IEEE Trans. Antennas Propag.*, vol. 58, no. 5 , pp.1579-1589.

Yoon, K.Y., Tateiba, M., Uchida, K. (2000). FVTD simulation for random rough dielectric surface scattering at low grazing angle, *IEICE Trans. Electron.*, E83-C(12), 1836-1843.

Yoon, K.Y., Tateiba, M., Uchida, K. (2002). A numerical simulation of low-grazing-angle scattering from ocean-like dielectric surfaces, *IEICE Trans. Commun.*, E85-B(10), pp.2344-2347.

14

Design of Radio-Frequency Transceivers for Wireless Sensor Networks

Bo Zhao and Huazhong Yang
Department of Electronic Engineering, TNLIST, Tsinghua University
Beijing, China
zhaobo06@mails.tsinghua.edu.cn
yanghz@tsinghua.edu.cn

1. Introduction

The SoC (System-on-Chip) design for the WSN (Wireless Sensor Networks) nodes is the most significant technology of modern WSN design. There are a large amount of nodes in a WSN system, the nodes are densely deployed either inside the environment or very close to it. Each node is equipped with a sensor, an ADC (Analog-to-Digital Converter), a MCU (Micro Controller Unit), a storage unit, a power management unit, and a RF (Radio-Frequency) transceiver, as shown in Fig. 1, so that it can sense, store, process, and communicate with other sensors using multi-hop packet transmissions. The basic specifications of WSN are reliability, accuracy, flexibility, expenses, the difficulty of development and power consumption. Because all the nodes are battery-powered, power consumption is the most important specification of WSN.

The core part of a WSN node is the RF transceiver, which is used to realize the wireless communication among the nodes. For a common used commercial chip, the distribution of power consumption is shown in Fig. 2, where TX and RX represent the transmitting mode and receiving mode of the transceiver. We can see that the RF part consumes the most power. Besides, modern RF design is composed of so many subjects that it requires IC designers to have sufficient knowledge. As a result, the IC design of RF transceivers becomes the most challenging research topic in the WSN field.

As the applications of WSN become more and more widespread, many companies have developed highly-integrated chips for RF transceivers. The main specifications of several commercial chips are shown in Table. 1. The common characters of these chips can be summarized into several aspects: 1)a low data rate, 2)low power consumption, 3)high sensitivity, 4)relatively low output power, and 5)a simple modulation scheme.

There are several ways to achieve low power consumption in WSN: 1)reduce the radiated power by using ad-hoc networks and multi-hop communication, 2)optimize the trade-off between communication and local computing, 3)design more power-efficient RF transceivers, and 4)develop more energy-efficient protocols and routing algorithms. And the third one is what we will talk about in this chapter.

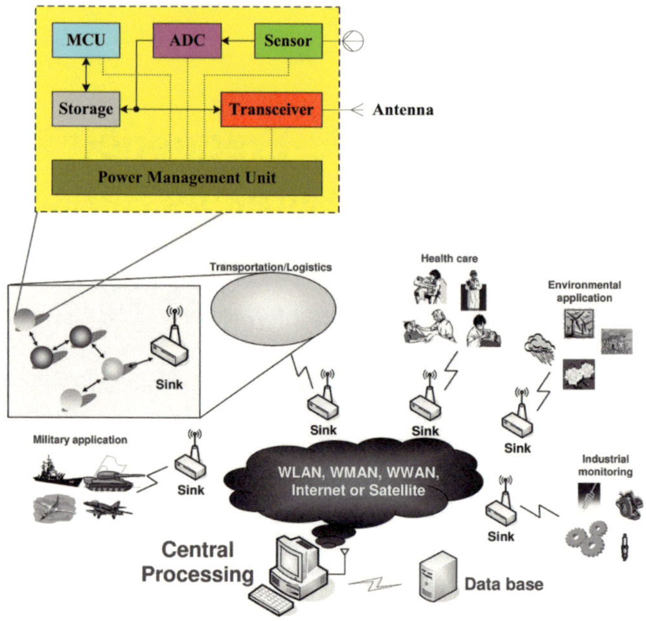

Fig. 1. The Wireless Sensor Networks.

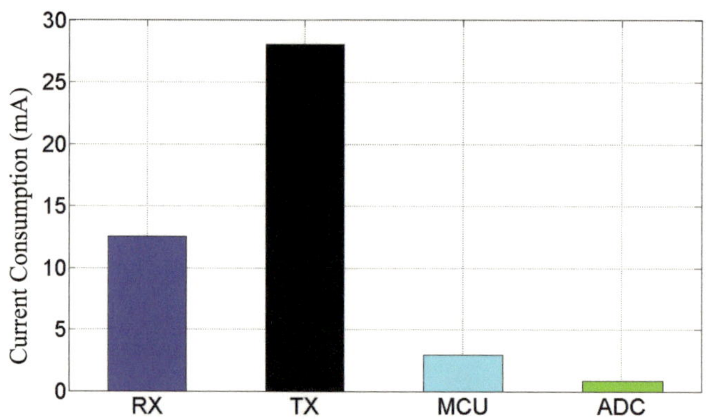

Fig. 2. The power distribution in a commercial WSN chip.

2. The classification of WSN transceivers

In this section, we make a classification of WSN transceivers, which are sorted by both modulation schemes and system architectures. Although some complex schemes such as OFDM (Orthogonal Frequency Division Multiplexing) can be adopted for the prospect of spectrum effective usage, RF designers are still inclined to simple schemes such as OOK (On-Off Keying), FSK (Frequency Shift Keying), UWB (Ultra-Wide Band), MSK (Minimum Shift Keying), BPSK (Binary Phase Shift Keying), QPSK (Quadrature Phase Shift Keying), and so on for low power consideration. From Table. 1, we can see that these simple modulation schemes are often adopted in the common-used commercial chips. In this section, we will analyze the advantages and disadvantages of the transceivers with these modulation schemes.

Chips	FB[a] (MHz)	DR[b] (kbps)	Power[c] (mW)	Sensitivity (dBm)	OTP[d] (dBm)	MS[e]
*TR*1000	916.5	115.2	14.4/36	-98	-1.2	OOK/ASK
*TRF*6903	300∼1000	19.2	60/111	-103	-12∼8	FSK/OOK
*CC*1000	300∼1000	76.8	30/87.8	-107	-20∼10	FSK
*CC*2420	2400	250	33.8/31.3	-95	0	O-QPSK
*nRF*905	433∼915	100	37.5/90	-100	-10∼10	GFSK
*nRF*2401	2400	0∼1000	75/39	-80	-20∼0	GFSK

[a]FB: The working Frequency Band of the chips.
[b]DR: The Data Rate of the transceivers.
[c]Power: The Power consumption for receiving/transmitting mode.
[d]OTP: The Output Transmitting Power.
[e]MS: The Modulation Scheme of the transceiver.

Table 1. The main specifications of several commercial RF chips common used in WSN.

2.1 The OOK transceiver

The OOK transceiver can often be realized by a simple architecture, since both the modulator and demodulator are easy to implement. As a result, the power consumption can be reduced. Another advantage is that a high data rate can be obtained by OOK. However, an AGC (Automatic Gain Control) with a wide dynamic range is often needed; a special coding is needed to avoid the saturation caused by long series of 0 or 1 in the receiver; it is spectrally inefficient; the most serious defect of OOK is that it is strongly susceptible to interferers, then the maximum communication distance of OOK transceivers is usually not long.
As a high data rate can be obtained by OOK transceivers, there are some works focused on multi-gigabit short-range wireless communications (Jri et al., 2010). The transceivers work at several GHz with low energy per bit. Besides, since the power consumption of OOK transceivers can be made very small, some recently works design OOK receivers for wake-up usages (Pletcher et al., 2009; Seungkee et al., 2010). The power consumption of these wake-up receivers can be as low as a few tens of μW.
One typical architecture of OOK transceivers is shown in Fig. 3. In the transmitter, the baseband data is modulated by a mixer with a carrier generated by an oscillator or a PLL (Phase-Locked Loop), and then the modulated signals are amplified by a PA (Power Amplifier) and emitted through an antenna. In the receiver, the signals received from an antenna is amplified by a LNA (Low-Noise Amplifier) firstly, and then detected by an envelop detector. At last, a simple DEM (DEModulator) can be used to demodulate the signals into 1-0 series. Therefore, in short-range and high-speed WSN applications, the OOK transceiver is a reasonable choice.

2.2 The FSK transceiver

As only the zero-crossing points of the signal contain useful information, FSK transceivers can work without an AGC, and special coding is not necessary. The modulator and demodulator are also easy to realize. As a result, the FSK transceiver has a simple architecture that ensures low power consumption and low cost. The spectrum efficient and ability to avoid interferences of FSK are both higher than that of OOK, so the communication distance of FSK transceivers can be longer. Besides, frequency hopping can be realized for FSK schemes. However, the complexity of FSK transceivers is relatively higher than that of OOK transceivers.

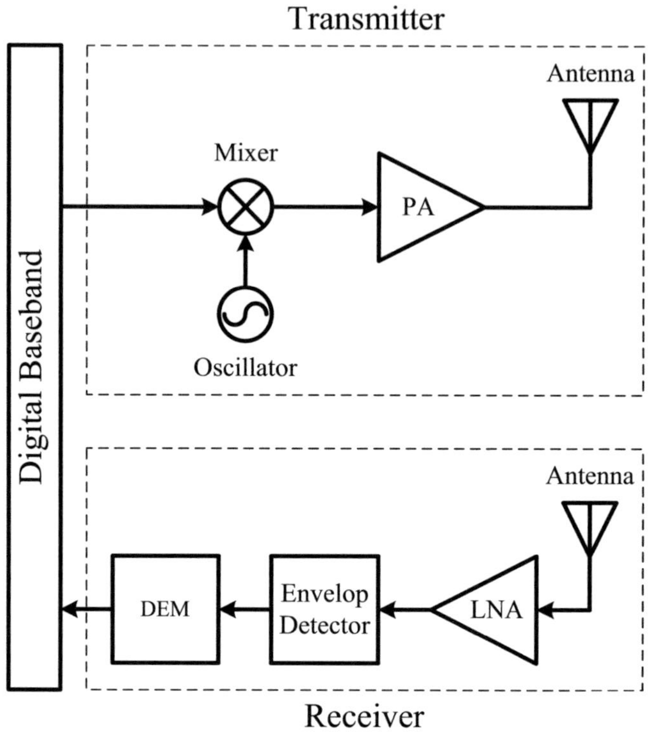

Fig. 3. A typical OOK transceiver.

A typical FSK transceiver is shown in Fig. 4. For the transmitter, the PLL directly digital modulation can be adopted (Perrott et al., 1997). This technology will be described in detail in section 4.4. The data stream from baseband is shaped firstly, and then input into a PLL to generate the FSK signals. Then the FSK signals are amplified by a PA and emitted by an antenna. In the receiver, the received signals are amplified by a LNA, and down-converted by a mixer that is followed by a filter to depress the interferences and the high-frequency component. The FSK demodulator is easy to realize since only the zero-cross points contain data information, and sometimes a zero-cross detector is enough. However, in many applications, the influence caused by frequency offset needs to be avoided. The power consumption of such transceiver is relatively low because the modulation and demodulation of FSK signals is easy to realize.
As we analyzed above, OOK can realize a high data rate, but the communication distance is short; FSK transceivers can be used for long-distance communication, but the data rate is often not high. Therefore, a multi-mode transceiver in which OOK and FSK can be compatible with each other may be a good choice. The OOK modulation can be implemented by directly modulating the output power of the PA in Fig. 4.

2.3 The UWB transceiver

UWB is defined as a signal that occupies a bandwidth wider than 500 MHz or a fractional bandwidth larger than 20%. Although UWB is not a modulation scheme, we list the UWB transceivers individually here because the advantages of UWB are remarkable: 1)a high data, 2)low cost and low power consumption, and 3)high accuracy in distance measuring due to

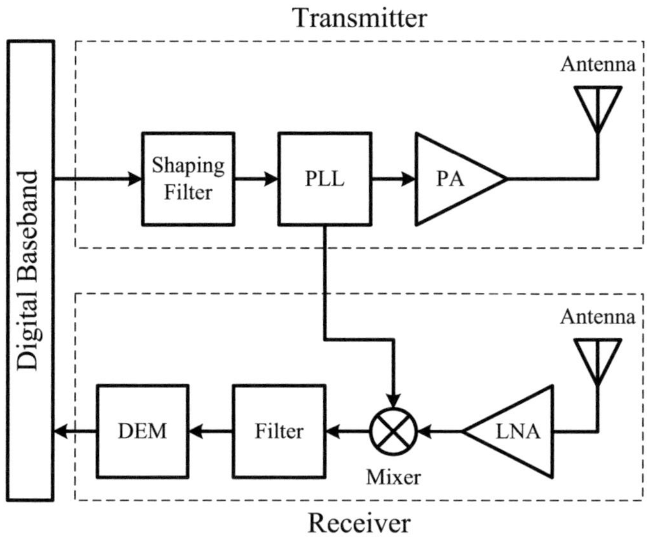

Fig. 4. An typical FSK transceiver.

the narrow pulse width of a few nanoseconds. However, the disadvantage is also serious: the spectrum is too wide and the emitting power is limited, so the single-hop transmission distance is very short (usually less than 10 m).

The IR (Impulse-Radio) UWB transceiver can be implemented with ultra-low power consumption. A typical example is shown in Fig. 5, where the architecture is similar to that of the typical OOK transceiver. The transmitter has a very simple structure, which is composed of an impulse generator and a PA. The impulse generator can be realized by a few logical gates. The receiver includes a LNA, a mixer as a squarer, an integrator and a comparator. In some applications, the architecture is even simpler than that of common OOK transceivers. Therefore, ultra-low power can be obtained with UWB transceivers, which fits the short-range WSN requirement very well.

2.4 Other Modulation Schemes

The spectrum efficiency of MSK is higher than FSK, but the demodulation is more complex and frequency hopping can not be realized for MSK. In 2003, IEEE released 802.15.4, which supports BPSK and O-QPSK (Offset-QPSK). The ability of BPSK to avoid interferences is better than that of FSK; and O-QPSK is more spectrally efficient. But the demodulation circuits of both BPSK and O-QPSK are complex and ADCs are usually needed. For general WSN applications that have a low data rate at several hundreds of Hz, the usage of these complex modulations is not recommended. However, in a few applications that have a high data rate and a strict spectrum limitation, MSK, BPSK, O-QPSK and even OFDM can be adopted.

2.5 The classification based on system architectures

In this section, we sort the transceivers according to the system architecture, and the types can be divided into superheterodyne, zero-IF (Intermediate Frequency), low-IF, slide-IF, super-regenerative, amplifier-sequenced hybrid architectures, and so on. These kinds of transceivers are summarized into Table 2. In WSN applications, low-IF transceivers are often adopted for high integration and low power consumption.

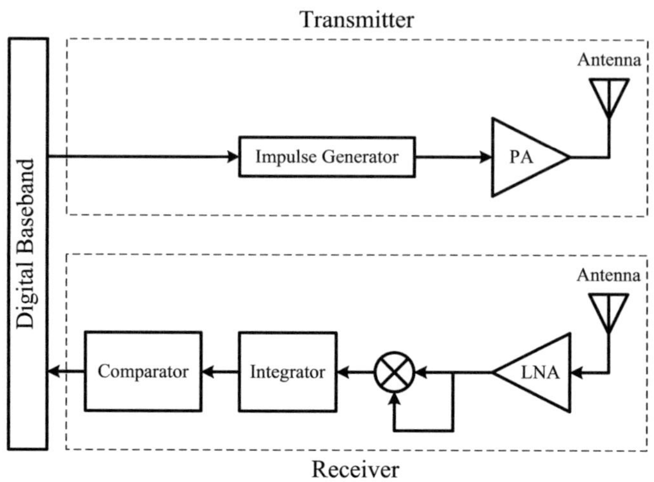

Fig. 5. A typical UWB transceiver.

A common transceiver architecture is shown in Fig. 6. In the transmitter, there is I/Q dual-path chains to realize the vector-signal transmitting. The coding, spreading, shaping, and so on are all finished in the digital domain. The data stream from digital baseband is converted to analog signals by two DACs (Digital-to-Analog Converters), and then the interferences and harmonics are filtered out by two filters. Two mixers are adopted for I/Q dual-path up-conversion and modulation. The phase difference of the two LO (Local Oscillation) signals input to the two mixers is 90°. At last, the I/Q signals are summed by an adder, amplified by a PA, and then emitted through an antenna. Assume the signal in the I path is a, the signal in the Q path is b, and then the transmission signal can be expressed as $a + jb$. Therefore, all the QAM (Quadrature Amplitude Modulation) signals in the constellation diagram can be generated by this transmitter. In the receiver, the received signals are amplified by a LNA firstly, and then down-converted by two mixers into I/Q dual-path signals. After being filtered by a filter, the analog signals are converted to digital domain by two ADCs. At last, the demodulating, de-spreading, synchronizing, decoding and so on are all done in digital domain. Such is the common architecture of SDR (Software Defined Radio), all the modulation schemes can be realized in this transceiver. Generally, the WSN usages have the characters of low cost, low power and a low data rate, so such complex architecture of SDR is not necessary in many WSN applications since simple analog circuits can be adequate. However, for specially WSN applications that need a complex modulation scheme, the SDR transceiver is indispensable.

3. The system design of transceivers

In this section, we will describe how to make system-level plans for the transceivers. The system-level specifications of a receiver are composed of sensitivity, maximum received signal, co-channel interferences, ACS (Adjacent/Alternate Channel Selectivity), image rejection, desensitization or blocking, and so on. And the specifications of a transmitter include ACPR (Adjacent Channel Power Ratio), EVM (Error Vector Magnitude), noise emission, and so on. These system-level specifications are decided by the protocols and network formation of WSN, and are affected by every submodule's performance, such as noise figure, gain, third intercept point, and so on. The main task for the system designers is to assign the system-

Architecture	Advantages	Disadvantages
Superheterodyne	Simple structure High sensitivity High selectivity	Poor integration level Large power consumption
Zero-IF[a]	Easy to be integrated Low power	DC offset Flicker noise
Low-IF	Easy to be integrated Immune to DC offset	Image rejection Difficult baseband design
Slide-IF	Small image interference Easy to be integrated	Difficult for LO design
super-regenerative	High sensitivity Low power	Low data rate Seriously affected by PVT[b] Poor stability
Amplifier-sequenced hybrid	Good stability Low power High Sensitivity	Low data rate

[a]IF: Intermediate Frequency
[b]PVT: Process, Voltage, and Temperature

Table 2. The classification based on system architecture.

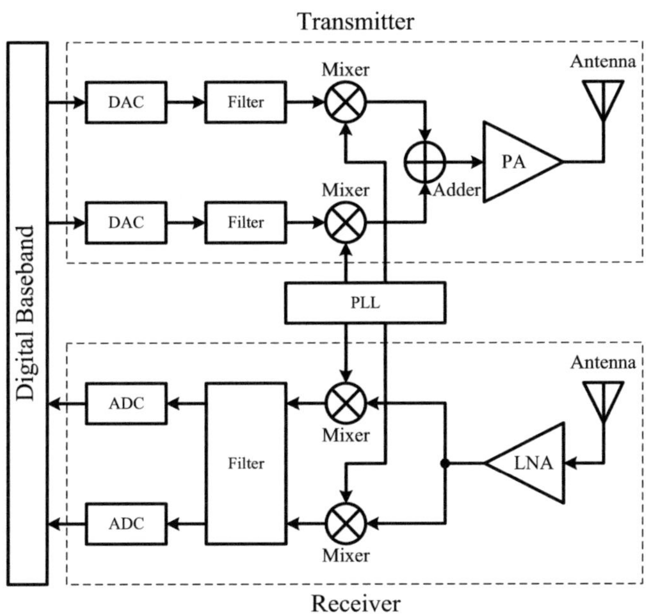

Fig. 6. A common architecture of transceivers.

level specifications to every individual submodules. Some of the conclusions in this section are referenced to Gu's book (Gu, 2005).

For the AWGN (Additive White Gaussian Noise) channel, the noise at the input of a receiver chain can be equivalent to thermal noise, and the sensitivity of a receiver can be expressed as:

$$R_{min}(dBm) = -174 + 10log(BW) + NF + CNR_{min} \tag{1}$$

where BW is the receiver noise bandwidth in Hz, NF is the overall noise figure of the receiver in dB, and CNR_{min} is the minimum CNR (Carrier-to-Noise Ratio) required for obtaining the required error rate.

As two different radio transmitters may use the same frequency in actual applications, RF receivers must have a certain ability to avoid the influence of co-channel interferences. Another important specification of receiver is the ACS, which measures the ability to receive a desired signal at the operating channel frequency in the presence of adjacent/alternate channel signals at a frequency offset from the assigned channel. The blocking characteristic is similar to the ACS, but it measures the anti-interference ability of an unwanted interferer at frequencies other than those of adjacent channels. The adjacent/alternate channel interference signal is usually modulated, while the blocking interferer is often defined as a continuous waveform tone. Both the adjacent/alternate channel selectivity and the blocking characteristics of a receiver are determined mainly by the characters of the channel filter, and the phase noise and spurs of the LOs in the adjacent/alternate channel bandwidth or around the unwanted interferer. For most wireless mobile systems, the desired signal level at receiver input for testing is defined as 3 dB above the reference sensitivity level R_{min} (Gu, 2005):

$$R_{d,i} = R_{min} + 3dB \tag{2}$$

For the single-conversion receiver, as shown in Fig. 6, it can be derived the ACS or the blocking characteristic $ACS_{adj/alt/block}$ to have the following form:

$$ACS_{adj/alt/block} = I_{adj/alt/block} - R_{d,i} \tag{3}$$

$$I_{adj/alt/block} = 10lg\left(\frac{10^{\frac{R_{d,i}-CNR}{10}} - 10^{\frac{-174+10lgBW+NF}{10}}}{10^{\frac{PN+10lgBW-\Delta R_{IF}}{10}} + 10^{\frac{SP-\Delta R_{IF}}{10}}}\right) \tag{4}$$

where CNR is the carrier-to-noise ratio for a given bit error rate, BW is the receiver noise bandwidth, NF is the noise figure of the RF front end, PN and SP are the phase noise and spurs of the LO signals, and ΔR_{IF} is the relative rejection of the IF channel filter to the adjacent/alternate channel signal or the blocking interferer.

The maximum received signal is decided by both the 1-dB compression point and the intermodulation behavior of the receiver. For a RF receiver chain with multi-stages, the input signal strength needs to be smaller than the 1-dB compression point of the first stage, and the output signal strength of every previous stage must be smaller than the 1-dB composition point of the following stage, otherwise the receiver will be saturated. The maximum received signal is also limited by the intermodulation because the nonlinearity will become more serious under larger signal strength. The intermodulation characteristics reflect the linearity of the receiver, and it is limited by the noise figure, the receiver noise bandwidth, the phase noise and spurs of the LOs, and the cross-modulation. For the single-conversion receiver, as shown in Fig. 6, the expression of the allowed maximum degradation of the desired signal caused by noise and interferences at the receiver input is:

$$D_{max} = R_{d,i} - CNR_{min} \tag{5}$$

The minimum third intercept point of a receiver can be expressed as:

$$IIP3_{min} = \frac{1}{2}\left[3I_{min} - 10lg(10^{\frac{D_{max}}{10}} - 10^{\frac{N_{nf}}{10}} - P_{pn} - P_{sp} - 10^{\frac{N_{cm}}{10}})\right] \tag{6}$$

where I_{min} is the minimum intermodulation interference tone level at receiver input. P_{pn} and P_{sp} are the phase noise and spurs contributions of LO, and they can be expressed as:

$$P_{pn} = 10^{\frac{N_{pn}+10lgBW+I_{in}-\Delta R}{10}} \tag{7}$$

$$P_{sp} = 10^{\frac{N_{sp}+I_{in}-\Delta R}{10}} \tag{8}$$

where N_{pn} is the average phase noise over the receiver bandwidth at a frequency offset equal to the space between the interference and the carrier, N_{sp} is the magnitude of the spurs at a frequency offset equal to or nearby the space between the interference and the carrier, I_{in} is the intermodulation interference tone, and ΔR is the rejection to the interference tone of a filter. N_{nf} denotes the receiver inherent noise converted to its input port:

$$N_{nf} = -174 + NF + 10lgBW \tag{9}$$

N_{cm} is the cross-modulation product, which can be approximately expressed as:

$$N_{cm} = I_{in} - 2IIP3_{lna} + 2(P_{Tx} + IL_{dRx} - R_{dTx}) + C \tag{10}$$

where $IIP3_{lna}$ is the LNA input third intercept point, P_{Tx} is the transmitter output power at the antenna port in dBm, IL_{dRx} is the receiver side insertion loss of the duplexer in dB, R_{dTx} is the duplexer receiver side filter rejection to the transmission in dB, and C is a correction factor associated with waveform magnitude fluctuation and interference tone offset frequency, which is approximately -3.8 dB and -5.8 dB for the cellular and PCS band CDMA mobile station receivers, respectively (Gu, 2005).

Then another specification can be introduced, that is the dynamic range, which is generally defined as the ratio of the maximum input level to the minimum input level at which the circuit satisfied the required bit error rate. A common used specification called SFDR (Spurious-Free Dynamic Range) bases the definition of the upper end of the dynamic range on the intermodulation behavior and the lower end on the sensitivity. The SFDR can be expressed as:

$$SFDR = \frac{2(IIP3 - NF)}{3} - SNR_{min} \tag{11}$$

where $IIP3$ is the input third intercept point of the RF front end, SNR_{min} is the required minimum SNR (Signal-to-Noise Ratio) of the IF circuits. The required $IIP3$ can be calculated by equation (6), then the $SFDR$ of a receiver can be obtained by equation (11).

Two strong in-band interference tones may cause a zero-IF or low-IF receiver to be completely jammed if the second-order intercept point of the receiver is not high enough. The two strong tones may generate in-channel interferences due to the second-order distortion of the receiver. The maximum level of the two equal blocking interference tones can be expressed as (Gu, 2005):

$$I_{block} = \frac{1}{2}\left[10lg\left(10^{\frac{R_{d,i}-CNR_{min}}{10}} - 10^{\frac{N_{nf}}{10}}\right) + IIP2\right] \tag{12}$$

where $IIP2$ is the input second intercept point of the receiver.

The main problem of the low-IF receiver architecture is the image rejection since the IF is so low that it is difficult to separate the image from the desired signal by filters. The imbalance between I and Q channel signals in the low-IF receiver determines the possible maximum image rejection. The image rejection IR and the imbalances of the I/Q amplitude and the phase have the following relationship:

$$IR = 10lg\frac{1+2(1+\Delta)cos\gamma+(1+\Delta)^2}{1-2(1+\Delta)cos\gamma+(1+\Delta)^2} \tag{13}$$

where γ is the I and Q phase imbalance from the nominal 90° offset in degree, Δ is the I and Q amplitude imbalance, which is usually expressed in dB by using the formula $10lg(1+\Delta)$.
For a full-duplex system, the receiver suffers from transmission leakage interferences particularly in the LNA. If a strong interference tone appears near the desired signal of the receiver, the amplitude modulation of the transmission leakage will cross-modulate the interference tone in the LNA. The spectrum of the cross-modulated tone may partially spread into the receiver channel bandwidth when the single-tone interferer is enough close to the desired signal. The receiver will be desensitized if the cross-modulation product getting into the receiver channel band is high enough. The expression of the allowed single-tone interferer is (Gu, 2005):

$$I_{st} = 10lg\left[\frac{10^{\frac{D_{max}}{10}} - 10^{\frac{N_{nf}}{10}}}{10^{\frac{-[2(IIP3_{lna}-IL_{dRx})-2(P_{Tx}-R_{dTx})-C]}{10}} + 10^{\frac{Npn+10lgBW}{10}} + 10^{\frac{Nsp}{10}}}\right] \tag{14}$$

where D_{max} can be got by equation (5), N_{nf} can be obtained by equation (9), P_{Tx} is the transmitter output power at the antenna in dBm, and C is the correction factor approximately equal to (Gu, 2005):

$$C = M_A + 6 + 10lg\frac{1.5\times BW - \Delta f}{2\times BW} \tag{15}$$

where Δf is the space between the interference tone and the carrier frequency of the desired receiving signal. M_A can be calculated by the probability density function $p(x)$ of the transmission signals and the normalized low-frequency product of the second-order distortion:

$$M_A = 10lg\left\{\int_{-\infty}^{\infty}\left[(10^{\frac{x}{10}})^2 - 10^{\frac{IM2_{dc}}{10}}\right]^2 p(x)dx\right\} \tag{16}$$

$$IM2_{dc} = 10lg\left[\int_{-\infty}^{\infty}(10^{\frac{x}{10}})^2 p(x)dx\right] \tag{17}$$

For the transmitter, the modulation accuracy is represented by EVM, which is defined as the mean square error between the samples of the actual and the ideal signals, normalized by the average power of the idea signal. The EVM of transmitters is influenced by the inter-symbol or inter-chip interference, the close-in phase noise of synthesized LO, the carrier leakage, the I and Q imbalance, the nonlinearity, the in-channel bandwidth noise, the reverse modulation of LO, and so on.
The influence caused by inter-symbol or inter-chip interferences can be obtained by:

$$EVM_{isi} = \sqrt{\sum_{k=-\infty}^{+\infty}\Delta I_{isi}^2(k)} \tag{18}$$

$$\Delta I_{isi}(k) = \left|\frac{h_{ir}(t_0+kT_s)}{h_{ir}(t_0)}\right| = \frac{\int_{t_0-\Delta t}^{t_0+\Delta t}|h_{ir}(t+kT_s)|dt}{\int_{t_0-\Delta t}^{t_0+\Delta t}|h_{ir}(t)|dt} \tag{19}$$

where $h_{ir}(t)$ is the impulse response of the pulse-shaping filter, k is equal to ±1, ±2, ±3, ..., Ts is the period of a symbol, and $2\Delta t$ is the during of a sampling pulse.
The influence caused by close-in phase noise of synthesized LO can be obtained by:

$$EVM_{pn} = \sqrt{2 \times 10^{\frac{N_{phase}}{10}} \times BW_{lf,pll}} \tag{20}$$

where N_{phase} is the average phase noise in dBc/Hz within the PLL loop bandwidth, and $BW_{lf,pll}$ is the bandwidth of the PLL loop filter in Hz.
The carrier leakage is mainly caused by the DC offset of the baseband, the LO-to-RF leakage and the IF-to-RF leakage. The influence on EVM caused by carrier leakage can be obtained by:

$$EVM_{cl} = \sqrt{\frac{CL_{offset}}{10} + \frac{CL_{lo}}{10} + \frac{CL_{if}}{10}} \tag{21}$$

where CL_{offset}, CL_{lo} and CL_{if} represents the leakage results from the DC offset, the LO-to-RF leakage and the IF-to-RF leakage, respectively.
The EVM caused by the I and Q imbalance can be expressed as:

$$EVM_{iq} = \sqrt{10^{\frac{IR}{10}}} \tag{22}$$

where IR is the image suppression, which can be calculated by equation (13).
It's assumed that only the signal amplitude equal to and greater than the output 1-dB compression of the power amplifier P_{-1} will affect the modulation accuracy, then the EVM caused by nonlinearity of the transmitter chain can be expressed as:

$$EVM_{nonlin} = \int_0^{\infty} P(\delta) \times \left(10^{\frac{\delta+1}{20}} - 1\right) d\delta \tag{23}$$

$$\delta = P_{Tx} - P_{-1} \tag{24}$$

where P_{Tx} is the output power level, and P_{δ} is the amplitude probability density function of the signal.
The EVM caused by in-channel bandwidth noise can be expressed by:

$$EVM_{ibn} = \sqrt{10^{\frac{N_{ibn}-P_{Tx}}{10}}} \tag{25}$$

where N_{ibn} is the integrated noise over the channel bandwidth, and P_{Tx} is the transmission power in dBm.
The transmission signals may be reflected from the load of the modulator, then the reflected signals and their harmonics may modulate the LO if the frequency of the carrier or the harmonics of the reflected signals is equal to LO frequency. Then reverse modulation occurs. The EVM caused by reverse modulation is:

$$EVM_{rm} = \sqrt{10^{\frac{N_{rm}}{10}}} \tag{26}$$

where N_{rm} is the integrated reverse modulation noise of the synthesized LO over the transmission signal bandwidth below the LO level.

The overall EVM of the transmission signal can be expressed as:

$$\begin{aligned} EVM_{total} &= EVM_{isi}^2 + EVM_{pn}^2 + EVM_{cl}^2 + EVM_{iq}^2 \\ &+ EVM_{nonlin}^2 + EVM_{ibn}^2 + EVM_{rm}^2 + \ldots \end{aligned} \tag{27}$$

The ACPR specification is generally defined as ratio of the power integrated over an assigned bandwidth in the adjacent/alternate channel to the total desired transmission power. The ACPR can be expressed as:

$$ACPR = \frac{\int_{f_a}^{f_a+\Delta B} SPD(f)df}{\int_{f_o-BW/2}^{f_o+BW/2} SPD(f)df} \tag{28}$$

where f_a is the start frequency of the adjacent/alternate channel, ΔB is the bandwidth of measuring adjacent/alternate channel power, which varies with different mobile systems. The general formula for the ACPR of a transmission signal at the output of the power amplifier with an output third intercept point $OIP3$ can be expressed as:

$$ACPR \approx 2(P_{Tx} - OIP3) - 9 + C_0 + 10lg(\frac{\Delta B}{BW}) \tag{29}$$

$$C_0 \approx 0.85 \times (PAR - 3) \tag{30}$$

where P_{Tx} is the transmission signal power at the output of the power amplifier, PAR is the peak-to-average ratio of the random noise. In the mobile communication systems, the adjacent/alternate power may be tested in a bandwidth ΔB that is different from the desired transmission signal bandwidth BW.

We only discuss the noise emissions that are those located outside of alternate channels here. In general, we like to have lower gain power amplifier for achieving low noise emission, but this is completely opposite to the gain setting of the power amplifier to obtain a good ACPR performance. The noise emission in mW/Hz of the transmitter has an expression as:

$$P_{nm} = G_{Tx} \times P_{n,in} + kT_0 G_{Tx}(NF_{Tx} - 1) \tag{31}$$

where $P_{n,in}$ in mW/Hz is the noise at the transmitter input, G_{Tx} is the overall transmitter gain, NF_{Tx} is the overall noise factor of the transmitter, and $kT_0 = 10^{-174/10}$mW/Hz.

4. The design of key modules in the transceiver

A common used transceiver is composed of a LNA, Mixers, filters, IF circuits, a PA, ADCs and DACs, a PLL and so on. The individual performance and the matching among these modules determine the performance of the whole transceiver system. A general description of these modules will be given in this section.

4.1 Low-Noise Amplifier

The specifications of a LNA can be summarized into:

- The working frequency;
- The noise figure;
- The third intercept point;
- The voltage or power gain;

- The reflection coefficient at the input port and the isolation between the output port and the input port;
- The power consumption;

The most important specification of the LNA is the noise figure since the first stage of a receiver chain decides the noise performance of the whole system. One of the common used LNAs is the inductively source degenerated type, and a typical design example is shown in Fig. 7(a). $M1$ is a common-source amplifier transistor, L_S is the source degenerated inductor, L_G is the gate inductor, V_{IN} is the input Port, and V_{OUT} is the output port; $M2$ is a common-gate transistor that is used for isolation and gain enhancement; the load Z_L can be a resistor, a inductor or a inductor-capacitor tank. This structure has a large gain and a low noise figure, but the input reflection is a problem. There is a trade-off between the noise figure and the input impedance matching.

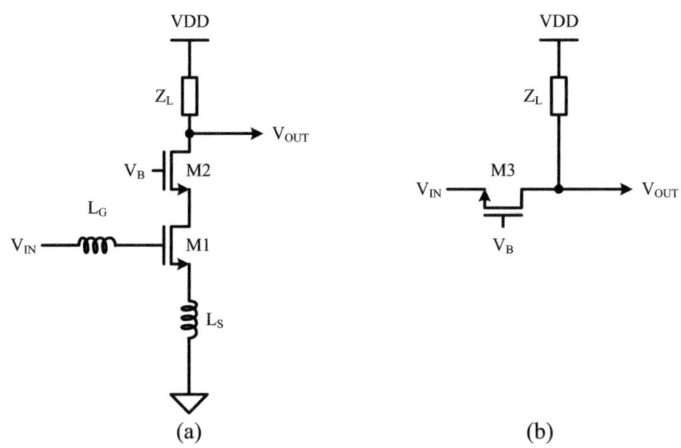

Fig. 7. (a)An inductively source degenerated LNA. (b)The common-gate input type.

Another common used LNA is the common-gate input type, as shown in Fig. 7(b). If the transconductance of the common-gate transistor $M3$ is g_m, then the input resistance is equal to $1/g_m$. Therefore, the input matching of common-gate LNA is easier to realize compared to the source degenerated LNA. However, the noise performance is poor, since the common-gate amplifier has a low gain.
The WSN receivers usually have a high sensitivity, as shown in Table 1. According to equation (1), the sensitivity is proportional to the noise figure of the RF front end. Therefore, in WSN applications, we often adopt the inductively source degenerated LNA assisted by some low-noise technologies.

4.2 Mixer

The mixers in transceivers can be divided into two types: 1)the up-converting mixers and 2)the down-converting mixers. The up-converting mixers are used in the transmitter, while the down-converting mixers are used for the receiver. The specifications of a mixer can be summarized into:

- The working frequency including RF frequency, LO frequency, and IF frequency;
- The noise figure;
- The third intercept point;

- The second intercept point (for zero-IF or low-IF receivers);
- The voltage or power conversion gain;
- The isolation between the RF port and the LO port, the RF port and the IF port, and the LO port and the IF port;
- The magnitude and phase imbalance between I and Q channel down converters (for the receivers and transmitters that use I and Q dual-path converters);
- The power consumption;

A classical mixer is known as the Gilbert cell, as shown in Fig. 8. I_B is a current source, $RF+$ and $RF-$ are the differential RF input ports, $LO+$ and $LO-$ are the differential LO input ports, and the output differential currents are $I_{OUT}+$ and $I_{OUT}-$; $M1$ and $M2$ convert the input RF voltage into current, and $M3$-$M6$ are used as switches for mixing.

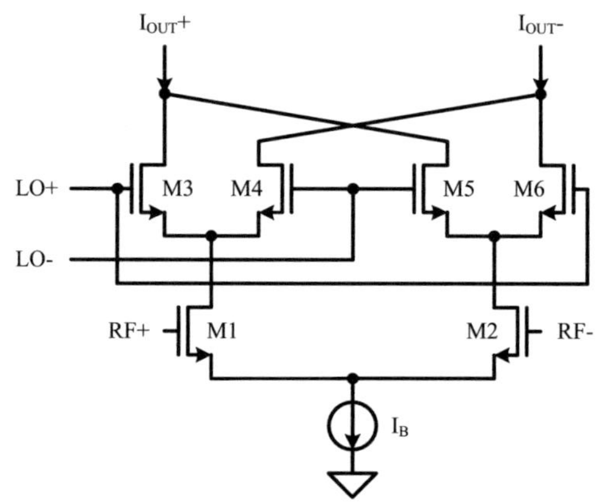

Fig. 8. The Gilbert cell.

The Gilbert mixer is a typical example of active mixers, which have high gain, low noise, but poor linearity. Another type of mixers are called passive mixers, which usually have low gain, large noise and high linearity. The passive mixers can be divided into two types: 1)voltage-mode mixers: the MOS switches are used for voltage switches, and loaded with high resistance. Because of the nonlinearity of the switches, distortions will be enlarged when the amplitudes of RF and IF signals are increased and the switches are modulated. 2)Current-mode mixers: the MOS switches are used for current switches, and loaded with low resistance. So the amplitudes of RF and IF signals are relatively low in current-mode mixers, then the linearity is improved. A typical example of current-mode passive mixers is shown in Fig. 9 (Valla et al., 2005). The input I_{IN} is a differential current signal, the output V_{OUT} is a voltage signal. Transistors $M1$-$M4$ are used for switches that are controlled by the LO signals $LO+$ and $LO-$. An OTA (OperaTional Amplifier) together with resistors $R1$, $R2$ and capacitors $C1$, $C2$ is adopted to amplify and filter the signals. Two additional capacitors $C3$ and $C4$ are used to generate an extra pole to suppress the amplitude. The LNA and mixers are often designed and tested together to realize an optimized trade-off among gain, noise figure and linearity for different applications.

The distribution of WSN nodes is random, and the distance between two nodes may be very short or very long, so the dynamic range of the transceivers is very important. According to equation (11), in order to improve the SFDR performance, the noise figure needs to be

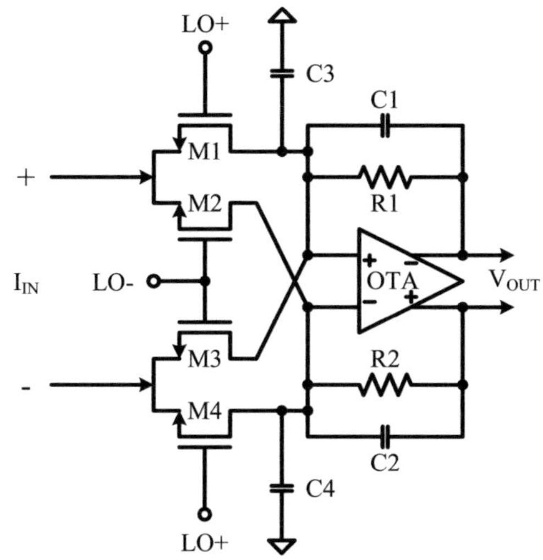

Fig. 9. A typical current-mode passive mixer.

decreased and the IIP3 needs to be enhanced. Therefore, a combination of a low-noise inductively source degenerated LNA and a current-mode passive mixer can be adopted for WSN usages.

4.3 Active Filter

According to the pattern of implement, the active filters usually used for transceivers can be summarized into three types: 1)switched-capacitor filters, in which the resistors are replaced by switched capacitors; 2)active-RC filters, which is composed of OTAs and resistor-capacitor networks; 3)gm-C filters, in which the resistors and inductors are replaced by transconductors.
For switched-capacitor filters, the advantages can be summarized here: 1)high precision without tuning, 2)small chip area and low power, and 3)insensitive to parasitics. However, there are several disadvantages: 1)affected by sampling, 2)requirement for extra clock generation circuit, and 3)not suitable for high-frequency applications.
For active-RC filters, the advantages can be summarized into: 1)high precision with tuning, 2)easy to design with classical RC structures, 3)insensitive to parasitics, 4)no sampling effect, and 5)large dynamic range. The disadvantages can be summarized into: 1)requirement for tuning circuits and 2)limited working frequency caused by OTAs.
For gm-C filters, the advantages can be summarized into: 1)high precision with tuning, 2)able to be realized based on simple open-loop OTAs, 3)lower power consumption than active-RC filters, 4)no sampling effect, and 5)good frequency performance. The disadvantages can be summarized into: 1)requirement for complex on-chip tuning circuits, 2)poor dynamic range, and 3)sensitive to parasitics.
According to the transfer characters, the filters can also be divided into four types: 1)Butterworth filters, which has the maximum flat amplitude in the pass band; 2)Chebyshev filters, which has the minimum ripples in the pass band; 3)Bessel filters, which has the maximum flat of group delay; 4)Ellipse filters, which has the minimum transition band. The other characters of these filters are summarized into Table 3.

Type	Amplitude-Frequency Characteristic			Phase-Frequency characteristic
	Pass Band	Stop Band	Transition Band	
Butterworth	Flat	Monotonic Decreasing	Gentle Monotonic Decreasing	Moderate
Chebyshev	Fluctuant	Monotonic Decreasing	Steep Monotonic Decreasing	Poor
Bessel	Flat	Monotonic Decreasing	Slowly Monotonic Decreasing	Excellent
Ellipse	Fluctuant	Fluctuant	Steep Monotonic Decreasing	Poor

Table 3. The characteristics of different filters.

In wireless transceivers, there is a special kind of filter named complex filter, which is usually used in low-IF receivers for image rejection. A classical complex filter is designed in 1995 (Crols & Steyaert, 1995). Fig. 10 shows the block diagram, which has I/Q dual-path inputs and I/Q dual-path outputs. The input of the Q path Q_{IN} is 90° delay of the input of the I path I_{IN}, and the output of Q path Q_{OUT} is also 90° delay of the output of the I path I_{OUT}. That is $Q_{IN} = -jI_{IN}$ and $Q_{OUT} = -jI_{OUT}$. The transfer function of this complex filter can be

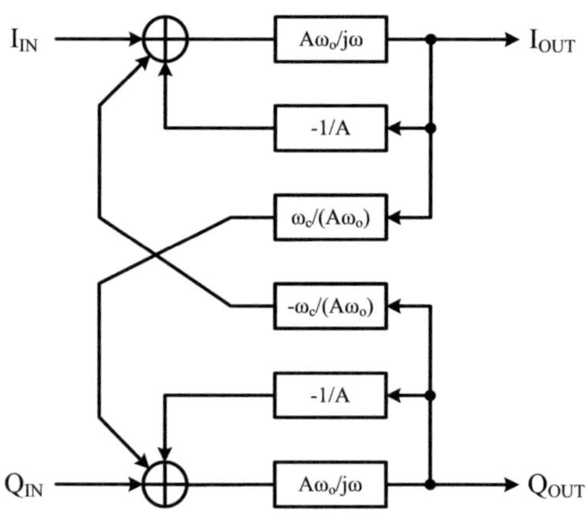

Fig. 10. The block diagram of a first-order complex filter.

expressed as:

$$H_{cf}(j\omega) = \frac{A}{1 + j(\omega - \omega_c)/\omega_o} \quad (32)$$

where ω_c is the central frequency, and $2\omega_o$ is the double-sideband bandwidth. It is equivalent to a low-pass filter's pass band moved by ω_c, and then the transfer curves of positive and negative frequency become asymmetric. As a result, the image can be rejected.

As shown in Table 2, low-IF transceivers has the advantages of easy to be integrated and immune to DC offset, so the low-IF SDR transceiver is adopted in many WSN applications. Besides, the data rate of WSN is usually not very high, then the IF can be relatively low and gm-C filters are not necessary. Therefore, an active-RC complex filter is suitable for such WSN receivers because of their low power, large dynamic range and image rejection function.

4.4 Phase-Locked Loop

The PLL is the core part of a transceiver system, as it's used for both the down-converting in receiver and the up-converting in transmitter. A typical sigma-delta charge-pump PLL is shown in Fig. 11 (Zhao et al., 2009), which is composed of a PFD (Phase-Frequency Detector), a charge pump, a loop filter, a VCO (Voltage-Controlled Oscillator), a multi-modulus frequency divider, and a sigma-delta modulator. Although the all-digital PLL has appeared in recent years, the classical charge-pump PLL is still widely used in the industrial community.

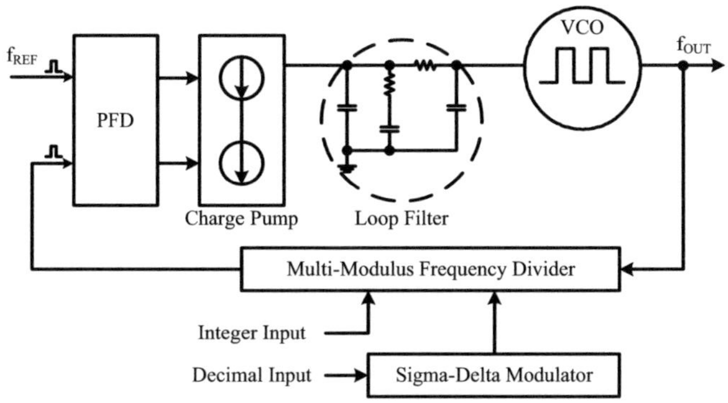

Fig. 11. A typical sigma-delta charge-pump PLL (Zhao et al., 2009).

For WSN transceivers, we tend to design low-power, full-integrated and fast-settling PLL. The power of the PLL is mainly limited by these modules: 1)VCO (Voltage-Controlled Oscillator), 2)prescaler, and 3)the buffer connected at the output of VCO. Therefore, the power reduction of these modules is significant to the low-power design of PLL.

For full-integrated design, the chip area needs to be decreased. In a typical PLL, the LF (Loop Filter) and the inductors in LC-tank VCO take up the largest chip area. In order to decrease the area of LF, some one has proposed a discrete-time architecture (Zhang et al., 2003). For the applications in low-frequency bands, the inductors will be large if the resonant frequency of the VCO is low, so the VCO is required to be designed at a high frequency with a frequency divider connected after it.

The settling speed of PLL is decided by the loop bandwidth. Too Large bandwidth brings not only fast settling, but also large in-band noise and spurs. As a result, there is a trade-off between the settling speed, and the phase noise and the spur performance. How to set the trade-off depends on the requirement of the transceiver system.

As we referenced in section 2.2, the PLL can be adopted for directly digital modulation, and such method is proposed by Perrott in 1997 (Perrott et al., 1997). The architecture of such PLL based transmitter is shown in Fig. 12. The data stream can be shaped by a filter firstly, then the shaped data are input into the sigma-delta modulator in order to change the dividing ratio. As a result, the output frequency can be modulated by the variation of dividing ratio according to the input data, and the FSK signals can be generated in this way. A PA is connected at the output of the PLL so that the FSK signals can be emitted through an antenna. Generally, the data rate can not be larger than the bandwidth of the PLL. Although some one has proposed a compensation technology with a digital filter whose transfer function is the reciprocal of PLL's (Perrott et al., 1997), mismatch and inaccuracy depress the performance in actual designs. As a result, we would rather enlarge the bandwidth of PLL to obtain a relatively high data rate.

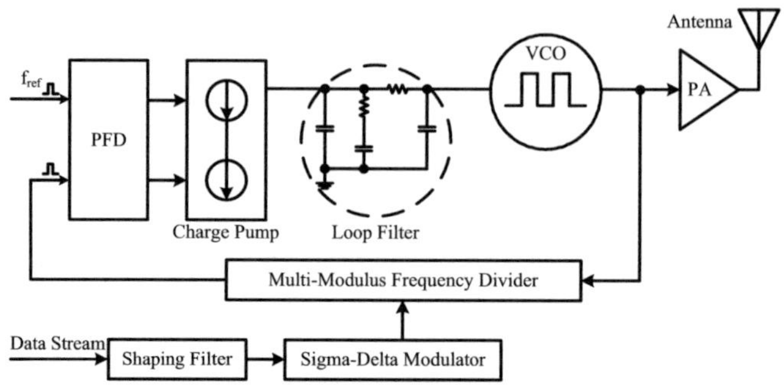

Fig. 12. A transmitter based on PLL directly digital modulation.

For WSN usages, the data rate is often not high, so such PLL directly digital modulation is a reasonable choice for common frequency and phase modulation schemes, such as FSK, MSK, and so on. As a result, mixers can be moved away, the cost and power consumption of the WSN transmitter can be saved a lot.

4.5 Power Amplifier

Generally, PA is the most power hungry module of a transceiver. Therefore, the output power of PA is usually relatively small for WSN usages, as shown in Table 1. The types of PA can be divided into class A, B, C, AB, D, E, F and F^{-1}.

For class-A PAs, the amplifier MOSFET is kept in the saturation region. The transistor always dissipates power because the product of drain current and drain voltage is always positive. It should be noticed that the maximum theoretical drain efficiency of class-A PAs is just 50%. However, drain efficiencies of 30∼50% are common for practical class-A PA designs. The normalized power output capability is about 1/8. The class-A amplifier provides high linearity at the cost of low efficiency and relatively large device stresses.

In a class-B PA, the device is shut off in half of every cycle. It should be mentioned that most practical class-B PAs are push-pull configurations of two MOSFETs. The peak drain current and maximum output voltage are the same as for the class-A PAs. The maximum drain efficiency for a class-B PA is 78.5%. The normalized power capability of the class-B PAs is 1/8, the same as for class-A PAs, since the output power, maximum drain voltage, and maximum drain current are the same.

In a class-C PA, the transistor conducts less than half the time. As the conduction angle shrinks toward zero, the efficiency approaches 100%, but the gain and output power unfortunately also tend toward zero at the same time. Furthermore, the normalized power capability of class-C PAs approaches zero as the conduction angle approaches zero. In one word, the efficiency can be large, but at the cost of normalized power capability, gain, and linearity.

The class-AB Pas conducts between 50% and 100% of a cycle. Both the conduction angle and efficiency of class-AB PA vary between that of class-A PA and class-B PA.

In a class-D PA, only one transistor is driven on at a given time, and one transistor handles the positive half-cycles and the other handles the negative half-cycles, just as a push-pull class-B PA. The difference between class-D and class-B is that the transistors are driven hard enough to make them act like switches for class-D PA, rather than as linear amplifiers. The normalized power capability of class-D PAs is about 0.32, which is better than a class-B push-pull and much better than a class-A PA. The MOS switches in class-D PAs function well only

at frequencies substantially below f_T, which is the cut-off frequency. Usually, one transistor fails to turn completely off before the other turns on, then the efficient is deteriorated.
The class-E PA uses a high-order reactive network that provides enough space to shape the switch voltage to have both zero value and zero slope at switch turn-on, then the switch loss is reduced. The efficiency can approach theoretically 100% with idea switches. The normalized power capability is about 0.098, which is worse than class-A PA. The class-E PA is more demanding of its switch performance than even class-A PAs because of the poor power capability and the reduced efficiency due to switch turn-off losses.
The termination of a class-F PA appears as an open circuit at odd harmonics of the carrier beyond the fundamental and as a short circuit at even harmonics, while the class-F^{-1} employs a termination that appears as an open circuit at even harmonics and as a short circuit at the odd harmonics. The class-F PA is capable of 100% efficiency in principle. The normalized power capability of class-F PAs is about 0.16, which is half that of the class-D PAs.
In summary, there is a trade-off between the efficiency and the linearity. For receivers with constant-envelope modulation, such as FSK, high-efficiency PAs can be adopted; for linear operation such as ASK (Amplitude Shift Keying), or systems with high ACPR requirement, high-linearity PAs can be adopted.
The high-efficiency PAs such as class-E are usually used in WSN transceivers, as the power consumption is the most significant specifications of WSN system.

4.6 IF circuits

The function of IF circuits includes demodulation, data decision, and clock recovery. There are two main kinds of IF circuits: 1)the digital scheme. In common receivers, an ADC is connected after the RF front end, and the frequency detecting, data decision and received signal strength indicating are all realized in digital domain, then the performance of the circuit can be easily improved in digital domain. Such is the general architecture for SDR as we described in section 2.5. However, the ADC usually consumes a large amount of power, and a high-linearity AGC circuit is required before the ADC. 2)The analog scheme. For low-power applications such as WSN, the CMOS analog resolution is sometimes a reasonable choice since the power consumption can be saved a lot.

4.7 ADC and DAC

There are two kinds of ADCs in a WSN node. One is connected after the sensor and used for data sampling, as shown in Fig. 1; and another is used in the transceiver, as shown in Fig. 6.
The main parameters of an ADC can be summarized into several aspects:

- Resolution: The minimum voltage level that can be discriminated by the ADC is $V_{ref}/2^N$ for a N-bit ADC with an input range from 0 to V_{ref}.
- DNL (Differential Non-Linearity): The maximum deviation between the actual conversion step and the idea conversion step.
- INL (Integrated Non-Linearity): The maximum deviation of actual center of bin from its idea location.
- Offset: The non-zero voltage or current at the output of the ADC when the input is zero since the OTAs or comparators have offset voltages and offset currents.
- Gain Error: The deviation of actual input voltage from the idea value when the ADC outputs the full-scale bits.

- SNR (Signal-to-Noise Ratio): The theory equation of SNR for a N-bit ADC can be expressed as:

$$SNR = 6.02N + 1.76 \tag{33}$$

- SNDR (Signal-to-Noise and Distortion Ratio): The power of the noise and harmonics divided by the power of signal.
- SFDR (Spurious Free Dynamic Range): The ratio of the signal's power to the maximum harmonic's power.
- ENOB (Effective Number of Bits): The ENOB can be calculated by:

$$ENOB = (SNDR - 1.76)/6.02 \tag{34}$$

- THD (Total Harmonic Distortion): The ratio of all the harmonics' power to the signal's power.

Besides the specifications above, another very important parameter is power consumption. In the aspect of architecture, the ADCs can be divided into flash, SAR (Successive AppRoximation), folding, pipeline, sigma-delta, and so on.

For the ADC connected after the sensor in a WSN node, SAR ADCs may be the best choice since the data collection is executed at most of the time. There are two reasons: 1)The only power hungry module in a SAR ADC is the comparator, so the overall power consumption is low; 2)the circuit structure of a SAR ADC is simple, so the cost of chip area is small. In the WSN transceivers, the data rate is not high and the modulation scheme is simple, so a low bandwidth and low precision ADC is usually enough. Therefore, a SAR ADC may also be a good choice for WSN transceivers.

The specifications of DAC also includes DNL, INL, SNR, SNDR, SFDR, power consumption, and so on. The DACs are often used in transmitters, as shown in Fig. 6, the shaped digital waves are converted to analog signals that are sent to up-converting mixers for modulation. As a result, the performance of the DACs will affect the EVM and ACPR of the transmitters.

5. Conclusion

The goal of this chapter is to give a brief manual for WSN transceiver design. Section 1 gives an introduction of WSN and the RF transceivers for WSN. The WSN transceivers are classified by both modulation schemes and architectures in section 2. How to calculate and assign the system specifications are described in section 3. The design of key modules is analyzed briefly in section 4. The readers is expected to master a top-to-down design method for WSN transceivers through the chapter.

6. References

Crols, J. & Steyaert, M. (1995). An analog integrated polyphase filter for a high performance low-IF receiver, *Symposium on VLSI Circuits, Digest of Technical Papers*, pp. 87–88.

Gu, Q. (2005). *RF System Design of Transceivers for Wireless Communications*, Springer Science+Business Media, LLC,.

Jri, L., Chen, Y. & Yenlin, H. (2010). A Low-Power Low-Cost Fully-Integrated 60-GHz Transceiver System With OOK Modulation and On-Board Antenna Assembly, *IEEE Journal of Solid-State Circuits, DOI-10.1109/JSSC.2009.2034806*, **45**(2): 264–275.

Perrott, M. H., Tewksbury, T. L. & Sodini, C. G. (1997). A 27-mW CMOS fractional-N synthesizer using digital compensation for 2.5-Mb/s GFSK modulation, *IEEE Journal of Solid-State Circuits, DOI-10.1109/4.643663*, **32**(12): 2048–2060.

Pletcher, N. M., Gambini, S. & Rabaey, J. (2009). A 52 μW Wake-Up Receiver With -72 dBm Sensitivity Using an Uncertain-IF Architecture, *IEEE Journal of Solid-State Circuits, DOI-10.1109/JSSC.2008.2007438,* **44**(1): 269–280.

Seungkee, M., Shashidharan, S., Stevens, M., Copani, T., Kiaei, S., Bakkaloglu, B. & Chakraborty, S. (2010). A 2mW CMOS MICS-band BFSK transceiver with reconfigurable antenna interface, *IEEE Radio Frequency Integrated Circuits Symposium,,* pp. 289–292.

Valla, M., Montagna, G., Castello, R., Tonietto, R. & Bietti, I. (2005). A 72-mW CMOS 802.11a direct conversion front-end with 3.5-dB NF and 200-kHz 1/f noise corner, *IEEE Journal of Solid-State Circuits, DOI-10.1109/JSSC.2004.842847,* **40**(4): 970– 977.

Zhang, B., Allen, P. E. & Huard, J. M. (2003). A fast switching PLL frequency synthesizer with an on-chip passive discrete-time loop filter in 0.25-μm CMOS, *IEEE Journal of Solid-State Circuits,* **38**(6): 855– 865.

Zhao, B., Mao, X., Yang, H. & Wang, H. (2009). A 1.41-1.72 GHz sigma-delta fractional-N frequency synthesizer with a PVT insensitive VCO and a new prescaler, *Analog Integrated Circuits and Signal Processing,* **59**(3): 265–273.

MAC & Mobility In Wireless Sensor Networks

Marwan Al-Jemeli[1], Vooi Voon Yap[2] and Fawnizu Azmadi Bin Hussin[1]
[1]*Universiti Teknologi PETRONAS*
[2]*Universiti Tuanku Abdul Rahman*
Malaysia

1. Introduction

The recent climate change has a significant impact on our planet environment. Therefore, deploying sensor networks to monitor the environment is becoming important. With sensor networks deployed in strategic location can provide the scientific communities useful data to be analyzed and take action if necessary. Typical environmental applications of sensor networks include, but not limited to, monitoring environmental conditions that affect crops and livestock, biological, Earth, and environmental monitoring and many more. Monitoring hazardous environment like volcanic activities is one of the important applications for Wireless Sensor Network (WSN) (Sohraby et al, 2007). WSN communicate wirelessly to pass and process information - see Figure 1.

These sensor networks are deployed far away from the nearest permanent energy source available which make them depending on their own energy source to provide the needed information.

WSNs usually consist of a large number of low-cost, low-power, multifunctional (or uni-functional) wireless devices deployed over a geographical area in an ad hoc fashion and with or without careful planning (this depends on the application mainly whether it is related to a real-time applications or non-real-time application). Individually, these devices have limited resources and have limited processing and communication capabilities. The cooperative operation behavior of these sensing devices gives a significant impact on a wide range of applications in several fields, including science and engineering, military settings, critical infrastructure protection, and environmental monitoring (Yu et al, 2006).
Networking distributed sensors are used in military and industrial applications and it dates back at least to the 1970s. back then the systems were primarily wired and small in scale. wireless technologies and low-power Very Large Scale of Integration (VLSI) design became feasible and emerged in 1990 and after that researchers began envisioning and investigating large-scale embedded wireless sensor networks for dense sensing applications (Krishnamachari, 2005).
However, wireless sensor networks have a major problem, that is, "network life time". Since WSN uses batteries, it does them in terms of storage, and processing power. Limited capabilities results in limited information efficiency. Current available technology on-shelf

allow us to produce sensors that consumes as little power as 100mW which means that the sensors can remain operational efficiently (depending on the application and the deployed nodes own capabilities) for about 10 months. Yet the life time of the network can be extended for further than 10 months.

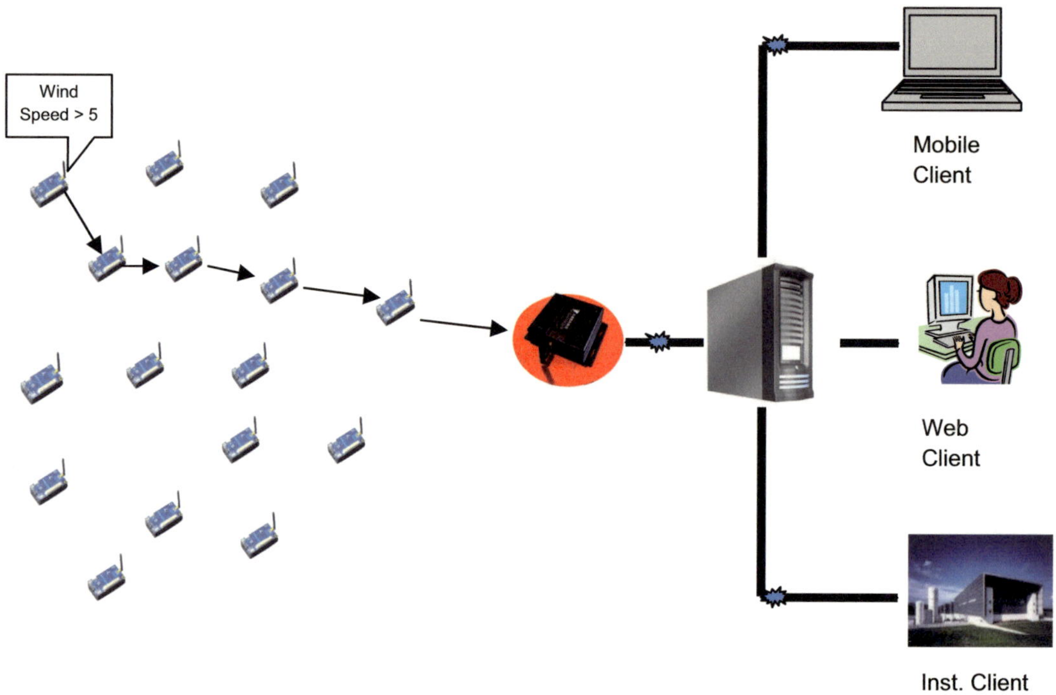

Fig. 1. Wireless Sensor Networks Example.

Some researchers proposed methods includes energy harvesting, solar energy and vibration energy. But these methods can only provide a small amount of energy to power these sensors, typically 20mw or less (Mainwaring et al,2002; Raghunathan et al, 2002). Maintenance and recharging these sensors is not a good option, and it will increase the expenses to keep the network alive and operational. Another alternative is to use energy efficient information processing and transacting algorithms to manage the network operation. We envisage that efficient routing and Medium Access Control (MAC) Protocols can help resolve this problem.
Information processing and routing is a technique used widely when it comes to provide a longer life time operation in wireless sensor networks however these techniques lacks the integrity as it has to compensate between either providing an energy efficient operation with the lack of high throughput or vice versa (Boukerche et al, 2005; Branyard et al, 2006).

One of the major levels of tweaking in networking systems is to manipulate the timing when to deliver particular packets at a precise times to achieve efficient operation. From the literature provided most of the available approaches consider the main purpose of manipulating information processing technique is to achieve better energy consumption in the nodes while sacrificing the system throughput quality and robustness (Branyard et al, 2006). the next section will discuss MAC theory and the related works that where done in this area of research follows it in section three our own theory and methodology. Section four and five will discuss the results that where obtained during our research . Section six will review Mobility issues in brief and section seven reviews our proposed future goals in this research.

2. MAC protocols effect in WSNs

MAC is the second layer after the physical layer in the Open System Interconnection (OSI) model in networking systems, MAC protocols controls when to send and receive distinguished packet between different nodes in a network. It controls the network interface when to establish the connection or the transaction between two or more hosts. Manipulating the operation of a MAC protocol can give its effect in terms of energy consumption and message delay between nodes (Van Hoesel, Havigna, 2004).

Different MAC protocols were defined for WSN because of its application dependency. MAC protocols have to compensate between providing energy efficient consumption with the availability of decent throughput to make the system dependable (Van Hoesel, Havigna, 2004; Law et al, 2005) .

An essential characteristic of wireless communication is that it provides an inherently shared medium. All MAC protocols for wireless networks manage the usage of the radio interface to ensure efficient utilization of the shared bandwidth. MAC protocols designed for wireless sensor networks have an additional goal of managing radio activity to conserve energy. Thus, while traditional MAC protocols must balance throughput, delay, and fairness concerns, WSN MAC protocols place an emphasis on energy efficiency as well (Krishnamachari, 2005). MAC layer affects the energy efficiency mainly through the adjustment of transmission scheduling and channel access. A common way to do that is via sleep scheduling from a long time scale, or time-division multiple access (TDMA), from a short time scale perspective. Similar to the shutdown technique of CPUs, sleep scheduling also explores the energy *vs.* response time tradeoffs in wireless communication. From previous studies, the response time is translated to network or application layer transmission delay or throughput.

(Mathioudakis et al, 2008) presented the most energy wastage sources in MAC protocols for WSNs:

The first source is caused by collisions, which occur when two or more nodes attempt to transmit simultaneously. The need to re-transmit a packet that has been corrupted by collision increases the energy consumption.

The second source of energy wastage is idle-listening, where a node listens for traffic that it is not sent. In a sample fetching operation, a silent channel can be high in several sensor applications.

The third source of waste is overhearing, which occurs when a sensor node receives packets that are destined for other nodes.

The fourth is caused by control packet overheads, which are required to regulate access to the transmission channel. Sending and receiving control packets consumes energy too, and less useful data packets can be transmitted.

The fifth source is over-emitting where the destination node is not ready to receive during the transmission procedure, and hence the packet is not correctly received.

Finally, the transition between different operation modes, such as sleep, idle, receive and transmit, can result in significant energy consumption. Limiting the number of transitions between sleep and active modes leads to a considerable energy saving.

2.1 Related Approaches

For wireless sensor networks the literature provided a lot of protocols and divided it into two major categories (Shukur et al, 2009):

1. **Contention Based MAC Protocols (CSMA carrier sense multiple access)**. The wireless nodes here contend to enter the medium of connectivity (which is the wireless medium in case of WSNs) and the winner node reserves the medium to itself until it finishes its operation. Examples for this kind of protocols are: IEEE 802.11, S-MAC (Ye et al, 2001), T-MAC (Van Dam, Longendean, 2003), R-MAC (Du et al, 2007) and others.
2. **TDMA (time division multiple access) Based MAC Protocols**. The medium here is divided into time slots each node knows its time slot when to enter the medium and do its operation. One popular TDMA based MAC protocol for WSNs is ALOHA (PARK et al, 2006).

Contention based MAC protocols offers a more scalability approach over TDMA based approaches because of the nature of TDMA approaches that requires slotting the time into slots to each node which is improper when deploying a large number of nodes. To list dome of the works in this area of research, follows are some approaches regarding CSMA based MAC protocols:

A popular contention based MAC protocol for wireless networks is the IEEE 802.11 which is the standard for WLAN applications. IEEE 802.11 performs well in terms of latency and throughput but it is not efficient in terms of energy consumption because of the idle listening problem. It has been shown that when the node is in idle listening state it consumes energy equivalent to the receiving energy and that is why this protocol is not suitable for WSNs applications (Ye et al, 2001).

Sensor-MAC, S-MAC is a contention based MAC protocol designed explicitly for wireless sensor networks proposed by (Ye et al, 2001). While reducing energy consumption is the primary goal of this design, the protocol also has good scalability and collision avoidance capability. It achieves good scalability and collision avoidance by utilizing a combined scheduling and contention scheme. It also achieves efficient energy consumption by using a scheme of periodic listening and sleeping which reduces energy consumption. In addition, it uses synchronization to form virtual clusters of nodes on the same sleep schedule. These schedules coordinate nodes to minimize additional latency. The protocol also uses the same mechanism to avoid the overhearing problem and hidden channel problem that is used in IEEE 802.11. But the S-MAC has a problem of latency because of periodic listen and sleep scheme which is dependent on the duty cycle.

WSNs applications have some unique operation characteristics, for example, low message rate, insensitivity to latency. These characteristics can be exploited to reduce energy consumption by introducing an active/sleep duty cycle. To handle load variations in time and location, (Van Dam, Langendeon, 2003) proposed the Timeout MAC T-MAC protocol. T-MAC can handle an adaptive duty cycle in a novel way: by dynamically ending the active part of it. This reduces the amount of energy wasted on idle listening, in which nodes wait for potentially incoming messages, while still maintaining a reasonable throughput. T-MAC uses *TA* (time out) packet to end the active part when there is no data to send/receive on the node. The protocol balances between energy efficient consumption and latency efficient throughput due to the scheme of burst data sending more effective in terms of energy consumption.

The concept of periodic listen and sleep approach was explored by (Suh, Ko, 2005). They proposed a novel MAC scheme named as TEEM (Traffic aware, Energy Efficient MAC) protocol. The proposed TEEM is based on the often cited contention-based MAC protocol S-MAC. The protocol achieves energy efficient consumption by utilizing 'traffic information' of each sensor node.

Thus, Suh and Ko show that the listen time of nodes can be reduced by putting them into sleep state earlier when they expect no data traffic to occur. In this method, they made two important modifications to the S-MAC protocol: the first modification was to make all nodes turn off the radio interface much earlier when no data packet transfer is expected to occur in the networks, and secondly eliminating communication of a separate RTS control packet even when data traffic is likely to occur. However, it lacks on latency efficiency to conserve energy.

The cross-layer approach protocol was investigated by (Pack et al, 2006) . They proposed a task aware MAC protocol for WSNs. The TA-MAC protocol determines the channel access probability depending on a node's and its neighbor nodes' traffic loads through the interaction with the data dissemination protocol. In this approach the TA-MAC protocol can reduce energy consumption and improve the throughput by eliminating unnecessary collisions. The TA-MAC protocol is feasible because it can be integrated with other energy efficient MAC protocol example, SMAC. The TA-MAC protocol focuses on the determination of channel access probability that is orthogonal to the previous MAC protocols for WSNs.

Another work that explores the cross-layer approach was presented by (Du et al, 2007) . The proposed scheme called Routing-enhanced MAC protocol (RMAC), exploits cross-layer routing information in order provide delay guarantee without sacrificing energy efficiency. Most importantly, RMAC can deliver a data packet *multiple* hops in a single operational cycle. During the SLEEP period in RMAC, a relaying node for a data packet goes to sleep and then wake up when its upstream node has the data packet ready to transmit to it. After the data packet is received by this relaying node, it can also immediately forward the packet to its next downstream node, as that node has just woken up and is ready to receive the data packet. The mechanism is implemented using a packet called Pioneer. This packet travels to all sensors in down-stream to synchronize the duty-cycles of the nodes to guarantee a multi-hop packet delivery. In this way the protocol achieved latency efficient operation.

(Erazo, Qain, 2007) developed the S-MAC to SEA-MAC, a protocol which aims for energy efficient operation for WSNs for environment monitoring. The protocol assumes only the base station node has the time synchronization schedule. Sensor nodes are active only when there is a sample to be taken from the environment which decreases the duty-cycle of the node and preserves energy. The packet which is responsible for initiating important data delivery in SEA-MAC is called TONE packet which is shorter in period than SYNC packet in S-MAC.

The literature trawl has revealed that few protocols use TDMA-based scheduling because of the overhead of time slot scheduling as sensor network deployment usually includes large number of sensors. A protocol that uses TDMA-based scheduling is the Energy and Rate (ER) proposed by (Kannan et al, 2003). The ER_MAC protocol has the ability of avoiding extra energy wastage.
The main advantages of ER-MAC are:

- packet loss due to collisions is absent because two nodes do not transmit in the same slot. Although packet loss may occur due to other reasons like interference, loss of signal strength etc.
- no contention mechanism is required for a node to start sensing its packets since the slots are pre-assigned to each node. No extra control overhead packets for contention are required.

ER-MAC uses the concept of periodic listen and sleep. A sensor node switches off its radio and goes into a sleep mode only when it is in its own time slot and does not have anything to transmit. It has to keep the radio awake in the slots assigned to its neighbors in order to receive packets from them even if the node with current slot has nothing to transmit.

Real-Time MAC (RT-MAC) proposed by (Sahoo, Baronia, 2007) is another TDMA-based MAC protocol that can provide delay guarantee. TDMA based MAC protocols suffers from latency caused by the assigning of time slots which takes up a lot of time because of the number of sensor nodes deployed. RT-MAC overcomes this problem by reutilizing the connection channel between two successive channel accesses of a sensor node. RT-MAC also allows sensors to go to sleep which preserves energy. Although it provides delay guarantee, the RT-MAC protocol requires a lot of computation that exhaust the sensor node itself in some cases like clock drifting problem.
There are other works on design of MAC protocol based on TDMA scheme (Ganeriwal et al, 2003; Egea-L'opez et al, 2006); they all share the same complexity in time slot assigning.
To summarize the investigated literature, we devised a table that illustrates the categories of MAC protocols proposed for WSNs showing their advantages and disadvantages. Refer to Table 2:

MAC Protocol	Category	Main Advantage	Main Disadvantage
IEEE 802.11	CSMA/CA	The Highest system throughput	Inefficient energy consumption
S-MAC	CSMA/CA	Scalable, energy efficient due to the sleep/listen scheme	Suffers from Latency issues
T-MAC	CSMA/CA	Energy efficient, Reasonable throughput	Requires extended control packet to achieve efficient operation
TEEM	CSMA/CA	Energy efficient due to the eliminating the use of RTS packet	Suffers from Latency issues
TA-MAC	CSMA/CA	Cross-Layer approach	Suffers from latency issues
R-MAC	CSMA/CA	Enhanced throughput	Control Packet Delivery overhead
SEA-MAC	CSMA/CA	Energy efficient operation	Suffers from Latency issues
ER-MAC	TDMA	Collision free environment	Scalability and latency issues
RT-MAC	TDMA	Increased the system throughput	Excessive calculation and clock drifting problems

Table 1. Summary of the related approaches for MAC Protocols.

3. Proposed solution and the methodology behind it

In this section we ill discuss our proposed solution and describe the operation on the protocol. It also discusses how it manages control packets and data packets exchanges between the network nodes. Energy consumption and packet exchange delay analysis are also discussed. To prove the method proposed we devised simulation experiments using the most common tool to simulate networking systems the Network Simulator 2 (NS2) (Issariyakul, Hossain, 2008). The analysis equations were based on the theory of S-MAC.

3.1 The Network Simulator 2 (NS2)

NS2 is the most widely used tool in researches involved in general networking systems analysis and wireless networking systems includes Mobile networking, Satellite networking, Wireless Sensor Networks, LAN networks and other network technologies. NS2 is built using C++ language and uses OTcl (Object Oriented Tcl) language as an interface with the simulator. The network topology is built using OTcl and the packet operation protocol is written in C++ (Issariyakul, Hossain, 2008).

3.1.2 Mobile Networking In NS2

The wireless model essentially consists of the MobileNode at the core, with additional supporting features that allows simulations of multi-hop ad-hoc networks, wireless LANs etc. A MobileNode thus is the basic Node object with added functionalities of a wireless and mobile node like ability to move within a given topology, ability to receive and transmit signals to and from a wireless channel.

3.1.3 Routing and MAC protocols provided in NS2

Two MAC layer protocols are implemented for mobile networks, which are IEEE 802.11 and TDMA, while S-MAC was added to NS2 as a Patch by (Ye et al,2001). The four different ad-hoc routing protocols currently implemented for mobile networking in NS2 are dsdv, dsr, aodv and tora.

3.2 The Proposed Scheme

The proposed scheme (Shukur, Yap, 2009, a; Shukur, Yap, 2009, b) considers the following:

1. Combining the functionality of SYNC packet with RTS packet will provide both energy and latency efficient operation which will eliminate the need of sending two packets and decrease control packet overhead. This packet from now on would be referred to as SEEK.
2. To increase the throughput of the system (SEEK) packet will be sent all the way to the down stream nodes before sending CTS packet to the upper stream node. This will open the way to DATA packet to move through the stream of nodes until DATA packet reaches the base station node. Figure 2 Describes the approach mentioned above.

3.2.1 Energy Consumption analysis

The first step is to analyze the proposed approach energy consumption for three nodes operation. The following assumptions are made for the analysis (using the scenario shown in Figure 3-3 below:

1. All nodes in the way are by all means available for any packet transmission.
2. The packet delivery direction is from node 1 to node 3.
3. No collision happens between nodes (assuming that Carrier Sense is successful in each transmission start).
4. SEEK packet follow this rule (SYNC<SEEK<SYNC+RTS).
5. DATA packet could be transmitted in one hop.
6. All control packets are fixed in size.
7. In a more realistic scenario upper-layer routing information provides the shortest route to the destination.
8. DATA packet can be transferred in one hop.
9. If the next node in the way is in sleep mode (SEEK) works as the signal that wakes up the node.

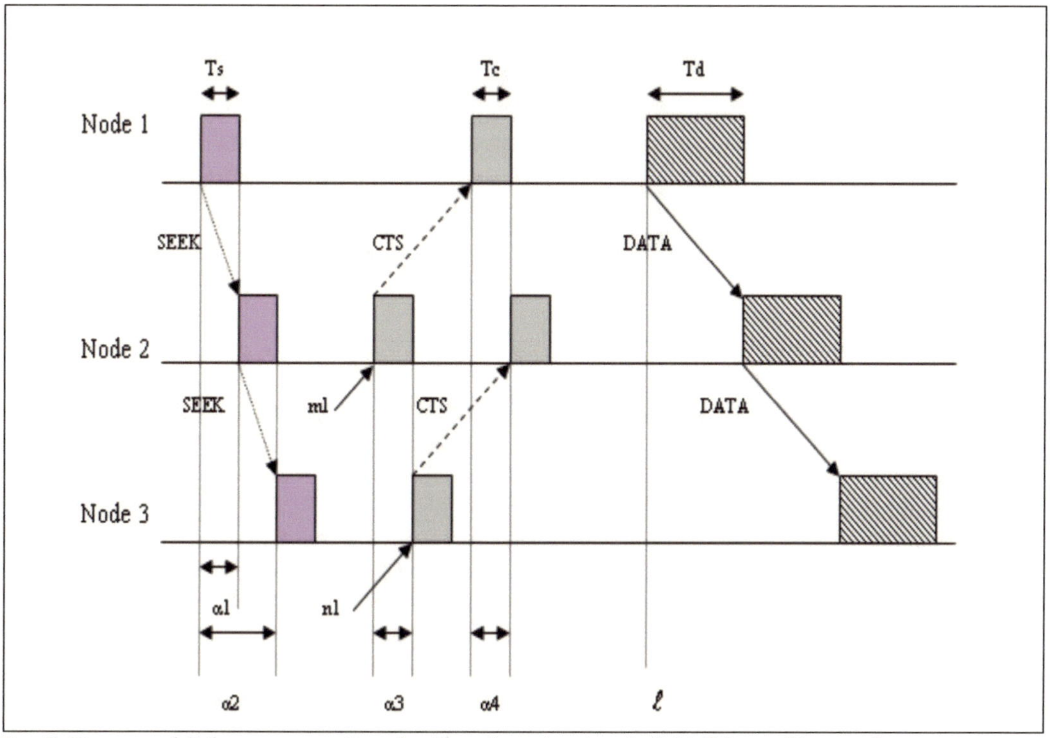

Fig. 2. Proposed Scheme operation for Synchronization in MAC layer Protocol.

The analysis scenario is described in Figure 3:

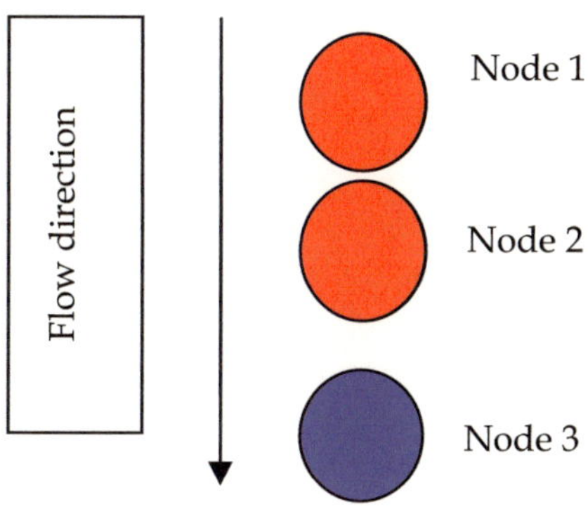

Fig. 3. Analysis Scenario.

Each Node operation is represented by a formula following the devised Synchronization timeline which is described in figure 2 above. Starting with the operation of Node (1) form the Scenario above (figure 3) the formula of Energy consumption can be represented as follow:

$$S_1(t) = X(t) + Y(t - \alpha_3) + Xd(t) \quad \text{.......(1)}$$

$$S_1(t) = P_t \times \text{rect}\left(\frac{t}{T_s}\right) + P \times \text{rect}\left(\frac{(t - \alpha_3 - m_1)}{T_c}\right) + P \times \text{rect}\left(\frac{(t - \ell)}{T_d}\right)$$

Where:

T_s: SEEK packet time length.
T_c : CTS packet time length.
T_d : DATA packet time length.
α : the delay in each state of transmitting SEEK packet and receiving CTS packet.
P_t : Transmission Power.
P : Reception Power.
X(t): rectangular function of delay for SEEK packet.
Y(t): rectangular function of delay for CTS packet out from the exact node.
Z(t): rectangular function of delay for CTS packet received from the down stream node.
X_d(t): rectangular function of delay for DATA packet.

Node (2) energy consumption is equal to the following equation:

$$S_2(t) = X(t - \alpha_1) + Y(t) + Z(t - \alpha_4) + Xd(t - \alpha_4) \quad \text{.......(2)}$$

$$S_2(t) = \begin{matrix} \left[P_t \times \text{rect}\left(\frac{(t - \alpha_1)}{T_s}\right) + P_t \times \text{rect}\left(\frac{(t - m_1)}{T_c}\right) + P \times \text{rect}\left(\frac{(t - \alpha_4)}{T_c}\right)\right] + \\ \left[P \times \text{rect}\left(\frac{(t - \alpha_4)}{T_c}\right) + P \times \text{rect}\left(\frac{(t - \ell - \alpha_4)}{T_d}\right)\right] \end{matrix}$$

Finally Node (3) energy consumption:

$$S_3(t) = X(t - \alpha_2) + Y(t) + Z(t) + X_d(t - \alpha_4) \quad \text{.......(3)}$$

$$S_3(t) = P_t \times \text{rect}\left(\frac{(t - \alpha_2)}{T_s}\right) + P_t \times \text{rect}\left(\frac{(t - n_1)}{T_c}\right) + P \times \left(\frac{(t - \ell - \alpha_4)}{T_d}\right)$$

From Equation (1,2 and 3) we can compute the energy consumed by following equation (4):

$$E_s = S_1(t) + S_2(t) + S_3(t) \quad \text{.......(4)}$$

Where (E_s) represents the energy consumed by the proposed analysis system in Figure (3-3). Substitute equations (1, 2 & 3) into (4) results in:

$$E_s(t) = \left[P_t * \mathrm{rect}\left(\frac{t}{T_s}\right) + P * \mathrm{rect}\left(\frac{t-\alpha_3-m_1}{T_c}\right) + P * \mathrm{rect}\left(\frac{t-1}{T_d}\right)\right] +$$
$$\left[P_t * \mathrm{rect}\left(\frac{t-\alpha_1}{T_s}\right) + P_t * \mathrm{rect}\left(\frac{t-m_1}{T_c}\right) + P * \mathrm{rect}\left(\frac{t-\alpha_4}{T_c}\right) + P * \mathrm{rect}\left(\frac{t-1-\alpha_4}{T_d}\right)\right] +$$
$$\left[P \times \mathrm{rect}\left(\frac{(t-\alpha_4)}{T_c}\right) + P \times \mathrm{rect}\left(\frac{(t-\ell-\alpha_4)}{T_d}\right)\right] +$$
$$\left[P_t \times \mathrm{rect}\left(\frac{(t-\alpha_2)}{T_s}\right) + P_t \times \mathrm{rect}\left(\frac{(t-n_1)}{T_c}\right) + P \times \left(\frac{(t-\ell-\alpha_4)}{T_d}\right)\right]$$

3.2.2 System Delay analysis

The proposed scheme deals with more that one node in a duty-cycle because of the concurrent (SEEK) packet transmission so the packet delay will only be counted as (extra SEEK packet) and (extra CTS packet) in the middle nodes, Below is the mathematical delay approach of the proposed scheme following the same parameters and the same assumptions made for energy consumption:

Node 1 delay:

$$D_1(t) = T_s + T_c + T_d \quad \ldots\ldots(5)$$

Node 2 delay:

$$D_2(t) = \alpha + T_s + 2 * T_c + T_d \quad \ldots\ldots(6)$$

Node 3 delay:

$$D_3(t) = T_s + T_c + T_d \quad \ldots\ldots(7)$$

From (5, 6 and 7) above a system delay equation can be derived:

$$D_s(t) = \sum_{1}^{N-2} \alpha * T_s + T_c \quad \ldots\ldots(9)$$

N: the number of nodes in the system.

While for S-MAC (Ye et al, 2001), because each node have to go through the same operation to send the data packet it is possible to describe S-MAC delay operation for the same system as:

$SYNC_t$: time length for SYNC packet.
RTS_t: time length for RTS packet.
CTS_t: time length for CTS packet.
$DATA_t$: time length for DATA packet.

Node 1 delay (S-MAC):

$$D_1\,(t) = SYNC_t + RTS_t + CTS_t + DATA_t. \quad\ldots\ldots\ldots\ldots\ldots\ldots (10)$$

Node 2 delay (S-MAC):

$$D_2\,(t) = D_1\,(t) + SYNC_t + RTS_t + CTS_t + DATA_t. \quad\ldots\ldots\ldots\ldots\ldots\ldots (11)$$

Node 3 delay (S-MAC):

$$D_3\,(t) = D_2\,(t) + SYNC_t + RTS_t + CTS_t + DATA_t. \quad\ldots\ldots\ldots\ldots\ldots\ldots (12)$$

From (10, 11 and 12) we can reach to a system delay equation using S-MAC:

$$D_{S\text{-}MAC}\,(t) = \sum_{1}^{N} D(N-1)(t) + SYNC_t + RTS_t + CTS_t + DATA_t \quad\ldots\ldots\ldots (13)$$

4. Implementation of the proposed solution

This section will discuss the implementaiton of the methodology and the simulation parameters used. Two simulation scenarios are devised and simulation parameters with a range of duty-cycles from (5% - 25%) for the first scenario and from (5%-40%) for the second scenario in three steps to cover most of operation environment that can a WSN suffer. The first simulation scenario (Figure 3) is represented by a straight line of nodes deployment .:

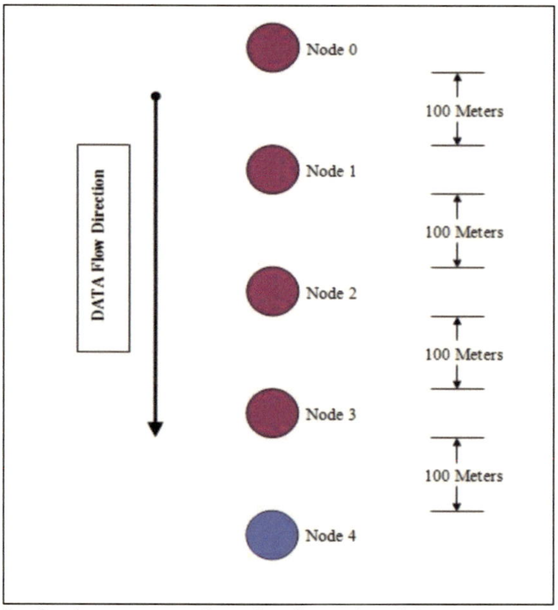

Fig. 3.a: A straight node deplyment used as the first simultion scene.

The simulation environment was built and made using NS2 version 2.33, the scenario consists of five nodes in one row Starts from node 0 to node 4 considering node 4 as the destination node in the simulation. Our main revals during where S-MAC protocol (as it is considered the base protocol to propose the Sleep-Listen Theory) and SEA-MAC because it is an inprovment over S-MAC in terms of the control packets handling. The proposed approach will be referred as Proposed Protocol (PP-) before or after any protocol name.

below is a table of the parameters that where allocated for the scenarion above:

Parameter	**Amplitude**
Simulation time	700 second
Duty-Cycle	5%, 10%, 25%
Routing Protocol	None
Node Idle power	100 mW
Node Rx Power	100 mW
Node Tx Power	100 mW
Node Sleep Power	1 mW
Transition Power	20 mW
Transition time	5 ms
Energy model	NS2 Energy model
Propogation model	TwoRayGround
Initial Energy for each node	1000 mJ

Table 3. the simulation parameters for the first scene

The second simulation (Figure 4) scenario consist of ten nodes. nine nodes 0-8 formed a square deployment and one node 9 was separated from the other as a base node. The simulations are conducted on a wide range of duty cycles from 5% - 40% in three steps (5, 25 and 40).

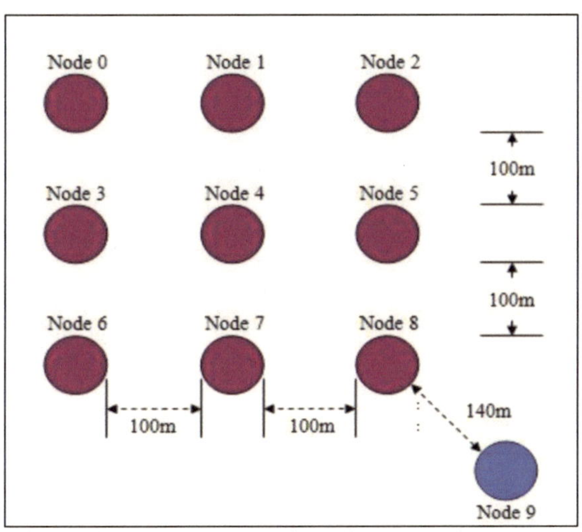

Fig. 4. square shape node deployment to simulate the second scene.

The Simulation parameters where allocated to this sceneario as in table 4 below:

Parameter	Amplitude
Simulation time	7000 seconds
Duty-Cycle	5%, 25%, 40%
Routing Protocol	DSR
Node Idle power	100 mW
Node Rx Power	100 mW
Node Tx Power	100 mW
Node Sleep Power	1 mW
Transition Power	20 mW
Transition time	5 ms
Energy model	NS2 Energy model
Propogation model	TwoRayGround
Initial Energy for each node	100000mJ

Table 4. the simulation parameters for the second scene.

The next section will discuss the results of the simulation and the effect of the proposed scheme on the operation of the MAC protocols and the Network itself.

4.1 First Scenario Results

To show the effect of the solution we propose we have to issue a comparison between our rivals S-MAC and SEA-MAC. Our criteria of comparison are Energy consumption, the number of collisions and the delay of packet transmission. Figure 5 shows the protocols operation in a 5% duty-cycle:

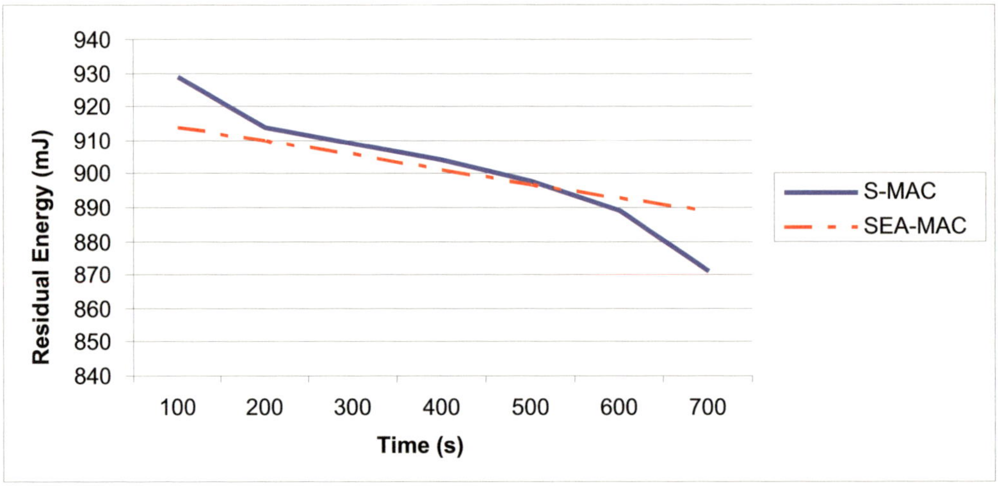

Fig. 5. S-MAC vs. SEA-MAC at 5% Duty-Cycle.

S-MAC operation is more prone to lose energy than SEA-MAC as it uses much more for (SYNC) packet than in SEA-MAC (TONE) packet. Figure 6-A shows the collisions occurance for both protocols. Figure 6-B show the delay effecincy of each protocol:

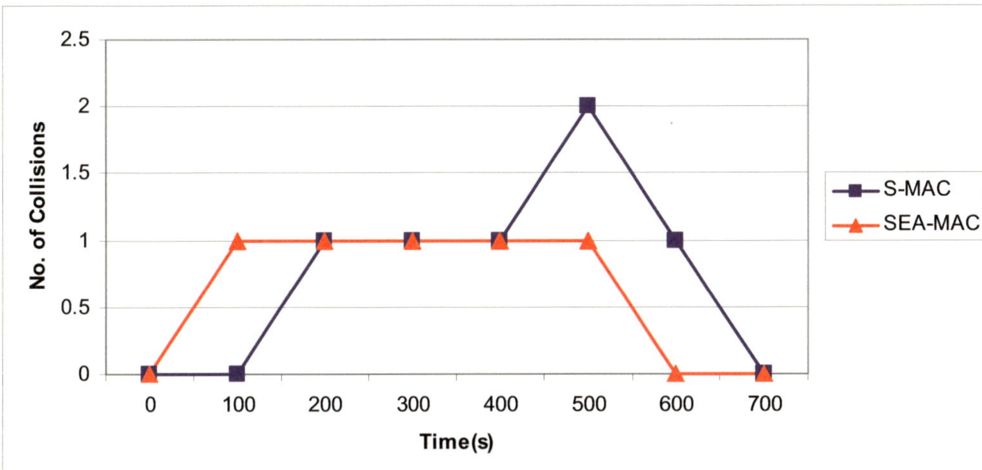

Fig. 6-A: S-MAC vs. SEA-Mac in terms of collision occurance at 5% Duty-Cycle

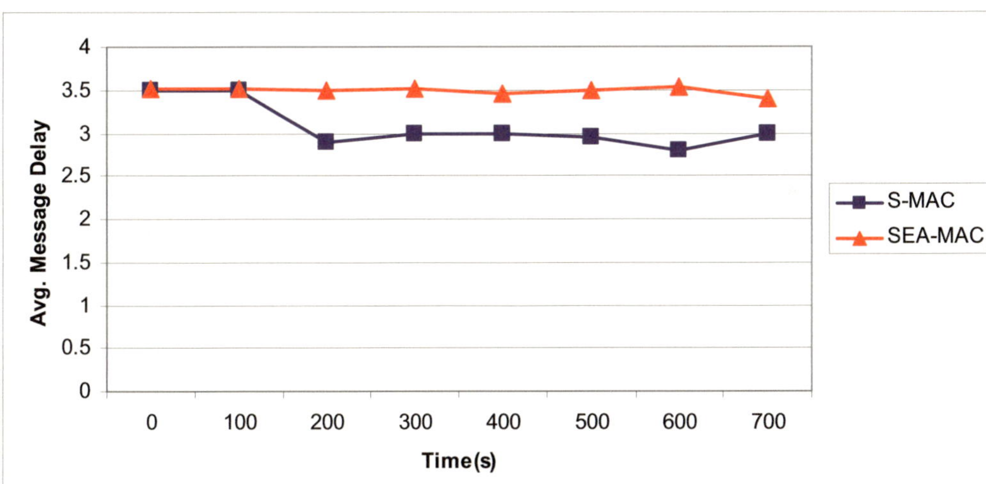

Fig. 6-B: S-MAC vs. SEA-MAC in terms of Message Delay at 5% Duty-Cycle

Overall, S-MAC offers better delay than SEA-MAC, if we increase the operation Duty-Cycle to 25%, we will observe that S-MAC a better performance in terms of Delay and collisions (Figure 7-A & B). But SEA-MAC offers better energy consumption over S-MAC because it uses a shorter activating packet called (TONE). Figure 8 shows the energy consumption effect.

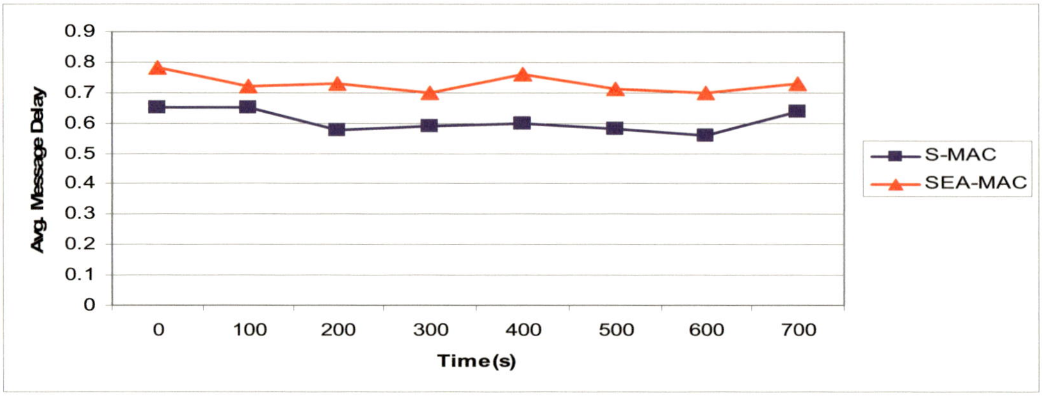

Fig. 7-A: S-MAC vs. SEA-MAC Message Delay at 25% operation Duty-Cycle.

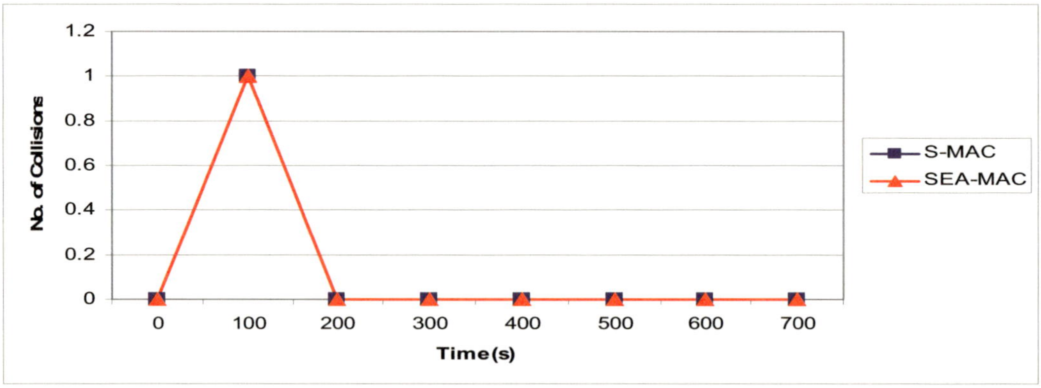

Fig. 7-B: S-MAC vs. SEA-MAC Collision Effect at 25% Duty-Cycle.

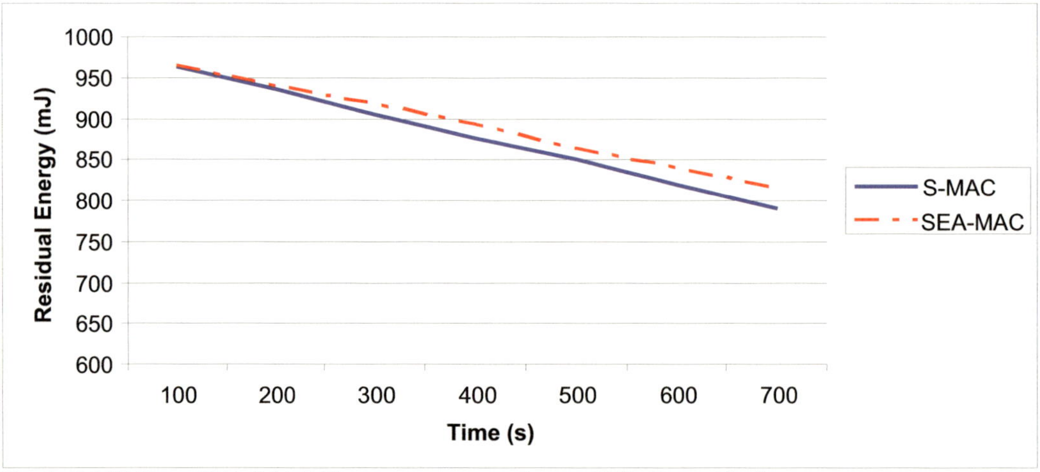

Fig. 8: S-MAC vs. SEA-MAC Energy consumption at 25% Duty-Cycle.

After impl;ementing our theory on both protocols (indicated by adding (PP-) before or after the protocol's name), S-MAC started to perform better than SEA-MAC in terms of energy consumption at low duty-cycle. While SEA-MAC kept the energy consumption better at higher Duty-Cycles. And to mention that both protocols after improvments provided Zero-collisions in the straight line simulation scenario. Figure 9-A,B,C&D shows the energy consumption and delay efficency at 5% & 25% duty-cycles.

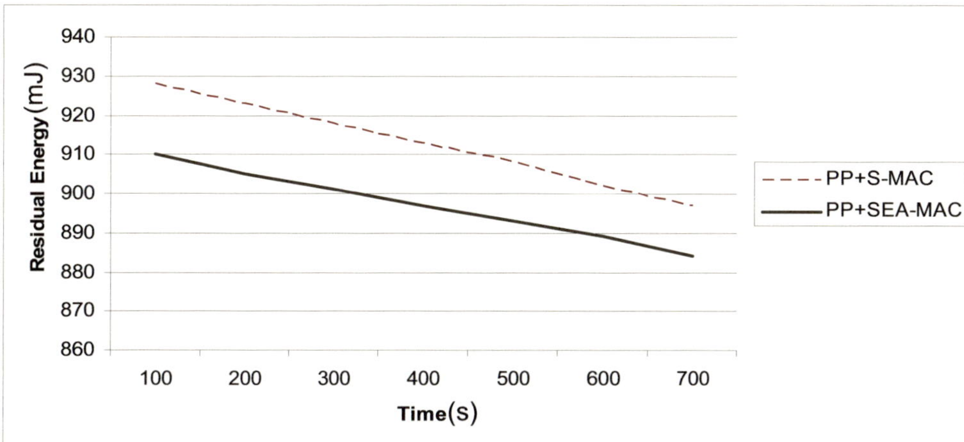

Fig. 9-A: PP+S-MAC vs. PP+SEA-MAC energy consumption at 5% Duty-Cycle.

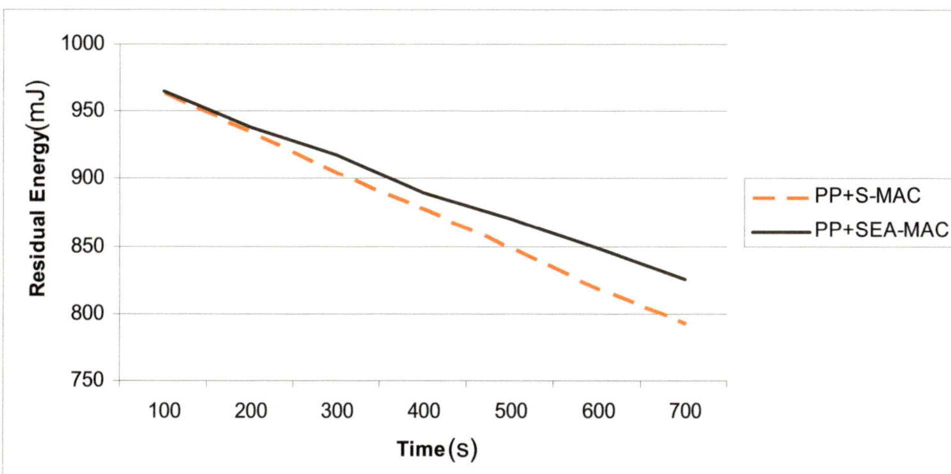

Fig. 9-B: PP+S-MAC vs. PP+SEA-MAC energy consumption at 25% Duty-Cycle.

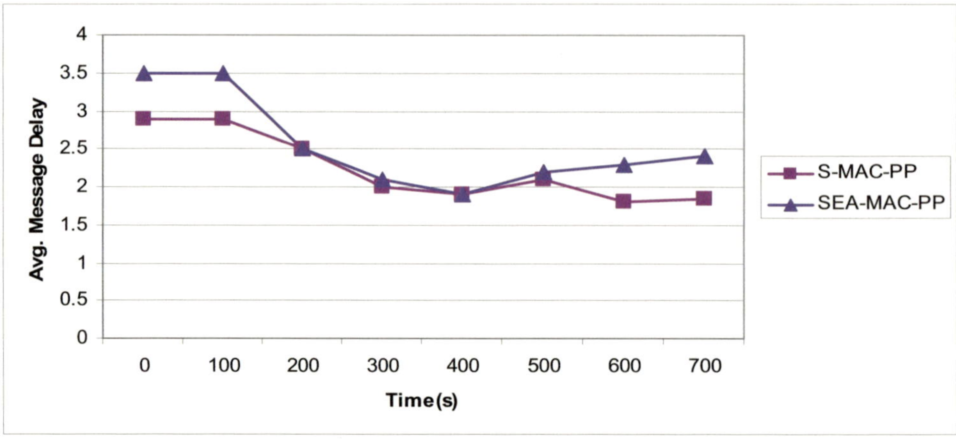

Fig. 9-C: PP+S-MAC vs. PP+SEA-MAC Delay effeciny at 5% Duty-Cycle.

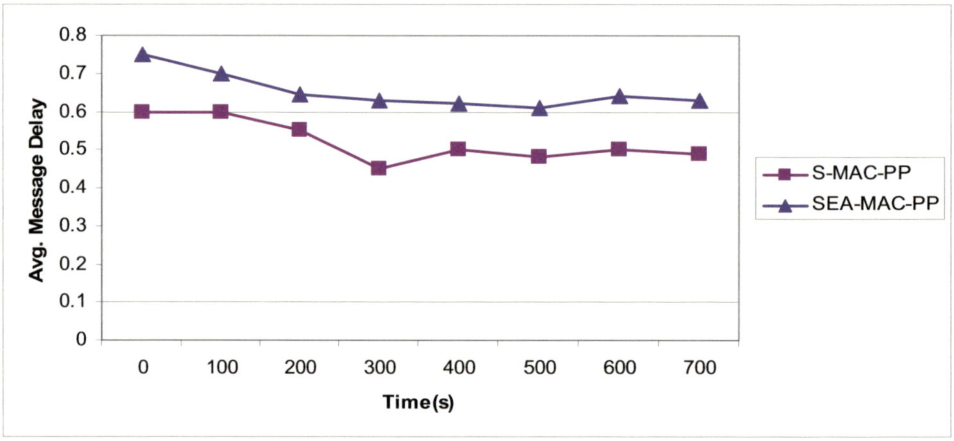

Fig. 9-D: PP+S-MAC vs. PP+SEA-MAC Delay comparison at 25% Duty-Cycle.

To summerize the result show above, The proposed scheme gave the effect on S-MAC and made the consumption in terms of energy at low Duty-Cycle operation better than the original scheme of S-MAC.
The proposed approach provided better operation in terms of energy consumption at high Duty-Cycle operation than the original SEA-MAC scheme.
Both protocols provided better throughput for most of the scenarios after adding the proposed scheme to the original scheme of the protocols.

4.2 The proposed Scheme effect for the second scenario

The second scenario has a new factor that gave an effect on the operation of both protocols S-MAC and SEA-MAC (with or without the implementation of the proposed theory). This is represented by the number of the deployed nodes. Increasing the number of the nodes can give a positive effect on the network operation as it will help to conduct the inquiry collection of the phenomena in a more fast paced operation. Figure 10-A,B &C shows the energy consumption, Delay and collisions occurrences. This effect is observed in Figure 10-A, where we can see the gap of consumption between SEA-MAC and S-MAC.

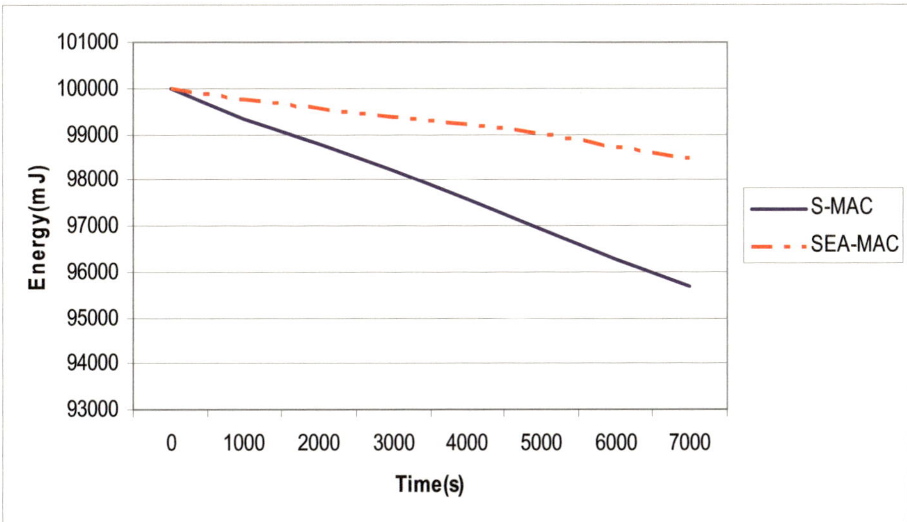

Fig. 10-A: S-MAC vs. SEA-MAC energy consumption at 5% Duty-Cycle.

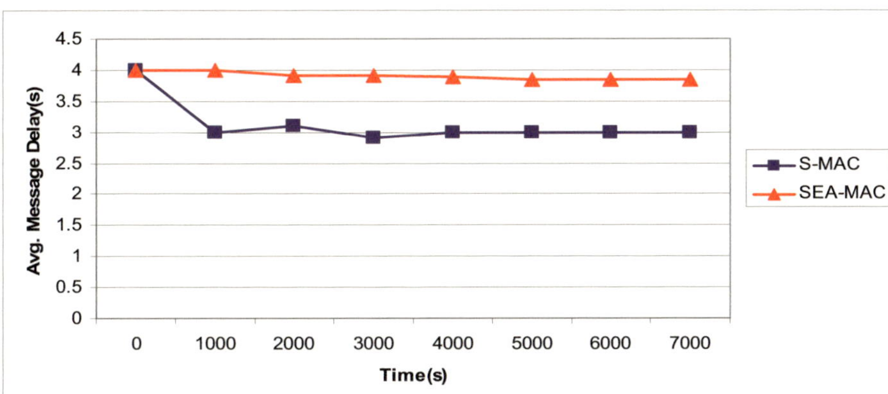

Fig. 10-B: S-MAC vs. SEA-MAC Delay average at 5% Duty-Cycle.

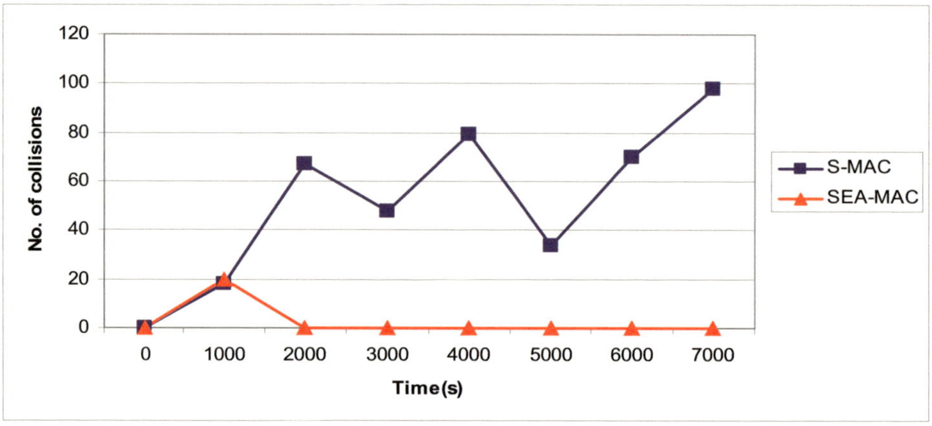

Fig. 10-C: S-MAC vs. SEA-MAC collisions ocurrances at 5% Duty-Cycle.

Adding the proposed approach to both protocols resulted in a different operation than the original ones. Figure 11-A,B&C shows that, it is observed that S-MAC was improved over SEA-MAC operation at low Duty-Cycle. This is due to the fact that S-MAC goes through four stages of operation (SYNC+RTS+CST+ACK) while SEA-MAC has (TONE+SYNC+RTS+CTS+ACK) which leads to a longer operation even with the compression of two control packets (SYNC&RTS), SEA-MAC has longer operation time than S-MAC at shorter duty-cycle.

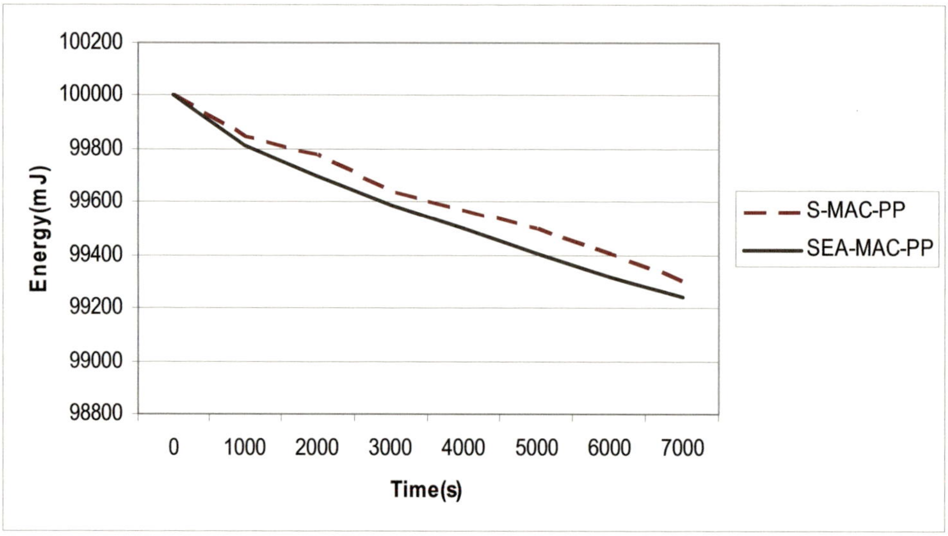

Fig. 11-A: PP+S-MAC vs. PP+SEA-MAC energy consumption at 5% Duty-Cycle.

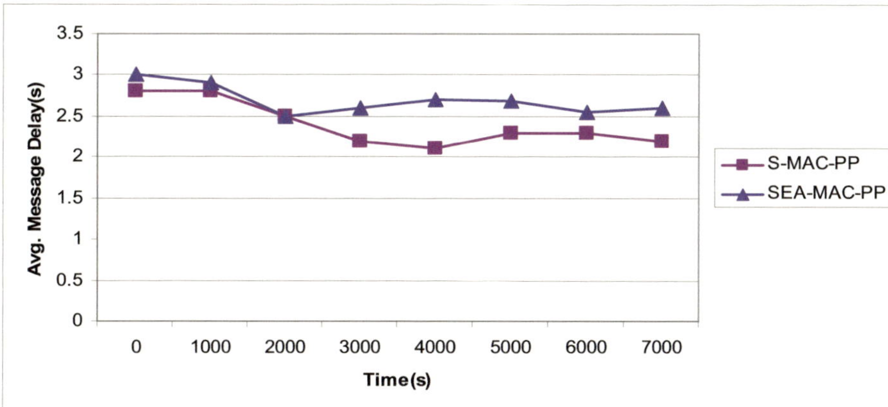

Fig. 11-B: S-MAC-PP vs. SEA-MAC-PP average Delay at 5% Duty-Cycle.

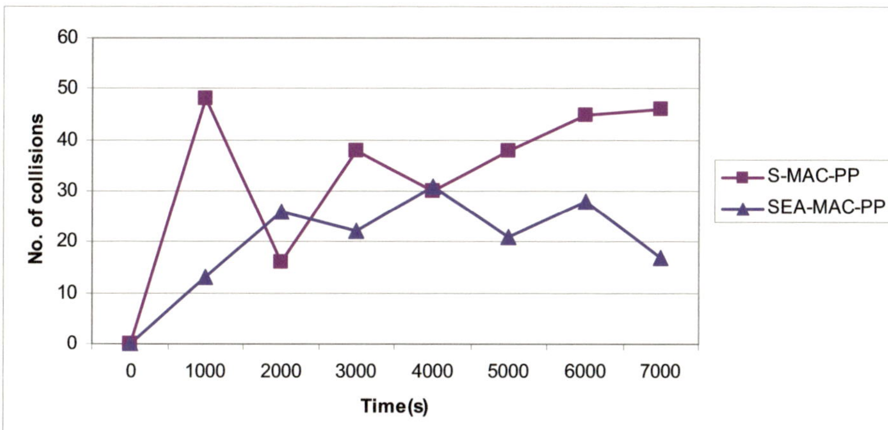

Fig. 11-C: S-MAC-PP vs. SEA-MAC-PP collisions at 5% Duty-Cycles.

4.3 Pros and Cons of the proposed theory

Overall, the proposed approached satisfied the quest as it does improve the operation of both protocols at different ranges of duty-cycle (we must note SEA-MAC with the proposed approach offered better energy consumption and delay operation at higher duty-cycles than S-MAC also implemented with the approach). Increasing the number of nodes result in collision occurance rather than the situation with the straight line deployment. Overall message delay is in favor of S-MAC at shorter duty-cycles and the advantage is to SEA-MAC at longer duty-cycle.

In the next section we will discuss breifly the mobility issues in WSN as it is considered an important part of this research area.

5. Mobility in WSN

Wireless sensor networks (WSN) offers a wide range of applications and it is also an intense area of research. However, current research in wireless sensor networks focuses on stationary WSN where they are deployed in a stationary position providing the base station with information about the subject under observation. However, a mobile sensor network is a collection of WSN nodes. Each of these nodes is capable of sensing, communication and moving around. It is the mobility capabilities that distinguish a mobile sensor network from the conventional 'fixed' WSN (Motari'c et al, 2002).

Mobile sensor networks offer many opportunities for research as these sensors involves: the estimate location of the node in a movement scenario, an efficient DATA and information processing schemes that can cope with the mobility measurements and requirements (this includes the routing theory and the potential MAC Protocol Used).

Most of the discussed approaches interms of routing theory, MAC and also allocation the location of the sensors are ment for stationary sensor nodes. Mobile sensor networks requiers extra care when it comes to design and implementing a network related protocols the conserns includes ad not exclusive to: energy consumption, message delay, location estimation accuracy and scurity of information traveled between the nodes to the base station.

To list some of the aspects that effects on designing an operapable Mobile sensor networks, the next sections will give a brief explination about routing theory, MAC approaches and Localaization scheme aimed for mobility applications.

5.1 Routing theory

Routing protocols are protocols aimed to offer transmitting the DATA through the network by utilizing the best available routes (not always the shortest ones) to the destination. When it comes to design routing protocols for mobile Sensor nodes, extra care should be taken in terms of timing the transportation between the nodes. Most of the routing protocol that are used and implemented for Wireless sensor networks (e.g. Ad hoc on demand Distance Vector (AODV) and Dynamic Source Routing (DSR)) are originally designed and optimized for ad hoc networks which utilizes devices like (Laptop computers and mobile phones) which has much powerful energy sources than the ones available in sensor nodes. And to the power issue mobility make the task even tougher.

5.2 MAC approaches

Even the approach discussed in this chapter does not satisfy the mobility issues in MAC protocols aimed for mobile sensor networks. The results from the current work suggest that the CSMA based MAC protocols has a better chance in overcoming this issue than TDMA based MAC protocols because of the time slotting issues that comes along with TDMA based systems. IEEE802.15.4 or best known as (Zigbee) is a MAC layer standard provided by IEEE organization aimed for low power miniatures. Still, it cannot be considered yet as a standard MAC protocols for mobile sensor networks as it is still in the development stages for such applications.

5.3 Localization Issues

Locating the sensor is an important task in WSN as it provides information about the phenomena monitored and what action should be taken at the occurrence of an action. Proposed localization schemes are aimed manly for stationary networks and partially for mobile networks. Some of the examples of localization techniques are (Boukerche et al, 2007):

RSSI: Received Signal Strength Indicator, which is the cheapest technique to establish a node location as the medium used is wireless medium and most of the wireless adapter are capable of capturing such information. The disadvantage of such approach is the accuracy of the information calculated by such approach.

GPS: Geo- Positioning System, the most used approach mobile nodes application and in some cases considered the easiest. The disadvantage of GPS systems is that it adds extra cost to systems in terms of financial cost and energy consumption costs and also accuracy issues.

TOA: Time On Arrival systems, the most accurate approach to achieve the location of the nodes. However there are some cons for this technique: first of all the cost is higher than GPS systems. Second the accuracy issue is dependent on how violent the environment being applied on as it requires a line-of-sight connection to capture the required information. And the last issue, because it is a mounted platform so it will consume energy like the issue with the GPS systems.

6. Future Research goals

The future research goal is to devise a template Network Model aimed for Mobile Wireless Sensor Networks. The template will take in consideration the concerns discussed in section five of this chapter. It is envisage that the proposed approach provided in this chapter can assist to devise a MAC approach that can be applied for various applications in WSN. The proposed template is designed for Habitat monitoring applications as they share some similarities in terms of the configurations and crucial guarantees. Future work would to utilize a Signal - to - noise Ratio estimator (Kamel, Jeoti, 2007) as a metric to define which route is the best to chose and on which nodes signal can estimate the location of the node. Cross-layer approach a definite approach and consideration that we aim utilize in our template.

7. Acknowledgments

We are greateful to both Dr. Brahim Belhaouari Samir, department of Electrical and Electronic Engineering in Universiti Teknologi PETRONAS, 31750 Tronoh, Perak for helping us with proposed scheme mathmatical analysis. and Mr. Megual A. Erazo, Computer science department in Florida international University for helping us to develop SEA-MAC protocol. Our thanks goes also to Universiti Teknologi PETRONAS for funding this research and achieve the aimed results.

8. References

Kazem Sohraby, Daniel Minoli and Taieb Znati "WIRELESS SENSOR NETWORKS Technology, Protocols, and Applications", 2007 by John Wiley & Sons, Inc.

Yang Yu, Viktor K Prasanna and Bhaskar Krishnamachari "Information processing and routing in wireless sensor networks", 2006 by World Scientific Publishing Co. Pte. Ltd.

Bhaskar Krishnamachari "Networking Wireless Sensors", Cambridge University Press 2005.

Alan Mainwaring, Joseph Polastre, Robert Szewczyk, David Culler and John Anderson "Wireless Sensor Networks for Habitat Monitoring", *WSNA'02,* September 28, 2002, Atlanta, Georgia, USA, ACM.

Vijay Raghunathan, Curt Schurgers, Sung Park, and Mani B. Srivastava "Energy Aware Wireless Sensor Networks", IEEE Signal Processing Magazine, 2002.

Azzedine Boukerche, Fernando H. S. Silva, Regina B. Araujo and Richard W. N. Pazzi "A Low Latency and Energy Aware Event Ordering Algorithm for Wireless Actor and Sensor Networks", MSWiM'05, October 10–13, 2005, Montreal, Quebec, Canada, ACM.

Rebecca Braynard, Adam Silberstein and Carla Ellis "Extending Network Lifetime Using an Automatically Tuned Energy-Aware MAC Protocol", Proceedings of the 2006 European Workshop on Wireless Sensor Networks, Zurich, Switzerland (2006).

Lodewijk van Hoesel and Paul J.M. Havinga "MAC Protocol for WSNs", SenSys'04, November 3-5, 2004, Baltimore, Maryland, USA, ACM.

Yee Wei Law, Lodewijk van Hoesel, Jeroen Doumen, Pieter Hartel and Paul Havinga "Energy Efficient Link Layer Jamming Attacks against Wireless Sensor Network MAC Protocols", SASN'05, November 7, 2005, Alexandria, Virginia, USA, ACM.

Ioannis Mathioudakis, Neil M.White, Nick R. Harris, Geoff V. Merrett, "Wireless Sensor Networks: A Case Study for Energy Efficient Environmental Monitoring", Eurosensors Conference 2008, 7-11 September 2008, Dresden, Germany.

Marwan Ihsan Shukur, Lee Sheng Chyan and Vooi Voon Yap "Wireless Sensor Networks: Delay Guarentee and Energy Efficient MAC Protocols", Proceedings of World Academy of Science, Engineering and Technology, WCSET 2009, 25-27 Feb. 2009, Penang, Malaysia.

Wei Ye, John Heidemann and Deborah Estrin "An Energy-Efficient MAC protocol for Wireless Sensor Networks", USC/ISI Technical Report ISI-TR-543, September 2001.

Tijs Van Dam and Keon Langendoen "An Adaptive Energy-Efficeint MAC Protocol for Wireless Sensor Networks", SenSys'03, November 5-7, 2003, ACM.

Shu Du, Amit Kumar Saha and David B. Johnson, "RMAC: A Routing-Enhanced Duty-Cycle MAC Protocol for Wireless Sensor Networks", INFOCOM 2007. 26th IEEE International Conference on Computer Communications. IEEE.

Jin Kyung PARK, Woo Cheol Shin and Jun HA "Energy-Aware Pure ALOHA for Wireless Sensor Networks", IEIC Trans. Fundamentals, VOL.E89-A, No.6 June 2006.

Changsu Suh and Young-Bae Ko, "A Traffic Aware, Energy Efficient MAC protocol for Wireless Sensor Networks", Proceeding of the IEEE international symposium on circuits and systems (IS CAS'05), May. 2005.

Sangheon Pack, Jaeyoung Choi, Taekyoung Kwon and Yanghee Choi, "TA-MAC: Task Aware MAC Protocol for Wireless Sensor Networks", Vehicular Technology Conference, 2006. VTC 2006-Spring. IEEE 63rd.

Miguel A. Erazo, Yi Qian, "SEA-MAC: A Simple Energy Aware MAC Protocol for Wireless Sensor Networks for Environmental Monitoring Applications", Wireless Pervasive Computing, 2007. ISWPC '07. IEEE 2nd international symposium 2007.

Rajgopal Kannan, Ram Kalidini and S. S. Iyengar "Energy and rate based MAC protocol for Wireless Sensor Networks" SIGMOD Record, Vol.32, No.4, December 2003.

Anirudha Sahoo and Prashant Baronia "An Energy Efficient MAC in Wireless Sensor Networks to Provide Delay Guarantee", Local & Metropolitan Area Networks, 2007. LANMAN 2007. 15th IEEE Workshop on.

Saurabh Ganeriwal, Ram Kumar and Mani B. Srivastava "Timing-sync Protocol for Sensor Networks", SenSys '03, November 5-7, 2003, Los Angeles, California, USA, ACM.

Esteban Egea-L´opez, Javier Vales-Alonso, Alejandro S. Mart´ınez-Sala, Joan Garc´ıa-Haro, Pablo Pav´on-Mari˜no, and M. Victoria Bueno-Delgado "A Real-Time MAC Protocol for Wireless Sensor Networks: Virtual TDMA for Sensors (VTS)", ARCS 2006, LNCS 3894, pp. 382–396, 2006, Springer-Verlag Berlin Heidelberg 2006.

Teerawat Issariyakul and Ekram Hossain "Introduction to Network Simulator NS2", SpringerLink publications-Springer US 2008.

Marwan Ihsan Shukur and Vooi Voon Yap "An Approach for efficient energy consumption and delay guarantee MAC Protocol for Wireless Sensor Networks", Proceedings of International Conference on Computing and Informatics, ICOCI 2009, 24-25 June 2009, Kuala Lumpur, Malaysia, a.

Marwan Ihsan Shukur and Vooi Voon Yap "Enhanced SEA-MAC: An Efficient MAC Protocol for Wireless Sensor Networks for Environmental Monitoring Applications", Conference on Innovative Technologies in Intelligent Systems and Industrial Applications, IEEE CITISIA 2009, 25 July 2009, MONASH University Sunway Campus, Malaysia, b.

Howard, A, Matari´c, M.J., and Sukhatme, G.S., "Mobile Sensor Network Deployment using Potential Fields: A Distributed, Scalable Solution to the Area Coverage Problem", *Proceedings of the 6th International Symposium on Distributed Autonomous Robotics Systems (DARS02)* Fukuoka, Japan, June 25-27, 2002.

Azzedine Boukerche, Horacio A. B. F. Oliveira and Eduardo F. Nakamura "Localization Systems For Wireless Sensor Networks", IEEE Wireless Communications Mag. Dec. 2007.

Nidal S. Kamel and Varun Jeoti "A Linear Prediction Based Estimation of Signal-to-Noise Ratio in AWGN Channel", ETRI journal, Volume 29, Number 5, October 2007.

16

Hybrid Optical and Wireless Sensor Networks

Lianshan Yan, Xiaoyin Li, Zhen Zhang, Jiangtao Liu and Wei Pan
Southwest Jiaotong University
Chengdu, Sichuan, China

1. Introduction

Wireless sensor network (WSN) has attracted considerable attentions during the last few years due to characteristics such as feasibility of rapid deployment, self-organization (different from ad hoc networks though) and fault tolerance, as well as rapid development of wireless communications and integrated electronics [1]. Such networks are constructed by randomly but densely scattered tiny sensor nodes (Fig. 1). As sensor nodes are prone to failures and the network topology changes very frequently, different protocols have been proposed to save the overall energy dissipation in WSNs [2-5]. Among them, Low-Energy-Adaptive-Clustering-Hierarchy (LEACH), first proposed by researchers from Massachusetts Institute of Technology [5], is considered to be one of the most effective protocols in terms of energy efficiency [6-7]. Another protocol, called Power-Efficient Gathering in Sensor Information Systems (PEGASIS), is a near optimal chain-based protocol [8].

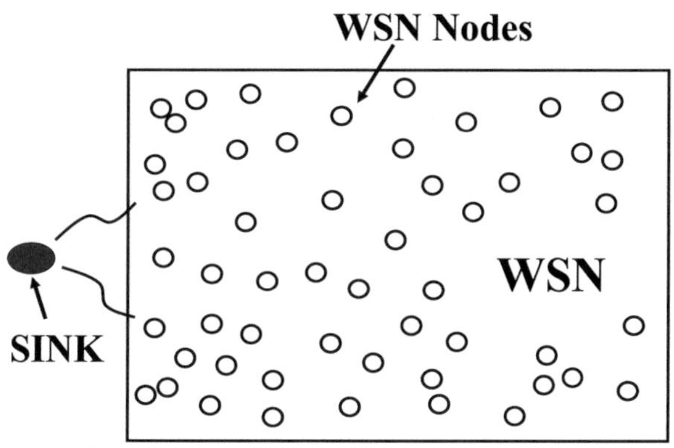

Fig. 1. Illustration of a wireless sensor network (WSN) with randomly scattered nodes (sink node: no energy restriction; WSN nodes: with energy restriction);

On the other hand, distributed fiber sensors (DFS) have been intensively studied or even deployed for analyzing loss, external pressure and temperature or birefringence distribution along the fiber link, ranging from hundreds of meters to tens of kilometers [9-13]. Mechanisms include Rayleigh, Brillouin or Raman scattering or polarization effects, through either time or frequency-domain analysis. Compared with conventional sensors including wireless ones, optical fiber sensors have intrinsic advantages such as high sensitivity, the immunity to electromagnetic interference (EMI), superior endurance in harsh environments and much longer lifetime.

Apparently it would be highly desirable to have integrated sensor networks that can take advantages of both WSNs and fiber sensor networks (FSNs). Such hybrid sensor networks can find major applications including monitoring inaccessible terrains (military, high-voltage electricity facilities, etc.), long-term observation of earthquake activity and large area environmental control with tunnels, and so on. So far hybrid sensor networks have been studied as well [14-16], while optical sensors in these networks are generally point-like (e.g. fiber-Bragg-grating based), and such nodes can be regarded as normal WSN nodes after optical to wireless signal conversion.

In this chapter, we first review typical WSN protocols, mainly about LEACH and PEGASIS, then evaluate the performance of LEACH protocols for different topologies, especially the rectangle one. We propose an improved algorithm based on LEACH and PEGASIS for the WSN, finally and most importantly, we propose an O-LEACH protocol for the hybrid sensor network that is composed of a DFS link and two separated WSNs. Most analyses about performance are done in terms of lifetime of the sensor networks.

2. Overview of WSN protocols

Wireless sensor networks (WSNs) generally are composed of small or tiny nodes with sensing, computation, and communication capabilities. Various routing, power management, and data dissemination protocols have been specifically designed for WSNs where energy awareness is an essential design issue. Among them, routing protocols might differ depending on the applications and network architectures. In general, routing protocols for the wireless sensor networks can be divided into flat-based, hierarchical-based, and location-based in terms of the underlying network structures [17-19]. As some protocols may be discussed intensively in other chapters of this book, here we give a brief review about major protocols.

(1) Flat-based routing protocol
Sensor nodes in flat-based routing protocols have the same role and collaborate together to perform the sensing task and multi-hop communication. Since the flat routing is based on flooding, it has several demerits, such as large routing overhead and high energy dissipation. Flat-based routing protocol is used in the early stage of WSNs, such as Flooding, Gossiping, SPIN, and Rumor.

(2) Hierarchical-based routing protocol
Hierarchical-based routing protocol is the main trend for WSN's routing protocols. In hierarchical-based routing protocols, the network is divided into several logical groups within a fixed area. The logical groups are called clusters. Sensor nodes collect the information in a cluster and a head node aggregates the information. Each sensor node delivers the sensing data to the head node in the cluster and the head node delivers the aggregated data to the base station which is located outside of the sensor network. Contrary to flat routing protocols, only a head node aggregates the collected information and sends it to the base station. Due to these advantages, sensor nodes can remarkably save their own

energy. In general, a hierarchical routing technique is regarded as superior to flat routing approaches. The classical Hierarchical-based routing protocols are LEACH, PEGASIS, H-PEGASIS, TEEN, and APTEEN. We will discuss the LEACH and PEGASIS protocols in more details later.

(3) Location-based routing protocol
Such protocol is based on the location information of senor nodes in WSNs. It assumes that each node would know its own location and its neighbor sensor nodes' location before sensor nodes sensing and collecting the peripheral information. The distance between neighbouring sensor nodes can be computed based on the incoming signal strength [17-18].

2.1 LEACH

Low Energy Adaptive Clustering Hierarchy (LEACH) was first introduced by Heinzelman, et al. in [5, 20] with advantages such as energy efficiency, simplicity and load balancing ability. LEACH is a cluster-based protocol, therefore the numbers of cluster heads and cluster members generated by LEACH are important parameters for achieving better performance.

In LEACH protocol, the sensor nodes in the network are divided into a number of clusters, the nodes organize themselves into preferred local clusters, a sensor node is selected randomly as the cluster head (CH) in each cluster and this role is rotated to evenly distribute the energy load among nodes of the network. The CH nodes compress data arriving from nodes that belong to the respective cluster, and send an aggregated packet to the BS in order to further reduce the amount of information that must be transmitted to the BS, thus reducing energy dissipation and enhancing system lifetime. After a given interval of time, randomized rotation of the role of CH is conducted to maximize the uniformity of energy dissipation of the network. Sensors elect themselves to be local cluster heads at any time with a certain probability. Generally only ~ 5% of nodes need to act as CHs based on simulation results. LEACH uses a TDMA/CDMA MAC to reduce intercluster and intracluster collisions. As data collection is centralized and performed periodically, this protocol is most appropriate when there is a need for constant monitoring by the sensor network.

The operation of LEACH is broken up into rounds, where each round begins with a set-up phase followed by a steady-state phase. In order to minimize overhead, the steady-state phase takes longer time compared to the set-up phase. In the setup phase, the clusters are organized and CHs are selected. In the steady state phase, the actual data transfer to the BS takes place. During the setup phase, each node decides whether or not to become a cluster head for the current round. A predetermined fraction of nodes, p, elect themselves as CHs. A sensor node chooses a random number between 0 and 1. If this random number is less than a threshold value *T(n)*, , the node becomes a cluster head for the current round. The threshold value is calculated based on Eq. (2-1):

$$T(n) = \begin{cases} \dfrac{p}{1 - p * (r \bmod \frac{1}{p})} & if \quad n \in G \\ 0 & otherwise \end{cases} \tag{2-1}$$

Where p is the desired percentage of the cluster heads (e.g. p=0.05), r is the current round, and G is the set of nodes that have not been cluster heads in the last 1/p rounds. Using this threshold, each node may be a cluster head sometime within 1/p rounds. All elected CHs broadcast an advertisement message to the rest of nodes in the network that they are the new CHs. After receiving the advertisement, all non-CH nodes decide on the cluster to which they want to belong based on the signal strength of the advertisement. The non-CH nodes then inform the appropriate CHs to be a member of the cluster. After receiving all the messages from the nodes that would like to be included in the cluster and based on the number of nodes in the cluster, the CH node creates a TDMA schedule and assigns each node a time slot when it can transmit information. This schedule is broadcast to all the nodes in the cluster. During the steady state phase, the sensor nodes can begin sensing and transmitting data to the CHs. The CH node must keep its receiver on to receive all the data from the nodes in the cluster. Each cluster communicates using different CDMA codes to reduce interference from nodes belonging to other clusters. After receiving all the data, the CH aggregates the data before sending it to the BS. This ends a typical round.
Advantages of LEACH include: (i) by using adaptive clusters and rotating cluster heads, LEACH allows the energy requirements of the system to be distributed among all the sensors; (ii) LEACH is able to perform local computation in each cluster to reduce the amount of data that must be transmitted to the base station. On the other hand, there are still some drawbacks about LEACH: (i) LEACH assumes that each node could communicate with the sink and each node has computational power to support different MAC protocols, which limits its application to networks deployed in large regions. (ii) LEACH does not determine how to distribute the CHs uniformly through the network. Therefore, there is the possibility that the elected CHs will be concentrated in one part of the network. (iii) LEACH assumes that all nodes begin with the same amount of energy capacity in each election round, assuming that being a CH consumes approximately the same amount of energy for each node. Hence LEACH is not appropriate for non-uniform energy nodes [17, 20].

2.2 PEGASIS

In [8], an enhancement over the LEACH protocol called Power-Efficient Gathering in Sensor Information Systems (PEGASIS) was proposed. The basic idea of the protocol is that nodes only receive from and transmit to the closest neighbours, and they take turns being the leader for communicating with the BS. This reduces the power required to transmit data per round as the power draining is spread uniformly over all nodes. Hence, PEGASIS has two main objectives: (i) to increase the lifetime of each node by using collaborative techniques; (ii) to allow only local coordination between nodes that are close together so that the bandwidth consumed in communication is reduced.

PEGASIS adopts a homogenous topology. In this topology, the BS lies far from sensors with the fixed position. The data is collected and compressed before sent to the next node. Hence the messages maintain ideally a fixed size when they are transmitted between sensors. To locate the closest neighbour node in PEGASIS, each node uses the signal strength to measure the distance to all neighbouring nodes and then adjusts the signal strength so that only one node can be heard. The chain in PEGASIS consists of those nodes that are closest to each other and form a path to the BS. The following describes the protocol briefly:

1) The chain starts with the furthest node from the BS to make sure that nodes father from the BS have close neighbours. Based on the greedy algorithm, the neighbour node joins into the chain with its distance increases gradually. When a node dies, the chain is reconstructed in the same manner to bypass the dead node.
2) To gather data in each round, a token is generated by the BS to set the aggregating direction after the token sent from the BS to an end node. Each node receives data from one neighbour, fuses with its own data, and transmits to the other neighbour on the chain.
3) Only one node transmits data to the BS in certain rounds, the leader is the node whose number is (i mod N) where N represents the number of the nodes in round i.

PEGASIS is better than LEACH in terms of energy saving due to following facts: (i) During the data localization, the distances that most of the nodes transmit information are much shorter compared to that in LEACH. (ii) The amount of data for the leader to receive is much less than LEACH. (iii) only one node transmits to the BS in each round.

Though PEGASIS has obvious advantages, it has some shortcomings. Firstly, though most sensors are joined on a chain to form a basically homogenous structure, a sensor with too much branches may perform many times of data receiving in a certain round thereby resulting in unbalanced energy problem. Secondly, all the nodes must keep active before the token arriving. This means there will be a large percentage of active nodes with nothing to do from the beginning, meaning a waste of energy and time. Thirdly, once a sensor on the chain was captured the whole net may be under the control by the attackers. The weak security could be a great threat [17, 20].

LEACH that is a cluster-based protocol and PEGASIS that is a chain-based protocol are the most classical Hierarchical-based routing protocols. They both have attracted intensive attention, and lots of routing protocols are based on these two. Next we will investigate some issues in details.

3. WSN Topologies

3.1 Shapes of different topologies

According to the shape of WSN monitoring area, application requirements and monitoring of different targets, different topologies should be chosen for deploying the WSN: circular topology is preferred for applications such as harbour, stadium etc. [21]; square topology is suitable for irrigation in agriculture, nature reserve area etc.; rectangular topology can be chosen for highway, railway, mine and other areas [22].

Here we study the life time of WSN in round, square, rectangular shapes of topology, and the three topologies are shown in Figs. 3.1 (a-c). In Fig. 3.1(a), the circular area is 10,000m^2 (same as square, rectangular areas) with the radius R= 56.419m and the base station is located at the center of the circle, i.e. (0, 0). In Fig. 3.1(b), the size of the square area is 100x100m^2 with the base station located on (0, 50) or (50, 175). In Fig.3.1(c), the size of the rectangular area is 50*200m with the base station located on (0, 25) or (100, 150).

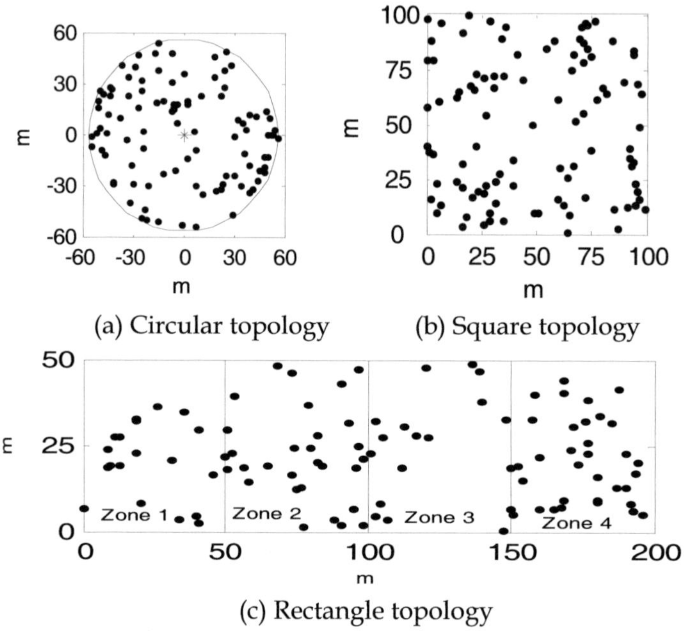

Fig 3.1 Three different types of WSN's topology:
Sink (a):(0,0); (b):(0,50) or (50,175); (c):(0,25) or (100,150)

The probability of cluster head node in the LEACH protocol has a certain impact on the WSN's lifetime. In our analysis, we divides the rectangle area into four smaller square areas, with the communication distance of nodes keeping short ($d_B < d_0$, where d_B is the broadcasting distance of the cluster head). From the first-order radio model, we can get the optimal cluster head probability formula as follows:

$$k_{opt} = \frac{\sqrt{N}}{\sqrt{2\pi}} \bullet \frac{M}{d_{toBS}} \tag{3-1}$$

Where N is the number of sensor nodes. In the rectangular region 100 nodes are scattered randomly, and the region is divided into four regions. In order to verify the difference of the number of nodes distributed in different regions, we simulate 100 independent iterations of the nodes' number in each region, and get the averages in the four regions as zone1(25.32)、zone2(24.83)、zone3(24.87) and zone4(24.98). It can be seen that the number of nodes in each region are around 25, there're almost no difference in the average nodes' numbers for the four regions, so in the text, the nodes are uniform distributed in the four regions, i.e. N=25. M is the side length of each small square region, here M=50. d_{toBS} is the distance between the sink and the node. As the distance of a node to the base station is different, we can change the percentage of cluster heads in different regions to reach the optimal value so that the lifetime of the whole WSN can be prolonged.

For the topology with a rectangular shape, we propose an improved LEACH algorithm. The main idea is described as follows:

(1) Divide the rectangular area into several small square areas with the same size;

(2) Elect the cluster heads separately in each region, and the optimal probabilities of the cluster heads for each region can be obtained from Eq. (3-1), i.e. the values are p_1=0.02 , p_2=0.03, p_3=0.03, p_4=0.02 ;

(3) After the cluster heads in each region are selected, the rest of the protocol is similar to LEACH.

The improved algorithm elects cluster heads in each region according to its probability of the cluster heads. In this way, it can make sure that there are clusters in every region and ensure the clusters distributed more uniformly in every region, which reduces the energy dissipation and improves the lifetime of the network.

3.2 Simulation results

We simulate the three shapes of topology that use (improved) LEACH as the routing protocol. Parameters used in simulation are listed in table 3.1. There are 100 sensor nodes randomly scattered with fixed position in each shape. We measure the round number when the first node died, 20% of nodes died and 50% of nodes died respectively as the criterion to estimate the lifetime of WSN.

Parameter	Value
Number of nodes	100
Initial energy (J)	0.5
Data packet length (bit)	4000
Control packet length(bit)	200
Energy dissipation of one Tx (nJ)	50
Energy dissipation of one Rx (nJ)	50
Energy aggregation energy (nJ)	5
Energy loss-free space (pJ/bit/m^2)	10
Energy loss-multipass fading (pJ/bit/m^4)	0.0013

Table 3.1 Parameters used in simulation

Table 3.2 shows the round numbers (the lifetime) of WSN for different BS locations and different percentage of dead nodes in circle, square and rectangular shapes of topology. As the sensor nodes distributed randomly in WSN that may statistically vary, we simulate every case for 100 iterations to get more accurate results. The percentage of the cluster heads is set to 5% in three shapes of topology. It can be seen from Table 3.2 that the longest lifetime of WSN is the circle shape of topology. The BS located in the center of the circle that is symmetric, so that the energy dissipation of nodes are more even and the lifetime of the network is prolonged. For the rectangle shape, the lifetime is different as the position of BS changes. Simulation results indicate that the lifetime for the BS in (0, 50) is longer that in (50,175). As the BS in (0, 50) is nearer to the sensor area, the energy for transmitting data to the BS is reduced.

Topology	Circle	Square		Rectangle	
Sink	(0, 0)	(0,50)	(50, 175)	(0, 25)	(100,150)
1%	757	740	639	470	555
20%	854	846	715	644	652
50%	923	919	788	785	717

Table 3.2 Lifetime comparisons of different shapes of topology

It can be seen form Table 3.2 that the lifetimes of the rectangular are poor for the BS both near and far away from the sensor area using conventional LEACH protocol. Then we use the improved LEACH algorithm that divides the rectangle region into four equal square regions. The cluster heads are elected separately in each region to make sure the cluster heads distribute uniformly in four regions.

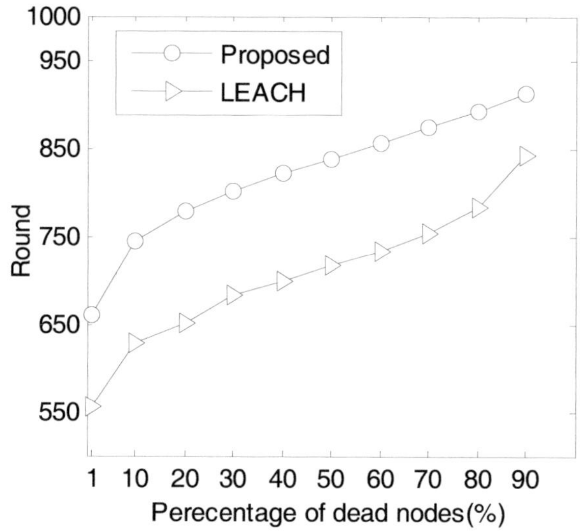

Fig 3.2 The relationship between percentage of dead nodes and number of rounds

Fig. 3.2 shows the improvement using the modified LEACH algorithm in terms of the lifetime of the network. It can be seen from the figure that the surviving round number of improved algorithm is increased under the same percentage of dead nodes, corresponding to the lifetime improvement of 19% (for the case of the first died node) compared with the conventional LEACH algorithm. This is due to the fact that the improved algorithm reduces the energy dissipation by dividing the rectangle region into sub-regions.
From the above analysis, when sensor nodes within a monitoring area can be manually deployed, one can select the circle topology of the WSN to maximize the lifetime. When the topology area is square and the base station location is variable, one can make the base station close to the WSN area to save energy and extend the network lifetime. When the monitoring area is a rectangular one, one can use the idea of partition to extend the lifetime.

4. LEACH & PEGASIS

4.1 Introduction of LEACH & PEGASIS algorithm

Brief introductions about LEACH and PEGASIS have been given in section 2. As pointed, there are three shortcomings for the LEACH protocol:

(1) The number of cluster heads is uncertain. If the number of cluster heads is large, cluster heads that need communicate directly with the base station will consume more energy. If the number of cluster heads is small, common nodes that need communicate with the remote cluster heads will consume more energy.

(2) When cluster heads are selected, the remaining energy of cluster heads is not considered. After a node is elected as the cluster head, maybe the remaining energy is not enough for the next round of communication, therefore may lead to failure of the entire cluster, and member nodes of the cluster will lose data. A blind spot will appear within the monitoring area.

(3) Many cluster heads communicate directly with the base station. Especially the cluster heads are far away from the base station. Transmission of data consumes a lot of energy. As the cluster head dies prematurely, the total energy of network consumes excessively.

To overcome the shortcomings of LEACH, we take the advantages of PEGASIS to construct a chain using greedy algorithm which only uses a node as the cluster head to communicate with the base station. Studies show that the approach of data fusion and multi-hop based on cluster can save the energy of nodes even better [23]. Here we propose an improved routing protocol called LEACH-P.

(1) The optimal number of cluster head is defined by Eq. (4-1).

$$m = \frac{\sqrt{N}}{\sqrt{2\pi}} \sqrt{\frac{\varepsilon_{fs}}{\varepsilon_{mp}} \frac{M}{d_{toBS}^2}} \tag{4-1}$$

Where m is the optimal number of cluster heads; N is the number of nodes; ε_{fs} is signal amplification factor in free space; ε_{amp} is signal amplification factor of the multipath fading channel; M is the side length; d_{toBS} is the distance between the cluster head and the base station.

(2) Cluster heads are decided by Eq. (2-1) in the LEACH algorithm without taking into account the residual energy of nodes. The new algorithm detects the residual energy of the cluster head to meet the required energy threshold E(r), i.e. the minimum energy to complete one round communication, which is sum of the energy for broadcasting information, receiving data packets and confirming messages from cluster members to, as well as communicating with its neighboring cluster heads.

In our simulation (parameters and definitions of acronyms are listed in Table. 4.1. We assume that the coverage of broadcasting is half of the diagonal of the field, which is less than d_0. Energy consumed through broadcasting is:

$$ETX * cPL + \varepsilon_{fs} * cPL * DB^2 \tag{4-2}$$

The energy that each cluster head needs to receive confirmation and data packets from its cluster members is:

$$19*((ERX + E_{DA})*PL + ERX*cPL \tag{4-3}$$

In each cluster, the average number of nodes that the cluster head needs to receive the information is 19.

The energy that the cluster head needs to send information to its neighboring cluster head is:

$$(ETX + E_{DA})*PL + \varepsilon_{fs}*PL*DB^2 \tag{4-4}$$

The distance between cluster heads is random, and we assume it as the broadcasting distance between nodes.

From above, we can obtain that the energy threshold value is ~0.0048J.

(3) The new algorithm (LEACH-P) randomly selects five cluster heads linked as a chain. The one with maximum residual energy is chosen to transfer information to the sink. Other cluster heads reduce energy dissipation through data fusion. LEACH-P uses the wireless communication model described in [24]. The energy dissipation that transmits K-bit data to the receiver over a certain distance d is:

$$E_t(k,d) = \begin{cases} kE_{elec} + k\varepsilon_{fs}d^2 & d<d_0 \\ kE_{elec} + k\varepsilon_{mp}d^4 & d\geq d_0 \end{cases} \tag{4-5}$$

Where E_{elec} is the energy dissipation of the transmitter; ε_{fs} and ε_{mp} are the energy dissipations of the power amplifiers; d_0 is the constant. The energy that a node needs to receive K-bit data is:

$$E_r(k) = kE_{elec} \tag{4-6}$$

The energy to fuse a number (L) of K-bit packets is:

$$E_f(L,k) = LkE_{DA} \tag{4-7}$$

Where E_{DA} is the energy dissipation of fusing 1-bit data.

The advantage of the PEGASIS link performance has been described and proved in [25]. The topology of the improved algorithm is shown in Fig 4.1. In the figure, the nodes within a cluster have the same symbol. There are five cluster heads with 5 symbols, one of them is elected as the leading cluster head that is responsible for the information exchange with the base station.

Parameter	Value
Sink location	(50, 175)
Sensing region	100x100
The number of nodes(N)	100
Initial energy(E_0) (J)	0.5
Data packet length(PL) (bit)	4000
Control packet length(cPL) (bit)	200
Energy dissipation of one (ETX) (nJ)	50
Energy dissipation of one (ERX) (nJ)	50
Energy aggregation energy(EDA) (nJ)	5
Energy loss-free space(Efs) ($pJ/bit/m^2$)	10
Energy loss-multipass fading (Emp) ($pJ/bit/m^4$)	0.0013

Table 4.1 Parameters used in simulation

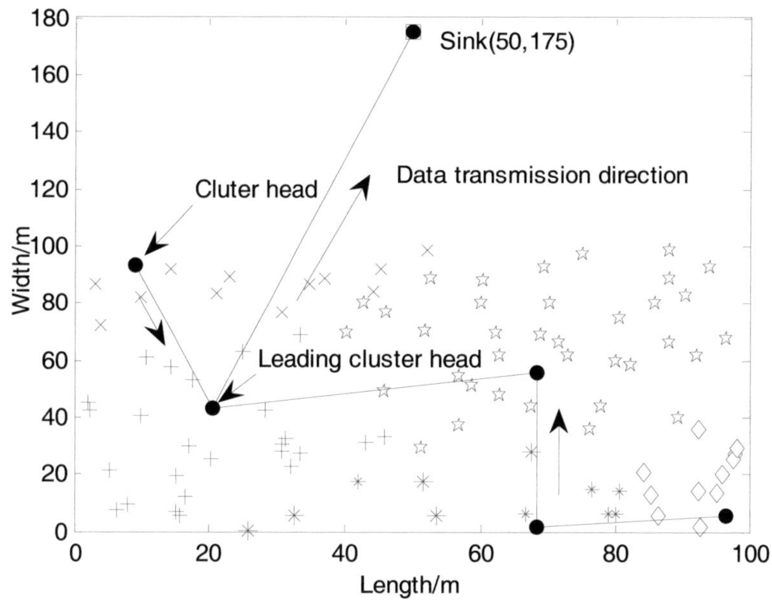

Fig. 4.1 Topology of the improved algorithm

The new algorithm not only solves the problem in the LEACH protocol that multiple cluster heads communicate directly with the base station to introduce more energy dissipation, but also overcomes the shortcoming of the long delay in the PEGASIS protocol. The flowchart of the algorithm is shown in Fig 4.2. In each round of communication, the improved algorithm is still divided into two stages : the setup stage and the stable stage.

In the setup stage, the election of cluster head depends on whether there is any dead node and the residual energy of cluster heads must be greater than E(r). After the election of cluster heads, the cluster heads broadcast information (advertisement message, ADV). Non-cluster head node chooses the cluster according to the signal strength after receiving the information, and sends a request that includes the cluster ID, its own ID, as well as its remaining energy state to join the cluster. Cluster heads are connected into a chain after the establishment of clusters, and the cluster head with maximum residual energy is assigned as the leading cluster head to communicate with the BS directly.

In the stable stage, common nodes of the cluster send information to the cluster head according to the TDMA time slot table. The cluster head receives data and integrates data into a packet. The packets are transmitted along the chain to the leading cluster head according to the Token. The leading cluster head receives and fuses data packets, then sends to the base station.

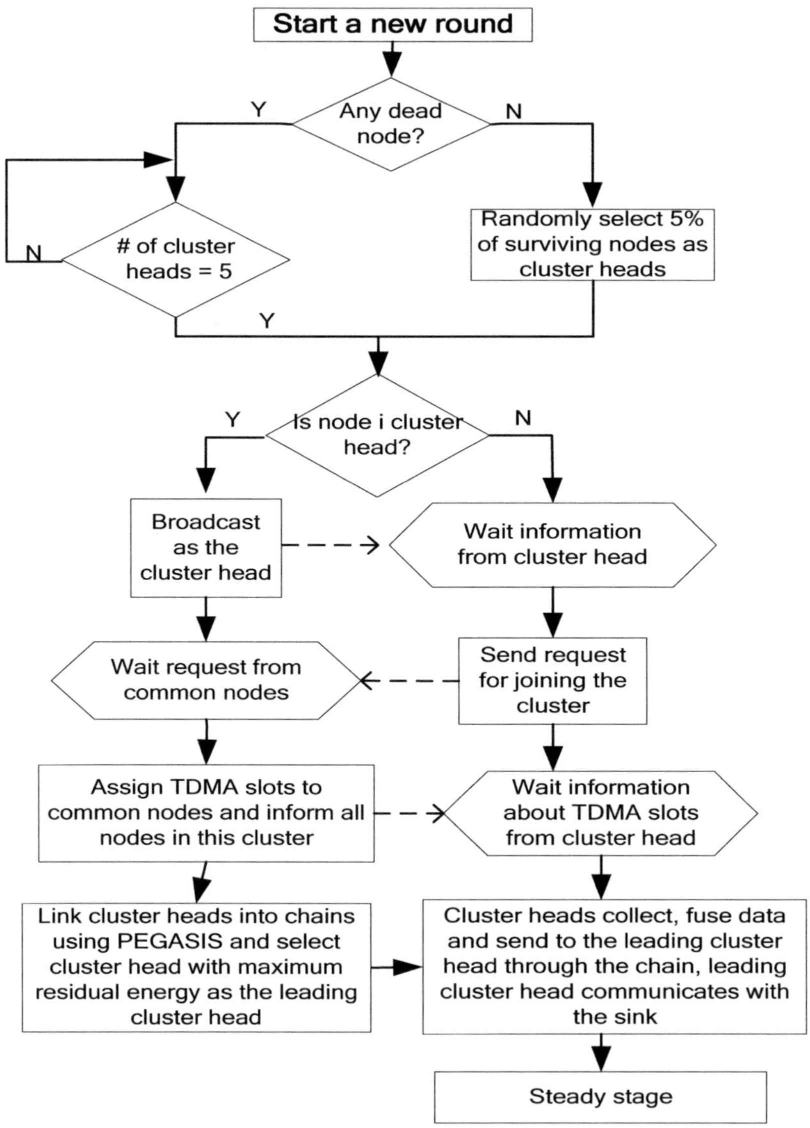

Fig. 4.2 Flow chart of proposed LEACH-P algorithm

4.2 Simulation results and analysis

The performance of improved algorithm (LEACH-P) is evaluated in terms of the lifetime of the network and data transmission delay. 100 sensor nodes are randomly distributed in the sensing area of 100x100m^2. Most simulation parameters are the same as those of the LEACH algorithm, with specific parameters listed in Table 4.1.

The lifetime in terms of rounds corresponding to 1%, 20%, 50%of nodes died are simulated for both the LEACH and improved algorithm. The results are compared in Table 4.2. The lifetime of the new algorithm improves by 17% for the case with 1% dead nodes. The numbers of rounds when 20% and 50% of nodes died are also improved compared with the LEACH protocol. Similar to previous simulation, we take 100 iterations for each case to reduce the statistical fluctuation.

Percentage of dead nodes	LEACH	LEACH-P
1%	639	751
20%	715	848
50%	788	919
100%	1296	1276

Table 4.2 Lifetime comparisons of different algorithms

Fig 4.3 also shows the lifetime comparison for the LEACH and improved algorithm. It can be seen that the round number corresponding to the first dead node and all dead nodes is ~ 639 and 1240 for the LEACH protocol, respectively. These two numbers are improved to ~751 and 1300 using the new algorithm. The Lower part of Fig 4.3 shows the variation of the cluster head number in the new algorithm with a fixed number as 5 until ~ 800 rounds (dead nodes appear). Compared with LEACH, the proposed algorithm can prolong the network lifetime and balance the energy dissipation of network nodes as well.

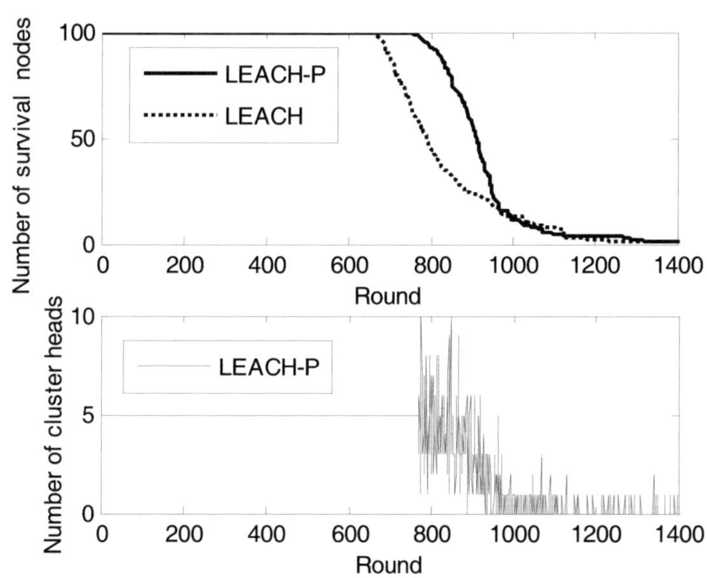

Fig. 4.3 Relationship between rounds and number of survival nodes

WSNs are generally deployed in harsh environments, where the base station is far away from the sensor field. The location of the base station has a great effect on the lifetime of the sensor network. When the distance between the base station and the monitored region is too large, some cluster heads will lead to excessive energy dissipation in the LEACH, shortening the lifetime of the network. On the other hand, the leading cluster head transmits data to the base station gathered from other cluster heads using multi-hop in the new algorithm. Therefore, the increasing distance between the base station and the monitored regions has less effect on the network lifetime. In Fig. 4.4, the network lifetime of LEACH and proposed algorithm are calculated as we vary the distance between the BS and sensor field. It can be seen from Fig. 4.4 that:

(1) When the base station location changes from 100 to 250, the lifetime of WSN for the LEACH-P algorithm keeps almost constant (~760). When using the LEACH algorithm, the lifetime is reduced from 737 to 328.

(2) When the base station location changes from 250 to 400, the lifetime of LEACH-P algorithm and LEACH algorithm are both reduced, but LEACH-P algorithm is still better than the conventional LEACH.

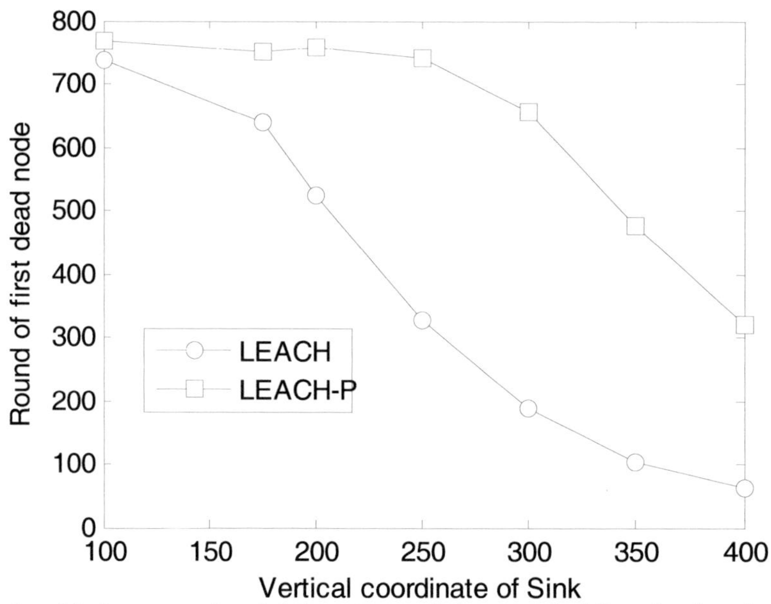

Fig. 4.4 Relationship between the sink location and the round of first dead node

Furthermore, Fig. 4.5 shows the numbers of live nodes using LEACH-P and LEACH protocols with the position of sink fixed at (50,300). Now the number of the first dead node for LEACH is only 187, but it is improved to 658 for the LEACH-P algorithm. Compared with LEACH, LEACH-P can prolong the network lifetime by 351%. Figs. 4.4 and 4.5 further indicate that LEACH-P is superior to LEACH with more evenly energy distribution among nodes even the sink location is far away from the sensor field.

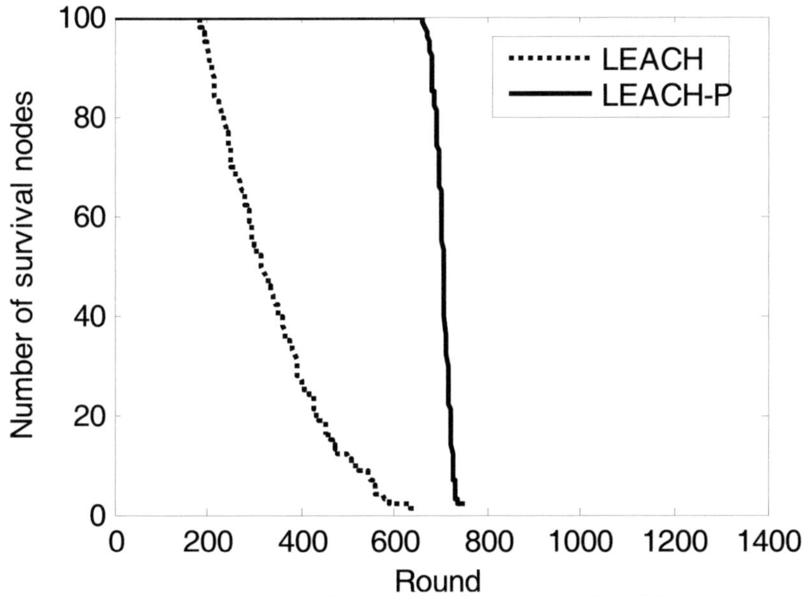

Fig. 4.5 Comparison of two algorithms for the sink location at (50,300)

Another significant advantage of LEACH-P is that it not only overcomes the issue of long delay of PEGASIS, but also inherits the idea of greedy into a chain. In order to verify the improvement of the transmission delay of the WSN, we calculate and compare the maximum distance that the data should be transmitted for every round in LEACH, PEGASIS and LEACH-P.

(1) When data is transmitted according to LEACH, all common nodes of the cluster send data to the cluster head in accordance with the TDMA time slots. The cluster head will fuse the data and send it to the sink. The longest distance of transmitting data in each round corresponds to the maximum distance of the sum of both common nodes to the cluster head and the cluster head to the sink.

(2) When data is transmitted using the token mechanism in PEGASIS, there is only one token which goes through the whole chain. The longest distance of transmitting data in each round corresponds to the sum of the distances of both the length of the entire chain and the leading cluster head to the sink.

(3) When data is transmitted using the TDMA and token mechanism in LEACH-P, in each cluster, the cluster head allocates TDMA time slot to common nodes. Common nodes send the data to the cluster head. Between cluster heads, the cluster heads are linked into chains and then transfer the data to the leading cluster head according to the token mechanism. The leading cluster head fuses the data and sends it to the sink. The longest distance of transmitting data in each round corresponds to the sum of the maximum distance of common nodes to cluster heads in two ends of the chain, the length of the chain and the distance between the leading cluster head and the sink.

Again, we simulate all cases with 100 iterations to get the statistical average value. The performance (network delay) is evaluated in terms of the average distance before the first dead node appearing in wireless sensor network.

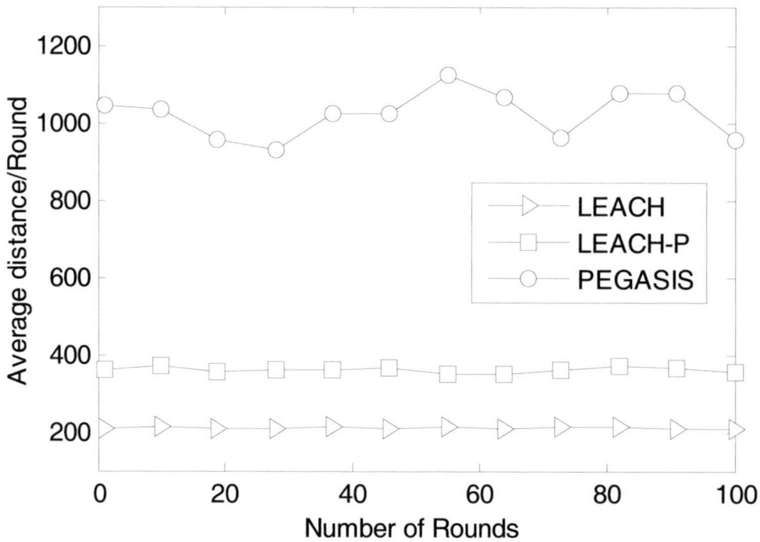

Fig. 4.6 Comparison of transmission delay in terms of maximum distance for three protocols

Fig. 4.6 illustrates the simulation results. In PEGASIS algorithm, the distribution of nodes and the chain length are quite different from round to round, so the average value is still fluctuating greatly. The average longest distances of transmitting data in each round for PEGASIS, LEACH and LEACH-P are 1026.7m, 209.9m and 357.33m, respectively. The real-time of LEACH-P is slightly lower than LEACH. However, compared with PEGASIS, LEACH-P increases by a dramatic value, i.e. ~290%.

Therefore, we can conclude for this section that LEACH-P combines the advantages of both LEACH and PEGASIS. It can not only reduce the energy dissipation of cluster heads compared to the LEACH algorithm in large-scale sensor networks, but also overcome the issue of poor real-time in the PEGASIS algorithm.

5. O-LEACH

Above sections are mainly about randomly scattered WSN nodes with different topologies. However, in some particular areas that are difficult to place wireless sensor nodes, we can lay distributed fiber sensor (DFS) along. DFS can achieve measurements such as temperature, strain /stress and so on, which associates with wireless sensor nodes to construct a new hybrid optical wireless sensor network. Incorporating distributed optical fiber sensor in rectangular topological region makes the WSN more suitable to work in harsh and large-scale regions. Meanwhile, the reliability and security of system and data are further protected.

5.1 O-LEACH algorithm description

We investigate an infrastructure of hybrid sensor network which is composed of a DFS link and two separated WSNs, as shown in Fig 5.1. The DFS link is located at the center of the whole sensor field and can cover a certain area. The two WSN fields are filled with randomly scattered nodes as usual. These nodes can communicate with each others. Unlike simple WSNs, since the DFS has to be powered on for data processing, we use one end of the DFS as the sink or the base station for all WSN nodes. We specially propose a new energy efficient communication protocol, optical LEACH (O-LEACH), based on the WSN LEACH protocol.

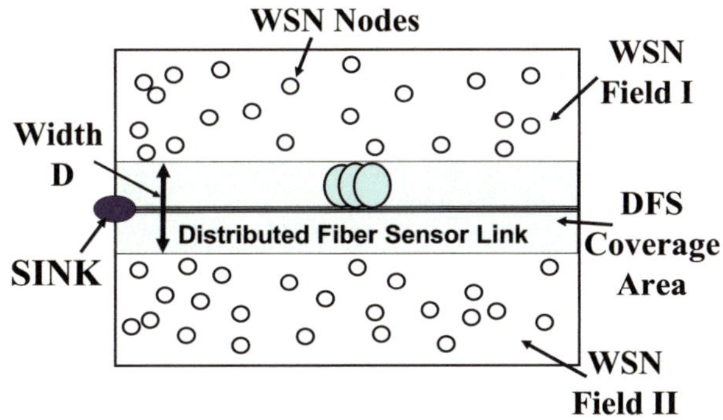

Fig. 5.1 Sensor field consisting of a distributed fiber sensor (DFS) link and two WSN fields (I and II), the sink node is located at one end of the DFS and the width of the DFS coverage area is D.

As a more general topology, Fig. 5.2 shows such hybrid sensor networks that have potential to cover much more broad areas under certain guidelines: (1) cascade of multiple rectangular regions in which the base station location of sensor node is (100,150), (2) the DFS is located in the monitor area with massive volume of data, harsh environment and poor security (located in the middle of the rectangular region for this paper) to link the rectangular region and the location of fiber's base station is (0, 25). The DFS can cover a certain area, for example 10m (vertical axis), to monitor the pressure or temperature information within the coverage area, and give the data back to the fiber base station.

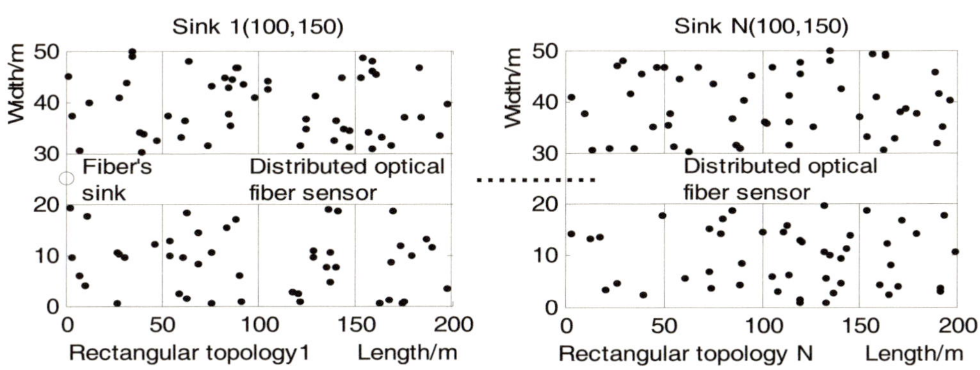

Fig. 5.2 The topology incorporating distributed optical fiber sensors

As nodes of two WSNs are power limited, the protocol is mainly dealing with these nodes. The flowchart of the O-LEACH protocol is shown in Fig. 5.3. As the operation of the standard LEACH protocol is separated into the setup phase and the steady phase, we also separate the O-LEACH operation into two phases. The steady phase is as same as the LEACH one. During the setup phase, there are two major differences between O-LEACH and LEACH: (1) nodes of WSNs cannot be deployed in the DFS coverage area; (2) the cluster head and the node should be within the same WSN field if two WSNs cannot communicate with each other (i.e. checking if $Position_{Cluster\_head}=Position_{Node\_I}$ in the flowchart). For most applications, it would be better to assume that two WSN fields are isolated due to the following reasons: (1) save information transfer energy, as transferring data over the DFS terrain would waste more energy; (2) wireless communication over the DFS area is not allowed for some applications. However, we simulate the case that nodes inside different WSN fields can communicate with each other as well for reference.

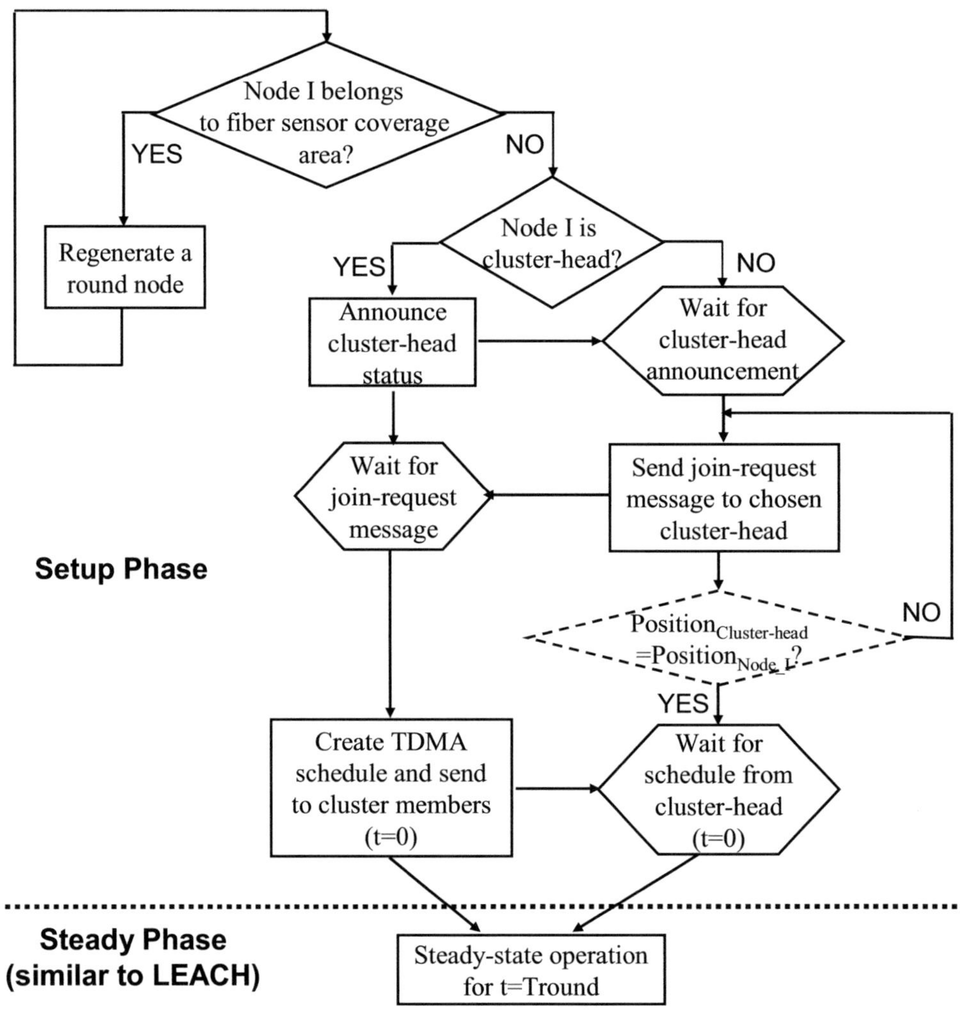

Fig. 5.3 Flowchart of the O-LEACH protocol

5.2 Simulation results

Based on the proposed O-LEACH protocol, we simulate the network performance in terms of the node lifetime. In our simulation model: (1) most of parameters (e.g. probability of a node to become a cluster head, data packet length, control packet length, etc.) are as the same as other LEACH-based simulation models (listed in Table 4.1); (2) the position of the sink in the LEACH model can be put in different places, while in our LEACH and O-LEACH ones, we put the sink of all the cases at the same position, i.e. in the middle of one edge of the sensor field; (3) as the network energy dissipation is a totally statistical behavior due to the random distribution of WSN nodes, we simulate every case for 1000 independent iterations (over days).

First we compare the network performance in terms of lifetime for the simple topology shown in Fig. 5.1. Three cases are simulated: original LEACH without DFS; O-LEACH with varying width of DFS coverage area (D) that two WSNs can either communicate with each other or not. As people use different parameters to evaluate the lifetime, i.e. the round number corresponding to the appearance of the first dead node, half of the dead nodes or the last survival node (refered as "first-dead", "half-dead" and "fully-dead" in the following part of the paper), we obtain all the three parameters and find that the network improvement may end up with quite different conclusions through these parameters.

For the LEACH case, we obtain that the average round number corresponding to "first-dead", "half-dead" and "fully-dead" are 731, 915, and 1741, respectively. Figs 5.4 (a) and (b) show the average lifetime evaluated by three "-dead" parameters for situations that the two WSNs can communicate with each other or not as we vary the value of D (from 5 to 50). We can see that: (1) the network performance in terms of lifetime keeps almost constant regardless the width of DFS coverage. (2) In the case that two WSNs cannot communicate with each other, nodes can save energy on broadcasting over smaller area (i.e. shorter distance), therefore, the average lifetime corresponding to either "first-dead" or "half-dead" is improved ~20% compared to the case that two WSNs are connected, while the last node's lifetime ("fully-dead") is more than doubled. (3) Compare to the conventional LEACH protocol, if two WSNs can communicate with each other, the improvement of O-LEACH is very limited (~2% and ~ 16% in terms of "first-dead" and "fully-dead", respectively). Therefore, it is expected and mostly required for such hybrid sensor networks to employ O-LEACH with two isolated WSNs. Furthermore, typical lifetime evolutions are compared as well in Fig. 5.4 (c) where D equals 20. These curves are specially chosen from thousands of simulated iterations with performance close to average ones. Results of LEACH protocol are also included.

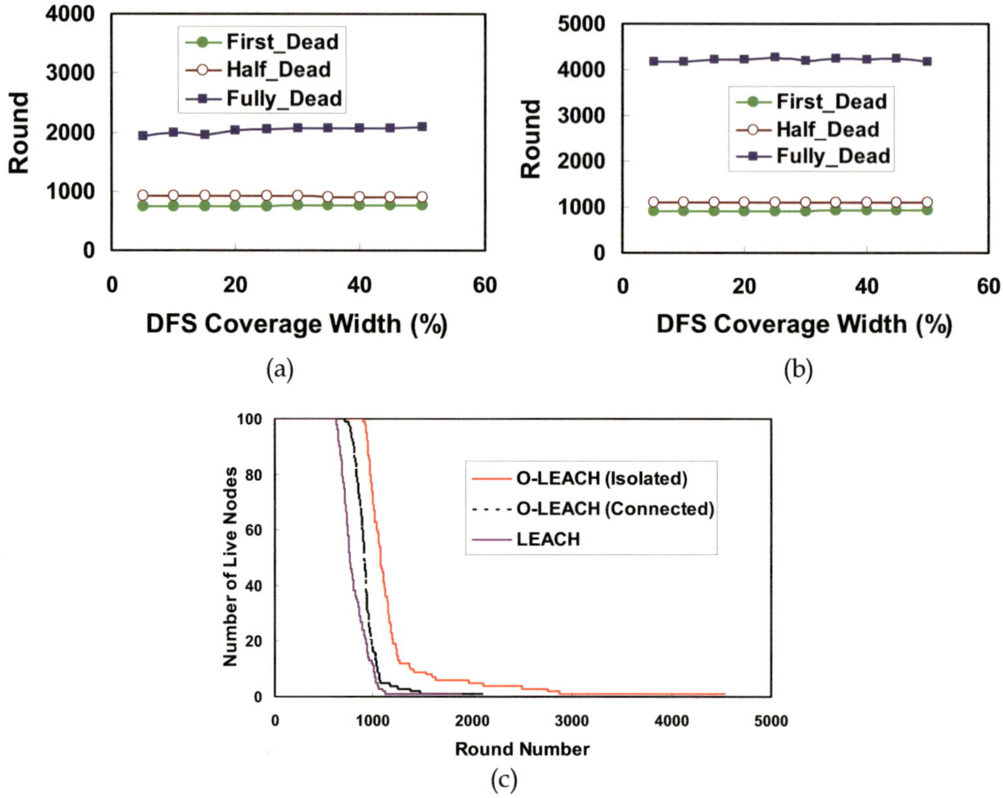

Fig. 5.4 Simulation results of network performance in terms of lifetime (round number) using O-LEACH protocol: (a) two WSNs can communicate with each other; (b) two WSNs are isolated; (c) typical lifetime evolution curves

A legitimate question for above network model is how close to the reality of the hybrid sensor network in terms of the coverage of DFS and WSN. As pointed in the introduction section, the distance of typical DFS link can vary from hundreds of meters to tens of kilometers, while the coverage diameter of WSN node is tens of meters. Therefore, it would be interesting to look into the case that the length of DFS link increases and the number of WSN nodes increases proportionally (to keep the approximate density).

To keep straightforward but simple, we fix the width of the whole sensor field to 100 and the coverage percentage of DFS to 20% (i.e. D equals to 20). Also we only consider the case that two WSNs are isolated. Fig. 5.5 illustrates the trend of normalized lifetime performance with increasing link length of DFS. The normalization is done using the ratio of the "-dead" parameter to the total node number. It is obvious that 100 is the optimum number for WSN nodes with parameters listed in Table I. As the length of the DFS link increases, the lifetime reduces dramatically, especially the "first-dead" parameter. More wireless sinks are required for longer DFS links, and the performance evaluation of various optical wireless sensor network topologies are of great interests for further investigation. On the other hand, our protocol and the simulation model can be adapted into networks with different parameters so that we can find the optimized network design.

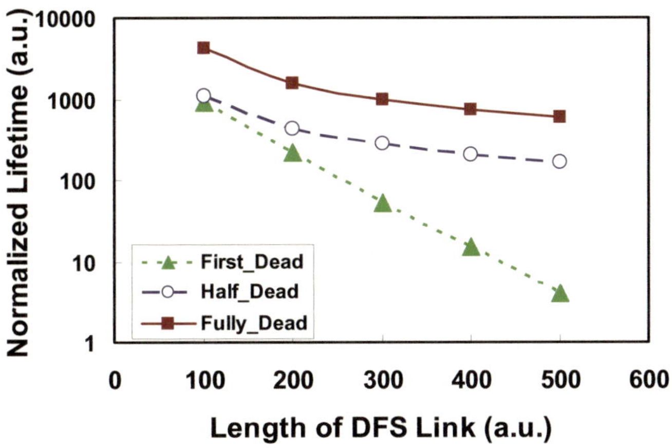

Fig. 5.5 Normalized lifetime as we increase the length of DFS link (number of WSN nodes are increased proportionally, the widths of the DFS coverage area and the whole sensor filed are fixed)

For the general topology shown in Fig.5.2, we simulate a 50*200m rectangular WSN region that is divided into four small regions. In the total area, we randomly scattered 100 nodes, the number of nodes in each region is: 21、 29、 23 and 27. The DFS is located in the middle of the total rectangular region. We assume that the nodes in the upper and the lower part of the optical region can communicate with each other, and the simulation results are shown in Fig 5.6.

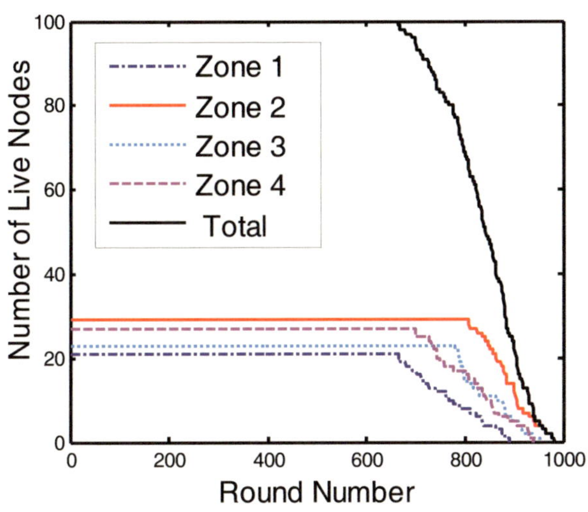

Fig. 5.6 Performance (# of live nodes vs. round) after introducing the DFS into the rectangular sensor area

We can see from Fig. 5.6 that the numbers of the first dead node in the four regions are 663, 805, 779 and 698, respectively. The first 20% nodes die slowly, but the remaining ones die rapidly in the total region. The results further demonstrate that the hybrid sensor network incorporating DFS with the O-LEACH protocol can evenly distribute the energy load among nodes, therefore prolong the overall lifetime of the network.

6. Conclusion

We discussed several improved algorithms (protocols) that can be used for WSNs or hybrid sensor networks with distributed fiber sensors involved. As sensor networks are much more complicated in real applications, more thorough and careful optimization of routing algorithms are required to meet specific requirements, such as real-time, long lifetime, security, and so on.

7. References

[1] I. F. Akyildiz, W. Su, Y. Sankarasubramaniam, E. Cayirci (2002). A survey on sensor networks, *IEEE Communication Magazine*, vol. 40, no.8, pp.102-114

[2] J. M. Kahn, R. H. Katz, and K. S. J. Pister (1999). Next century challenges: mobile networking for smart dust, *Proc. ACM MobiCom '99*, Washington DC, pp. 271–78

[3] V. Rodoplu and T. H. Meng (1999). Minimum energy mobile wireless networks, *IEEE JSAC*, vol. 17, no. 8, pp.1333–1344

[4] K. Sohrabi et al. (2000). Protocols for self-organization of a wireless sensor network, *IEEE Pers. Commun.*, pp.16–27

[5] W. R. Heinzelman, A. Chandrakasan, and H. Balakrishnan (2000). Energy-efficient communication protocol for wireless microsensor networks, *IEEE Proc. Hawaii Int'l. Conf. Sys. Sci.*, pp. 1–10

[6] X. Fan, Y. Song (2007). Improvement on LEACH protocol of wireless sensor network, *IEEE SENSORCOMM*, pp.260-264

[7] H. Jeong, C.-S. Nam, Y.-S. Jeong, D.-R. Shin (2008). A mobile agent based LEACH in wireless sensor network, *Conf. on Advanced Comm. Technol. (ICACT)*, pp. 75-78

[8] Stephanie Lindsey and Cauligi S. Raghavendra (2002). PEGASIS: Power-Efficient Gathering in Sensor Information System, *2002 IEEE Aerospace Conference*, vol. 3, pp.1125-1130

[9] X. Bao, D. J. Webb, and D. A. Jackson (1993). 32-km distributed temperature sensor using Brillouin loss in optical fiber, *Opt. Lett.*, vol. 18, pp.1561–1563.

[10] D. Garus, T. Gogolla, K. Krebber, F. Schliep (1997). Brillouin optical-fiber frequency-domain analysis for distributed temperature and strain measurements, *J. Lightwave Technol.*, vol.15, no.4, pp.654-662

[11] S.M. Maughan, H. H. Kee, T. P. Newson (2001). A calibrated 27-km distributed fiber temperature sensor based on microwave heterodyne detection of spontaneous Brillouin scattered power, *IEEE Photon. Technol. Lett.*, vol. 13, no 5, pp. 511-513

[12] J. C. Juarez, E. W. Maier, K. N. Choi, H. F. Taylor (2005). Distributed fiber-optic intrusion sensor system, *J. Lightwave Technol.* vol.23, no.6, pp.2081-2087

[13] D. Iida, F. Ito (2008). Detection sensitivity of Brillouin scattering near Fresnel reflection in BOTDR measurement, *J. Lightwave Technol.*, vol. 26, no.4, pp.417-424

[14] D. Kedar and S. Arnon (2003). Laser 'Firefly' Clustering; a New Concept in Atmospheric Probing, *IEEE Photon. Tech. Lett.*, vol.15, no.1 pp. 1672–1624

[15] S. Teramoto, and T. Ohtsuki (2004). Optical wireless sensor network system using corner cube retroreflectors (CCRs), *IEEE Globecom'04*, pp.1035-1039

[16] D. Kedar, S. Arnon (2005). Second generation laser firefly clusters: an improved scheme for distributed sensing in the atmosphere, *Appl. Opt.*, vol. 44, no.6, pp.984-992

[17] Jamal N. AL-Karaki, Ahmed E. Kamal (2004). Routing Techniques in Wireless Sensor Networks: A Survey, *IEEE Wireless Communications*, Dec.

[18] W. Heinzelman, J. Kulik, and H. Balakrishnan (1999). Adaptive Protocols for Information Dissemination in Wireless Sensor Networks, *Proc. 5th ACM/IEEE Mobicom*, Seattle, WA, pp. 174–85.

[19] J. Kulik, W. R. Heinzelman, and H. Balakrishnan (2002). Negotiation-Based Protocols for Disseminating Information in Wireless Sensor Networks, *Wireless Networks*, vol. 8, pp. 169–85.

[20] Wendi Beth Heinzelman (2000). Application-Specific Protocol Architectures for Wireless Networks (PhD), Boston: Massachusetts Institute of Technology

[21] Vivek Mhatre, Catherine Rosenberg (2004). Design guidelines for wireless sensor networks: Communication, clustering and aggregation, *Ad Hoc Networks*, vol.2, no.1, pp. 45-63

[22] Ning Xu, Sumit Rangwala, Krishna Kant Chintalapudi, Deepak Ganesan, Alan Broad, Ramesh Govindan, Deborah Estrin (2004). A Wireless sensor network for structural monitoring, *Proc. 2nd international conference on Embedded networked sensor systems*, Baltimore, MD, USA, pp.13-24.

[23] Katayoun Sohrabi, Jay Gao, Vishal Ailawadhi , Gregory J.Pottie (2000). Protocols for Self-organization of a Wireless Sensor Network, *IEEE Personal Communications*, vol.7, no.5, pp.16-27

[24] ISO.16484-5, Building automation and control systems part 5 data communication protocol, 2003.

[25] Stephanie Lindsey, Cauligi Raghavendra, Krishna M. Sivalingam (2002). Data Gathering Algorithms in Sensor Networks Using Energy Metrics, *IEEE Transactions on Parallel and Distributed Systems*, vol.13, no.9, pp.924-935

Range-free Area Localization Scheme for Wireless Sensor Networks

Vijay R. Chandrasekhar[1], Winston K.G. Seah[2], Zhi Ang Eu[3] and Arumugam P. Venkatesh[4]
[1]*Stanford University, USA**
[2]*Victoria University of Wellington, New Zealand**
[3]*National University of Singapore*
[4]*National University of Singapore*

Abstract

For large wireless sensor networks, identifying the exact location of every sensor may not be feasible and the cost may be very high. A coarse estimate of the sensors' locations is usually sufficient for many applications. In this chapter, we describe an efficient Area Localization Scheme (ALS) for wireless sensor networks. ALS is a range-free scheme that tries to estimate the position of every sensor within a certain area rather than its exact location. Furthermore, the powerful sinks instead of the sensors handle all complex calculations. This reduces the energy consumed by the sensors and helps extend the lifetime of the network. The granularity of the areas estimated for each node can be easily adjusted by varying some system parameters, thus making the scheme very flexible. We first study ALS under ideal two-ray physical layer conditions (as a benchmark) before proceeding to test the scheme in more realistic non-ideal conditions modelled by the two-ray physical layer model, Rayleigh fading and lognormal shadowing. We compare the performance of ALS to range-free localization schemes like APIT (Approximate Point In Triangle) and DV (Distance Vector) Hop, and observe that the ALS outperforms them. We also implement ALS on an experimental testbed and, show that at least 80% of nodes lie within a one-hop region of their estimated areas. Both simulation and experimental results have verified that ALS is a promising technique for range-free localization in large sensor networks.

Keywords: Localization, Wireless Sensor Network, Positioning, Range-free

1. Introduction

Deployment of low cost wireless sensors is envisioned to be a promising technique for applications ranging from early warning systems for natural disasters (like tsunamis and

---

*This work done by these authors in the Institute for Infocomm Research, Singapore.

wildfires), ecosystem monitoring, real-time health monitoring, and military surveillance. The deployment and management of large scale wireless sensor networks is a challenge because of the limited processing capability and power constraints on each sensor. Research issues pertaining to wireless sensor networks, from the physical layer to the application layer, as well as cross-layer issues like power management and topology management, have been addressed[1]. Sensor network data is typically interpreted with reference to a sensor's location, e.g. reporting the occurrence of an event, tracking of a moving object or monitoring the physical conditions of a region. Localization, the process of determining the location of a sensor node in a wireless sensor network, is a challenging problem as reliance on technology like GPS [2] is infeasible due to cost and energy constraints, and also physical constraints like indoor environments.

In very large and dense wireless sensor networks, it may not be feasible to accurately measure the exact location of every sensor and furthermore, a coarse estimate of the sensor's location may suffice for most applications. A preliminary design of the Area Localization Scheme (ALS) [3] has been proposed, which can only function in an (unrealistic) ideal channel and definitely not in a real environment with fading, shadowing and other forms of interference. In this chapter, we describe algorithms and techniques that will enable the Area Localization Scheme (ALS) to be deployable in a real environment. ALS is a centralized range-free scheme that provides an estimation of a sensor's location within a certain area, rather than the exact coordinates of the sensor. The granularity of the location estimate is determined by the size of areas which a sensor node falls within and this can be easily adjusted by varying the system parameters. The advantage of this scheme lies in its simplicity, as no measurements need to be made by the sensors. Since ALS is a range-free scheme, we compare its performance to other range-free schemes like APIT (Approximate Point In Triangle) [4], DV-Hop[5] and DHL (Density-aware Hopcount-based Localization) [6]. To validate our schemes, we first use simulations developed in Qualnet[31] to evaluate the performance of ALS and show that it outperforms other range-free localization schemes. We then follow with an implementation of ALS on a wireless sensor network test bed and conduct tests in both indoor and outdoor environments. We observe that at least 90% of nodes lie within a 1-hop region of their estimated areas, i.e. within their individual transmission radius.

The rest of the paper is organized as follows. Section 2 provides a survey of related work on wireless sensor network localization. Section 3 then describes the key aspects of the basic Area Localization Scheme. Section 4 describes the simulation environment and evaluates the performance of the ALS and compares it to other range-free schemes. Section 5 discusses the performance of the ALS evaluated on a wireless sensor network test bed for both indoor and outdoor environments. This section also discusses how the ALS scheme is extended to a generic physical layer model from the two-ray model used in the simulation studies. Section 6 presents our conclusions and plans for future work.

2. Related Work

A number of localization schemes have been proposed to date. The localization schemes take into account a number of factors like the network topology, device capabilities, signal propagation models and energy requirements. Most localization schemes require the location of some nodes in the network to be known. Nodes whose locations are known are referred to as *anchor nodes* or *reference nodes* in the literature. The localization schemes that use reference nodes can be broadly classified into three categories: range-based schemes, range-free schemes and schemes that use signal processing or probabilistic techniques (hereafter referred to as probabilistic schemes). There also exist schemes that do not require such reference locations in the network.

A. Range-based Schemes

In range-based schemes, the distance or angle measurements from a fixed set of reference points are known. Multilateration, which encompass atomic, iterative and collaborative multilateration techniques, are then used to estimate the location of each sensor node. Range-based schemes use ToA (Time of Arrival), TDoA (Time Difference of Arrival), AOA (Angle of Arrival) or RSSI (Received Signal Strength Indicator) to estimate their distances to anchor nodes. UWB based localization schemes [7][8], GPS [2], Cricket [9] and other schemes [11][12][13] use ToA or TDoA of acoustic or RF signals from multiple anchor nodes for localization. However, the fast propagation of RF signals implies that a small error in measurement could lead to large errors. Clock synchronization between multiple reference nodes or between the sender and the receiver is also an extremely critical issue in schemes that use ToA or TDoA. AOA allows sensor nodes to calculate the relative angles between neighbouring nodes [14][15]. However, schemes that use AOA entail sensors and reference nodes to be equipped with special antenna configurations which may not be feasible to embed on each sensor. Complex non-linear equations also need to be solved[15]. Schemes that use RSSI [16][17][18] have to deal with problems caused by large variances in reading, multi-path fading, background interference and irregular signal propagation.

B. Range-free Schemes

Range-free localization schemes usually do not make use of any of the techniques mentioned above to estimate distances to reference nodes, e.g. centroid scheme [19] and APIT [4]. Range quantization methods like DV-Hop [5] and DHL [6] associate each 1-hop connection with an estimated distance, while others apply RSSI quantization [20]. These schemes also use multilateration techniques but rely on measures like hop count to estimate distances to anchor nodes. Range-free schemes offer a less precise estimate of location compared to range-based schemes.

C. Probabilistic Schemes

The third class of schemes use signal processing techniques or probabilistic schemes to do localization. The fingerprinting scheme [21], which uses complex signal processing, is an example of such a scheme. The major drawback of fingerprinting schemes is the substantial effort required for generating a signal signature database, before localization can be performed. Hence, it is not suitable for adhoc deployment scenarios in consideration.

D. Schemes without Anchor/Reference Points

The fourth class of schemes is different from the first three in that it does not require anchor nodes or beacon signals. In [22], a central server models the network as a series of equations representing proximity constraints between nodes, and then uses sophisticated optimization techniques to estimate the location of every node in the network. In [23], Capkun *et al.* propose an infrastructure-less GPS-free positioning algorithm.

E. Area-based Localization

Most of the localization schemes mentioned above calculate a sensor node's exact position, except for [4], which uses an area-based approach. In [4], anchor nodes send out beacon packets at the highest power level that they can. A theoretical method, based on RSSI measurements, called Approximate Point in Triangle (APIT), is defined to determine whether a point lies inside a triangle formed by connecting three anchor nodes. A sensor node uses the APIT test with different combinations of three audible anchor nodes (audible anchors are anchor nodes from which beacon packets are received) until all combinations are exhausted. Each APIT test determines whether or not the node lies inside a distinct triangular region. The intersection of all the triangular regions is then considered to estimate the area in which the sensor is located. The APIT algorithm performs well when the average number of audible anchors is high (for example, more than 20). As a result, a major drawback of the algorithm is that it is highly computationally intensive. An average of 20 audible anchors would imply that the intersection of ${}^{20}C_3 = 1140$ areas need to be considered. Furthermore, the algorithm performs well only when the anchor nodes are randomly distributed throughout the network, which is not always feasible in a real deployment scenario.

3. Area Localization Scheme Fundamentals

In ALS, the nodes in the wireless sensor network are divided into three categories according to their different functions: reference nodes, sensor nodes and sinks.

A. Reference/Anchor nodes

The main responsibility of the reference/anchor (both terms will be used interchangeably) nodes is to send out beacon signals to help sensor nodes locate themselves. Reference nodes are either equipped with GPS to provide accurate location information or placed in pre-determined locations. In addition, the reference nodes can send out radio signals at varying power levels as required. For an Ideal Isotropic Antenna, the received power at a distance *d* from the transmitter is given by:

$$P_r = P_t G_t G_r \left(\frac{\lambda}{4\pi d} \right)^2 \tag{1}$$

while the two-ray ground reflection model considers both the direct path and a ground reflection path, and the received power at a distance *d* is given by:

$$P_r = \frac{h_r^2 h_t^2 G_r G_t P_t}{d^4} \quad \text{for} \quad d > \frac{4\pi h_t h_r}{\lambda} \tag{2}$$

where P_r is the received power, P_t is the transmitted power, d is the distance between the transmitter and receiver, λ is the wavelength and, h_t and h_r are the heights of the transmitter and receiver respectively. G_t and G_r represent the gains of the transmitter and receiver respectively in equations (1) and (2).

From the above equations, it can be clearly seen that if the received power is fixed at a certain value, the radio signal with a higher transmitted power reaches a greater distance. Using one of the physical layer models described above and the threshold power that each sensor can receive, the reference node can calculate the power required to reach different distances. Each reference node then devises a set of increasing power levels such that the highest power level can cover the entire area in consideration. The reference nodes then broadcast several rounds of radio signals. The beacon packet contains the ID of the reference node and the power level at which the signal is transmitted (which can be simply represented by an integer value, as explained below.)

Let *PS* denote the set of increasing power levels of beacon signals sent out by a reference node. For now, let us assume that all the reference nodes in the system send out the same set *PS* of beacon signals. In the ALS scheme, a sensor node simply listens and records the power levels of beacon signals it receives from each reference node. In real environments, small scale fading and shadowing can cause the power levels received by the sensor nodes to vary significantly from the expected power levels calculated by the path loss models in equations (1) and (2). Sending out beacon signals in the set *PS* only once might lead to inaccurate beacon reception by sensor nodes. As a result, the reference nodes send out the beacon signals in set *PS* multiple times. The sensor nodes can then calculate the statistical average (mode or mean) of the received power levels from each reference node.

Let the number of power levels in set *PS* be denoted by N_p and the N_p power levels in set *PS* be represented by $P_1, P_2, P_3, \ldots, P_{Np}$. The power levels $P_1, P_2, P_3, \ldots, P_{Np}$ can be represented by simple integers, e.g. increasing values corresponding to increasing power levels; therefore sensor nodes only need to take note of these integer values that are contained in the beacon packets and the hardware design can be kept simple as there is no need for accurate measurement of the received power level. Let the number of times that the same set of beacon signals *PS* are sent out be denoted by N_r, also referred to as the number of rounds. The power *MP* in dB required to cover the entire area is calculated from equation (1) or (2), based on the physical layer model in consideration. The power *LP* in dB required to cover a small distance δ (say 10 m) is also calculated. The values $P_1, P_2, P_3, \ldots, P_{Np}$ are then set to be N_p uniformly distributed values in the range [*LP*, *MP*] in the dB scale. The simple procedure followed by the reference nodes is shown below:

```
for i = 1: Nr
        for j=1: Np
                Send beacon signal at power level Pj
        end for
end for
```

The transmissions by the different reference nodes do not need to be synchronized. However, the reference nodes schedule the beacon signal transmissions so to avoid collisions. The transmitted set of power levels *PS* need not be the same for all the reference nodes, and can be configured by the network administrator. Also, the set of power levels *PS* need not be uniformly distributed too. It is also not necessary for the reference nodes to know each other's position and levels of transmitted power, but there should be at least one sink or a central agent that stores the location information of all the reference nodes.

B. Sensor node

A sensor node is a unit device that monitors the environment. Sensors typically have limited computing capability, storage capacity, communications range and battery power. Due to power constraints, it is not desirable forsensor nodes to make complex calculations and send out information frequently.

1) Signal Coordinate Representation:

In the ALS scheme, the sensors save a list of reference nodes and their respective transmitted power levels and forward the information to the nearest sink when requested or appended to sensed data. The sinks use this information to identify the area in which the sensors reside in. However, if the number of reference nodes is large, the packets containing location information may be long, which might result in more traffic in the network. A naming scheme is hence designed.

The sensor nodes use a signal coordinate representation to indicate their location information to the sinks. Power contour lines can be drawn on an area based on the set of beacon signal power levels *PS* transmitted by each reference node, and their corresponding distances travelled. The power contour lines divide the region in consideration into many sub-regions (which we refer to as areas) as shown in Figure 1 below. Each area in the region can be represented by a unique set of n coordinates, hereafter referred to as the *signal coordinate*.

Suppose there are n reference nodes, which are referred to as $R_1, R_2, \ldots,$ and R_n. For a sensor in an area, let the lowest transmitted power levels it receives from the n reference nodes be $S_1, S_2, \ldots,$ and S_n respectively. $S_1, S_2, \ldots,$ and S_n are simple integer numbers indicating the different power levels rather than the actual signal strengths. The mappings between integer levels and the actual power values are saved at the reference nodes and sinks. The signal coordinate is defined as the representation $< S_1, S_2, \ldots, S_n >$ such that each S_i, the ith element, is the lowest power level received from R_i.

For example, consider a square region with reference nodes at the four corners, as shown in Figure 1. In this case, the set of power levels *PS* is the same for all the four reference nodes and there are three power levels in the set *PS*. The smallest power level in the power set *PS* is represented by the integer 1 while the highest power level is represented by the integer 3. For each node, the contour lines drawn on the region represent the farthest distances that the beacon signals at each power level can travel. Contour lines for beacon power levels 1 and 2 are drawn. The power level 3 for each corner reference node can reach beyond the corner that is diagonally opposite to it and so, its corresponding contour line is not seen on the square region. Thus, for each reference node, the two contour lines corresponding to power levels 1 and 2 divide the region into three (arc) areas.

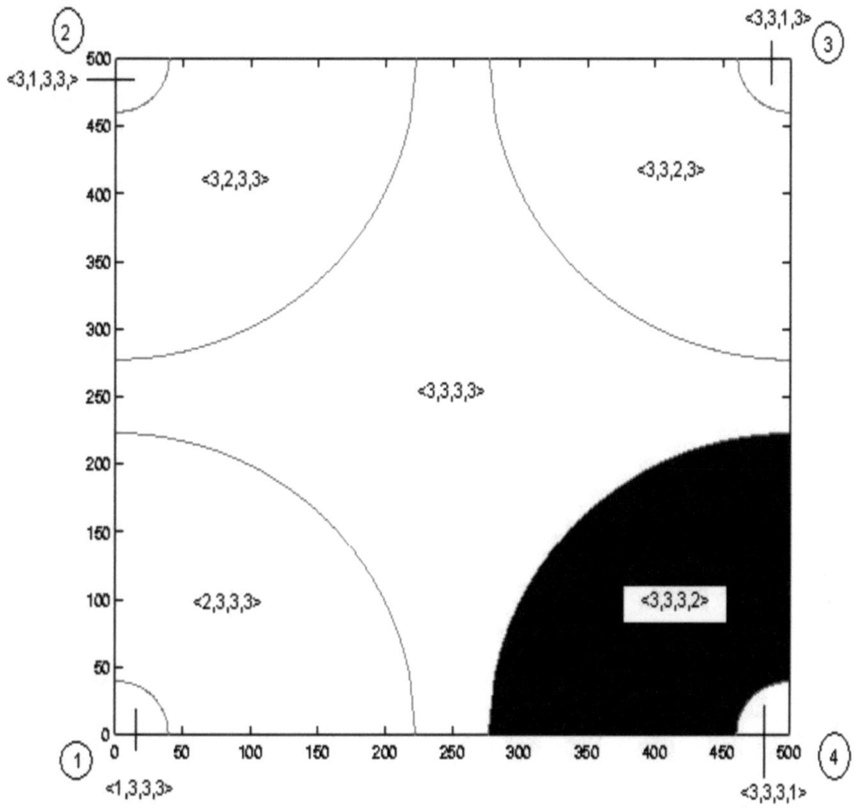

Fig. 1. Example of ALS under ideal isotropic conditions; shaded region is <3, 3, 3, 2>

For a sensor node in the shaded area(lower right) in Fig. 1, the lowest power level received from reference nodes 1, 2 and 3 is 3.The sensor node in the shaded area receives beacon signals at power levels 2 and 3 from reference node 4. So, the lowest power level received by the sensor from reference node 4 is 2. As a result, the shaded area in the figure can be represented by the unique signal coordinate <3,3,3,2>. Similarly, every other area in the square region can be represented by a unique signal coordinate, as shown in the figure. As stated in the signal coordinate definition, the lowest power level received from reference node i forms the ith element of the signal coordinate. Sensors use this unique signal coordinate to identify the area in which they are located.
Thus, if all the sensors and sinks agree in advance on the set(s) of beacon power levels *PS* transmitted by each reference node, the sensor nodes can use the signal coordinate $<S_1,S_2,\ldots,S_n>$ to indicate their area location information to the sinks. Similarly, when a sink need to get information from sensors specific to a certain area, it includes the signal coordinate in its request and the sensors simply compare the incoming signal coordinate to their own to see if they lie in the relevant area.

2) Algorithm

In the ALS scheme, the sensor node simply listens to signals from all reference nodes and records the information that it receives from them. A sensor node at a particular location

may receive localization signals (beacon messages)at different power levels from the same reference node, as explained above. The sensor records its signal coordinate and forwards the information to the sink(s) using the existing data delivery scheme, as and when requested.
Let the signal coordinate of a node be denoted $<S_1, S_2,...,S_n>$ where n is the number of reference nodes. A sensor node uses variables $L_{11}, L_{12,} ...,L_{1Nr}$ to represent the lowest power levels received by the sensor from reference node 1 during rounds 1 to $N_{r.}$ Similarly, let $L_{i1}, L_{i2},...,L_{iNr}$ represent the lowest power levels received by the sensor from reference node i during rounds 1 to $N_{r.}$ Let the number of reference nodes be n. Initially, all the values $L_{11}, L_{12,} ..., L_{1Nr}, L_{21}, L_{22}, ..., L_{2Nr}, ..., L_{n1}, L_{n2}, ..., L_{nNr}$ are set to zero. The zeros imply that the sensor nodes have received no signals from the reference nodes.
The pseudo-code running on each sensor node is shown below. After initialization, the sensor nodes start an infinite loop to receive beacon messages from reference nodes and follow the algorithm shown below. Since a reference node sends out several rounds of beacon signals, the sensor node may hear multiple rounds of beacon signals from the same reference node. If the sensor receives a signal from reference node i for the first time during round j, it sets L_{ij} to be the lowest received power level for that round; otherwise, if the received power level from reference node i in round j is lower than the current value in L_{ij}, L_{ij} is set to the latest received power level. After all the reference nodes have completed sending out beacon messages, the power levels L_{i1} to L_{iNr} on each sensor represent the lowest power levels received from reference node i during rounds 1 to N_r respectively.

```
Initialization:
    for i=1 to n
            for j = 1 to Nr
                    Lij = 0
            end for
    end for
Loop:
    Receive a message
    if (the message is from reference nodei during round j)
            if (Lij= 0 || received power level <Lij) ; received power level ⇐ integer
representation
                    Lij = received power level
            end if
    end if
```

Each reference node sends out beacon signals at all the power levels in the set PS N_r times (N_r rounds). In real conditions, fading and shadowing can cause the power levels to vary erratically about the expected signal strength predicted by the large scale fading model. Hence, the lowest signal power level received by a sensor from a reference node need not be the same for all the rounds 1 to N_r, i.e. all the values L_{i1} to L_{iNr} need not be the same. One is then faced with the problem of deciding which value L_{ix} to pick as $S_{i,}$ the ith element of the signal coordinate.
Hence, a threshold value *CONFIDENCE_LEVEL* is defined. This parameter represents the confidence level with which the values $S_1, S_2, ..., S_n$ can be estimated, and is an operational

value that end users can specify to suit their requirements. For example, this value was set to 80% of N_r in our performance studies. If there is a power level L_{ix} that occurs with frequency greater than *CONFIDENCE_LEVEL* in the set $\{L_i, \ldots, L_{iNr}\}$, then L_{ix} is set to be the ith element in the node's signal coordinate, i.e. $S_i = L_{ix}$. If there is no power level with frequency greater than *CONFIDENCE_LEVEL*, then all the distinct power levels in the set $\{L_{i1}..L_{iNr}\}$ are considered possible candidates of the ith element of the signal coordinate, and further refinement may be necessary.

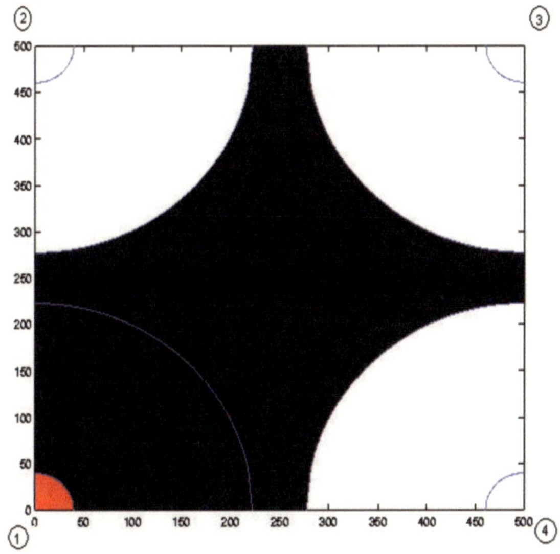

(a) Black region ≡<{2,3}, 3, 3, 3>.
(Black ∪Red) regions ≡<{1,2,3}, 3,3,3>

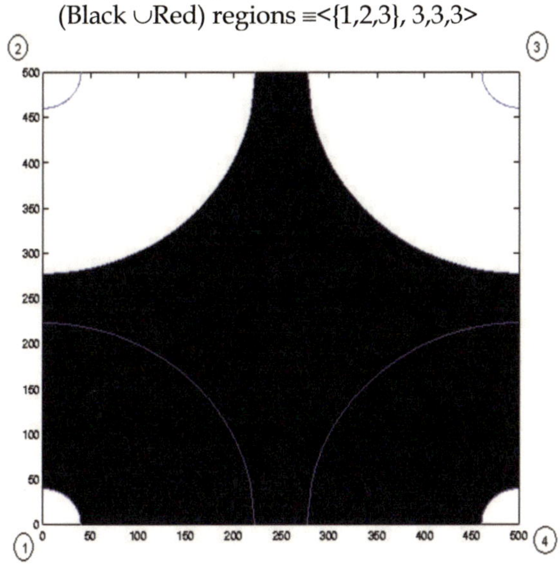

(b) Black region ≡<{2,3}, 3, 3,{2,3}>

Fig. 2. Illustration of Signal Coordinate Representation

This concept is further illustrated by a couple of examples and we assume the same scenario as in Fig. 1. In Fig. 1, we have assumed ideal isotropic channel conditions and each element in the signal coordinate has been ascertained with a high confidence level.

Fig. 2 illustrates scenarios of non-ideal channel conditions where beacons messages may be lost. Fig. 2(a) shows the case <{2,3},3,3,3>, where the first element of the signal coordinate is either 1 or 2. This happens when the lowest power level received from reference 1 during the N_r rounds of beacon messages oscillates between 1 and 2. Both values (1 and 2) can be considered as possible candidates for S_1, if no power level L_{1x} occurs with frequency greater than *CONFIDENCE_LEVEL* in the set $\{L_{11}, ..., L_{1Nr}\}$. The union of the black and red regions in Fig. 2(a) represents the region <0, 3, 3, 3>, where the value of 0 implies that there is no information available on the first element. This could happen in the case when no beacon packets are received from reference node 1, and the signal coordinate region <{1,2,3},3,3,3> is considered as a result. Thus, every element S_i in the set $<S_1, S_2, ..., S_n>$ need not be a unique value, but could be a set of values as shown in Fig. 2(b). While more than one element of a signal coordinate may have multiple values, we consider a signal coordinate to be valid only if at least half of its values have been determined with a high confidence level. From the above description, it can be clearly seen that the sensor nodes do not perform any complicated calculations to estimate their location. Neither do they need to exchange information with their neighbours.

C. Sink

In wireless sensor networks, data from sensor nodes are forwarded to a sink for processing. From a hardware point of view, a sink usually has much higher computing and data processing capabilities than a sensor node. In ALS, a sensor node sends its signal coordinate (location information) to a sink according to the data delivery scheme in use. The sensor itself does not know the exact location of the area in which it resides nor does it know what its signal coordinate represents. It is up to the sink(s) to determine the sensor's location based on the signal coordinate information obtained from the sensor. One assumption of the ALS scheme is that the sink knows the positions of all the reference nodes and their respective transmitted power levels, whether by directly communicating with the reference nodes, or from a central server, which contains this information. Therefore, with the knowledge of the physical layer model and signal propagation algorithms, the sink is able to derive the map of areas based on the information of the transmitted signals from the reference nodes. With the map and the signal coordinate information, the sink can then determine which area a sensor is in from the received data, tagged with the signal coordinate.

In the ALS scheme, the choosing of the signal propagation model plays an important part in the estimation accuracy. For different networks, different signal propagation models can be used to draw out the signal map according to the physical layer conditions. An irregular signal model could divide the whole region into many differently shaped areas, as shown in Fig. 3. Any adjustments made to the underlying physical layer model will have no impact on the sensor nodes, which just need to measure their signal coordinates and forward them to the sink. An immediate observation is the diverse area granularity, which affects the accuracy of the location estimation. The granularity issue will be discussed in the next section.

A key advantage of ALS is its simplicity for the sensors with all the complex calculations done by the sink. Thus, the localization process consumes little power at the sensor nodes, helps to extend the life of the whole network. Furthermore, it has a covert feature whereby anyone eavesdropping on the transmission will not be able to infer the location of sensors from the signal coordinates contained in the packets.

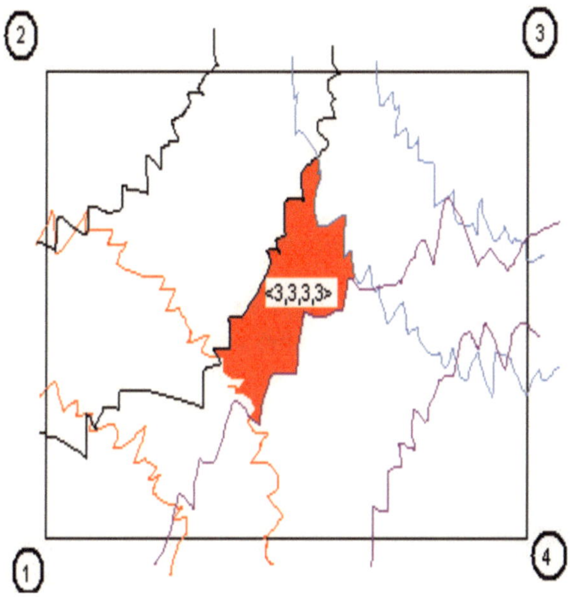

Fig. 3. Irregular contour lines arising from a non-ideal signal model

4. Performance Evaluation of ALS

We evaluate ALS using simulations as well as field experimentation using commercially available wireless sensor nodes.

A. Performance metrics for ALS

The metrics, accuracy and granularity, are used to evaluate the performance of the scheme. High levels of accuracy and granularity are desired; however, accuracy begins to suffer as granularity increases, since the probability of estimating the location of a node correctly in a smaller area decreases. Hence, in order to have a fair evaluation of ALS, we normalize the accuracy with respect to the granularity or average area estimate, that is, *normalized accuracy* = accuracy / average area estimate.

Another metric, *average error*, is defined to compare the performance of ALS to other range free schemes. The Center of Gravity (COG) or centroid of the final area estimate is assumed to be location of the node. Average error is then defined to be the average of the Euclidian distances between the original and estimated locations for all the nodes in the network.

B. Simulation scenario and parameters

The QUALNET 3.8 simulation environment is used to evaluate the performance of ALS. The system parameters used in our simulations are described below.

- *Region of deployment*: Square of size 500m × 500m.
- *Physical layer*: For the ideal case, it is modelled by the two-ray model given in equation (2). In the non-ideal case, Rayleigh fading and lognormal shadowing are also factored into the two-ray model.
- *Node placement*: A wireless sensor network with 500 nodes (eight of which are reference nodes) is used. The sensors are placed randomly throughout the region, and the eight reference nodes are positioned at the four corners and the four mid points of the sides of the square region. Although there are eight reference nodes, only four transmit beacon signals during each round of ALS. The sensor nodes in the network are assumed to be static, and the maximum velocity of objects in the surrounding is set to 1 m/s.
- *Reference-to-Node Range ratio* (RNR): This parameter refers to the average distance a reference beacon signal travels divided by the average distance a regular node signal travels. The radio range of sensors is set to 50 m, while the radio range of reference nodes is set to 1000 m, which is large enough for the beacon signals to cover the entire area. Therefore, the RNR value is 20.
- *Node Density* (ND): The node density refers to the average number of nodes within a node's radio transmission area. This value is close to 13 for the network scenario in consideration.
- *Reference Node Percentage* (RNP): The reference node percentage refers to the number of reference nodes divided by the total number of nodes. In our case, the system has a low RNP of 1.6% (8/500).
- *Receiver Threshold Power*: The receiver threshold power refers to the lowest signal strength of a packet that a node can receive. The value is set to -85 dBm.
- N_r: Number of times each beacon signal is sent out by a reference node. This parameter is set to 20.
- *CONFIDENCE_LEVEL*: 80%.

C. Simulation studyof ALS under ideal conditions

LP is set to -13 dBm and *MP* is set to 17 dBm. The number of power levels is then increased from 3 to 7 and the performance of the scheme is observed. All the sensors lie in their estimated areas as the experiment is carried out under ideal conditions. On the other hand, the granularity increases as the average area estimate decreases (Table 1), and as a result, the normalized accuracy metric improves, shown in Fig 4.

				Ideal conditions	
Iteration No.	No. of power levels	LP (dBm)	*MP*(dBm)	Avg. Area Est. as % of area size	% nodes that lie in their estimated area
1	3	-13	17	58.5	100
2	4	-13	17	17.4	100
3	5	-13	17	8.3	100
4	6	-13	17	5.8	100
5	7	-13	17	4.6	100

Table 1. Ideal case – granularity increases as the number of power levels increases.

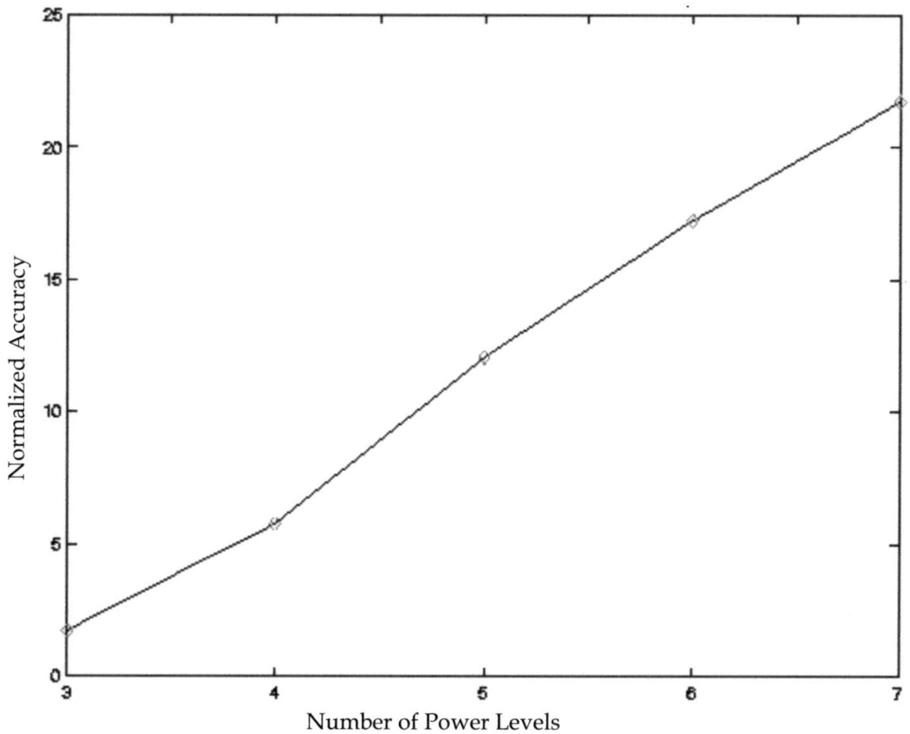

Fig. 4. Ideal case: Normalized Accuracy (accuracy/granularity) vs. Number of power levels.

D. Simulation study of ALS under non-ideal conditions

We first demonstrate the impact of decreasing the difference in adjacent power levels on the signal coordinates measured by the sensors. A signal coordinate $<S_1, S_2, S_3, S_4>$ is considered to be valid only if at least two of the four elements S_i can be measured with a confidence level of 80%. The measured signal coordinate is considered wrong if any valid element, S_i, differs from the actual value.

LP is set to -13dBm, while *MP* is set to 17dBm, as in the ideal case, and the number of power levels is increased from 3 to 7. The difference in adjacent power levels is $(MP-LP)/(N_p-1)$. For example, when N_p is set to 3, the three power levels are -13 dBm, 2 dBm and 17 dBm, and the difference in adjacent power levels is 15 dBm.

It is observed that the percentage of nodes that measure their signal coordinate correctly decreases from 96% to 28% as the number of power levels increases from 3 to 7. Fading and shadowing can cause the received signal strength to vary by as much as +10 dBm to -30 dBm of the expected value. The variance in measured signal coordinate increases, as the fading effect causes the received signal strength to vary by much more than the difference in adjacent power levels. As a result, fewer signal coordinates are measured correctly with a high confidence level (Fig. 5.). Nodes that were close to the edges of regions in the area were more prone to error than the nodes that are in the centre a region.

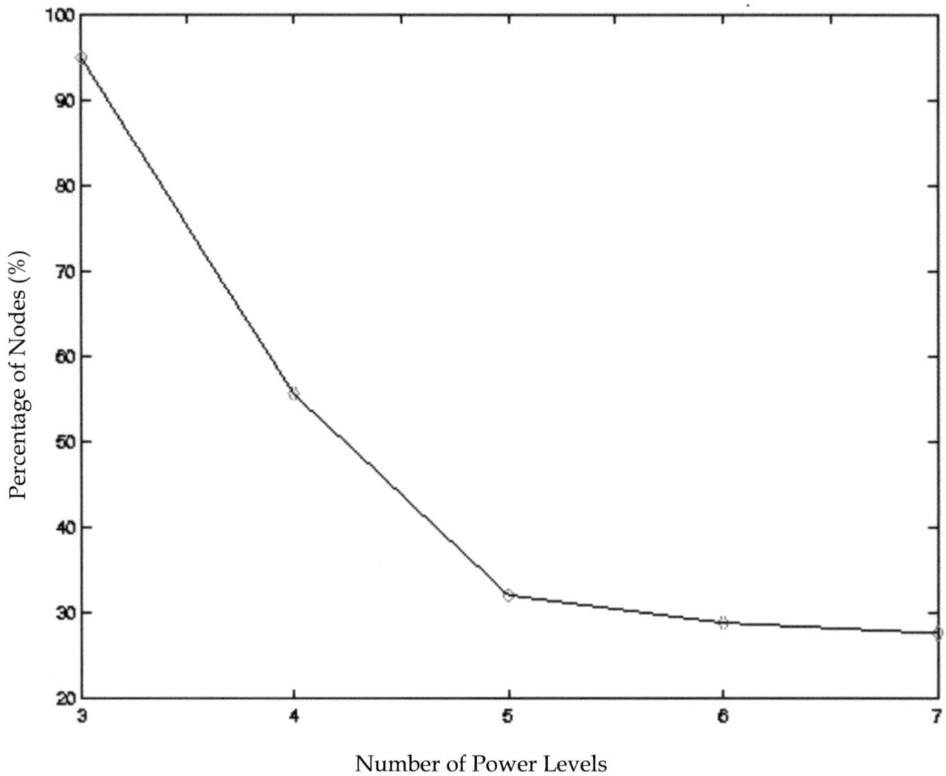

Fig. 5. Percentage of nodes that measure signal coordinate correctly vs Number of power levels

From the above discussion, it is evident that the difference in adjacent power levels should be set as large as possible. We therefore use a different set of power levels in each round of ALS, with the difference between adjacent power levels being set to 15 to 20 dBm, with each set *PS* having three distinct power levels. The *LP* and *MP* for each round are shown in Table 2. We consider 10 rounds in our simulations. For the first five rounds, the reference nodes at the four corners send out beacon signals, and for the next five, the reference nodes at the mid-points of the four sides send out beacon signals. For example, Fig. 6 shows the region after the reference nodes send out six sets of beacon signals. Each colour represents a distinct set of power contour lines. The final area estimate of a sensor is the intersection of the areas obtained from each round of the beaconing process. If the areas obtained from all the rounds completed do not intersect, the largest intersecting area obtained is considered as the area estimate. Thus, the final area estimate of each sensor node is one small region or a combination of many small regions in the final area shown in Fig. 6. The experiment is then carried out under both ideal and non-ideal conditions. The ideal conditions scenario serves as a benchmark to compare how well the ALS scheme performs under non-ideal conditions. The results obtained are shown in Table 2 and Fig. 7.

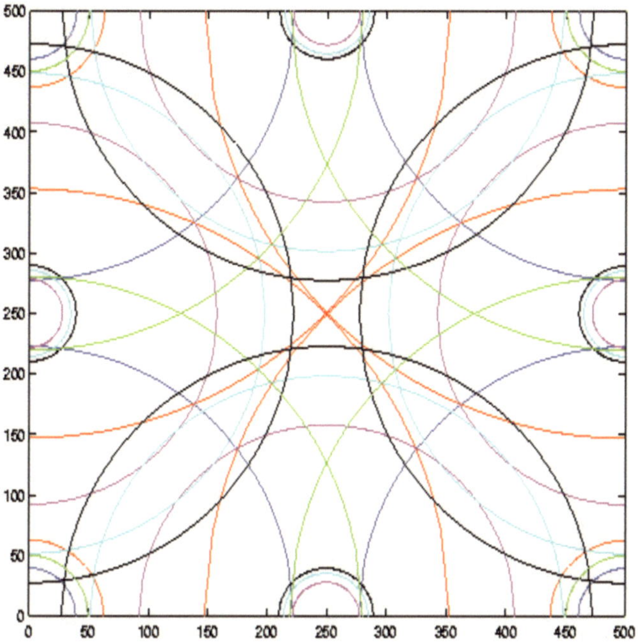

Fig. 6. Region after six rounds of ALS; each color represents a set of power levels.

For the non-ideal case, as the number of rounds increases from 1 to 10, the accuracy decreases from 98.47% to 69.9% (Table 2). The accuracy drops because a wrong signal coordinate measured in any one round of ALS would result in the final area being estimated incorrectly, as the intersection of areas from all rounds is considered in the final area estimate. On the other hand, granularity increases as the average area estimate decreases from 59.99% of the area size to 1.33% of the area size (Table 2) and this is a consequence of the intersection area for each sensor becoming smaller and smaller as the number of rounds increases.

				Ideal two-ray conditions		Non-ideal two-ray conditions		
Number of rounds finished	No. of power levels	*LP* (dBm)	*MP* (dBm)	Avg. Area Est. as % of area size	%nodes correctly localized	Avg. Area Est. as % of area size	% nodes correctly localized	% of nodes that lie 1-hop away
1	3	-16	30	49.76	100	59.99	98.47	1.53
2	3	-14	30	30.06	100	42.17	94.39	5.61
3	3	-13	30	19.26	100	31.18	89.29	10.71
4	3	-11	30	5.10	100	11.30	84.18	15.82
5	3	-9	30	2.80	100	5.58	82.65	9.69
6	3	-16	30	1.55	100	3.97	81.63	10.71
7	3	-14	30	1.17	100	3.09	77.55	10.71
8	3	-13	30	0.96	100	2.49	75.51	12.76
9	3	-11	30	0.71	100	1.72	72.96	18.37
10	3	-9	30	0.60	100	1.33	69.90	21.9

Table 2. Data and results for the non-ideal case

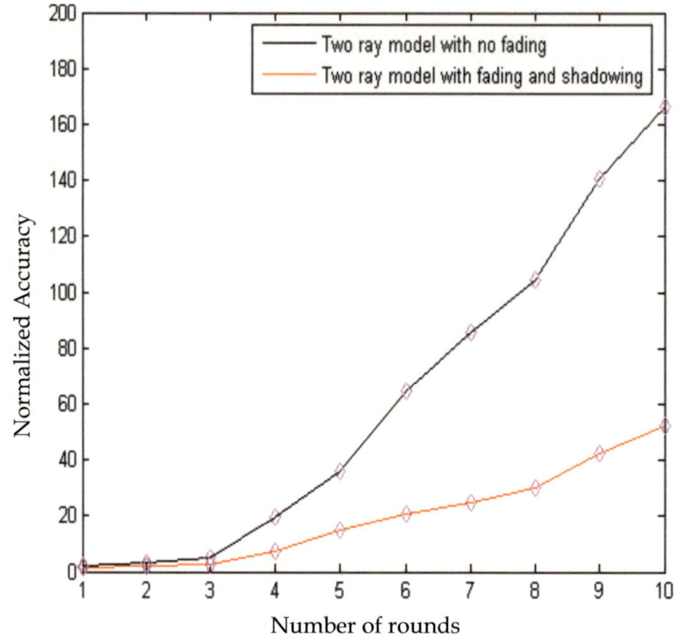

(a) Normalized accuracy starts to flatten out as the number of rounds increases

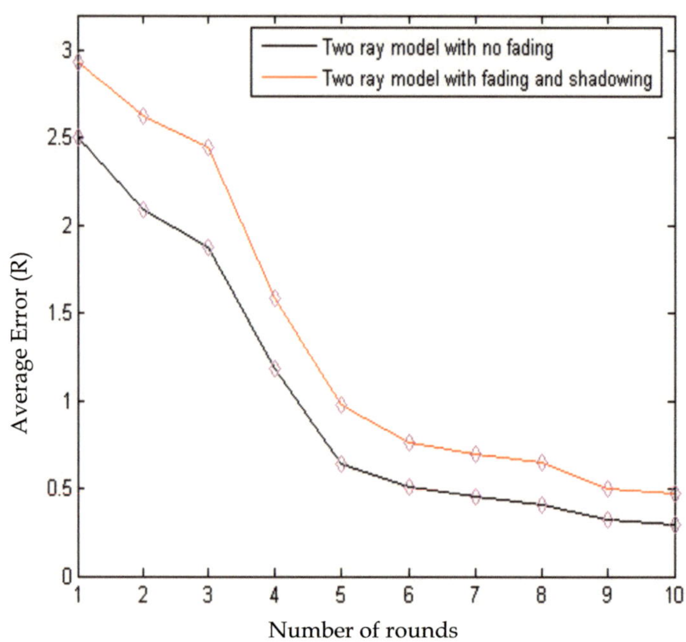

(b) Average error decreases as the number of rounds increases.

Fig. 7. ALS performance after multiple rounds

For the non-ideal case, the normalized accuracy metric improves, and starts to flatten out as the number of rounds increases. The performance metric increases as the decrease in average area estimate is greater than the decrease in accuracy after each additional round of ALS. The performance flattens out because of the quantization of power levels, and the constraint of maintaining a significant difference between adjacent power levels. ALS can be stopped once desired accuracy levels and granularity are obtained. The desired average area estimate and accuracy level, as well as the computational complexity of performing an extra round, with the increased overhead in beacon messages should all be taken into account before an additional round is executed. After 10 rounds, the average error drops below 0.5*R (where R is the Radio Range of a sensor) for both the ideal and non-ideal conditions (Fig. 7(b)).

E. One-hop Neighbourhood

Nodes that are closer to contour line boundaries are more prone to have their signal coordinates measured wrongly. An analysis was carried out to investigate the error patterns of nodes that did not lie in their estimated areas. It was observed that nodes, whose locations were estimated incorrectly, very often lie in an adjacent area to their actual location area.

Let the average area estimate of the nodes in the sensor network be denoted by A (for example, A = 1.33% of region size at the end of 10 rounds in our simulation). The area estimate of each node can then be approximated by a circle of area A (of radius $\sqrt{(A/\pi)}$). Circles with radius $\sqrt{(A/\pi)}$ and $2\sqrt{(A/\pi)}$ are drawn from the estimated location of the node. The circular ring between radii $\sqrt{(A/\pi)}$ and $2\sqrt{(A/\pi)}$ is defined as the one-hop neighbourhood region of the node. This concept of one-hop neighbourhood is illustrated with an example in Fig. 8. Referring to Table 2, we observe that the average area estimate is large for the first four rounds. As a result, all nodes lie within their estimated area or in the one-hop neighbourhood. As more rounds of ALS are executed, the accuracy decreases and the number of nodes that fall in the one-hop neighbourhood increases from 9.69% to 21.9%. It can be seen that, when A = 1.33%, more than 90% of nodes either lie in their estimated areas or in an area one-hop away.

The significance of the one-hop neighbourhood lies in various application scenarios that ALS can be applied to. Consider an application scenario where a particular sensor in the network detects an event and an unmanned vehicle is sent to the area (estimated by ALS) to investigate. If the vehicle fails to find the sensor in the estimated area, it would then expand its search in the surrounding areas that are one-hop away, two-hop away and so on. The chance of finding the sensor within a one-hop range of the estimated area is very high (> 90%), as evident from Table 2.

F. Comparison with other range-free schemes

In this section, we compare ALS against other range-free localization schemes proposed for wireless sensor networks. The range-free area localization schemes and range-free distance vector based localization schemes chosen for comparison with ALS are the following: PIT (Point in Triangle) and APIT (Approximate Point in Triangle) schemes [4], DV-Hop [5], and DHL [6]. For a fair comparison, the chosen algorithms share a common set of system parameters described in Section 4.B. The results obtained after ten rounds of ALS are compared to the other two categories of range-free schemes.

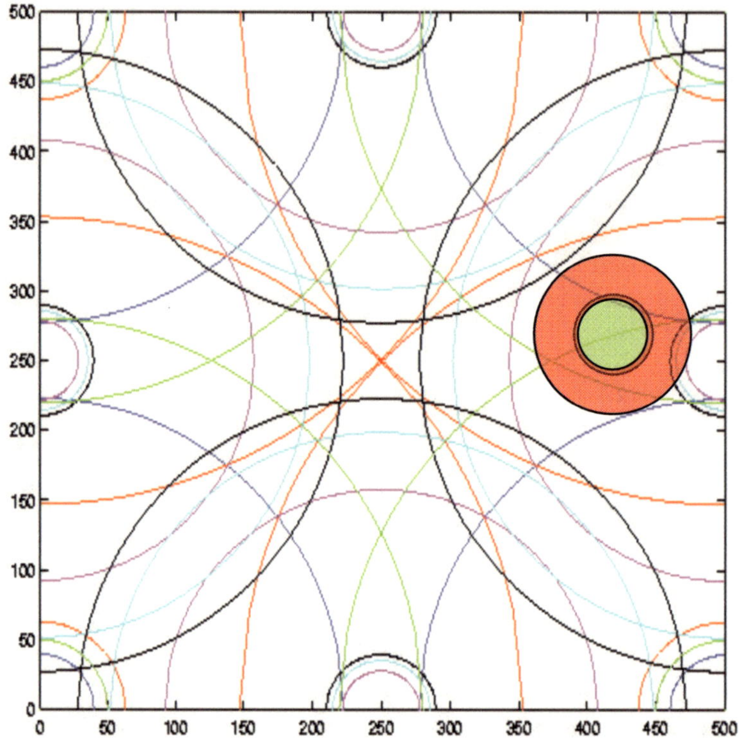

Fig. 8. One-hop neighbourhood: green area represents the estimated area of a node in the final area, while the surrounding red area represents the corresponding one-hop neighbourhood

1) Comparison with area based scheme:APIT (Approximate Point in Triangle)

In the PIT and APIT schemes[4], a node chooses three reference nodes from all audible reference nodes (reference nodes from which a beacon was received) and tests whether it is inside the triangle formed by connecting these three reference nodes. The theoretical method used to determine whether a point is inside a triangle or not is called the Point-In-Triangle (PIT) test. The PIT test can be carried out only under ideal physical layer conditions, when every node in the network is mobile can move around its own position. Due to the infeasibility of conducting such a test, an APIT (Approximate Point in Triangle) test is proposed. The APIT uses RSSI information of beacon signals to determine whether it is inside or outside a given triangle. The PIT or APIT tests are carried out with different audible reference node combinations until all combinations are exhausted. The information is then processed by a central server to narrow down the possible area that a target node resides in. An area scan aggregation algorithm is used to determine the intersection of the areas and determine the final area estimate of the node.

Fig. 9 shows all the possible triangles for the given configuration of the eight reference nodes. There are 52 triangles in total ($^8C_3$ - 4). The sensor nodes determine whether they are in or out of each of the 52 triangles, and the final area estimate computed is a small region or combination of regions on the area. Since PIT and APIT are area localization schemes, their performance are compared with ALS using the normalized accuracy metric. The following five cases are compared and results shown in Fig. 10:

i) ALS under ideal physical layer conditions after six rounds
ii) PIT under ideal physical layer conditions
iii) APIT under ideal physical layer conditions
iv) ALS under non-ideal physical layer conditions after six rounds
v) APIT under non-ideal physical layer conditions

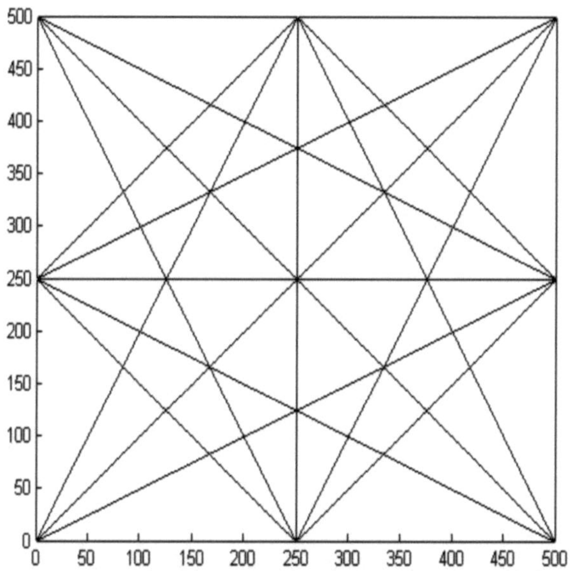

Fig. 9. All possible triangles for PIT and APIT schemes with 8 reference nodes: 4 at the corners and 4 at the mid-points of sides. There are 52 triangles in total.

The PIT and APIT schemes are carried out under ideal conditions to establish the performance limits that can be achieved with the APIT algorithm under non-ideal conditions. For the given scenario, it is observed (as shown in Fig. 7(a)) that ALS under ideal conditions outperforms both PIT and APIT after just six rounds.

Not all APIT tests yield correct results, even under ideal physical layer conditions. As a result, the performance of APIT under ideal conditions is slightly lower than PIT, due to lower accuracy levels. Under non-ideal conditions, it is observed that ALS performs much better than APIT. This is primarily because fluctuating RSSI values causes a number of APIT tests to be incorrect. It is also observed that only around 60% of the 52 APIT tests are correct for each sensor. This results in large area estimates on the network area. Thus, lower accuracy levels and higher area estimates cause the performance of the APIT scheme to suffer. ALS, on the other hand, is more resilient to fading and shadowing due to the significant difference in adjacent beacon power levels.

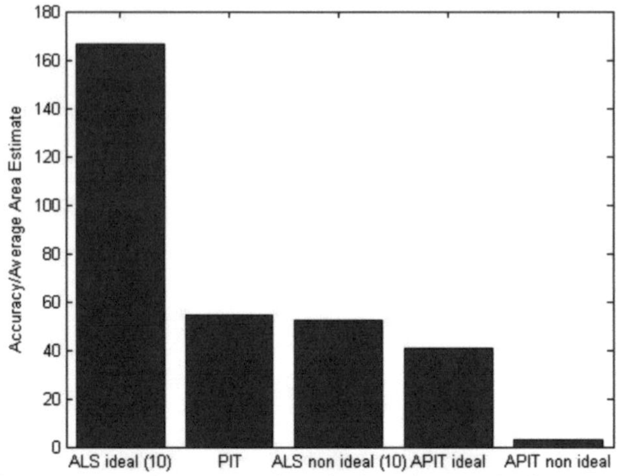

Fig. 10. ALS outperforms PIT and APIT under ideal and non-ideal scenarios respectively

The ALS scheme is much more computationally efficient than APIT. For the scenario in consideration, the area estimate obtained from the intersection of just 10 regions for ALS, one from each round, results in a better performance than APIT, which considers the intersection of 52 regions. Thus, ALS achieves the desired performance level as APIT at a much lower computational cost. The computational complexity in number of areas is given by $O(N_r)$ for ALS and $O({}^{N}C_3)$ for APIT.

2) Comparison with distance vector based schemes: DV-Hop and DHL

Distance Vector based localization schemes estimate the point location of a node. The location estimation error is then defined to be the Euclidian distance between the actual position and the estimated position of the node. The average of the location estimation errors of all the nodes in the network is used to compare the performance of the three localization schemes. The location errors are normalized with respect to the transmission range of the node. For ALS and APIT, the Center of Gravity (COG) of the final predicted region is used as the estimated position of the node. Again, localization using PIT, APIT and ALS schemes are carried out under ideal conditions to establish the performance limits that can be achieved by the algorithms under non-ideal conditions.

DV-Hop localization uses a mechanism that is similar to classical distance vector routing. Each reference node broadcasts a beacon, which contains its location information and a hop-count parameter initialized to one. The beacon is flooded throughout the network. Each sensor node maintains the minimum counter value per reference node of all beacons it receives and ignores those beacons with higher hop-count values. Beacons are flooded outward with hop-count values incremented at every intermediate hop. Through this mechanism, all nodes in the network (including other reference nodes) get the shortest distance, in hops, to every reference node. In order to convert hop-count into physical distance, the system estimates the average distance per hop without range-based techniques. Once a node can calculate the distance estimation to more than three reference nodes in the plane, it uses triangulation (or multilateration) to estimate its position. The DV-Hop scheme performs well in networks with uniform node density, as the size of each hop is assumed to be constant. The DHL scheme is an enhancement to the DV-Hop scheme for networks with non-uniform node density. In the DHL scheme, the size of each hop is not assumed to be constant. Instead, the size of each hop depends on the density of nodes in the neighbourhood.

In the simulations, the RNR parameter is set to 1 for DV-Hop and DHL, i.e. the radio range for both the reference and sensor nodes is set to 50m. From the results shown in Fig. 11, it can be observed that even under non-ideal physical layer conditions, ALS performs better than the other range-free schemes.

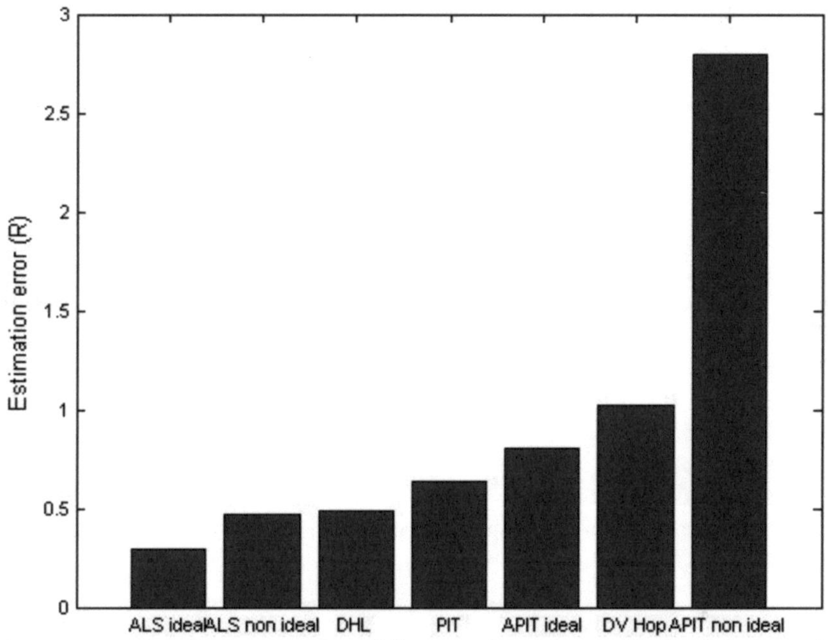

Fig. 11. Average estimation error for different algorithms. Under ideal and non-ideal conditions, ALS (after 10 rounds) outperforms all the other schemes.

The average estimation error of DV-Hop and APIT under non-ideal conditions is greater than R (radio range of sensor node.) The APIT scheme under non-ideal conditions is severely affected by the fluctuations in RSSI of the beacon packets. APIT would perform better if there were more reference nodes, but the improved performance would be at the expense of a much higher computational cost. For example, it is observed from the APIT simulations in [4] that achieving an average estimation error of close to 0.5R, for a similar set of system parameters used here, would require more than 15 audible reference nodes. This would entail computing the intersection of more than 455 ($^{15}C_3$) areas and hence, is highly computation intensive. The performance of DV-Hop is contingent on the density distribution of nodes in the network and the estimate used for the average distance of a single hop. The performance of DV-Hop suffers if the distribution of nodes in the network is non-uniform. For both schemes, viz. DV-Hop and DHL, we observe that the localization error of the nodes along the sides is higher than nodes at the centre of the region [24].

5. Performance of ALS on a WSN Test bed

The performance of the ALS depends on the model used for the physical layer. The radio environment can be modelled by the empirical log-distance path loss model, as shown below:

$$PL(d)[dB] = PL(d_0) + 10n\log(\frac{d}{d_0}) + X_\sigma(dB) \quad (3)$$

where n is the path loss exponent which indicates the rate at which the path loss increases with distance, d_0 is the reference distance (typically set to 1 m), d is the distance between the transmitter and receiver, $PL(d_0)$ is the power received at distance d_0, and X_σ is a zero-mean Gaussian distributed random variable (in dB) with standard deviation σ. X_σ describes the random shadowing effects over a large number of measurements for the same transmitter-receiver separation.

To extend ALS to any generic physical layer model, we implement a testing phase where the parameters n and X_σ are estimated, which can be achieved by the reference nodes mutually measuring the received signal strength from one another's beacons. The model in equation (3) can then be used to determine the different transmit power levels and to draw out the signal map.

We implemented the ALS on a wireless sensor network test bed [25]. MicaZ motes by Crossbow Technology Inc. [32] are used as both reference and sensor nodes. The MicaZ motes allow transmission of signals at only 8 power levels: -25, -15, -10, -7, -5, -3, -1 and 0 dBm, corresponding to MicaZ transmission power settings of 3, 7, 11, 15, 19, 23, 27 and 31 respectively. As a result, we were constrained to use only these eight power levels for the reference nodes. However, for a real deployment, we anticipate the anchor nodes to be more sophisticated devices with the ability to finetune the power at which they transmit beacon signals.

Fig. 12. ALS experiment in an obstacle free environment – the experiment was carried out in a 30m×30m region in a soccer field. There were no obstacles present inside the region.

Fig. 13. ALS experiments in an obstacle ridden environment - the experiment was carried out in a 30m×30m region in a park with several trees in the region. Several of the sensors were placed behind trees, as seen in the picture.

The experiments were carried out in both indoor and outdoor environments. For the outdoor scenario, the experiment was first carried out in an environment with no obstacles (Fig. 12), and subsequently in an obstacle-ridden environment (trees, park benches, etc) (Fig. 13). RSSI measurements were made to estimate the path loss exponent of radio signals in each environment. The path loss exponent is calculated using regression analysis on the RSSI measurements, and was determined to be 2.92 for the indoor environment and 2.96 for the outdoor environment. The path loss exponents were then used to estimate the ranges of the beacon signals sent out at different power levels by reference nodes. The measured and estimated range measurements for the indoor and outdoor environments are shown in Table 3, and we observed that the estimated range values tally with the measured range values for different power levels.

Power Level	Indoor (Reference Node Height = 12 cm Estimated (m)	Indoor (Reference Node Height = 12 cm Measured (m)	Outdoor(Reference Node Height = 190 cm) Estimated (m)	Outdoor (Reference Node Height = 190 cm) Measured (m)
3	2.2	2.5	2.5	2.5
7	4.9	5.5	13.0	13.5
11	7.3	8.0	19.2	17.5
15	9.3	9.0	24.3	24.5
19	10.8	13.0	28.3	30.0
23	12.0	13.0	33.1	33.5
27	14.0	15.0	38.7	37.0
31	15.0	15.0	50.0	50.0

Table 3. Estimated and Measured Range measurements for different MicaZ power levels for indoor and outdoor environments. The slightly greater differences in indoor estimated and measured values are due multipath effects.

We observed that the range measurements vary with the height at which the reference nodes are placed, as also noted in [18]. In particular, the communication ranges of the nodes increased when the reference nodes were raised above the ground. We, therefore, raised the height of the reference nodes for our experiments. For the indoor environment, the reference nodes are placed on plastic pots at a height of 12 cm above ground level. Similarly, for the outdoor environment, the reference nodes are mounted on wooden easels at a height of 190 cm above ground level, while the sensors are placed on plastic pots, 12 cm above ground level (as shown in Fig. 14.)
We also observed that the range measurements depend on the relative orientation between the transmitter and receiver antennas, and the radio patterns of the MicaZ antennas are not circular. Both these observations are confirmed by the results obtained by Lymberopoulos *et al.* [26] and Tan *et al.* [27]. Therefore, we run each ALS experiment four times with four different relative orientations between the transmitter and receiver antennas. For example, all sensors are placed facing a certain direction initially (North). The ALS experiment is then carried out four times, with reference nodes facing North, South, East and West directions respectively. The results are then combined to obtain the final area estimate of each sensor. This helps alleviate the problems caused by the radio irregularity of the sensor antennas.

Fig. 14. Reference nodes mounted 190 cm above ground; sensor nodes are mounted on plastic pots 12 cm above ground level.

The experimental set up is similar to the set up described in the simulation scenario. An area size of 10m × 10m is chosen for the indoor environment while the area size of 30m × 30m is chosen for the outdoor environment. Between 30 and 35 sensors are deployed randomly throughout the area, with eight reference nodes positioned at the four corners and the four mid-points of the sides of the square region. Since packets are often lost due to varying channel conditions, the *CONFIDENCE_LEVEL* parameter is set to a relatively low value of 30%. The results of the experiments are summarized in the Table 4 below.

			Ideal results				Experimental results				
Environment	Description	No. of power levels	Avg. Error	Accuracy	Avg. Area Estimate	Accuracy / Avg. Area Estimate	Avg. Error	Accuracy	One hop accuracy	Avg. Area Estimate	Accuracy / Avg. Area Estimate
Indoor	10m×10m	3	0.71	35/35 (100%)	4.28%	23.36	1.09	21/35 (60%)	9/35 (25.71%)	3.48%	17.24
Outdoor (No obstacles)	30m×30m	4	1.45	30/30 (100%)	1.70%	58.82	2.04	17/30 (56.67%)	13/30 (43.33%)	2.76%	20.53
Outdoor (With obstacles)	30m×30m	4	1.45	30/30 (100%)	1.70%	58.82	3.37	16/30 (53.33%)	9/30 (30.00%)	2.58%	20.67

Table 4. Summary of Experimental Results

We also plot the actual versus point estimated locations for the sensors in the different environments, namely, Fig. 15 for indoor, Fig. 16 for outdoor (open field/no obstacles) and Fig. 17 for outdoor (park/with obstacles). The crosses are the actual locations of the sensors, the circles are the predicted point estimated location of the sensors (the approximate centre of the region) and the squares are the locations of the reference nodes. A shorter line connecting a cross-circle pair denotes higher accuracy.

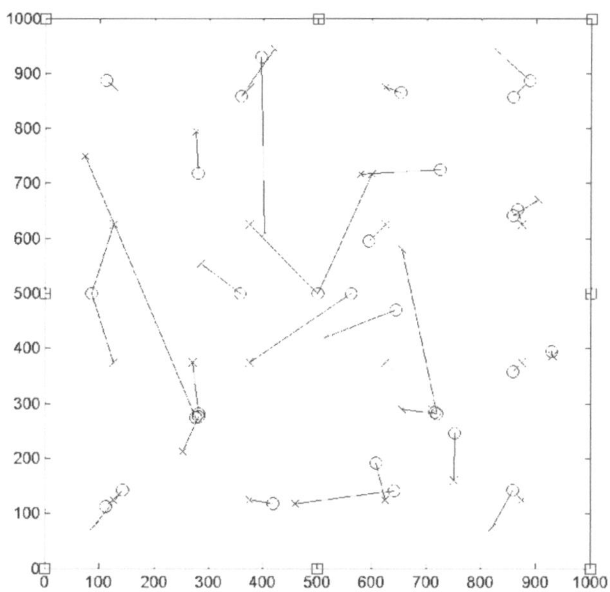

Fig. 15. Actual versus Estimated Locations of Sensors in Indoor Environment

The set of ideal results refers to the case where every sensor node measures its signal coordinate correctly and hence the area in which it is located, correctly. The average error increases from 2.04 m to 3.37 m as we move from an obstacle-free outdoor environment to an obstacle-ridden environment. Moving from an obstacle-free to an obstacle-ridden environment outdoors, we notice that the accuracy drops but the number of nodes present in the one-hop neighbourhood increases. For all the scenarios, more than 80% of sensor nodes lie within their predicted area or within a one-hop region of the predicted areas with the average area size estimate of less than 3.5%. This means that in real deployments, sensors can be located quickly once the predicted region is calculated. The performance of

ALS becomes worse as we move from the open field to the park because multipath effects from obstacles cause fluctuations in signal strength resulting in sensor nodes receiving incorrect signal coordinates. The accuracy of localization also depends on many other factors, such as, type of hardware, environment, number of reference nodes, and the size of the deployment area.

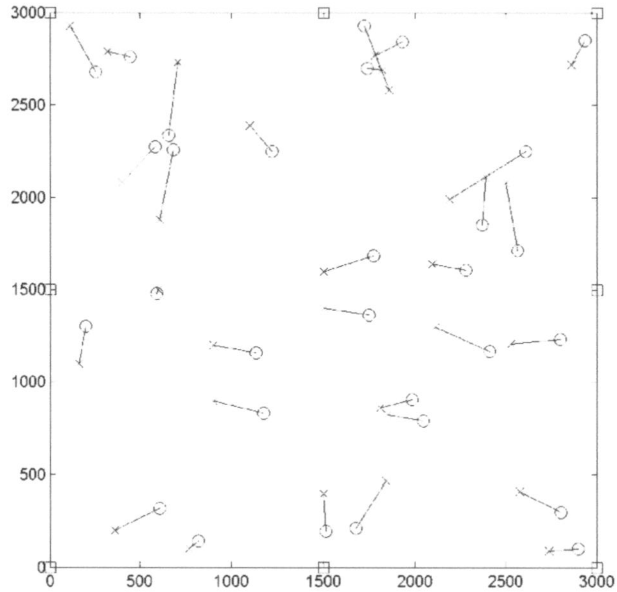

Fig. 16. Actual versus Estimated Locations of Sensors in Outdoor Environment (no obstacles)

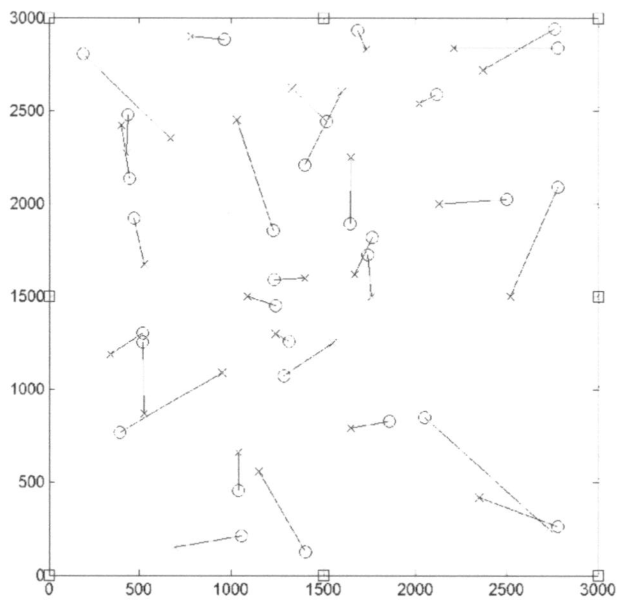

Fig. 17. Actual versus Estimated Locations of Sensors in Outdoor Environment (with obstacles)

In addition, we also compared ALS with other localization schemes that utilize radio signals, which have also been implemented and tested experimentally. From Table 5, show the results of ALS with such schemes, and it is clear that the performance of ALS in the field trials is comparable or better than these other localization schemes.

Scheme	Hardware Platform	Environment	No of reference nodes	Region Size	Error
ALS	micaZ	Indoor	8	10m x 10m	1.09m
ALS	micaZ	Outdoor (no obstacles)	8	30m x 30m	2.04m
ALS	micaZ	Outdoor (with obstacles)	8	30m x 30m	3.37m
Ranging-based [18]	mica2dot	Outdoor (open field)	(not mentioned; 49-node network)	50m x 50m	4.91m
Mote Track [28]	mica2	Indoor	20	1742 m^2	50th percentile: 2m 90th percentile: 3m
Ecolocation [29]	mica2	Outdoor	10	12m x 12m	around 1m
Probability Grid [30]	mica2	Outdoor	3	5x5 grid; approx 12m apart	79% of radio range where radio range is about 15m

Table 5. Comparisons of Field Test results against other localization schemes

6. Conclusion

The ALS is a range free localization scheme that provides a coarse estimation of the location of a sensor within a certain area. While the sensors simply record the signal levels received from reference nodes, the sinks carry out most of the complicated computations. The granularity of the area estimates can be increased easily by modifying certain system parameters. The simulations results in Qualnet show that ALS is a promising scheme as more than 90% of nodes are located in their estimated areas or in a region one-hop away. We implemented the ALS algorithm on a wireless sensor network testbed and tested it in both indoor and outdoor scenarios. We observed that 100% of nodes lie within a one-hop region in an outdoor, obstacle-free environment, and 83% (or more) of nodes of lie within a one-hop region in an indoor or outdoor, obstacle-ridden environment. With high quality transmission device and antennas implemented at the reference nodes (as compared to the motes used in the experiment), it is highly foreseeable that the accuracy of ALS will increase. As part of our ongoing and future work, we have first addressed the issue of non-uniform areas (e.g. as shown in Fig. 6) by aggregating areas of different sizes to create more uniformity [33]. Subsequently, we will be improving the reference nodes and also develop routing protocols that is able to utilize the location information provided by the ALS algorithm. A sensor can therefore estimate whether it is nearer or further away from the destination, compared to its previous hop, based on the signal coordinate information of its neighbour, the destination and itself, and this information can be used for developing fast and efficient routing protocols. Another benefit is the covert nature of the scheme, which can be exploited to meet privacy needs.

7. References

[1] I. Akyildiz, W. Su, Y. Sankarasubramaniam and E. Cayirci, "A Survey on Sensor Networks", *IEEE Communications Magazine*, Vol. 40, No. 8, pp 102-114, Aug2002.

[2] Global Positioning System standard Positioning Service Specification, 2nd Edition, June 2, 1995.

[3] Q. Yao, S. K. Tan, Y. Ge, B.S. Yeo, and Q. Yin, "An Area Localization Scheme for Large Wireless Sensor Networks",*Proceedings of the IEEE 61st Semiannual Vehicular Technology Conference (VTC2005-Spring)*, May 30 - Jun 1, 2005, Stockholm, Sweden.

[4] T. He, C. Huang, B. Blum, J. Stankovic and T. Abdelzaher, "Range-Free Localization Schemes for Large Scale Sensor Networks", *Proceedings of the 9th ACM International Conference on Mobile Computing and Networking (Mobicom 2003)*, Sep 14-19 2003, San Diego, CA, USA.

[5] D. Niculescu and B. Nath, "DV Based Positioning in Ad Hoc Networks", *Telecommunication Systems*, Vol. 22, No. 1-4, pp 268-280, 2003.

[6] S.Y. Wong, J.G. Lim, S.V. Rao and Winston K.G. Seah, "Density-aware Hop-count Localization (DHL) in wireless sensor networks with variable density", *Proceedings of the IEEE Wireless Communications and Networking Conference (WCNC 2005)*, 13-17 Mar 2005, New Orleans, L.A.,USA.

[7] S. Gezici, Z. Tian, G. Giannakis, H. Kobayashi, A. Molisch, V.Poor and Z. Sahinoglu, "Localization via Ultra Wide Band Radios", *IEEE Signal Processing Magazine*, Vol. 22, No. 4,Jul 2005, pp. 70-84.

[8] Y. Xu, J. Shi and X. Wu, "A UWB-based localization scheme in wireless sensor networks", *Proceedings of the IET Conference on Wireless, Mobile and Sensor Networks 2007 (CCWMSN07)*, Dec 12-14, 2007, Shanghai, China.

[9] N. B. Priyantha, A. Chakraborty and H. Balakrishnan, "The Cricket Location-Support system", *Proceedings of the 6th ACM International Conference on Mobile Computing and Networking (Mobicom 2000)*, Aug 6-11, 2000, Boston, MA, USA.

[10] Y. Kwon, K. Mechitov, S. Sundresh, W. Kim and G. Agha,"Resilient Localization for Sensor Networks in Outdoor Environments", *Proceedings of 25th IEEE International Conference on Distributed Computing Systems* (ICDCS 2005), Jun 6-10, 2005, Columbus, Ohio, USA.

[11] P. Bahl and V. Padmanabhan, "RADAR: an in-building RF-based user location and tracking system", *Proceedings of the 19th Annual Joint Conference of the IEEE Computer and Communications Societies* (INFOCOM 2000),Mar 26-30, 2000, Tel Aviv, Israel.

[12] X. Cheng, A. Thaeler, G. Xue and D. Chen, "TPS: A Time-Based Positioning Scheme for Outdoor Sensor Networks", *Proceedings of the 23rd Annual Joint Conference of the IEEE Computer and Communications Societies (INFOCOM 2004)*, Mar 7-11, 2004, Hong Kong.

[13] A. Savvides, C. C. Han and M. B. Srivastava, "Dynamic Fine-grained Localization in Ad-Hoc networks of Sensors",*Proceedings of the 7th ACM International Conference on Mobile Computing and Networking (Mobicom 2001)*, Jul 16-21, 2001, Rome, Italy.
[14] D. Niculescu and B. Nath, "Ad Hoc Positioning System (APS) Using AOA", *Proceedings of the 22ndAnnual Joint Conference of the IEEE Computer and Communications Societies (INFOCOM 2003)*, Mar 30-Apr 3, 2003, San Francisco, CA, USA.
[15] N. Malhotra, M. Krasniewski, C. Yang, S. Bagchi, and W. Chappell, "Location Estimation in Ad-hoc networks with Directional Antennas",*Proceedings of 25th IEEE International Conference on Distributed Computing Systems (ICDCS 2005)*, Jun 6-10, 2005, Columbus, Ohio, USA.
[16] L. Girod and D. Estrin, "Robust Range Estimation Using Acoustic and Multimodal Sensing", *Proceedings of the International Conference on Intelligent Robots and Systems (IROS 2001)*, Oct 29-Nov 3, 2001, Maui, HI, USA.
[17] L.Evers, S. Dulman and P. Havinga, "A Distributed Precision Based Localization Algorithm for Ad-Hoc Networks", *Proceedings of the 2nd International Conference on Pervasive Computing (PERVASIVE 2004)*, Apr 21-23, 2004, Linz, Vienna, Austria.
[18] K. Whitehouse, C. Karlof and D. Culler, "A practical evaluation of radio signal strength for ranging-based localization", *ACM SIGMOBILE Mobile Computing and Communications Review*, Special Issue on Localization, Vol. 11 , No. 1, pp. 41-52, Jan 2007.
[19] N. Bulusu, J. Heidemann and D. Estrin, "GPS-less Low Cost Outdoor Localization for Very Small Devices", *IEEE Personal Communications Magazine*,Vol. 7, No. 5, pp. 28-34, Oct 2000.
[20] X. Li, H. Shi and Y. Shang, "Sensor network localisation based on sorted RSSI quantisation", *International Journal of Ad Hoc and Ubiquitous Computing*, Vol. 1, No. 4, pp. 222-229, 2006.
[21] R. Battiti, M. Brunato, and A. Villani, "Statistical learning theory for location fingerprinting in wireless LANs" *Tech. Rep. DIT-02-0086*, Dipartimento di Informatica e Telecomunicazioni, Universita di Trento, 2002.
[22] L. Doherty, K. Pister, and L. Ghaoui, "Convex Position Estimation in Wireless Sensor Networks", *Proceedings of the 20th Annual Joint Conference of the IEEE Computer and Communications Societies (INFOCOM 2001)*, Apr 22-26, 2001, Anchorage, AK, USA.
[23] S. Capkun, M. Hamdi and J. Hubaux, "GPS-free positioning in mobile ad-hoc networks", *Proceedings of the 34th Annual Hawaii International conference on System Sciences*, Jan 3-6, 2001, Hawaii, USA.
[24] Jeffrey Tay, Vijay R. Chandrasekhar and Winston K.G. Seah, "Selective Iterative Multilateration for Hop Count Based Localization in Wireless Sensor Networks". *Proceedings of the 7th International Conference on Mobile Data Management (MDM'06)*, May 13-16, Nara, Japan, 2006.
[25] Vijay R. Chandrasekhar, Z.A. Eu, Winston K.G. Seah and Arumugam P. Venkatesh, "Experimental Analysis of Area Localization for Wireless Sensor Networks", *Proceedings of the IEEE Wireless Communications and Networking Conference* (WCNC2007), Mar 11-15, 2007, Hong Kong.
[26] D. Lymberopoulos, Q. Lindsey and A. Savvides, "An Empirical Analysis of Radio Signal Strength Variability in IEEE 802.15.4 Networks using Monopole Antennas", *Proceedings of the Second European Workshop on Sensor Networks (EWSN 2006)*, Feb 13-15, 2006, ETH, Zurich, Switzerland.

[27] Eddie B.S. Tan, J.G. Lim, Winston K.G. Seah and S.V. Rao, 'On the Practical Issues in Hop Count Localization of Sensors in a Multihop Network', *Proceedings of the 63rd IEEE Vehicular Technology Conference* (VTC2006-Spring), May 8-10, 2006, Melbourne, Victoria, Australia.
[28] K. Lorincz and M. Welsh, "Motetrack: A Robust, Decentralized Approach to RF-Based Location Tracking", *Proceedings of the International Workshop on Location- and Context-Awareness* (LoCA2005), May 12-13, 2005, Munich, Germany.
[29] K. Yedavalli, B. Krishnamachari, S. Ravula and B. Srinivasan, "Ecolocation: A Sequence Based Technique for RF Localization in Wireless Sensor Networks", *Proceedings of Information Processing in Sensor Networks* (IPSN2005), Apr 25-27, 2005, Los Angeles, CA, USA.
[30] R. Stoleru and J. A. Stankovic, "Probability Grid: A Location Estimation Scheme for Wireless Sensor Networks", *Proceedings of Sensor and Ad Hoc Communications and Networks Conference* (SECON2004), Oct 4-7, 2004, Santa Clara, CA, USA.
[31] Scalable Networks Inc., QualNet Simulator, available from: http://www.scalable-networks.com/.
[32] Crossbow Technology Inc., homepage: http://www.xbow.com.
[33] V.A. Pillai, Winston K.G. Seah and Y.H. Chew, "Improved Area Estimates for Localization in Wireless Sensor Networks", *Proceedings of the 16th Asia-Pacific Conference on Communications* (APCC), Auckland, New Zealand, Nov 1-3, 2010.

Part 3

Information and Data Processing Technologies

18

Data Fusion Approach for Error Correction in Wireless Sensor Networks

Maen Takruri
Centre for Real-Time Information Networks (CRIN)
University of Technology, Sydney
Australia

Subhash Challa
NICTA Victoria Research Laboratory
The University of Melbourne
Australia

1. Introduction

Wireless Sensor Networks (WSNs) emerged as an important research area (Estrin et al., 2001). This development was encouraged by the dramatic advances in sensor technology, wireless communications, digital electronics and computer networks, enabling the development of low cost, low power, multi-functional sensor nodes that are small in size and can communicate over short distances (Akyildiz et al., 2002). When they work as a group, these nodes can accomplish far more complex tasks and inferences than more powerful nodes in isolation. This led to a wide spectrum of possible military and civilian applications, such as battlefield surveillance, home automation, smart environments and forest fire detection.

On the down side, the wireless sensors are usually left unattended for long periods of time in the field, which makes them prone to failures. This is due to either sensors running out of energy, ageing or harsh environmental conditions surrounding them. Besides the random noise, these cheap sensors tend to develop drift in their measurements as they age. We define the drift as a slow, unidirectional long-term change in the sensor measurement. This poses a major problem for end applications, as the data from the network becomes progressively useless. An early detection of such drift is essential for the successful operation of the sensor network. In this process, the sensors, which otherwise would have been deemed unusable, can continue to be used, thus prolonging the effective life span of the sensor network and optimising the cost effectiveness of the solutions.

A common problem faced in large scale sensor networks is that sensors can suffer from bias in their measurements (Bychkovskiy et al., 2003). The bias and drift errors (systematic errors) have a direct impact on the effectiveness of the associated decision support systems. Calibrating the sensors to account for these errors is a costly and time consuming process. Traditionally, such errors are corrected by site visits where an accurate, calibrated sensor is used to calibrate other sensors. This process is manually intensive and is only effective when the number of sensors deployed is small and the calibration is infrequent. In a large scale sensor

network, constituted of cheap sensors, there is a need for frequent recalibration. Due to the size of such networks, it is impractical and cost prohibitive to manually calibrate them. Hence, there is a significant need for auto calibration (Takruri & Challa, 2007) in sensor networks.
The sensor drift problem and its effects on sensor inferences is addressed in this work under the assumption that neighbouring sensors in a network observe correlated data, i.e., the measurements of one sensor is related to the measurements of its neighbours. Furthermore, the physical phenomenon that these sensors observe also follows some spatial correlation. Moreover, the faults of the neighbouring nodes are likely to be uncorrelated (Krishnamachari & Iyengar, 2004). Hence, in principle, it is possible to predict the data of one sensor using the data from other closely situated sensors (Krishnamachari & Iyengar, 2004; Takruri & Challa, 2007). This predicted data provides a suitable basis to correct anomalies in a sensor's reported measurements. At this point, it is important to differentiate between the measurement of the sensor or the reported data which may contain bias and/or drift, and the corrected reading which is evaluated by the error correction algorithms. The early detection of anomalous data enables us not only to detect drift in sensor readings, but also to correct it.
In this work, we present a general and comprehensive framework for detecting and correcting both the systematic (drift and bias) and random errors in sensor measurements. The solution addresses the sparse deployment scenario of WSNs. Statistical modelling rather than physical modelling is used to model the spatio-temporal cross correlations among sensors' measurements. This makes the framework presented here likely to be applicable to most sensing problems with minor changes. The proposed algorithm is tested on real data obtained from the Intel Berkeley Research Laboratory sensor deployment. The results show that our algorithm successfully detects and corrects drifts and noise developed in sensors and thereby prolongs the effective lifetime of the network.
The rest of the chapter is organised as follows. Section 2 presents the related work on error detection and correction in WSNs literature. We present our network structure and the problem statement in Section 3. Sections 4 and 5 formulate the Support Vector Regression and Unscented Kalman Filter framework for error correction in sensor networks. Section 6 evaluates the proposed algorithm using real data and section 7 concludes with future work.

2. Related Work

The sensor bias and drift problems and their effects on sensor inferences have rarely been addressed in the sensor networks literature. In contrast, the bias correction problem has been well studied in the context of the multi-radar tracking problem. In the target tracking literature the problem is usually referred to as the *registration problem* (Okello & Challa, 2003; Okello & Pulford, 1996). When the same target is observed by two sensors (radars) from two different angles, the data from those two sensors can be fused to estimate the bias in both sensors. In the context of image processing of moving objects, the problem is referred to as *image registration*, which is the process of overlaying two or more images of the same scene taken at different times, from different viewpoints, and/or by different cameras. It geometrically aligns two images: the reference and sensed images (Brown, 1992). Image registration is a crucial step in all image analysis tasks in which the final information is gained from the combination of various data sources like in image fusion (Zitova & Flusser, 2003). That is, in order to fuse two sensor readings, in this case two images, the readings must first be put into a common coordinates systems before being fused. The essential idea brought forth by the solution to the registration problem is the augmentation of the state vector with the bias components. In other words, the problem is enlarged to estimate not only the states of the targets, using the radar measurements for example, but also the biases of the radars. This is the approach we consider in the case of sensor networks. Target tracking filters, in conjunction with sensor drift models are used to estimate the sensor drift in real time. The estimate is used for correction and as a feedback to the next estimation step. The presented methodology is a robust framework for auto calibration of sensors in a WSN.

A straightforward approach to bias calibration is to apply a known stimulus to the sensor network and measure the response. Then comparing the ground truth input to the response will result in finding the gain and offset for the linear drifts case (Hoadley, 1970). This method is referred to by (Balzano & Nowak, 2007) as non-blind calibration since the ground truth is used to calibrate the sensors. Another form of non-blind calibration is manually calibrating a subset of sensors in the sensor network and then allowing the non-calibrated sensors to adjust their readings based on the calibrated subset. The calibrated subset in this context form a reference point to the ground truth (Bychkovskiy, 2003; Bychkovskiy et al., 2003). The above mentioned methods are impractical and cost prohibitive in the case of large scale sensor networks.

The calibration problem of the sensor network was also tackled by (Balzano & Nowak, 2007; 2008) in a different fashion. They stated that after sensors were calibrated to the factory settings, when deployed, their measurements would differ linearly from the ground truth by certain gains and offsets for each sensor. They presented a method for estimating these gains and offsets using subspace matching. The method only required routine measurements to be collected by the sensors and did not need ground truth measurements for comparison. They referred to this problem as blind calibration of sensor networks. The method did not require dense deployment of the sensors or a controlled stimulus. However, It required that the sensor measurements are at least slightly correlated over space i.e. the network over sampled the underlying signals of interest. The theoretical analysis of their work did not take noise into consideration and assumed linear calibration functions. Therefore, the solution might not be robust in noisy conditions and will probably result in wrong estimates if applied in a scenario where the relationship between the measurement and the ground truth is nonlinear. The evaluations they presented showed that the method worked better in a controlled environment.

An earlier work on blind calibration of sensor nodes in a sensor network was presented in (Bychkovskiy, 2003; Bychkovskiy et al., 2003). They assumed that the sensors of the network under consideration were sufficiently densely deployed that they observed the same phenomenon. They used the temporal correlation of signals received by neighbouring sensors when the signals were highly correlated to derive a function relating the bias in their amplitudes. Another method for calibration was considered by (Feng et al., 2003). They used geometrical and physical constraints on the behaviour of a point light source to calibrate light sensors without the need for comparing the measurement with an accurate sensor (ground truth). They assumed that the light sensors under consideration suffered form a constant bias with time.

The authors in (Whitehouse & Culler, 2002; 2003) argued that calibrating the sensors in sensor networks is a problematic task since it comprises large number of sensor that are deployed in partially unobservable and dynamic environments and may themselves be unobservable. They suggested that the calibration problem in sensor/actuator networks should be expressed as a parameter estimation problem on the network scale. Therefore, instead of calibrating each sensor individually to optimise its measurement, the sensors of the network are calibrated to optimise the overall response of the network. The joint calibration method they presented calibrated sensors in a controlled environment. The method was tested on an ad-hoc localisation system and resulted in reducing the error in the measured distance from 74.6% to 10.1%. The authors claimed that the joint calibration method could be transformed into an auto calibration technique for WSNs in an uncontrolled environment i.e. some form of blind calibration where the value of the ground truth measurement (here the distance) is unknown. They formulated the problem as a quadratic programming problem. Similar to (Whitehouse & Culler, 2002; 2003), blindly calibrating range measurements for localisation purposes between sensors using received signal strength and/or time delay were considered in (Ihler et al., 2004; Taylor et al., 2006).

The work of (Elnahrawy & Nath, 2003) aimed to reduce the uncertainties in the sensors readings. It introduced a Bayesian framework for online cleaning of noisy sensor data in WSNs. The solution was designed to reduce the influence of random errors in sensors measurements on the inferences of the sensor network but did not address systematic errors. The framework was applied in a centralised fashion and on synthetic data set and showed promising results.

The author of (Balzano, 2007) described a method for in-situ blind calibration of moisture sensors in a sensor network. She used the Ensemble Kalman Filter (EnKF) to correct the values measured by the sensors, or in other words, to estimate the true moisture at each sensor. The state equation was governed by a physical model of moisture used in environmental and civil engineering and the measurements were assumed to be related to the real state by a certain offset and gain. The state (moisture) vector was augmented with the calibration parameters (gain and offset) and then the gains and offsets were estimated to recover the correct state from the measurements.

Another method for detecting a single sensor failure that is a part of an automation system (a sort of wired sensor network) was proposed by (Sallans et al., 2005). Using the incoming sensor measurement, a model for the sensor behaviour was constructed and then optimised using an online maximum likelihood algorithm. Sensor readings were compared with the model. In event that the sensor reading deviated from the modelled value by a certain threshold, the system labelled this sensor as faulty. On the other hand, when the difference was small, the system automatically adapted to it. This made the system capable of adapting to slow drifts.

A neural network-based instrument surveillance, calibration and verification system for a chemical processing system (a sort of wired sensor network) was introduced in (Xu et al., 1998). The neural network used the correlation in the measurements of the interconnected sensors to correct the drifting sensors readings. The sensors that were discovered to be faulty were replaced automatically with the best neural network estimate thus restoring the correct signal. The performance of the system depended on the degree of correlation of the sensors readings. It was also found that the robustness of the monitoring network was related to the amount of signal redundancies and the degree of signal correlations. The authors concluded that their system could be used to continuously monitor sensors for faults in a plant. However, they noted that retraining the entire network may be necessary for major changes in plant operating conditions

Support Vector Machines (SVM) were used in (Rajasegarar et al., 2007) to detect anomalies and faulty sensors of a sensor network. The data reported by the sensors were mapped from the input space (the space where the features are observed) to the feature space (higher dimensional space) using kernels. The projected data were then classified into clusters and the data points that did not lie in a normal data cluster were considered anomalous. The sensor that always reported anomalous data was considered faulty.

The authors of (Guestrin et al., 2004) presented a method for in-network modelling of sensor data in a WSN. The method used kernel linear regression to fit functions to the data measured by the sensors along a time window. The basis functions used were known by the sensors. Therefore, if a sensor knew the weights of its neighbour, it would be able to answer any query about the neighbour within the time window. So instead of sending the measured data of the whole window period from one sensor to another, sending the weights would considerably reduce the communication overhead. This was one of the aims of the method. The other aim was to enable any sensor in the network to estimate the measured variable at points within the network where there were no sensors using the spatial correlation in the network. An application for the introduced method is computing contour levels of sensor values as in (Nowak & Mitra, 2003). Even that the work in (Guestrin et al., 2004) considered the unreliable communication between distant sensors and the noise in sensor readings, it did not address the systematic errors (drift and bias) which can build up along time and propagate among sensors causing the continuously modelled functions to produce estimates that deviate from the ground truth values.

In addition to its superb capabilities in generalisation, function estimation and curve fitting, Support Vector Machines (SVR) is used in other applications such as forecasting and estimating the physical parameters of a certain phenomenon. In (Wang et al., 2003), SVR was utilised in medical imaging for nonlinear estimation and modelling of functional magnetic resonance imaging (fMRI) data to reflect their intrinsic spatio-temporal autocorrelations. Moreover, SVR was used in (Gill et al., 2006) to successfully predict the ground moisture at a site using meteorological parameters such as relative humidity, temperature average solar radiation, and moisture measurements collected from spatially distinct locations. A similar experiment to predict ground moisture was reported in (Gill et al., 2007). In addition to using the SVR to predict the moisture measurements ahead in time, they introduced the use of an EnKF to correct or match the predicted values with the real measurements at certain points of time (whenever measurements are available) to keep the predicted values close to the measurements taken on site and eventually reduce the prediction error.

The above survey, has introduced most of the work undertaken in the area of fault detection and fault detection/correction in wireless sensor networks. This research approaches the problem in a more comprehensive manner resulting in several novel solutions for detecting and correcting drift and bias in WSNs. It does not assume linearity of the sensor faults (drift) with time and addresses smooth drifts and drifts with sudden changes and jumps. It also considers the cases when the sensors of the network are densely and sparsely (non densely) deployed. Moreover, it introduces recursive online algorithms for the continuous calibration of the sensors. In addition to all of that, the solutions presented are decentralised to reduce the communication overhead. Some of the papers that have arisen from this research are surveyed below: (Takruri & Challa, 2007) introduced the idea of drift aware wireless sensor network which detects and corrects sensors drifts and eventually extends the functional life time of the network. A formal statistical procedure for tracking and detecting smooth sensors drifts using decentralised Kalman Filter (KF) algorithm in a densely deployed network was introduced in (Takruri, Aboura & Challa, 2008; Takruri, Challa & Chacravorty, 2010). The sensors of the network were close enough to have similar temperature readings and the average of their measurements was taken as a sensible estimate to be used by each sensor to self-assess. As an upgrade for this work, the KFs were replaced in (Takruri, Challa & Chacravorty, 2010; Takruri, Challa & Chakravorty, 2008) by interacting multiple model (IMM) based filters to deal with unsmooth drifts. A more general solution was considered in (Takruri, Rajasegarar, Challa, Leckie & Palaniswami, 2008). The assumption of dense sensor deployment was relaxed. Therefore, each sensor in the network ran an SVR algorithm on its neighbours'

corrected readings to obtain a predicted value for its measurements. It then used this predicted data to self-assess its measurement, detect (track) its drift using a KF and then correct the measurement.
A more robust and reliable decentralised algorithm for online sensor calibration in sparsely deployed wireless sensor networks was presented in (Takruri, Rajasegarar, Challa, Leckie & Palaniswami, 2010). The algorithm represents a substantial improvement of method in (Takruri, Rajasegarar, Challa, Leckie & Palaniswami, 2008). By using an Unscented Kalman Filter (UKF) instead of the KF, the bias in the estimated temperature (system error) was dramatically reduced compared to that reported in (Takruri, Rajasegarar, Challa, Leckie & Palaniswami, 2008). This is justified by the fact that UKF is a better approximation method for propagating the mean and covariance of a random variable through a nonlinear transformation than the KF is. The algorithm was then upgraded in (Takruri et al., 2009) to become more adaptable for under sampled sensor measurements and consequently, allowing for reducing the communication between sensors and maintain the calibration. This led to reducing the energy consumed from the batteries. Unlike the work in (Balzano, 2007), statistical modelling rather than physical relations was used to model the spatio-temporal cross correlations among the sensors measurements. Similar to (Takruri, Rajasegarar, Challa, Leckie & Palaniswami, 2008), statistical modelling was achieved by applying SVR. This in principal made the framework applicable to most sensing problems without needing to find the physical model that describes the phenomenon under observation, and without the need to abide by the constraints of that physical formulation. The algorithm runs recursively and is fully decentralised. It does not make assumptions regarding the linearity of the drifts as opposed the work in (Balzano & Nowak, 2007). The implementation of the algorithm on real data obtained from the Intel Berkeley research laboratory (IBRL) showed a great success in detecting and correcting sensors drifts and extending the functional lifetime of the network.
In this chapter, we present another model for error detection and correction in sparsely deployed WSNs. Similar to (Takruri, Rajasegarar, Challa, Leckie & Palaniswami, 2010), SVR is used to model the spatio-temporal cross correlations among the sensors measurements to obtain a predicted value for the actual ground truth measurements and Unscented Kalman Filter is used to estimate the corrected sensors readings. However, both algorithms are substantially different in terms of the training data set used for training the SVR framework, the dynamic equations that govern the models and the estimated variables. The state transition function in the new model is taken to be linear resulting in much lower computational complexity than (Takruri, Rajasegarar, Challa, Leckie & Palaniswami, 2010) and comparable results.

3. Network Structure and Problem Statement

Consider a wireless sensor network with a large number of sensors distributed randomly in a certain area of deployment such as the one shown in Figure 1. The sensors are grouped in clusters (sub-networks) according to their spatial proximity. Each sensor measures a phenomenon such as ambient temperature, chemical concentration, noise or atmospheric pressure. The measurement, say temperature, is considered to be a function of time and space. As a result, the measurements of sensors that lie within the same cluster can be different from each other. For example, a sensor closer to a heat source or near direct sunlight will have readings higher than those in a shaded region or away from the heat source. An example of a cluster is shown using a circle in Figure 1. The sensors within the cluster are considered to be capable of communicating their readings among each other.

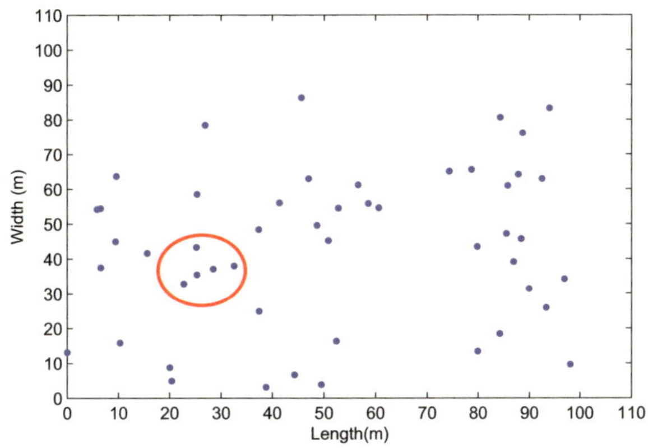

Fig. 1. Wireless sensor area with encircled sub-network

As time progresses, some nodes may start experiencing drift in their readings. If these readings are collected and used from these nodes, they will cause the users of the network to draw erroneous conclusions. After some level of unreliability is reached, the network inferences become untrustworthy. Consequently, the sensor network becomes useless. In order to mitigate this problem of drift, each sensor node in the network has to detect and correct its own drift using the feedback obtained from its neighbouring nodes. This is based on the principle that the data from nodes that lie within a cluster are correlated, while their faults or drifts instantiations are likely to be uncorrelated. The ability of the sensor nodes to auto-detect and correct their drifts helps to extend the effective (useful) lifetime of the network. In addition to the drift problem, we also consider the inherent bias that may exist within some sensor nodes. There is a distinct difference between these two types of errors. The former changes with time and often becomes accentuated, while the latter, is considered to be a constant error from the beginning of the operation. This error is usually caused by a possible manufacturing defect or a faulty calibration.

The sensor drift that we consider in this work is slow smooth drift that we model as linear and/or exponential function of time. It is dependent on the environmental conditions, and strongly relate to the manufacturing process of the sensor. It is highly unlikely that two electronic components fail in a correlated manner unless they are from the same integrated circuit. Therefore, we assume that the instantiations of drifts are different from one sensor to another in a sensor neighbourhood or a cluster. Figure 2 shows examples of the theoretical models for smooth drift.

Consider a sensor sub-network that consists of n sensors deployed randomly in a certain area of interest. Without loss of generality, we choose a sensor network measuring temperature, even though this is generally applicable to all other types of sensors that suffer from drift and bias problems. Let T be the ground truth temperature. T varies with time and space. Therefore, we denote the temperature at a certain time instance and sensor location as $T_{i,k}$ where i is the sensor number and k is the time index. At each time instant k, node i in the sub-network measures a reading $r_{i,k}$ of $T_{i,k}$. It then estimates and reports a *drift corrected* value $x_{i,k}$ to its neighbours. The corrected value $x_{i,k}$ should ideally be equal to the ground truth temperature $T_{i,k}$. If all nodes are perfect, $r_{i,k}$ will be equal to the $T_{i,k}$, and the reported values will ideally be equal to the readings, i.e., $x_{i,k} = r_{i,k}$.

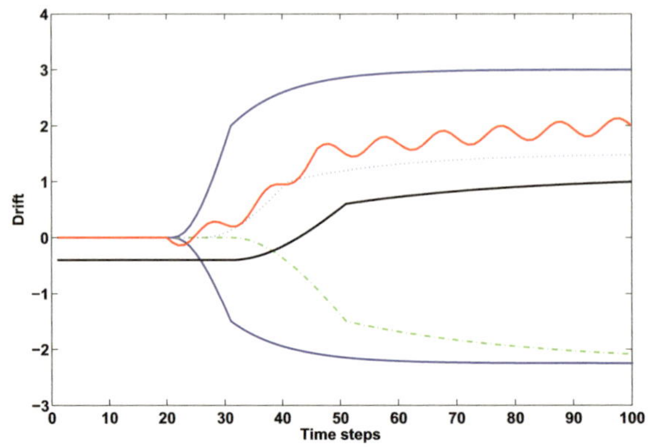

Fig. 2. Examples of smooth drifts

To estimate the corrected value $x_{i,k}$, each node i first finds a predicted value $\widetilde{x}_{i,k}$ for its temperature as a function of the corrected measurements collected from its neighbours in the previous time step using $\widetilde{x}_{i,k} = f(\{x_{j,k-1}\}_{j=1,j\neq i}^{n})$. Then it fuses this predicted value together with its measurement $r_{i,k}$ and the projected drift $d_{i,k}$ to result in an error corrected sensor measurement $x_{i,k}$. In practice, each sensor reading comes with an associated random reading error (noise), and a drift $d_{i,k}$. This drift may be null or insignificant during the initial period of deployment, depending on the nature of the sensor and the deployment environment. The problem we address here is how to account for the drift in each sensor node i, using the predicted value $\widetilde{x}_{i,k}$, so that the reading $r_{i,k}$ is corrected and reported as $x_{i,k}$.
In the following sections, $\widetilde{x}_{i,k}$ is computed using a support vector regression (SVR) modelled function that takes into account the temporal and spatial correlations of the sensor measurements. In this work, SVR approximates $\widetilde{x}_{i,k}$ using the previous corrected readings of all the sensors in the neighbourhood (cluster) excluding the sensor itself $\widetilde{x}_{i,k} = f(\{x_{j,k-1}\}_{j=1,j\neq i}^{n})$.

4. Modelling and predicting measurements using Support Vector Regression

The purpose of using Support Vector Regression (SVR) is to predict the actual sensor measurements $\widetilde{x}_{i,k}$ of a sensor node i at time instant k using the corrected measurements from neighbouring sensors. The intention is that each sensor learns a model function $f(.)$ that can be used for predicting its subsequent actual (error free) measurements through out the whole period of the experiment. SVR implements this in two phases, namely the *training phase* and the *running phase*. During the training phase, sensor measurements collected during the initial deployment period (training data set) are used to model the function $f(.)$. During the *running phase*, the trained model $f(.)$ is used to predict the subsequent actual sensor measurements $\widetilde{x}_{i,k}$.
We assume that the training data (collected during the initial periods of deployment) is void of any drift and can be used for training the SVR at each node. This is a reasonable assumption in practice, as the sensors are usually calibrated before deployment to ensure that they are working in order. Similar to our work in (Takruri, Rajasegarar, Challa, Leckie & Palaniswami, 2010), we use the widely used Gaussian kernel SVR for our evaluations (Scholkopf & Smola, 2002). However, the training data set used here is slightly different in that it comprises the

corrected readings of the neigbours and does not take into consideration the corrected reading of node i itself. The training data set at each node i is given by $X_s = (TrX, TrZ)$, where $TrX = \{x_{j,k-1} : j = 1...n, k = 1...m, j \neq i\}$, $TrZ = \{x_{i,k} : k = 1...m\}$ and m is number of training data vectors. A detailed explanation of our implementation of the SVR can be found in (Takruri, Rajasegarar, Challa, Leckie & Palaniswami, 2010).

The model obtained via SVR training is then used during the *running phase* for predicting subsequent actual measurements $\widetilde{x}_{i,k}$. The difference between the sensor reading $r_{i,k}$ and the SVR modelled value $\widetilde{x}_{i,k}$, $y^{(2)}_{i,k}$, which we refer to as the drift measurement of node i at time instant k, is used by an *Unscented Kalman Filter* together with $r_{i,k}$ to estimate the corrected reading $x_{i,k}$ and the drift $d_{i,k}$ as will be shown in the following section.

5. Iterative measurement estimation and correction using an SVR-UKF framework

The solution to the smooth drift problem consists of the following iterative steps. At stage k, a reading $r_{i,k}$ is made by node i. The node also has a prediction for its corrected measurement (actual temperature at this sensor), $\widetilde{x}_{i,k} = f(\{x_{j,k-1}\}^n_{j=1,j\neq i})$, as a function of the corrected measurements of all neighbouring sensors in the cluster from the previous time step. Using this predicted value ($\widetilde{x}_{i,k}$) together with $r_{i,k}$, the corrected reading $x_{i,k}$ and the drift value $d_{i,k}$ are estimated. The node then sends the corrected sensor value $x_{i,k}$ to its neighbours. After that, each node collects the neighbourhood corrected measurements and computes $\widetilde{x}_{i,k}$ and so on. It is important here to emphasise that our main objective is to estimate $x_{i,k}$ the corrected reading which represents our estimate for the ground truth value $T_{i,k}$ at node i. Assuming that $x_{i,k}$ and $d_{i,k}$ change slowly with time the dynamics of $x_{i,k}$ and $d_{i,k}$ are mathematically described by:

$$x_{i,k} = x_{i,k-1} + \eta^{(1)}_{i,k} \quad \eta^{(1)}_{i,k} \sim N(0, Q^{(1)}_{i,k}) \tag{1}$$

$$d_{i,k} = d_{i,k-1} + \eta^{(2)}_{i,k} \quad \eta^{(2)}_{i,k} \sim N(0, Q^{(2)}_{i,k}) \tag{2}$$

where $\eta^{(1)}_{i,k}$ and $\eta^{(2)}_{i,k}$ are the process noises. They are taken to be uncorrelated Gaussian noises with zero means and variances $Q^{(1)}_{i,k}$ and $Q^{(2)}_{i,k}$, respectively.

The value $x_{i,k}$ is never sensed or measured. What is really measured is $r_{i,k}$, the reading of the sensor. As we argued earlier, $r_{i,k}$ deviates from $x_{i,k}$ by both systematic and random errors. The random error is taken to be a Gaussian noise $w_{i,k} \sim N(0, R_{i,k})$ with zero mean and variance $R_{i,k}$ (measurement noise variance). The systematic error is referred to as the drift $d_{i,k}$. This leads to (3).

$$y^{(1)}_{i,k} = r_{i,k} = x_{i,k} + d_{i,k} + w_{i,k} \quad w_{i,k} \sim N(0, R_{i,k}) \tag{3}$$

We also define $y^{(2)}_{i,k}$ as the difference between the measurement $r_{i,k}$ and the SVR modelled value $\widetilde{x}_{i,k}$ and refer to $y^{(2)}_{i,k}$ as the drift measurement of node i at time instant k.

$$\begin{aligned} y^{(2)}_{i,k} &= y^{(1)}_{i,k} - f(\{x_{j,k-1}\}^n_{j=1,j\neq i}) \\ &= x_{i,k} + d_{i,k} + w_{i,k} - f(\{x_{j,k-1}\}^n_{j=1,j\neq i}) \\ &= x_{i,k} + d_{i,k} + w_{i,k} - \widetilde{x}_{i,k} \quad w_{i,k} \sim N(0, R_{i,k}) \end{aligned} \tag{4}$$

The model is expressed in vector notation as follows:

$$X_{i,k} = \begin{bmatrix} x_{i,k} \\ d_{i,k} \end{bmatrix} = \begin{bmatrix} 1 & 0 \\ 0 & 1 \end{bmatrix} \begin{bmatrix} x_{i,k-1} \\ d_{i,k-1} \end{bmatrix} + \begin{bmatrix} \eta_{i,k}^{(1)} \\ \eta_{i,k}^{(2)} \end{bmatrix} \quad (5)$$

$$Y_{i,k} = \begin{bmatrix} y_{i,k}^{(1)} \\ y_{i,k}^{(2)} \end{bmatrix} = \begin{bmatrix} 1 & 1 \\ 1 & 1 \end{bmatrix} \begin{bmatrix} x_{i,k} \\ d_{i,k} \end{bmatrix} + \begin{bmatrix} w_{i,k} \\ w_{i,k} \end{bmatrix} - \begin{bmatrix} 0 \\ \widetilde{x}_{i,k} \end{bmatrix} \quad (6)$$

The noise component associated with $X_{i,k}$ is Gaussian with mean vector $\mu_{X_{i,k}} = [0 \;\; 0]^T$ and covariance matrix $Qx_{i,k} = \begin{bmatrix} Q_{i,k}^{(1)} & 0 \\ 0 & Q_{i,k}^{(2)} \end{bmatrix}$. The noise component associated with $Y_{i,k}$ has a mean vector $\mu_{Y_{i,k}} = [0 \;\; 0]^T$ and covariance matrix $Ry_{i,k} = \begin{bmatrix} R_{i,k} & R_{i,k} \\ R_{i,k} & R_{i,k} \end{bmatrix}$ which indicates that it is not White Gaussian. The system is clearly observable when $\widetilde{x}_{i,k} = x_{i,k}$, i.e. when $\widetilde{x}_{i,k}$ is a true, bias free, representation of $x_{i,k}$ and the difference between $x_{i,k}$ and $\widetilde{x}_{i,k}$ is zero.

Since the noise component associated with $Y_{i,k}$ is not White Gaussian, the KF cannot be used (Lu et al., 2007) to estimate $x_{i,k}$ and $d_{i,k}$. Another filter that can be used for solving such a problem is the Particle Filter. Unfortunately, the high computational complexity of the Particle Filter makes it unsuitable for the use in WSNs, where the sensors are limited in their energy and computational capabilities. A better alternative is to use the UKF. The Unscented Transformation (UT) was introduced by Julier et al. in (Julier et al., 1995) as an approximation method for propagating the mean and covariance of a random variable through a nonlinear transformation. This method was used to derive UKF in (Julier & Uhlmann, 1997). UKF can deal with versatile and complicated nonlinear sensor models and non-Gaussian noise that are not necessarily additive (Challa et al., 2008) with a comparable computational complexity to the Extended Kalman Filter (EKF) (Wan & van der Merwe, 2000). It also outperforms the EKF since it provides better estimation for the posterior mean and covariance to the third order Taylor series expansion when the input is Gaussian, whereas, the EKF, only achieves the first order Taylor series expansion (Wan & van der Merwe, 2000). Below, we explain the UKF algorithm in detail.

The UT as mentioned before is a method for finding the statistics of a random variable $Z = g(X)$ which undergoes nonlinear transformation. Let X of dimension L be the random variable that is propagated through the nonlinear function $Z = g(X)$. Assume that X has a mean $\hat{X}$ and a covariance P. According to (Challa et al., 2008), to find the statistics of Z using the scaled unscented transformation, which was introduced in (Julier, 2002), the following steps must be followed: First, $2L + 1$ (where L is the dimension of vector X) weighted samples or *sigma points* $\sigma_i = \{\mathcal{W}_i, \mathcal{X}_i\}$ are deterministically chosen to completely capture the true mean and covariance of the random variable X. Then, the sigma points are propagated through the function $g(X)$ to capture the statistics (mean and covariance) of Z. A selection scheme that satisfies the requirement is given below:

$$\begin{aligned}
\mathcal{X}_0 &= \hat{X}, \;\; \mathcal{W}_0^m = \frac{\lambda}{\lambda + L} \\
\mathcal{W}_0^c &= \frac{\lambda}{\lambda + L} + (1 - \alpha^2 + \beta) \\
\mathcal{X}_i &= \hat{X} + (\sqrt{(L+\lambda)P})_i, \;\; \mathcal{W}_i = \frac{1}{2(\lambda + L)} \\
\mathcal{X}_{L+i} &= \hat{X} - (\sqrt{(L+\lambda)P})_i, \;\; \mathcal{W}_{L+i} = \frac{1}{2(\lambda + L)}
\end{aligned} \tag{7}$$

where $i = 1, ..., L$ and $\lambda = \alpha^2(L + \kappa) - L$ is a scaling parameter. α determines the spread of the sigma points around the mean $\hat{X}$ and is usually set to a small positive value (*e.g.*, 0.001). κ is a secondary scaling parameter which is usually set to 0, and β is used to incorporate prior knowledge of the distribution of X. The optimal value of β for a Gaussian distribution is $\beta = 2$ as stated in (Wan & van der Merwe, 2000). The term $(\sqrt{(L+\lambda)P})_i$ is the ith row of the matrix square root of matrix $(L + \lambda)P$. In our work here α, κ and β are taken to be equal to 0.001, 0, 2, respectively. The UKF is used to estimate $X_{i,k}$ for sensor i at time step k. The dimension L of $X_{i,k}$ is equal to 2. This means that we only have five sigma points for each node i. The steps of the UKF algorithm are given below as in (Challa et al., 2008):

Let $\hat{X}_{i,k-1|k-1}$ be the prior mean of the state variable and $P_{i,k-1|k-1}$ be the associated covariance for node i. To simplify the notation we write the prior mean of the state variable and the associated covariance as $\hat{X}_{k-1|k-1}$ and $P_{k-1|k-1}$ (without showing the sensor number i) keeping in mind that they refer to a certain sensor node i. This also applies for all the other parameters we use in describing the UKF algorithm.

The sigma points are calculated from (7) and then propagated through the state equation function $g(.)$. This results in $\mathcal{X}_{0,k|k-1}, \mathcal{X}_{1,k|k-1}, \mathcal{X}_{2,k|k-1}, \mathcal{X}_{3,k|k-1}$ and $\mathcal{X}_{4,k|k-1}$ as shown in (8).

$$\mathcal{X}_{k|k-1} = g(\mathcal{X}_{k-1}) = \mathcal{X}_{k-1} \tag{8}$$

The predicted mean and covariance of the state variable are given by (9) and (10), respectively.

$$\hat{X}_{k|k-1} = \mathcal{W}_0^m \mathcal{X}_{0,k|k-1} + \sum_{i=1}^{2L} \mathcal{W}_i \mathcal{X}_{i,k|k-1} \tag{9}$$

$$\begin{aligned}
P_{k|k-1} &= \mathcal{W}_0^c(\mathcal{X}_{0,k|k-1} - \hat{X}_{k|k-1})(\mathcal{X}_{0,k|k-1} - \hat{X}_{k|k-1})^T \\
&+ \sum_{i=1}^{2L} \mathcal{W}_i(\mathcal{X}_{i,k|k-1} - \hat{X}_{k|k-1})(\mathcal{X}_{i,k|k-1} - \hat{X}_{k|k-1})^T + Qx_k
\end{aligned} \tag{10}$$

The propagated sigma points are then passed through the measurement function $h(.)$ as shown in (11).

$$\mathcal{Y}_{k|k-1} = h(\mathcal{X}_{k|k-1}) \tag{11}$$

Then the predicted mean and covariance of each sensor measurement are given by (12) and (13), respectively.

$$\hat{Y}_{k|k-1} = \mathcal{W}_0^m \mathcal{Y}_{0,k|k-1} + \sum_{i=1}^{2L} \mathcal{W}_i \mathcal{Y}_{i,k|k-1} \tag{12}$$

$$P_{Y_kY_k} = \mathcal{W}_0^c(\mathcal{Y}_{0,k|k-1} - \hat{Y}_{k|k-1})(\mathcal{Y}_{0,k|k-1} - \hat{Y}_{k|k-1})^T + \sum_{i=1}^{2L} \mathcal{W}_i(\mathcal{Y}_{i,k|k-1} - \hat{Y}_{k|k-1})(\mathcal{Y}_{i,k|k-1} - \hat{Y}_{k|k-1})^T + Ry_k \quad (13)$$

The cross covariance of the predicted state and sensor measurement is found by (14).

$$P_{X_kY_k} = \mathcal{W}_0^c(\mathcal{X}_{0,k|k-1} - \hat{X}_{k|k-1})(\mathcal{Y}_{0,k|k-1} - \hat{Y}_{k|k-1})^T + \sum_{i=1}^{2L}[\mathcal{W}_i(\mathcal{X}_{i,k|k-1} - \hat{X}_{k|k-1})(\mathcal{Y}_{i,k|k-1} - \hat{Y}_{k|k-1})^T] \quad (14)$$

where

$$K_k = P_{X_kY_k}P_{Y_kY_k}^{-1} \quad (15)$$

The updated posterior mean and covariance of the state are then estimated by (16) and (17), respectively.

$$\hat{X}_{k|k} = \hat{X}_{k|k-1} + K_k(Y_k - \hat{Y}_{k|k-1}) \quad (16)$$

$$P_{k|k} = P_{k|k-1} + K_kP_{Y_kY_k}K_k^T \quad (17)$$

where $\hat{X}_{k|k}$ and $P_{k|k}$ are the mean and covariance of the state of node i at time step k.
Figure 3 shows a block diagram of our drift correction algorithm. It clearly summarises the stages of the error detection and correction framework in one of the nodes in the cluster. The steps of the algorithm are stated below:

Decentralised error correction algorithm using the SVR-UKF framework
At time step k

- Each node i finds its predicted corrected measurement $\tilde{x}_{i,k}=f(\{x_{j,k-1}\}_{j=1,j\neq i}^{n})$.
- Each node i obtains its reading $y_{i,k}^{(1)} = r_{i,k}$.
- Each node i calculates the drift measurement $y_{i,k}^{(2)}$.
- Each node i finds the sigma points $\sigma_i = \{\mathcal{W}_i, \mathcal{X}_i\}$ from $\hat{X}_{i,k-1|k-1} = [\hat{x}_{i,k-1|k-1} \; \hat{d}_{i,k-1|k-1}]^T$.
- For each node i, the sigma points are propagated through the state equation function $g(.)$.
- The UKF estimates the corrected measurement and the drift using (9)-(17) and then sends the result to the neighbouring nodes.
- The algorithm reiterates.

6. Evaluation

Our aim is to evaluate the ability of our proposed framework to correct the drift experienced in sensor nodes and to extend the functional life of the sensor network. The data in our evaluation are a set of real sensor measurements gathered from a deployment of wireless sensors in the IBRL (2006, Accessed on 07/09/2006).
In 2004, a set of wireless sensors with 55 sensor nodes (including a gateway node) were deployed in the IBRL lab for monitoring the lab environment (refer to Figure 4). They recorded

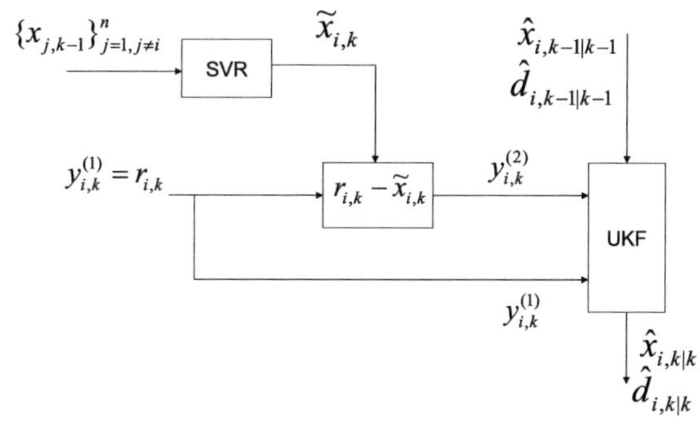

Fig. 3. The SVR-UKF Measurement correction framework at node i.

temperature, humidity, light and voltage measurements at 30 seconds intervals during the period starting from 28th February 2004 to 5th April 2004.

The data from the sensor nodes are re-sampled at seven minute intervals and the first 2000 samples are used for our evaluation purposes. This corresponds to the data collected during a ten day period from 28th February 2004 to 9th March 2004. We use the first 1000 samples (this corresponds to the first five days' data) as the training set for use in the *training phase*. An exponential drift is introduced to the real data in each node, starting randomly after the first 1000 samples. The data after 1000 samples and up to 2000 samples are used in the *running phase* for testing our algorithm for drift correction. These samples correspond to the next five days of the IBRL data. Temperature measurements are used in all our evaluations.

We formed a network of sensors using nodes selected from the IBRL deployment using sixteen sensor nodes. The node IDs used are {1,2,3,4,6,7,8,9,10,31,31,33,34,35,36,37}. Each sensor communicates only with its closest 8 neighbours.

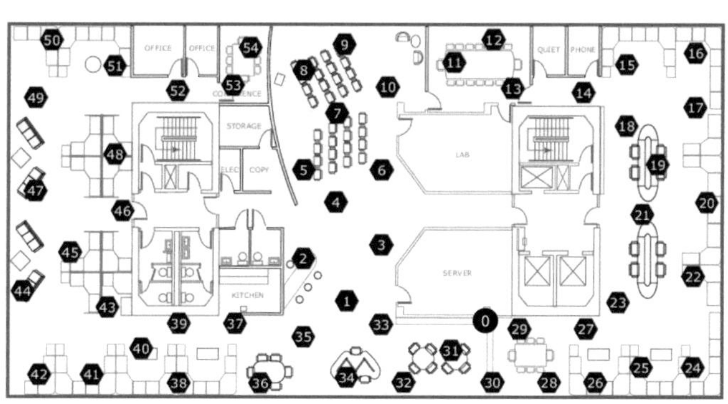

Fig. 4. Sensor nodes in the IBRL deployment. Nodes are shown in black with their corresponding node-IDs. Node 0 is the gateway node (2006, Accessed on 07/09/2006).

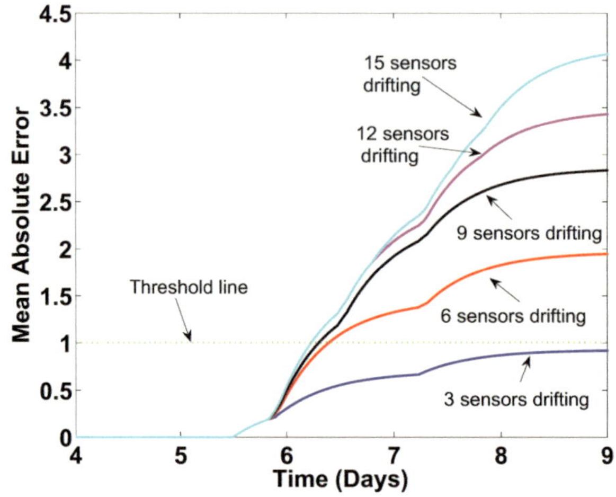

Fig. 5. Mean Absolute Error for the network without correction.

Our algorithm is implemented in MatLab, utilising the SVR toolbox from (Canu et al., 2005) and the UKF toolbox from (Särkkä & Hartikainen, 2007). For comparison purposes, we run the algorithm on two data sets. One, the data *without* the introduced drift (WOD), and the other, the data *with* drift introduced (WD). Initially, the SVR of each node is trained on the first 1000 samples of its neighbours readings.

The UKF parameters μ, κ and β are set to the default values as explained in Section 5. Through out our evaluations, we take $Q^{(1)}_{i,k} = Q^{(2)}_{i,k} = Q_{i,k}$. $Q_{i,k}$ and $R_{i,k}$ are tuned using trial and error for both cases. The values used in our evaluation are $Q_{i,k} = 0.001$ and $R_{i,k} = 0.02$. If $R_{i,k}$ is set to a high value, the estimated temperature will follow the reading (which may have drift) whereas if $R_{i,k}$ is set to a small value, the estimated temperature will not be able to follow the real temperature. Thus, it will not totally correct the error. On the other hand, a high value for $Q_{i,k}$ will result in oscillatory estimates and lead to an unstable state. Hence, a trade off has to be considered in selecting the values for $Q_{i,k}$ and $R_{i,k}$ to obtain the best results.

We have conducted two simulations using two data sets. One data set has no drifts introduced. We denote this data set by *R-WOD*, which stands for *'Readings Without Drift'* and represents the sensor measurements that only suffer from noise. The other is the same data set with drifts introduced in several scenarios. We denote the readings of this data set by *R-WD*, which stands for *'Readings With Drift'* and represents the sensor measurements that suffer from both drift and noise. The drift scenarios considered in *R-WD* are as follows: scenario 1 (SCN 1) being one node drifting, scenario 2 (SCN 2) being two nodes drifting and so on until the last scenario (SCN 16) having all nodes drifting. The resulting corrected measurements obtained when the algorithm is run on the *R-WD* data sets are denoted by *DCM-WD*, which stands for *'Drift Corrected Measurement for readings With Drift'*. Similarly, the corrected measurements obtained using data set *R-WOD* are denoted by *DCM-WOD*, which stands for *'Drift Corrected Measurement for readings WithOut Drift'*.

To evaluate the performance of our algorithm from the network's point of view, we compare the average absolute error of all the sensors of the network with and without implementing our drift correction algorithm.

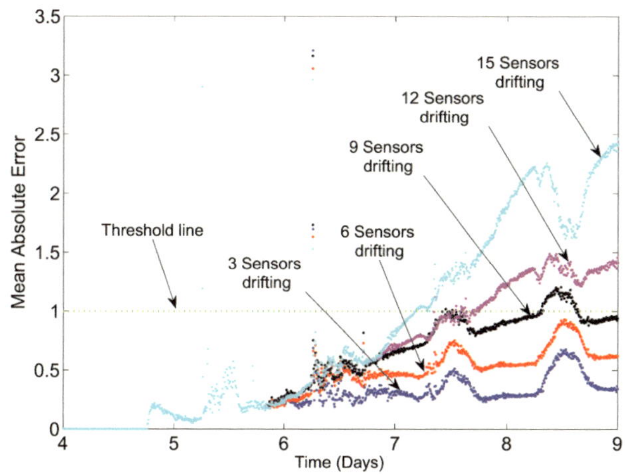

Fig. 6. Mean Absolute Error for the network with correction for 2001 samples in 10 days.

Figure 5 shows the mean absolute error between the true temperatures (R-WOD) and the values reported by the sensors (R-WD) for the whole network, for five different scenarios. The mean absolute error of the network is computed for each scenario as follows: for each node, at each instant of time, the absolute error between the true temperature (R-WOD) and the value reported by the sensors (R-WD) is computed. The average for all these nodes' absolute errors is then found. This gives the mean absolute error of the network. Similarly, the mean absolute error between the true temperatures (R-WOD) and the drift corrected measurements (DCM-WD) is calculated at each instant of time and plotted in Figure 6. By comparing Figures 5 and 6 it is evident that applying the drift correction algorithm results in less measurements error for all of the scenarios. For our evaluation purposes we assumed that the maximum mean absolute error that can be tolerated in the network is 1^oC. If the mean absolute error of the network exceeds that limit, the network is deemed to be useless or has broken down. This maximum limit is shown by a horizontal threshold line in Figures 5 and 6. The choice of the threshold is dependent on the error tolerance allowed by the application.

In Figure 5, it is evident that the curves for scenarios 6, 9, 12 and 15 cross the threshold line after the 6th day of the experiment. In contrast, in Figure 6, the curves for scenarios 6 and 9 do not cross the threshold line at all for the whole period of the experiment, while the curves of scenarios 12 and 15 cross the threshold line on the 8th day and the 7th day, respectively. This demonstrates that our algorithm extends the operational life of the network for all of the scenarios.

In another simulation we repeated the experiment after doubling the sampling rate. This resulted in 4001 samples for the 10 days experiment. Figure 7 shows the mean absolute error between the true temperatures (R-WOD) and the drift corrected measurements (DCM-WD) for the whole network for five different scenarios. The error is computed in a similar way to the method used for the Figures 5 and 6. By comparing Figures 5 and 7, it is evident that the application of the error correction algorithm results in less measurements errors for all of the scenarios.

Looking at Figures 6 and 7, we can notice that the performance when using 4001 samples is better for scenario 9 since the absolute error curve does not cross the 1^oC threshold line as it does in the 2001 samples case. This means that the operational lifetime has been extended from around 8 days in the case of 2001 samples, to more than 9 days for the case of 4001 samples. Moreover, we notice in Figure 7 that the curves for each scenario are smoother than the corresponding curves in Figure 6 and that the observed occasional jumps and peaks are smaller. The jumps in the curves are caused by the fast changes in the readings or the ambient temperature at some instants of time. An effective way of reducing the size of the jumps is to increase the sampling rate as we noticed in Figure 7. However, that would be at a cost of the increased communication overhead due to the increased data transmissions among the sensors. This means that a trade off between the smoothness of the curves and the communication overhead has to be made. Another important thing to note in both Figures 6 and 7 is that the mean absolute error of the network's estimated temperatures is proportional to the number of sensors developing drift.

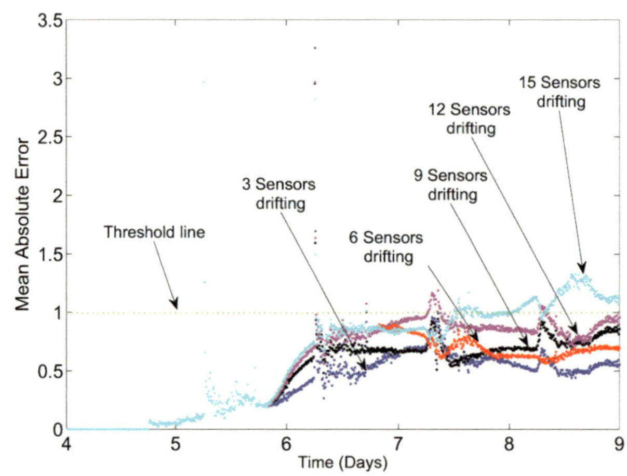

Fig. 7. Mean Absolute Error for the network with correction for 4001 samples in 10 days.

The choice of $R_{i,k}$ and $Q_{i,k}$ is crucial. It affects the accuracy of estimating the temperature and the induced error. In general, we can say that increasing $R_{i,k}$ improves the tracking of drift in the drifting sensors. However, it also increases both the induced error in drift estimation in the non drifting sensors and the fluctuations in the drifting sensors. Since, the error caused by the fast changes in temperature in the case of 4001 samples is less than that for the case of 2001 samples (as explained previously), $R_{i,k}$ is taken to be 0.05 for the case of 4001 and 0.02 for the 2001 samples case. This way, the drift tracking is improved for the 4001 samples case keeping error levels comparable to the case of 2001 samples. On the other hand, increasing $Q_{i,k}$ increases the fluctuations in the estimated drift in both drifting and non-drifting sensors and causes the response to become less stable. The $Q_{i,k}$ used in all our simulations is equal to 0.001.

It is important here to note that for comparison purposes, the data set used in the evaluations of this chapter is the same as the data set used in (Takruri, Rajasegarar, Challa, Leckie & Palaniswami, 2010) evaluations. Comparing figure 8, which is quoted from (Takruri, Rajasegarar, Challa, Leckie & Palaniswami, 2010), with figure 6, where both of them are for 2001 samples, it can be clearly noticed that SVR-UKF framework presented in (Takruri, Rajasegarar, Challa, Leckie & Palaniswami, 2010) outperforms the algorithm presented in this chapter in correcting sensor readings errors as it manages to keep the absolute error for the case of 9 sensors drifting below the threshold line. However, this is at the cost of substantially increased computational complexity. The state transition function used in this chapter is linear and the number of sigma points is 5 whereas the state transition function used in the algorithm in (Takruri, Rajasegarar, Challa, Leckie & Palaniswami, 2010) is the SVR modelled function which is highly nonliner and the number of sigma points is 19. This results in a computational complexity of several orders of magnitude.

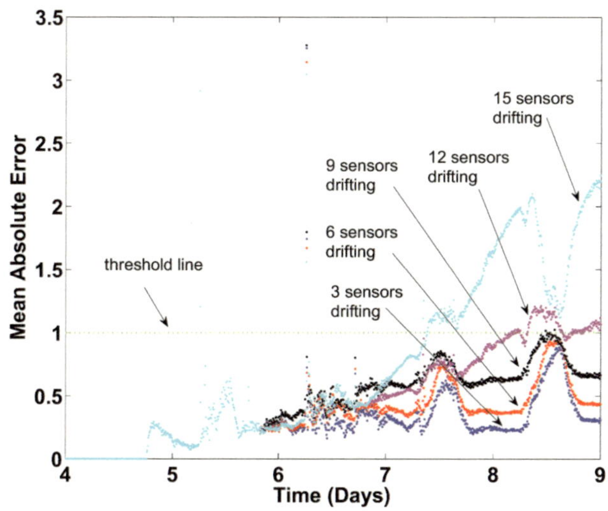

Fig. 8. Mean Absolute Error for the network with correction for 2001 samples in 10 days. Quoted from (Takruri, Rajasegarar, Challa, Leckie & Palaniswami, 2010).

The error correction performance for a sensor node in a cluster is dependent on the correlation of the actual temperature at the sensor under consideration with the actual temperatures at the neighbours. The SVR at a sensor predicts the actual temperature at the sensor $\tilde{x}_{i,k}$ using previous estimates of the neighbourhood $\{\hat{x}_{j,k-1|k-1}\}_{j=1,j\neq i}^{n}$. Therefore, low correlation will lead to a poor prediction, and result in poor estimate of the actual temperature at the sensor under consideration. In practice, the correlation among the nodes may change depending on their spatial proximity within the cluster and with the change in the observed phenomenon along time.

It can be observed in the IBRL sensor deployment that not all the sensors were subject to the same conditions. This is because of their physical locations. Some of the nodes were closer to air conditioning. Some were closer to windows and hence were affected by the sun. Some were closer to the kitchen and thus affected by the heat and humidity coming from there. Furthermore, the patterns followed by sensor measurements changed seasonally. As an example, during a week period, the pattern followed in week days was different than that followed in weekends. That was because the air conditioning was reduced or turned off in the laboratory on the weekends. This caused the interrelationship among the sensors to vary not only with their spatial locations, but also with time.

A solution to overcome such a problem is to choose the neighbour sensors of each node so that they are physically close and subject to similar conditions. An alternative solution is to upgrade the model to become incremental with time to account for phenomenal changes. This can be achieved using *incremental learning* of the SVR. The learning process can then be performed at each time step (incrementally) or at predefined short intervals, depending on how severe the change is. Incremental SVR learning algorithms (Platt, April 1998; Shilton et al., Jan. 2005) can be utilised with the UKF to perform adaptive drift correction in the network. Devising an adaptive drift correction framework by incorporating incremental learning is a direction for future work.

7. Conclusion

In this chapter we have proposed a formal statistical algorithm for detecting and correcting sensor measurement errors in sparsely deployed WSN based on the assumption that sensor nodes in a neighbourhood observing a certain physical phenomenon have correlated measurements and uncorrelated drifts and biases. We have used SVR to model the spatio-temporal correlations in the neighbouring sensor measurements to obtain predictions of the future sensor measurements. The predicted data have then been used by a UKF to estimate the actual value of the measured variable at the sensor under consideration. The algorithm runs recursively and is fully decentralised. Extensive evaluations of the presented algorithm on real data obtained from the IBRL have proved that it is effective in detecting and correcting sensor errors and extending the effective life of the network.
In future, we plan to upgrade our algorithm to become more adaptive to any phenomenal changes that may occur in the network deployment area by implementing an incremental SVR to periodically re-train the SVR. This will also be tested on sensor networks deployed both in a controlled Lab environment and uncontrolled outdoor environments.

8. References

2006 (Accessed on 07/09/2006). http://db.lcs.mit.edu/labdata/labdata.html, *[online]*.
URL: *http://db.lcs.mit.edu/labdata/labdata.html*

Akyildiz, I. F., Su, W., Sankarasubramaniam, Y. & Cayirci, E. (2002). Wireless sensor networks: a survey, *Comp. Networks* **38**: 393–422.

Balzano, L. (2007). *Addressing fault and calibration in wireless sensor networks*, Master's thesis, University of California, Los Angeles, California.

Balzano, L. & Nowak, R. (2007). Blind calibration of sensor networks, *Information Processing in Sensor Networks* .

Balzano, L. & Nowak, R. (2008). Blind calibration of networks of sensors: Theory and algorithms, *Networked Sensing Information and Control*, Springer US, pp. 9–37.

Brown, L. (1992). A survey of image registration techniques, *ACM Computing Surveys* **24**(4): 326–376.

Bychkovskiy, V. (2003). *Distributed in-place calibration in sensor networks*, Master's thesis, University of California, Los Angeles, California.

Bychkovskiy, V., Megerian, S., Estrin, D. & Potkonjak, M. (2003). A collaborative approach to in-place sensor calibration, *Int. Workshop on Information Processing in Sensor Networks* pp. 301–316.

Canu, S., Grandvalet, Y., Guigue, V. & Rakotomamonjy, A. (2005). Svm and kernel methods matlab toolbox, Perception Systï£¡mes et Information, INSA de Rouen, Rouen, France.
URL: *http://asi.insa-rouen.fr/enseignants/ arakotom/toolbox/index.html*

Challa, S., Evans, R., Morelande, M. & Musicki, D. (2008). *Fundamentals of Object Tracking*, Cambridge University Press.

Elnahrawy, E. & Nath, B. (2003). Cleaning and querying noisy sensors, *in Proceedings of ACM WSNAï£¡03*.

Estrin, D., Girod, L., Pottie, G. & Srivastava, M. (2001). Instrumenting the world with wireless sensor networks, *Int. Conference on Acoustics, Speech, and Signal Processing)* .

Feng, J., Megerian, S. & Potkonjak, M. (2003). Model-based calibration for sensor networks, *Sensors* pp. 737 – 742.

Gill, M. K., Asefa, T., Kemblowski, M. W. & McKee, M. (2006). Soil moisture prediction using support vector machines1, *J. of the American Water Resources Association* **42**(4): 1033–1046.
URL: *http://www.blackwell-synergy.com/doi/abs/10.1111/j.1752-1688.2006.tb04512.x*

Gill, M. K., Kemblowski, M. W. & McKee, M. (2007). Soil moisture data assimilation using support vector machines and ensemble kalman filter, *J. of the American Water Resources Association* **43**(4): 1004–1015.

Guestrin, C., Bodik, P., Thibaux, R., Paskin, M. & Madden, S. (2004). Distributed regression: an efficient framework for modeling sensor network data, *In IPSN04*, pp. 1–10.

Hoadley, B. (1970). A baysian look at inverse linear regression, *J. of the American Stats. Association* **65**(329): 356–369.

Ihler, A., Fisher, J., Moses, R. & Willsky, A. (2004). Nonparametric belief propagation for self-calibration in sensor networks, *In Proceedings of the Third international Symposium on Information Processing in Sensor Networks*.

Julier, S. (2002). The scaled unscented transformation, *American Control Conference* **6**: 4555–4559.

Julier, S. J., Uhlmann, J. K. & Durrant-Whyte, H. F. (1995). A new approach for filtering non-linear systems, *American control Conference* pp. 1628 – 1632.

Julier, S. & Uhlmann, J. (1997). A new extension of the Kalman filter to nonlinear systems, *Int. Symp. Aerospace/Defense Sensing, Simul. and Controls*.

Krishnamachari, B. & Iyengar, S. (2004). Distributed bayesian algorithms for fault-tolerant event region detection in wireless sensor networks, *IEEE Tran. Computers* **53**(3): 241–250.

Lu, S., Cai, L., Lu, D. & Chen, J. (2007). Two efficient implementation forms of unscented kalman filter, *IEEE Int. Conference on Control and Automation* pp. 761 – 764.

Nowak, R. & Mitra, U. (2003). Boundary estimation in sensor networks: Theory and methods, *In IPSN*, pp. 80–95.

Okello, N. & Challa, S. (2003). Simultaneous registration and track fusion for networked trackers, *conference on information fusion*, Montrï£¡al, Canada.

Okello, N. & Pulford, G. (1996). Simultaneous registration and tracking for multiple radars with cluttered measurements, *IEEE Signal Processing Workshop on Statistical Signal and Array Processing* pp. 60–63.

Platt, J. (April 1998). Sequential minimal optimization: A fast algorithm for training support vector machines, *Technical Report 98-14, Microsoft Research, Redmond, Washington.* .
URL: *citeseer.ist.psu.edu/platt98sequential.html*

Rajasegarar, S., Leckie, C., Palaniswami, M. & Bezdek, J. (2007). Quarter sphere based distributed anomaly detection in wireless sensor networks, *Proceedings of the IEEE International Conference Communications (IEEE ICC '07)*, UK.

Sallans, B., Bruckner, D. & Russ, G. (2005). Statistical model-based sensor diagnostic for automation systems, *in* M. L. Chavez (ed.), *Fieldbus systems and their applications*, Elsevier, pp. 239–246.

Scholkopf, B. & Smola, A. (2002). *Learning with Kernels*, MIT Press.

Shilton, A., Palaniswami, M., Ralph, D. & Tsoi, A. C. (Jan. 2005). Incremental training of support vector machines, *IEEE Tran. on Neural Networks* **16**(1): 114–131.

Särkkä, S. & Hartikainen, J. (2007). Ekf/ukf toolbox for matlab v1.2, Centre of Excellence in Computational Complex Systems Research, Helsinki University of Technology (HUT), Finland. http://www.lce.hut.fi/research/mm/ekfukf/.

Takruri, M., Aboura, K. & Challa, S. (2008). Distributed recursive algorithm for auto calibration in drift aware wireless sensor networks, *in* K. Elleithy (ed.), *Innovations and Advanced Techniques in Systems, Computing Sciences and Software Engineering*, Springer, pp. 21–25.

Takruri, M. & Challa, S. (2007). Drift aware wireless sensor networks, *Proceedings of the 10th international conference on information fusion*, Quebec City, Canada.

Takruri, M., Challa, S. & Chacravorty, R. (2010). Recursive bayesian approaches for auto calibration in drift aware wireless sensor networks, *Journal of Networks* **5**(7): 823–832.

Takruri, M., Challa, S. & Chakravorty, R. (2008). Auto calibration in drift aware wireless sensor networks using the interacting multiple model algorithm, *Mosharaka International Conference on Communications, Computers and Applications (MIC-CCA 2008)*, Amman, Jordan.

Takruri, M., Challa, S. & Yunis, R. (2009). Data fusion techniques for auto calibration in wireless sensor networks, *Proceedings of the 12th International conference on information fusion*, Seattle, USA.

Takruri, M., Rajasegarar, S., Challa, S., Leckie, C. & Palaniswami, M. (2008). Online drift correction in wireless sensor networks using spatio-temporal modeling, *Proceedings of the 11th international conference on information fusion*, Cologne, Germany.

Takruri, M., Rajasegarar, S., Challa, S., Leckie, C. & Palaniswami, M. (2010). Spatio-temporal modelling based drift aware wireless sensor networks, *IET Wirel. Sens. Syst. (under review)* .

Taylor, C., Rahimi, A., Bachrach, J., Shrobe, H. & Grue, A. (2006). Simultaneous localization, calibration, and tracking in an ad hoc sensor network, *In Proceedings of 5th International Conference on Information Processing in Sensor Networks (IPSN06)*, pp. 27–33.

Wan, E. & van der Merwe, R. (2000). The unscented kalman filter for nonlinear estimation, *IEEE Symposium 2000 (AS-SPCC)* .

Wang, Y. M., Schultz, R. T., Constable, R. T. & Staib1, L. H. (2003). Nonlinear estimation and modeling of fmri data using spatio-temporal support vector regression, *Information Processing in Medical Imaging* **2732**: 647–659.

Whitehouse, K. & Culler, D. (2002). Calibration as parameter estimation in sensor networks, *ACM International Workshop on Wireless Sensor Networks and Applications (WSNA'02)*.

Whitehouse, K. & Culler, D. (2003). Macro-calibration in sensor/actuator networks, *Mobile Networks and Applications Journal (MONET), Special Issue on Wireless Sensor Networks* .

Xu, X., Hines, J. W. & Uhrig, R. E. (1998). On-line sensor calibration monitoring and fault detection for chemical processes, *Maintenance and Reliability Conference (MARCON 98)*, pp. 12–14.

Zitova, B. & Flusser, J. (2003). Image registration methods: a survey, *Image and Vision Computing* **21**(1): 977–1000.

19

Target Tracking in Wireless Sensor Networks

Jianxun LI* and Yan ZHOU**,*
* *Department of Automation, Shanghai Jiao Tong University, Shanghai 200240, China*
** *College of Information Engineering, Xiangtan University, Xiangtan 411105, China*

1. Introduction

Wireless sensor networks (WSNs) have gained worldwide attention in recent years, particularly with the proliferation in Micro-Electro-Mechanical Systems (MEMS) technology which has facilitated the development of smart sensors (Akyildiz et al., 2006; Akyildiz et al., 2007; Yick et al., 2008). Target tracking in WSNs is an important problem with a large spectrum of applications (Akyildiz et al., 2006; Zhao et al., 2002), such as surveillance (Valera & Velastin, 2005), natural disaster relief (Wang et al., 2003), traffic monitoring (Li et al., 2009), pursuit evasion games, etc.

1.1 Opportunities and challenges

A target tracking system through WSNs can have several advantages (Veeravalli & Chamberland, 2007): (i) qualitative and fidelity observations; (ii) signal processing accurately and timely; and (iii) increased system robustness and tracking accuracy. However, the use of sensor networks for target tracking presents a number of new challenges. These challenges include limited energy supply and communication bandwidth, distributed algorithms and control, and handling the fundamental performance limits of sensor nodes, especially as the size of the network becomes large. Unlike traditional networks, a WSN has its own design and resource constraints. Resource constraints include a limited amount of energy, short communication range, low bandwidth, and limited processing and storage in each node. Design constraints are application dependent and are based on the monitored environment. The environment plays a key role in determining the size of the network, the deployment scheme, and the network topology.
Power consumption is the most important design factor for WSNs (Shorey et al., 2006). Commonly, saving power during the operation of the electronic device could be achieved on more than one protocol level. Plenty of research work is dedicated to the design of power efficient schemes for target tracking which try to explore good trade-off between power consumption and tracking accuracy (see e.g. Lee et al. 2007; Xu & Lee, 2003; Walchli et al., 2007; Tsai et al., 2007, and the references therein).
Besides, the traditional target tracking methodologies make use of a centralized approach. As the number of sensors rise in the network, more messages are passed on towards the sink and will consume additional bandwidth. Thus traditional approaches are not fault tolerant as there is single point of failure and does not scale well. However, in sensor networks,

hundreds, and in the extreme, hundreds of thousands of sensors are deployed in a large geographical area. In some cases dropped from airplanes, or deployed using artillery shells. Requiring that every node must work in order for the network to operate is difficult to achieve. The network must have a high level of fault tolerance in order to be of any practical value (Hoblos et al., 2000).
In a word, target tracking algorithm considering the tradeoff between the tracking accuracy and network resources such as energy, bandwidth, and communication/computation burden is challenging.

1.2 Contributions and chapter organization

In this chapter, a self-contained overview of tracking approaches through a WSN is given according the architecture of the networks. As can be seen soon, WSNs are typically classified into two main categories (Sohraby et al., 2006): hierarchical network and peer-to-peer network. For the former, naïve, tree, cluster, and hybrid network based methods are reviewed in detail. While for the latter, average consensus is usually adopted to achieve network-wide agreement on target estimate and two approaches commonly used including the dynamic consensus filter and alternating direction method. Then advantage and limitations of these approaches are compared.
Considering the stringent energy and bandwidth limitation, quantized messages based tracking method is discussed separately. Local data quantization/compression is usually adopted so as to reduce the required expenditure of resources. Then the signal processing unit, i.e. the fusion center (FC), or cluster head (CH), or any node in the network, combines the quantized messages from local sensors to produce a final estimation of the target state. In the quantized scenario, Shannon's "rate-distortion" bound that notionally is a curve of possible and impossible points on distortion (e.g. MSE versus bit-rate axes) is important and practical. In the literature, target tracking algorithms using quantized information can be categorized to mainly quantized measurement based and quantized innovation based tracking. Both categories will be overview in tail with the characteristics discussed.
The remainder of this chapter is organized as follows. Section 2 gives a taxonomy on target tracking approaches through a WSN. In Section 3, details of target tracking approaches in hierarchical network are highlighted focusing on: naïve activation based tracking, tree-based tracking methods, cluster-based tracking approaches, and hybrid tracking methods. In Section 4, target tracking in peer-to-peer sensor networks are reviewed with both embedded filter based consensus and alternating direction based consensus methods discussed. Comparison criterions are given in Section 5 and the methods aforementioned are compared. Section 6 discusses the methods based on quantized information in detail, which followed by Section 7 that gives the future research directions and concluding remarks.

2. A taxonomy on tracking methods

In the literature, various algorithms and approaches for target tracking are presented but there is a deficiency of well defined classification of solutions of this well known application of WSN. This chapter presents a taxonomy of target tracking approaches as well as discussing each method under the appropriate category.
The taxonomy on the tracking techniques presented is based on the network architecture. In this chapter, we classify the WSNs into two categories (Veeravalli & Chamberland, 2007):

hierarchical (category 1 WSNs) and peer-to-peer (category 2 WSNs). For the category 1 WSNs (see Fig. 1 (a)), almost invariably mesh-based systems with multihop radio connectivity among or between wireless nodes are employed. The sensors in the vicinity of an event must be able to monitor the event of interest and report back to the sink. A sink sensor node has capability to communicate with outside world such as laptop, base station. The important characterizations of the category 1 WSNs are that (i) sensor nods can support communications on behalf of other sensor nodes by acting as repeaters; (ii) the forwarding node can support data processing or information fusion on behalf of the senor nodes. The category 2 WSNs (see Fig. 1 (b)), which are also referred as flat, or point-to-point network, generally are with single-hop radio connectivity to wireless nodes utilizing static routing over the wireless network. The main features of the category 2 WSNs are that (i) the forwarding node only supports static routing; (ii) each node only communicates with its neighbouring node(s) and network-wide consensus can be achieved through information exchange between neighbours.

As illustrated in Fig. 2, we taxonomize the tracking algorithms into two aspects according to the aforementioned two categories of network architecture. One is hierarchical network based tracking, the other is peer-to-peer network based tracking. The former can be further classified into four schemes, which are: Naïve activation based tracking, tree-based tracking, cluster-based tracking, and hybrid methods. In tree-based target tracking, nodes in a network may be organized in a hierarchical tree or represented as a graph in which vertices represent sensor nodes and edges are links between nodes that can directly communicate with each other.

The cluster-based methods provide scalability and better usage of bandwidth than other types of methods. If CH is formed via local network processing, extra messages are reduced and fewer messages are transmitted towards base station thus providing security as well as less usage of bandwidth (Rapaka & Madria, 2007). In the conventional cluster architecture, clusters are formed statistically at the time of network deployment and the properties of each cluster are fixed such as number of members, area covered, etc. Static clustering has several drawbacks regardless of its simplicity, for example, static membership is not robust

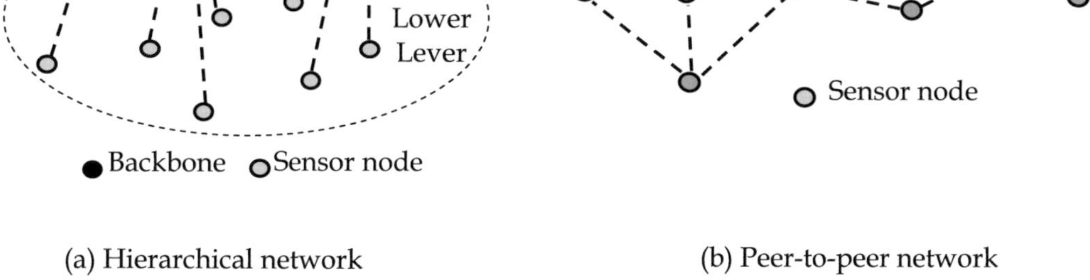

Fig. 1. Architecture of wireless sensor networks

from fault-tolerance point of view and it prevents sensor in different clusters from sharing information. In contrast, dynamic clustering offers several advantages where clusters are formed dynamically depending on occurrence of certain events, for instance, when a node with sufficient battery and computational power detects an event, it comes forward to act as a CH. To make sure only one CH remains active for target tracking, some decentralized mechanism is adapted. The CH invites nearby sensor nodes and makes them members of that cluster. Since sensors don't statistically form a cluster, they may belong to altered cluster at different timings. As only one cluster is active at a time, redundant data and interference is reduced.

According to the methods approaching consensus, the peer-to-peer networks based target tracking systems can be further classified into embedded filter based tracking and alternating-direction based tracking. Since network-wide consensus can be achieved through only information exchange between neighbours, considerable communication energy is reduced and this makes the peer-to-peer networks based tracking scalable for large-scale sensor network.

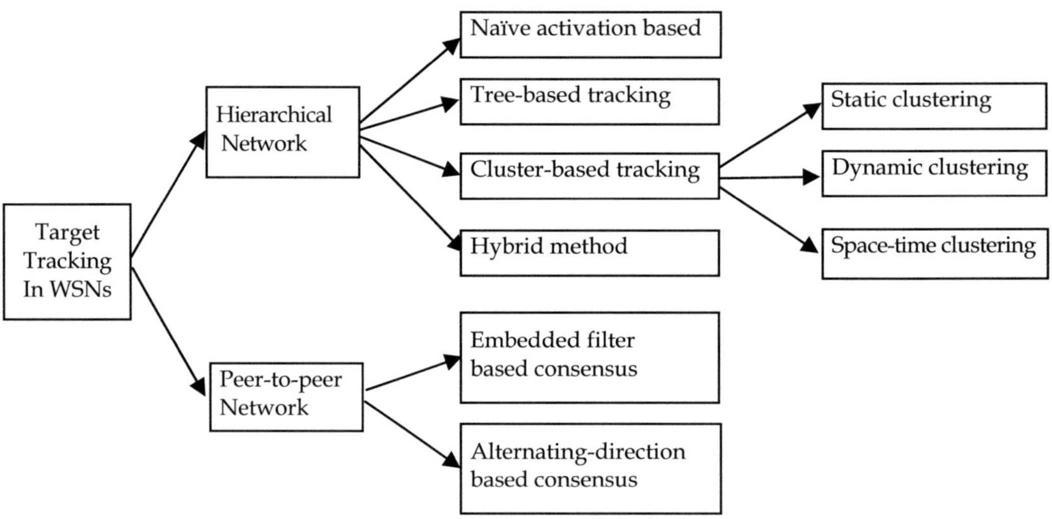

Fig. 2. Taxonomy of target tracking algorithms in WSNs.

3. Tracking methods for hierarchical networks

3.1 Naïve activation based tracking

Naïve activation (or direct communication) based tracking scheme (Guo et al., 2003) is the simplest approach, for which all nodes are in tracking mode all the time. Each node sends the local measurement to the sink node or base station. Then the base station estimates and predicts the target state according to the received local measurements. Since it offers the best tracking results, it is a useful baseline for comparison. However, this strategy offers the worst energy efficiency and it inflects heavy communication and computation burden on the base station of sink node. This makes the naïve approach not robust against base station failure especially for the case of link failure and channel congestion.

3.2 Tree-based tracking

Centralized target tracking approaches are both time and energy consuming, to avoid this limitation tree-based tracking methods are proposed, for example, Scalable Tracking Using Networked sensors (STUN) (Kung & Vlah, 2003), Dynamic Convoy Tree-Based Collaboration (DCTC) (Zhang & Cao, 2004a; Zhang & Cao, 2004b), Deviation Avoidance Tree (DAT) (Lin et al., 2004), and Dynamic Object Tracking (DOT) approach (Tsai et al., 2007).

Specifically, STUN (Kung & Vlah, 2003) is a tree-based approach in which a cost is assigned to each link calculated by Euclidean distance between the two nodes. The leaf nodes are used for tracking the moving object and then sending collected data to the sink through intermediate nodes. The intermediate nodes keep a record of detected object and whenever there is a change in that record, they send updated information to the sink. However, STUN has two drawbacks. First, drain and balance tree does not replicate physical sensor network as it is a logical tree, hence an edge may consist of multiple communication hops and may raise communication cost. Second, the construction of DAB tree does not consider query cost.

In (Zhang & Cao, 2004a; Zhang & Cao, 2004b), a spanning tree rooted at the sensor node close to a target is used for target tracking, with the target position estimated by the location of the root sensor. More statistically-oriented algorithms for mobile target identification and localization are proposed therein, which allows the designer to directly model the distributional properties of sensor signals.

A network aggregation model by organizing sensor nodes in logical tree is intended in (Lin et al., 2004). As physical topology of the network is considered, thus reducing the total communication cost. The object tracking involves two steps: update and query. In first step; location update cost is reduced by Deviation Avoidance Tree (DAT) algorithm and in second step; query cost is reduced by query cost reduction algorithm.

DOT (Tsai et al., 2007), a unique protocol reports the tracking information of moving object to moving source. First of all, the face neighbours are identified by Gabriel graph. In target discovery step, source sends request to sensor nodes and the node close to the target replies back. To detect moving target continuously, the spatial neighbours of near sensor node are waken up. In target tracking step, source send query to beacon node (node keeping track information), which reply back target's next location and the source moves towards next beacon node. The process is repeated until the source catches the target.

Lin et al. propose a CLOUD framework to track the region-based event (Lin et al., 2005). The basic idea is to dynamically form a tree-based collective structure for each event region in each time slot. However, both of their approaches in (Lin et al., 2005; Jiang et al., 2004) are limited by the dependence on the tree structure for the network topology.

In (Jin & Nittel, 2006), an R-tree sensor network topology is adopted for the detection and tracking of region-based targets. Two approaches: forward-all and forward-description methods are proposed for the detection of event regions. Furthermore, the authors describe three detailed algorithms: boundary detection algorithm, merging algorithm and description improvement algorithm to deal with the problems of how to detect an event boundary, how to merge the event region obtained from the child nodes in the R-tree, and how to simplify and smooth the event boundary.

3.3 Cluster-based tracking

To facilitate collaborative data processing in target tracking-centric sensor networks, the cluster architecture is usually used in which sensors are organized into clusters, with each cluster consisting of a CH and several slave nodes (members). Hierarchical (clustering) techniques can aid in reducing useful energy consumption (Heinzelman et al., 2002). Clustering is particularly useful for applications that require scalability to hundreds or thousands of nodes. Scalability in this context implies the need for load balancing and efficient resource utilization. Clustering can be extremely effective in one-to-many, many-to-one, one-to-any, or one-to-all (broadcast) communication. For example, in many-to-one communication, clustering can support data fusion and reduce communication interference (Younis & Fahmy, 2004).

3.3.1 Static clustering

Conventionally, clusters are formed statically at the time of network deployment. The attributes of each cluster, such as the size of a cluster, the area it covers, and the members it possesses, are static. In spite of its simplicity, the static cluster architecture suffers from several drawbacks. First, fixed membership is not robust from the perspective of fault tolerance. If a CH dies of power depletion, all the sensors in the cluster render useless. Second, fixed membership prevents sensor nodes in different clusters from sharing information and collaborating on data processing. Finally, fixed membership cannot adapt to highly dynamic scenarios in which sensors in the region of high (low) event concentration may be instrumented to stay awake (go to sleep).

3.3.2 Dynamic clustering

Dynamic cluster architectures, on the other hand, offer several desirable features (Chen et al., 2003). Formation of a cluster is triggered by certain events of interest (e.g., detection of an approaching target with acoustic sounds). When a sensor with sufficient battery and computational power detects (with a high signal-to-noise ratio, SNR) signals of interest, it volunteers to act as a CH. No explicit leader (CH) election is required and, hence, no excessive message exchanges are incurred. As more than one "powerful" sensors may detect the signal, multiple volunteers may exist. A judicious, decentralized approach has to be applied to ensure that only one CH is active in the vicinity of a target to be tracked with high probability. Sensors in the vicinity of the active CH are "invited" to become members of the cluster and report their measurements to the CH. Compared with the static clustering approaches, dynamic clustering networked sensors do not statically belong to a cluster and may support different clusters at different times. Moreover, as only one cluster is active in the vicinity of a target with high probability, redundant data is suppressed and potential interference and contention at the MAC level is mitigated.

Examples of dynamic cluster-based tracking are information-driven sensor querying (IDSQ) (Zhao et al., 2002), DELTA (Walchli et al., 2007), and RARE (Olule et al., 2007).

Zhao et al. addressed the dynamic sensor collaboration problem in distributed tracking to determine dynamically which sensor is most appropriate to perform the sensing, what needs to be sensed, and to whom to communicate the information (Zhao et al., 2002). They developed the IDSQ approach, enabling collaboration based on resource constraints and the const of transmitting information. Information utility functions employed include entropy,

Mahalanobis distance, and a measure on expected posterior distribution. This approach assumes that each node in the network can locally estimate the cost of sensing, processing and communicating data to another node. Although the approach is power efficient (since only few nodes are active at any given time), it is applied for tracking a single object only.

Walchli et al. present DELTA (Walchli et al., 2007), a distributive algorithm for tracking a person moving at constant speed by dynamically making a cluster and selecting CH based on light measurement. The CH is responsible to reliably monitor moving object and collaborate with sensor nodes. The limitation of DELTA algorithm is that it can only deal with constant speed, whereas, varying speed is not considered.

Energy aware probabilistic target localization algorithm for a single target using cluster-based WSN is proposed in (Zou & Chakrabarty, 2003), where a two step protocol for communication between CH and sensors in the cluster is put forward. In the first step, sensors detecting the target report to the CH by a short message. Then the CH executes localization procedure to determine the subset of sensors in the vicinity of target and query detailed target information from them.

Olule et al. investigate an energy efficient target tracking protocol based on two algorithms, ARE-Area (Reduced Area Reporting) and RARE-Node (Reduction of Active node Redundancy) via static clustering (Olule et al., 2007). RARE-Area reduces number of nodes participating in tracking by inhibiting far away nodes from taking part in tracking. RARE-node reduces redundant information by identifying overlapping sensors. Cluster is formed dynamically by prediction during target tracking (Jin et al., 2006), thus reducing number of nodes involved in tracking. Although the method consumes low energy, the missing target recovery procedure is not well defined.

Quantized measurements are usually adopted in such a network to attack the problem of limited power supply and communication bandwidth. Very recently, the problem of target tracking in a WSN that consists of randomly distributed range-only sensors is considered in (Zhou et al., 2010)). The posterior Cramér-Rao lower bounds (CRLB) on the mean squared error (MSE) on target tracking with quantized range-only measurements are derived. Due to the analytical difficulties, particle filter is applied to approximate the theoretical bounds. In this paper, recursion of posterior CRLB on tracking based on both constant velocity (CV) and constant acceleration (CA) model for target dynamics and a general range-only measuring model for local sensors are obtained. More details on tracking using quantized messages can be found in Section 6.

3.3.3 Space-time clustering

In order to present the event processing with high accuracy, Phoha et al. propose the dynamic space-time clustering (DSTC) (Phoha et al., 2003a). In this architecture, clusters of space-time neighbouring nodes are dynamically organized to present the event around by combining the local information among nodes in the inner space-time cluster. The type and track of the target then are estimated by the CH.

Phoha et al. propose two methods by combining the DSTC and beamforming: one is DSTC beamforming controlled, the other is DSTC logic controlled beamforming (Phoha et al., 2003b). The former is composed of hundreds of low-cost DSTC nodes and a few beamforming nodes, which estimate the target position through triangulation. In the case of failure of beamforming nodes, the DSTC nodes are activated to localize the target. The latter determine a cluster to track the target according to DSTC logic, while the member nodes run the beamforming algorithm to estimate the target state.

3.4 Hybrid method

Hybrid methods are referred to the tracking algorithms that fulfill the requirements of more than one types of target tracking. Examples include distributed predictive tracking (DPT) (Yang & Sikdor, 2003), DCAT (Chen et al., 2003), and Hierarchical prediction strategy (HPS) (Wang et al., 2008).

The DPT adopts a clustering based approach for scalability and a prediction based tracking mechanism to provide a distributed and energy efficient solution (Yang & Sikdor, 2003). The protocol is proven to be robust against node or prediction failures which may result in temporary loss of the target and recovers from such scenarios quickly and with very little additional energy use.

A decentralized dynamic clustering algorithm for single target tracking (Here we referred as dynamic clustering for acoustic tracking, DCAT) is proposed in (Chen et al., 2003). Using Voronoi Diagrams, clusters are formed and only one CH becomes active when the acoustic signal strength detected by CH exceeds a pre-determined threshold. The CH then asks the sensors in its vicinity to join cluster by sending a broadcast packet. The sensor based on the probabilistic distance estimates between itself and target, decides whether it should reply to CH. Afterwards, CH executes a localization method to estimate location of target based on sensor replies and sends result to the sink.

In HPS, cluster is formed using Voronoi division and a target next location is predicted via Least Square Method but overheads are not well defined. HVE protocol uses cluster structure and prediction for estimating shape and size of forwarding zone and delivering mobicast messages.

A location model determines the granularity of location information and the prediction model processes the historical data to predict next movement of mobile object. An interesting example of multiple targets tracking using prediction is given in (Chong et al., 2003).

4. Tracking methods for peer-to-peer networks

For the tree- or cluster-based methods, sensing task is usually performed by several nodes at a time and inflicts heavy computation burden on the root node or the CH. This makes the tree- or cluster-based WSN tracking systems lack of robustness in case of root node or the CH failures. On the contrary, another architecture for target tracking is the peer-to-peer WSN. As it can guarantee that sensors obtain the desired estimates and rely only on single-hop communications between neighbouring nodes, the limitations mentioned above are not encountered in peer-to-peer WSN based target tracking systems.

On the other hand, the well-known strategy concerning estimation and tracking is decentralized Kalman filtering or nonlinear filtering scheme, e.g. extended Kalman filtering (EKF), unscented Kalman filtering (UKF), and particle filtering (PF), which involve state estimation using a set of local filters that communicate with all other nodes (see e.g. Li & Wang, 2000; Mutambara, 1998; Vercauteren & Wang, 2005, and the references therein). The information flow in the traditional decentralized Kalman filtering (see e.g. Mutambara, 1998) or unscented Kalman filtering scheme (Vercauteren & Wang, 2005) is all-to-all with communication complexity of O(N2) (here N is the number of sensors in the network), which is not scalable for sensor networks (Speyer et al., 2004). On the contrary, the peer-to-peer network tracking is usually based on average consensus algorithms that have proven to be effective tools for performing network-wide distributed computation task ranging from flocking to robot rendezvous as in the papers (Olfati-Saber & Murry, 2004; Tanner et al., 2007; Kar & Moura, 2009), and the references therein. Hence, we refer this kind of methods as average consensus based tracking (AC tracking).

4.1 Embedded filter based consensus

Distributed estimation using peer-to-peer WSNs is based on successive refinements of local estimates maintained at individual sensors. In a nutshell, each iteration of the algorithm comprises a communication step where the sensors interchange information with their neighbours, and an update step where each sensor uses this information to refine its local estimate. In this context, estimation of deterministic parameters in linear data models, via decentralized computation of the BLUE or the sample average estimator, was considered in (Olfati-Saber & Murry, 2004; Scherber & Papadopoulos, 2005; Xiao & Boyd, 2004) using the notion of consensus averaging. Decentralized estimation of Gaussian random parameters was reported in (Delouille et al., 2004) for stationary environments, while the dynamic case was considered in (Spanos et al., 2005).

Olfati-Saber introduces a distributed Kalman filtering (DKF) algorithm that uses dynamic consensus strategy in (Olfati-Saber, 2005; Olfati-Saber, 2007). The DKF algorithm consists of a network of micro-Kalman filters each embedded with a high-gain high-pass consensus filter (or consensus protocol). The role of consensus filters is to estimate of global information contribution using only local and neighbouring information. Recently, the problem of estimating a simpler scenario with a scalar state of a dynamical system from distributed noisy measurements based on consensus strategies is considered in (Carli et al., 2006), the focuses are with the interaction between the consensus matrix, the number of messages exchanged per sampling time, and the Kalman gain for scalar systems.

Very recently, the distributed and scalable robust filtering problem using average consensus strategy in a sensor network is investigated in (Zhou & Li, 2009a). Specifically, based on the information form robust filter, every node estimates the global average information contribution using local and neighbours' information rather than using the information from whole network. Due to the adoption of iterations of robust filter, the proposed algorithm relaxes the necessity to have the prior knowledge of the noise statistics. Moreover, the proposed algorithm is applicable to large-scale sensor network since each node broadcasts message only to its neighbouring nodes.

The aforementioned embedded filter based consensus for distributed target tracking is proposed for linear systems with Gaussian or energy bounded noises, there is little result on tracking algorithm for nonlinear dynamic systems and/or nonlinear observations. In (Zhou & Li, 2009b), a distributed scalable Sigma-Point Kalman filter (DS2PKF) is proposed for distributed target tracking in a sensor network based on the dynamic consensus strategy. The main idea is to use dynamic consensus strategy to the information form sigma-point Kalman filter (ISPKF) that derived from weighted statistical linearization perspective. Each node estimates the global average information contribution by using local and neighbours' information rather than by the information from all nodes in the network. Therefore, the proposed DSPKF algorithm is completely distributed and applicable to large-scale sensor network. A novel dynamic consensus filter is proposed, and its asymptotical convergence performance and stability are discussed.

4.2 Alternating-direction based consensus

Alternating-direction method of multipliers (Bertsekas & Tsitsiklis, 1999) is proven to be efficient in solving the distributed estimation (Schizas et al., 2008a; Schizas et al., 2008b). Recently, decentralized estimation of random signals in arbitrary nonlinear and non-Gaussian setups was considered in (Schizas & Giannakis, 2006), while distributed estimation of stationary Markov random fields was pursued in (Dogandzic & Zhang, 2006).

Adaptive algorithms based on in-network processing of distributed observations are well-motivated for online parameter estimation and tracking of (non)stationary signals using peer-to-peer WSNs. To this end, a fully distributed least mean-square (D-LMS) algorithm is developed in (Schizas et al., 2009), offering simplicity and flexibility while solely requiring single-hop communications among sensors. The resultant estimator minimizes a pertinent squared-error cost by resorting to i) the alternating-direction method of multipliers so as to gain the desired degree of parallelization and ii) a stochastic approximation iteration to cope with the time-varying statistics of the process under consideration. Information is efficiently percolated across the WSN using a subset of "bridge" sensors, which further tradeoff communication cost for robustness to sensor failures. For a linear data model and under mild assumptions aligned with those considered in the centralized LMS, stability of the novel D-LMS algorithm is established to guarantee that local sensor estimation error norms remain bounded most of the time.

Forero et al. develop a decentralized expectation-maximization (EM) algorithm to estimate the parameters of a mixture density model for use in distributed learning tasks performed with data collected at spatially deployed wireless sensors (Forero et al., 2008). The E-step in the novel iterative scheme relies on local information available to individual sensors, while during the M-step sensors exchange information only with their single hop neighbours to reach consensus and eventually percolate the global information needed to estimate the wanted parameters across the WSN.

5. Analysis and comparison

All the methods mentioned above are compared in Table 1 in terms of tracking accuracy, communicational burden, scalability, computational complexity, and fault tolerance, etc. In Table 1, we rate the method into four levels, i.e. A-D, according to the performance criterions mention above. We note that criterions, such as communicational burden, tracking accuracy and fault tolerance, are proportional to energy utilization for target tracking through WSNs. If communicational burden is high for cluster formation, more energy is consumed. High tracking accuracy demand will ultimately end with additional energy usage. Similarly fault tolerance will increase overheads and energy consumption. The total energy consumption and bandwidth usage during target tracking is the key concern in the majority of the methods since the network is with strictly limited energy and bandwidth. The energy consumption of a sensor node can be divided into three main domains, radio communication, sensing and data processing.

It is also worth pointing out that all the rating levels are relative since different methods are proposed within different network scenarios. For example, the AC tracking is mainly for the peer-to-peer network to improve the scalability. However, the cluster-based tracking such as IDSQ is mainly for the energy consumption and the lifetime.

Method	Tracking accuracy	Scalability	Computational complexity	Communicational burden	Fault tolerance
DC	A	D	B	D	A
STUN	C	C	C	C	B
DCTC	B	C	C	B	C
DAT	D	C	C	C	C
DOT	D	C	C	C	C
R-tree	B	B	D	C	C
IDSQ	C	B	A	A	B
DELTA	C	B	B	B	B
RARE	C	B	C	B	C
DSTC	B	B	B	B	B
DPT	D	C	B	C	C
DCAT	A	C	A	C	B
HPS	C	C	B	C	C
AC tracking	B	A	B	A	A

Table 1. Comparison of target tracking methods in WSNs

6. Quantized scenario

In the WSN tracking system, ach sensor node acquires measurements which are noisy linear or nonlinear transformations of the target state. The sensors then transmit measurements to the fusion center (for the FC-based WSNs) or the neighbouring nodes (for the distributed peer-to-peer WSNs) in order to form a state estimate. If measurements were available at a common location, minimum mean-square error (MMSE) estimates could be obtained using a Kalman filter, or nonlinear estimation methods, such as UKF and PF. However, since measurements are distributed in space and there is limited communication bandwidth, the measurements have to be quantized before transmission. Thus, the original estimation problem is transformed into decentralized state estimation based on quantized measurements. The problem is further complicated by the harsh environment typical of WSNs; see e.g., Chong & Kumar, 2003, and Culler et al., 2004.

The problem of decentralized estimation based on quantized measurements has been studied in early works such as Gubner, 1993, and Lam & Reibman, 1993. Recently, universal decentralized estimation taking into account local signal-to-noise ratio (SNR) and the channel path loss in sensor network is studied (Xiao et al., 2005). When the noise probabilistic density function (PDF) is unknown, the problem of estimation based on severely quantized data has been also addressed in (Luo, 2005).

In this section, we category the tracking methods based on quantized information into quantized measurements and quantized innovations. The latter is usually with higher accuracy when using the same quantization bit rate. It is because that the range of innovations is commonly little that causes little quantization noise.

6.1 Quantized measurement based tracking

Quantizing measurements to estimate a parameter of interest is not the same as quantizing a signal for later reconstruction (Gray, 2006). Instead of a reconstruction algorithm, the objective is finding, e.g., MMSE optimal, estimators using quantized observations (Papadopoulos et al., 2001; Ribeiro & Giannakis, 2006). Furthermore, optimal quantizers for reconstruction are, generally, different from optimal quantizers for estimation. State estimation using quantized observations is a nonlinear estimation problem that can be solved using e.g., EKF, UKF, or PF.

From the measurement fusion perspective, the problem for target tracking using quantized information in WSNs is investigated in (Zhou & Li, 2009c) and (Zhou et al., 2009a). Due to the limited energy and bandwidth, each activated node quantizes and then transmits the local measurements by probabilistic quantization strategy. The FC estimates the target state in a dimension compression way instead of merging all the quantized messages to a vector (augmented scheme). A closed-form solution to the optimization problem for bandwidth scheduling is given, where the total energy consumption measure is minimized subject to a constraint on the mean square error (MSE) incurred by quasi-best linear unbiased estimation (Quasi-BLUE) fusion. The results are extended to the case of tracking maneuvering target and correlation noise in (Zhou & Li, 2009d) and (Zhou et al., 2009b), respectively.

Quantizing measurements is an efficient way that gives tradeoff between the bandwidth/energy constraints and tacking accuracy. However, if the values of measurements are large, quantizing measurements will bring large information loss under the limited bandwidth, which means that the variance of the quantization noise is large. In this scenario, the quantized measurements based tracking will have a low filtering accuracy. To reduce the information loss and improve the filtering accuracy, quantized innovations based tracking has been extensively investigated recently. Since the values of innovation data are smaller than those of measured data, quantizing innovations will bring smaller information loss than quantizing measurements under the same bandwidth constraint.

6.2 Quantized innovation based tracking

Surprisingly, for the case where quantized observations are defined as the sign of the innovation (SOI) sequence, it is possible to derive a filter with complexity and performance very close to the clairvoyant KF based on the analog-amplitude observations (Ribeiro et al., 2006). Even though promising, the approach of (Ribeiro et al., 2006) is limited to a particular 1-bit per observation quantizer. Msechu et al. introduce two novel decentralized KF estimators based on quantized measurement innovations (Msechu et al., 2008). In the first quantization approach, the region of an observation is partitioned into contiguous, non-overlapping intervals where each partition is binary encoded using a block of bits. Analysis and Monte Carlo simulations reveal that with minimal communication overhead, the mean-square error (MSE) of a novel decentralized KF tracker based on 2-3 bits comes stunningly close to that of the clairvoyant KF. In the second quantization approach, if intersensor communications can afford bits at time , then the th bit is iteratively formed using the sign of the difference between the observation and its estimate based on past observations (up to time 1) along with previous bits (up to 1) of the current observation.

Recently, by optimizing the filter with respect to the quantization levels, a multiple-level quantized innovation Kalman filter (MLQ-KF) for estimation of linear dynamic stochastic systems is proposed in (You et al., 2008). Furthermore, Sukhavasi and Hassibi propose a particle filter that approximates the optimal nonlinear filer and observe that the error covariance of the particle filter follows the modified Riccati recursion (Sukhavasi, & Hassibi, 2009).

Very recently, Zhou et al. investigate the decentralized collaborative target tracking problem in a WSN from the fusion of quantized innovations perspective (Zhou et al., 2009c). A hierarchical fusion structure with feedback from the FC to each deployed sensor is proposed for tracking a target with nonlinear Gaussian dynamics. Probabilistic quantization strategy is employed in the local sensor node to quantize the innovation. After the FC received the quantized innovations, it estimates the state of the target using the Sigma-Point Kalman Filtering (SPKF). To attack the energy/power source and communication bandwidth constraints, the tradeoff between the communication energy and the global tracking accuracy is considered in (Zhou et al., 2009d). By Lagrange multiplier, a closed-form solution to the optimization problem for bandwidth scheduling is given, where the total energy consumption measure is minimized subject to a constraint on the covariance of the quantization noises. Simulation example is given to illustrate the proposed scheme obtains average percentage of communication energy saving up to 41.5% compared with the uniform quantization, while keeping tracking accuracy very closely to the clairvoyant UKF that relies on analog-amplitude measurements. In (Ozdemir et al., 2009), a new framework for target tracking in a wireless sensor network using particle filters is proposed. Under this framework, the imperfect nature of the wireless communication channels between sensors and the FC along with some physical layer design parameters of the network are incorporated in the tracking algorithm based on particle filters. It is call "channel-aware particle filtering" that derived for different wireless channel models and receiver architectures. Furthermore, the posterior CRLBs for the proposed channel-aware particle filters are also given.

7. Concluding remarks and open research directions

The extensively research of target tracking through WSNs inspired us to present a literature survey. In this chapter, we have explored the categories of target tracking methods, including tree-based, cluster-based, hybrid, and consensus-based tracking algorithm. Considering the stringent limitation on energy supply, the quantized messages based tracking has been discussed separately.

The emergence of WSN in the variety of application areas brought many open issues to researchers. The open research issues for target tracking in WSNs include, channel-aware tracking, mobile node aided tracking, multitarget association & tracking, cross-layer design, and fault tolerant tracking methods, etc.

First, wireless communication channels between sensors and the FC or base station are not perfect. Incorporating the statistics of the channel imperfection to the tracking algorithm is expected to improve the tracking accuracy. Second, the scenario becomes complicated in the presence of multiple targets and their tracking with mobile sensors which leads to intend more realistic solutions. Message transmission consumes more energy than local processing, thus, well organized computing and nominal transmission of messages without degradation of performance must be considered while designing a target tracking method (Rapaka & Madria, 2007). Data association is an important problem when multiple targets are present in a small region. Each node must associate its measurements of the environment with individual targets. Combining the track association and tracking becomes more complicated, especially in circumstance of low cost sensor network with limited computation capacity and communication bandwidth (Li et al., 2010).

Another interesting issue for target tracking is the consideration of node failure. The sensor nodes are usually deployed in harsh environments so various nodes may fail, may be attacked or node energy may be depleted due to obstacles. Therefore, fault tolerant target tracking algorithms and protocols must be designed for wireless sensor networks as the fault tolerant approaches developed for traditional wired or wireless networks are not well suited for WSN because of various differences between these networks (Ding & Cheng, 2009).

The cross-layered approach in WSN is more effective and energy efficient than in traditional layered approach. While traditional layered approach endures more transfer overhead, cross-layered approach minimizes these overhead by having data shared among layers (Melodia et al., 2006; Kwon et al., 2006; Song & Hatzinakos, 2007). In the cross-layered approach, the protocol stack is treated as a system and not individual layers, independent of each other. Layers share information from the system. The development of various protocols and services in a cross-layered approach is optimized and improved as a whole.

In last decades, the problem of decentralized information fusion has been discussed extensively in the literature. However, the algorithms developed are free of energy and communication constraints, see e.g. Sun & Deng, 2004; Li & Wang, 2000; Zhou & Li, 2008a; Zhou & Li, 2008b. Novel fusion approaches include practical constraints in WSNs while keeping high fusion performance must be investigated (Ruan et al., 2008). Moreover, tracking with adaptive quantization thresholds and/or allocated bandwidth is another promising research direction since the communicational condition dependent quantization will definitely improve the estimation accuracy while using less communicational energy (Zhou et al., 2011; Xu & Li, 2010).

Finally, WSNs have the potential to enhance and change the way people interact with technology and the world (Aboelaze & Aloul, 2005). The direction of future WSNs also lies in identifying real business and industry needs. Interactions between research and development are necessary to bridge the gap between existing technology and the development of business solutions. Applying sensor technology to different applications will improve business processes as well as open up more problems for researchers.

8. Acknowledgements

The work was jointly supported by the National Natural Science Foundation of China (Under Grant 60874104, 60935001); 973Project (2009CB824900, 2010CB734103); Shanghai Key Basic Research Foundation (08JC1411800).

9. References

Aboelaze, M. & Aloul, F. (2005). Current and future trends in sensor networks: a survey, *Proceedings of the 14th IEEE Intl. Conf. on Wireless and Optical Communication*, New York, USA, pp. 133-138

Akyildiz, I.F.; Su, W. Sankarasubramaniam, Y. & Cayirci, E. (2002). Wireless Sensor Network: A Survey. *Computer Networks*, vol. 38, no. 4, pp. 393–422

Akyildiz, I.F.; Melodia, T. & Chowdury, K.R. (2007). Wireless multimedia sensor networks: A survey. *IEEE Wireless Communications*, vol. 14, no. 6, pp. 32-39

Bertsekas, D. P. & Tsitsiklis, J. N. (1999). *Parallel and Distributed Computation: Numerical Methods*, 2nd ed. Belmont, MA: Athena Scientific

Carli, R. Chiuso, A. Schenato, L. & Zampieri, S. (2006). Distributed Kalman filtering based on consensus strategies. *IEEE J. Selected Areas in Communications*, vol. 26, no. 4, pp. 622-632

Chen, W.P.; Hou, J.C. & Sha, L. (2003). Dynamic clustering for acoustic target tracking in wireless sensor networks, *Proceedings of 11th IEEE International Conf. Network Protocols*, Atlanta, Georgia, USA, pp. 284–294

Chen, W.P.; Hou, J.C. & Sha, L. (2004). Dynamic clustering for acoustic target tracking in wireless sensor networks. *IEEE Transactions on Mobile Computing*, vol. 3, no. 3, pp. 258-273

Chong, C.Y. & Kumar, S. (2003). Sensor networks: Evolution, opportunities, and challenges. *Proc. IEEE*, vol. 91, pp. 27–41

Chong, C.Y.; Zhao, F. Mori, S. & Kumar, S. (2003). Distributed tracking in wireless Ad Hoc sensor networks, *Proceedings Sixth Intl. Conf. on Information Fusion*, Cairns, Australia, pp. 431–438

Culler, D.; Estrin, D. & Srivastava, M. (2004). Overview of sensor networks. *Computer*, vol. 37, no. 8, pp. 41–49

Delouille, V.; Neelamani, R. & Baraniuk, R. (2004). Robust distributed estimation in sensor networks using the embedded polygons algorithm, *Proceedings of the 3rd Int. Symp. Info. Processing Sensor Networks*, Berkeley, CA, pp. 405–413

Ding, M. & Cheng, X. (2009). Fault tolerant target tracking in sensor networks, *Proceedings of the tenth ACM international symposium on Mobile ad hoc networking and computing*, New Orleans, LA, USA

Dogandzic, A. & Zhang, B. (2006). Distributed estimation and detection for sensor networks using hidden Markov random field models. *IEEE Trans. Signal Process.*, vol. 54, no. 8, pp. 3200–3215

Forero, P.A.; Cano, A. & Giannakis, G.B. (2008). Consensus-based distributed expectation-maximization algorithm for density estimation and classification using wireless sensor networks, *Proceedings of the IEEE Int'l Conf. Acoustics, Speech and Signal Processing*, pp. 1989-1992

Gray, R. M. (2006). Quantization in task-driven sensing and distributed processing, *Proceedings Int. Conf. Acoustics, Speech, Signal Processing*, Toulouse, France, vol. 5, pp. V-1049–V-1052

Gubner, J. (1993). Distributed estimation and quantization. *IEEE Trans. Information Theory*, vol. 39, no. 5, pp.1456-1459

Guo, W.H.; Liu, Z.Y. & Wu, G.B. (2003). An energy-balanced transmission scheme for sensor networks, *Proceedings of the 1st Intl. Conf. Embedded Networked Sensor Systems*. Los Angeles, CA, USA, pp. 300-301

Heinzelman, W. R.; Chandrakasan, A. & Balakrishnan, H. (2002). An application-specific protocol architecture for wireless microsensor networks. *IEEE Transactions on Wireless Communications*, vol. 1, no. 4, pp. 660–670

Hoblos, G.; Staroswiecki, M. & Aitouche, A. (2000). Optimal design of fault tolerant sensor networks, *Proceedings of the IEEE Int'l Conf. on Control Applications*, pp. 467-472

Jiang, C.; Dong, G. & Wang, B. (2005). Detection and tracking of region-based evolving targets in sensor betworks, *Proceedings of 14th Int. Conf. on Computer Communications and Networks, ICCCN 2005*, pp:563 - 568

Jin, G. & Nittel, S. (2006). NED: An efficient noise-tolerant event and event boundary detection algorithm in wireless sensor networks, *Proceedings of the 7th Int. Conf. on Mobile Data Management*

Jin, G.Y.; Lu, X.Y. & Park, M.S. (2006). Dynamic clustering for object tracking in wireless sensor networks. *Ubiquitous Computing Systems*, pp.200-209

Kar, S. & Moura, J.M.F. (2009). Distributed consensus algorithms in sensor networks with imperfect communication: link failures and channel noise. *IEEE Trans. Signal Processing*, vol. 57, no. 1, pp. 355-369

Kung, H.T. & Vlah, D. (2003). Efficient location tracking using sensor networks, *Proceedings of the IEEE Wireless Communications and Networking Conference*, New Orleans, Louisiana, USA

Kwon, H.; Kim, T.H. Choi, S. & Lee, B.G. (2006). A cross-layer strategy for energy-efficient reliable delivery in wireless sensor networks. *IEEE Trans. on Wireless Communications*, vol. 5, no. 12, pp. 3689-3699

Lam, W. & Reibman, A. (1993). Quantizer design for decentralized systems with communication constrains. *IEEE Trans. Communications*, vol. 41, no. 8, pp.1602-1605

Lee, S.M.; Cha, H. & Ha, R. (2007). Energy-aware location error handling for object tracking applications. *Wireless Sensor Networks, Computer Communication*, vol. 30, pp. 1443-1450

Li, X.; Shu, W. Li, M. Huang, H.Y. Luo P.E. & Wu, M.Y. (2009). Performance evaluation of vehicle-based mobile sensor networks for traffic monitoring. *IEEE Trans. Vehic. Tech.*, vol. 58, no. 4, pp. 1647-1653

Li, X. R. & Wang, J. (2000). Unified optimal linear estimation fusion – Part II: Discussions and examples, *Proceedings of the 3rd ISIF Conf. On Information Fusion*, MoC2/18-MoC2/25

Li, X.R.; Zhu, Y. & Han, C. (2000). Unified optimal linear estimation fusion, *Proceedings of the 39th IEEE Conf. On Decision and Control*, Sydney, Australia, 10-25

Li, Z.; Chen, S. Leung, H. & Bosse, E. (2010). Joint data association, registration, and fusion using EM-KF. *IEEE Trans. Aerospace and Electronic Systems*, Vol. 46, no. 2, pp. 496-507

Lin, C.; King, C. & Hsiao, H. (2005). Region abstraction for event tracking in wireless Sensor networks, *Proceedings of the 8th Int. Symposium on Parallel Architectures, Algorithms, and Networks*

Lin, C.Y.; Peng, W.C. & Tseng, Y.C. (2004). Efficient in-network moving object tracking in wireless sensor networks. Department of Computer Science and Information Engineering - National Chiao Tung University

Luo, Z. (2005). Universal decentralized estimation in a bandwidth constrained sensor network. *IEEE Trans. Information Theory,* vol. 51, pp.2210-2219

Melodia, T.; Vuran, M.C. & Pompili, D. (2006). The state of art in cross-layer design for wireless sensor networks. in *Network Architect. In Next Generation Internet* (M. Cesana, L. Fratta, eds.) Berlin Heidelberg: Springer-Verlag

Msechu, E.J.; Roumeliotis, S.I. Ribeiro, A. & Giannakis, G.B. (2008). Decentralized quantized Kalman filtering with scalable communication cost, *IEEE Trans. Signal Process.*, vol. 56, no. 8, pp. 3727–3741

Mutambara, A.G.O. (1998). *Decentralized estimation and control for multisensory systems*, Boca Raton, FL: CRC Press

Olfati-Saber, R. (2005). Distributed Kalman filter with embedded consensus filters, *Proceedings of the 44th IEEE Conf. on Decision and Control, and European Control Conf.*, pp. 8179-8184

Olfati-Saber, R. (2007). Distributed Kalman filtering for sensor networks. *46th IEEE Conf. on Decision and Control*, New Orleans, USA

Olfati-Saber, R. & Murry, R.M. (2004). Consensus problems in network of agents with switching topology and time-delays. *IEEE Trans. Automat. Control*, vol. 49, no.9, pp. 101-115

Olule, E.; Wang, G. Guo, M. & Dong, M. (2007). RARE: An energy efficient target tracking protocol for wireless sensor networks, *Proceedings of the Intl. Conf. on Parallel Processing*, pp. 1298-1306

Ozdemir, O.; Niu, R. & Varshney, P.K. (2009). Tracking in wireless sensor networks using particle filtering: physical layer considerations. *IEEE Trans. Signal Processing*, bol. 57, no. 5, pp. 1987-1999

Papadopoulos, H.; Wornell, G. & Oppenheim, A. (2001). Sequential signal encoding from noisy measurements using quantizers with dynamic bias control. *IEEE Trans. Inf. Theory*, vol. 47, no. 3, pp. 978–1002

Phoha, S.; Javobson, N. & Friedlander, D. (2003a). Sensor network based localization and target tracking hybridization in the operational domains of beamforming and dynamic space-time clustering. *Proceedings of the Global Telecommunications Conference*. Michigan, USA, pp. 2952- 2956

Phoha, S.; Jacobson, N. Friedlander, D. & Brooks, R. (2003b). Sensor network based localization and tracking through hybridization in the operational domains of beamforming and dynamic space-time clustering. *Proceedings of the Conf. Global Telecommunications*. Michigan, USA, pp. 1137- 1140

Rapaka, A. & Madria, S. (2007). Two energy efficient algorithms for tracking objects in a sensor network. *Wireless Communication Mobile Computing*, vol. 12, no. 7, pp.809-819

Ribeiro, A. & Giannakis, G. B. (2006). Bandwidth-constrained distributed estimation for wireless sensor networks, Part II: Unknown PDF. *IEEE Trans. Signal Process.*, vol. 54, no. 3, pp. 1131–1143

Ribeiro, A.; Giannakis, G. B. & Roumeliotis, S. I. (2006). SOI-KF: Distributed Kalman filtering with low-cost communications using the sign of innovations. *IEEE Trans. Signal Process.*, vol. 54, no. 12, pp. 4782–4795

Ruan, Y.; Willett, P. Marrs, A. Palmieri, F. & Marano, S. (2008). Practical fusion of quantized measurements via particle filtering. *IEEE Trans. Aerosp. Electron. Syst.*, vol. 44, no. 1, pp. 15-29

Scherber, D. & Papadopoulos, H. C. (2005). Distributed computation of averages over *ad hoc* networks. *IEEE J. Sel. Areas Commun.*, vol. 23, no. 4, pp. 776–787

Schizas, I. D. & Giannakis, G. B. (2006). Consensus-based distributed estimation of random signals with wireless sensor networks, *Proceedings of the 40th Asilomar Conf. Signals, Systems, Computers*, Monterey, CA

Schizas, I.D.; Mateos, G. & Giannakis, G.B. (2009). Distributed LMS for consensus-based in-network adaptive processing. *IEEE Trans. Signal Process.*, vol. 57, no. 6, pp. 2365–2382

Shorey, R.; Ananda, A. L. Chan, M. C. & Ooi, W. T. (2006). *Mobile, Wireless, and Sensor Networks: Technology, Applications, and Future Directions*, IEEE Press, Piscataway, pp. 173-196

Schizas, I.D.; Ribeiro, A. & Giannakis, G.B. (2008a). Consensus in Ad Hoc WSNs with noisy links—part I: distributed estimation of deterministic signals. *IEEE Trans. Signal Processing*, vol. 56, no. 1, pp. 350-364

Schizas, I.D.; Ribeiro, A. & Giannakis, G.B. (2008b). Consensus in Ad Hoc WSNs with noisy links—part II: distributed estimation and smoothing of random signals. *IEEE Trans. Signal Processing*, vol. 56, no. 4, pp. 1650-1666

Sohraby, K.; Minoli, D. Znati, T. (2006). *Wireless sensor networks: Technology, protocols, and applications*. Wiley-interscience, New Jersey

Song, L. & Hatzinakos, D. (2007). A cross-layer architecture of wireless sensor networks for target tracking. *IEEE/ACM Transactions on networking*, vol. 15, no. 1, pp. 145-158

Spanos, D. P.; Olfati-Saber, R. & Murray, R. J. (2005). Distributed sensor fusion using dynamic consensus, *Proceedings of the 16th IFAC World Congr.*, Prague, Czech Republic

Speyer, J.L. (2004). Computation and transmission requirements for a decentralized linear-quadratic-Gaussian control problem. *IEEE Trans. Automatic Control*, vol. 49, no. 9, pp.1453-1464

Sukhavasi, R.T. & Hassibi, B. (2009). Particle filtering for quantized innovations, *Proceedings of the IEEE Int. Conf. Acout., Speech and Signal Process.*, Taiwan, China

Sun, S.L. & Deng, Z.L. (2004). Multi-sensor optimal information fusion Kalman filter. *Automatica*, vol. 40, no. 5, pp. 1017-1023

Tanner, H.G.; Jadbabaie, A. & Pappas, G.J. (2007). Flocking in fixed and switching networks. *IEEE Trans. Automat. Control*, vol. 52, no. 5, pp. 863-868

Tsai, H.W.; Chu, C.P. & Chen, T.S. (2007). Mobile object tracking in wireless sensor networks. *Computer Communications*, vol. 30, pp. 1811–1825

Valera M. & Velastin, S.A. (2005). Intelligent distributed surveillance systems: a review. *IEE Proc. Visual Image Signal Process.*, vol. 152, no. 2, pp. 192-204

Veeravalli V.V. & Chamberland, J.F. (2007). Detection in sensor networks. in Swami, A. Zhao, Q. Hong, Y.W. & Tong, L. Eds, *Wireless Sensor Networks: Signal Processing And Communications Perspectives*, pp. 119-148, John Wiley & Sons

Vercauteren, T. & Wang, X. (2005). Decentralized sigma-point information filters for target tracking in collaborative sensor networks. *IEEE T. Signal Processing*, vol. 53, no. 8, pp. 2997-3009

Walchli, M.; Skoczylas, P. Meer M. & Braun, T. (2007). Distributed event localization and tracking with wireless sensors, *Proceedings of the 5th Intl. Conf. on Wired/Wireless internet Communications*

Wang, Q.X.; Chen, W.P. Zheng, R. Lee, K. & Sha, L. (2003). Acoustic target tracking using tiny wireless sensor devices, *Proceedings of the Int. Workshop on Information Processing in Sensor Networks*

Wang, Z.; Li, H. Shen, X. Sun, X. & Wang, Z. (2008). Tracking and predicting moving targets in hierarchical sensor networks, *Proceedings of the IEEE Intl. Conf. on Networking, Sensing and Control*, pp. 1169-1173

Xiao, J.J.; Cui, S. Luo, Z. Q. & Goldsmith, A.J. (2006). Power scheduling of universal decentralized estimation in sensor networks. *IEEE Trans. Signal Process.*, vol. 54, no. 2 , pp. 413-421

Xiao, L. & Boyd, S. (2004). Fast linear iterations for distributed averaging. *Syst. Control Lett.*, vol. 53, pp. 65–78

Xu, J. & Li, J. (2010). State estimation with quantized sensor information in wireless sensor networks. *IET Signal Processing*, (peer reviewed, to appear).

Xu Y. & Lee, W.C. (2003). On localized prediction for power efficient object tracking in sensor networks, *Proceedings of the 1st Intl. Workshop on Mobile Distributed computing*, Providence RI, pp. 434–439

Yang, H. & Sikdor, B. (2003). A protocol for tracking mobile targets using sensor network, *Proceedings of the First IEEE International Workshop on Sensor Network Protocols and Applications*, Anchorage, Alaska, pp. 71–81

Yick, J.; Mukherjee, B. & Ghosal, D. (2008). Wireless sensor network survey. *Computer Networks*, vol. 52, no. 12, pp. 2292-2330

You, K.; Xie, L. Sun, S. & Xiao, W. (2008). Multiple-level quantized innovation Kalman filter, *Proceedings of the 17th IFAC World Congress*, Seoul, Korea

Younis, O. & Fahmy, S. (2004). Distributed clustering in ad-hoc sensor networks: a hybrid, energy-efficient approach, *Proceedings of the 23rd Annual Joint Conf. IEEE Computer & Communications Societies*, INFOCOM

Zhang, W. & Cao, G. (2004a). Optimizing tree reconfiguration for mobile target tracking in sensor networks. IEEE INFOCOM, HongKong, China

Zhang, W. & Cao, G. (2004b). DCTC: Dynamic convoy tree-based collaboration for target tracking in sensor networks. *IEEE Transactions on Wireless Communications*, vol. 3, no. 5, pp. 1689-1701

Zhao, F.; Shin, J. & Reich, J. (2002). Information-driven dynamic sensor collaboration for tracking applications. *IEEE Signal Proces. Mag.*, vol.19, no. 2, pp. 61-72

Zhou, Y. & Li, J. (2008a). Data fusion of unknown correlations using internal ellipsoidal approximation, Proceedings 17th International Foundation Automatic Control World Congress, Seoul, Korea

Zhou, Y. & Li, J. (2008b). Robust decentralized data fusion based on internal ellipsoid approximation, *Proceedings of the 17th International Foundation Automatic Control World Congress*, Seoul, Korea

Zhou, Y. & Li, J. (2009a). Average consensus based scalable robust filtering for sensor network, *Proceedings of the 5th Intl. Conf. Wireless Communications, Networking and Mobile Computing*, Beijing, China

Zhou, Y. & Li, J. (2009b). Scalable distributed sigma-point Kalman filtering for sensor networks: dynamic consensus approach, *Proceedings of the IEEE Intl. Conf. on Systems, Man, and Cybernetics*, San Antonio, Taxas, USA

Zhou, Y. & Li, J. (2009c). Quantized measurement fusion for target tracking in wireless sensor networks, *Proceedings of the 48th IEEE Conf. on Decision and Control and 28th Chinese Control Conference*, pp. 6835-6840, Shanghai, China

Zhou, Y. & Li, J. (2009d). Collaborative maneuvering target tracking in wireless sensor network with quantized measurements, *Proceedings of the IEEE Intl. Conf. on Systems, Man, and Cybernetics*, San Antonio, Taxas, USA

Zhou, Y. Li, J. & Wang, D. (2009a). Weighted average approach to quantized measurement fusion in wireless sensor network, *Proceedings of the 5th Intl. Conf. Wireless Communications, Networking and Mobile Computing*, Beijing, China

Zhou, Y.; Li, J. & Wang, D. (2009b). Quantized measurement fusion in wireless sensor network with correlated sensor noises, *Proceedings of the 7th IEEE Intl. Conf. on Control & Automation*, pp. 1868-1873, Christchurch, New Zealand

Zhou, Y.; Li, J. & Wang, D. (2009c). Collaborative Target Tracking in Wireless Sensor Networks using Quantized Innovations and Sigma-Point Kalman Filtering, *Proceedings of the IEEE Intl. Symp. Industrial Electronics*, Seoul, Korea

Zhou, Y.; Li, J. & Wang, D. (2009d). Unscented Kalman filtering based quantized innovation fusion for target tracking in WSN with feedback, *Proceedings of the Intl. Conf. Machine Learning and Cybernetics*, Baoding, China

Zhou, Y.; Li, J. & Wang, D. (2010). Posterior Cramér-Rao lower bounds for target tracking in sensor networks with quantized range-only measurements. *IEEE Signal Processing Letters* , vol. 17, no. 2, pp. 157-160

Zhou, Y.; Li, J. & Wang, D. (2011). Target tracking in wireless sensor networks using adaptive measurement quantization. *Science In China – Series F: Information Sciences* (peer reviewed, to appear).

Zou, Y. & Chakrabarty, K. (2003). Target localization based on energy considerations in distributed sensor networks, *Proceedings of the 1st IEEE Intl. Workshop on Sensor Network Protocols and Applications*, Anchorage, Alaska, pp.51–58

20

A Gaussian Mixture Model-based Event-Driven Continuous Boundary Detection in 3D Wireless Sensor Networks

Jiehui Chen[1,2], Mariam B.Salim[2] and Mitsuji Matsumoto[2]
[1]*Global COE Program International Research and Education Center for Ambient SoC sponsored by MEXT,Japan*
[2]*Graduate School of Global Information and Telecommunication Studies Waseda University, Tokyo, Japan*

1. Introduction

Wireless sensor networks (WSNs) may consist of tiny, energy efficient sensor nodes communicating via wireless channels, performing distributed sensing and collaborative tasks for a variety of monitoring applications. One of the critical problems in sensor applications is detecting boundary sensors in a complex sensor network environment where sensed data is often required to be associated with spatial coordinates. In (Zhong, et al, 2007) a COBOM protocol that monitors the boundary of a continuous object was proposed. Sensor nodes are assigned with a Boundary sensor Node (BN) array to store BN information. The boundary monitoring is based on the changes to the observations in the BN array. As a updated version, (Kim,J.H. et al,2008) presented the DEMOCO protocol that enhanced COBOM by considering sensor nodes on one side of the boundary line called the "IN" range, and ignoring those on the other side of the boundary line called the "OUT" range which theoretically reduces approximately by half of the number of the selected BNs. Others like (Basu, et al, 2006; Eren,T. et al, 2004; He,T. et al,2003; Nissanka,B. et al, 2003) also involve two-dimensional (2D) sensor localizations. To address the issues of adaptive sensor coverage and tracking for dynamic network topology, the authors of (Guo, et al, 2008) utilized a Gaussian mixture model to characterize the mixture distribution of object locations and proposed a novel methodology to adaptively update sensor node placement according to the ML estimates of mass object locations with a distributed implementation of an EM algorithm to reduce communication costs. Moreover, (Olfati-Saber, et al, 2007) discussed a flocking-base mobility model for Distributed Kalman Filtering (DKF) in mobile sensor networks and (Funke, et al, 2006; Funke, et al, 2007) demonstrated efficient boundary detection algorithms with only the connectivity information.

In fact, the boundary detection problem has been mostly considered for 2D sensor networks and the case of 3D sensor networks has gone practically unnoticed. Despite the fact that difference between the normal 2D and the more realistic 3D scenario is only one extra dimension, network topology could be much more complex and the location scheme has to be more robust towards network irregularities. Taking a step further to expand from 2D to 3D sensor applications, several neighborhood-measurement (Peng, et al, 2006) based 3D range-free boundary detection models (Lance, et al, 2001; Hu, et al, 2004; Ji, et al, 2004; Yi, et al, 2003; Yi, et al, 2004, Andreas, et al, 2002) have been proposed. However, their tight dependence on sensor node densities and availability of sufficient neighbors are too optimistic for real 3D sensor applications due to their non-uniform sensor node densities and topology randomization. On the other hand, a range-based model such as in (Zhang, et al, 2006) does not make any assumption about sensor node densities and network topology. Instead it introduced a strong entity called mobile location assistants (LAs) that enables each location-unaware sensor node to easily estimate its own position using the measurable AOAs (Peng, et al, 2006) and RSS (Samitha, et al, 2010). Similar approaches like (He, et al, 2003; Nissanka, et al, 2003; Bulusu, et al, 2001) assume that a small fraction of sensor nodes called anchors or beacons have a priori knowledge of their location and (Liu&Wang, et al, 2009) proposed a range-based positioning method using beacon signals, that doesn't require time synchronization since the beacon sensor nodes estimate the range based on frequency differences instead of time differences. To conclude, all the aforementioned approaches either introduced strong entities or made irrational assumptions. Furthermore, (Liu&Manli, et al, 2009) presented a new high precision WSN positioning method with reasonable implementation cost for a 3D case. Reference sensor nodes with known locations transmit linear frequency modulation continuous waves (FMCWs), while other sensor nodes estimate the range difference to them based on the received signals' frequency difference, called time frequency difference arrival (TFDA).

Motivated by all above observations, instead of introducing miraculous assisting entities, our range-free Gaussian Mixture Model (GMM)-based approach performs a connectivity information-based segmentation algorithm (Zhu, et al, 2009) that partitions an irregular sensor field into nicely shaped pieces, associated with an enhanced BN Array and efficient distributed in-network information extraction virtual Thick Section Model (TSM); to the best of our knowledge, this is the first work that presents a principled algorithmic approach integrating computational geometry constructs adopted simultaneously for boundary detection in both 2D and 3D network areas. It is promising that our new statistical Gaussian mixture model (Mclachlan, et al, 2000)-based method in this paper is capable of fusing multivariate real-valued sensor inputs to detect boundaries of events in a mathematically principled manner. More precisely, the distribution of sensor readings within each sensor node's spatial neighborhood is mathematically formulated using most popular finite GMMs. The model selection techniques (Figueiredo, et al, 2002;Akaike, et al, 1973; Schwarz, et al, 1978; Solla, et al, 2000) can then effectively identify the correct number of modes for finite mixture models. Therefore, Boundary and Non-Boundary sensor nodes can be consequently distinguished from their neighboring sensor node data distributions.

The remainder of this paper is organized as follows: the next section details enhancement to the BN Array concept; Section 3 simply describes general problems in boundary detection; Section 4 presents the proposed robust Boundary Detection scheme for 3D (BD3D) sensor networks in detail; Section 5 proves BD3D by simulation results; Finally, Section 6 concludes the paper with future work.

2. Enhancement to BN Array

In (Chintalapudi, et al, 2003) three different schemes which can only take inputs of the 0/1 decision predicates from neighboring sensor nodes are proposed. (Jin, et al,2006) presents a noise-tolerant algorithm named NED for event and event boundary detection. In NED, the moving mean of the readings of the neighboring sensor node set is used as the estimate for a certain sensor node. The authors of (Min, et al, 2005) propose Median-based approaches for outlying classification and event frontline detection, where the median is a useful and robust estimator which works directly with continuous numbers, rather than binary 0/1 readings. An extra description of the BN-Array of COBOM (Zhong, et al, 2007) and DECOMO (Kim, et al, 2008) is given in this section. Suppose we have a sensor node v (N_v) and its neighbors $\xi(N_v) = \sum_{i=0}^{k} N_{u_i}$(k is the potential number of neighbors) (k = 6 in Figure 1). Let us consider the BN array in (Kim, et al, 2008; Zhong, et al, 2007):

1	1	1	1	0	0
N_{u_1}	N_{u_2}	N_{u_3}	N_{u_4}	N_{u_5}	N_{u_6}

Table 1. BN Array of N_v [Note: "0" and "1" are sensor readings (sample)].

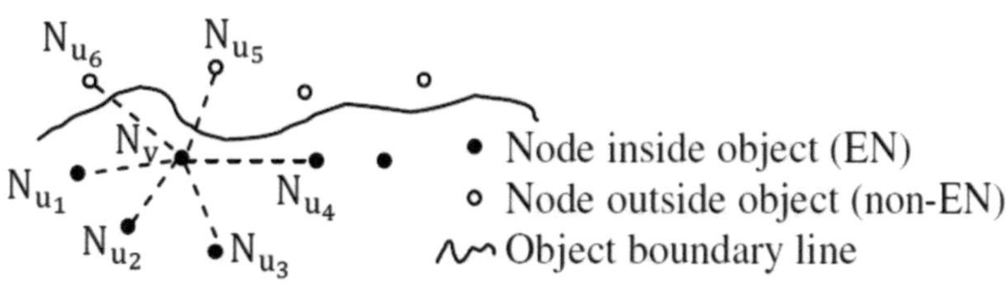

Fig. 1. Readings of neighbors in BN Array of N_v.

In Figure 1, the sensor readings of $\xi(N_v)$ only indicate the relative locations of its neighbors only. Correspondingly, there is no own sensor reading, as a result, N_v judges itself by inquiring $\xi(N_v)$ in a time and energy consuming way. In our model, we applied a head with 1 byte more space for the BN Array to store its own sensor reading as well (see Table 2) for self-judgment as a EBN or non-EBN. Here, we denote a BN inside object as Event BN (EBN), and a BN outside object as non-EBN. That is very important for monitoring applications in the sensor network because an Event sensor Node (EN) is usually highly responsible for sending and receiving the aggregated data should be constantly aware of own status.

1(head)	1	1	1	1	0	0
N_v	N_{u_1}	N_{u_2}	N_{u_3}	N_{u_4}	N_{u_5}	N_{u_6}

Table 2. BD3D BN Array of N_v.

Figure 2(a,b) show the expected boundary lines in COBOM and DEMOCO, respectively. Despite the fact that the shape of the expected boundary line in the 2D model of BD3D (see

Figure 3) is similar to that of DEMOCO, the knowledge about Boundary sensor Nodes (BNs) promises to be different because we can clearly distinguish EBN and non-EBN as well.

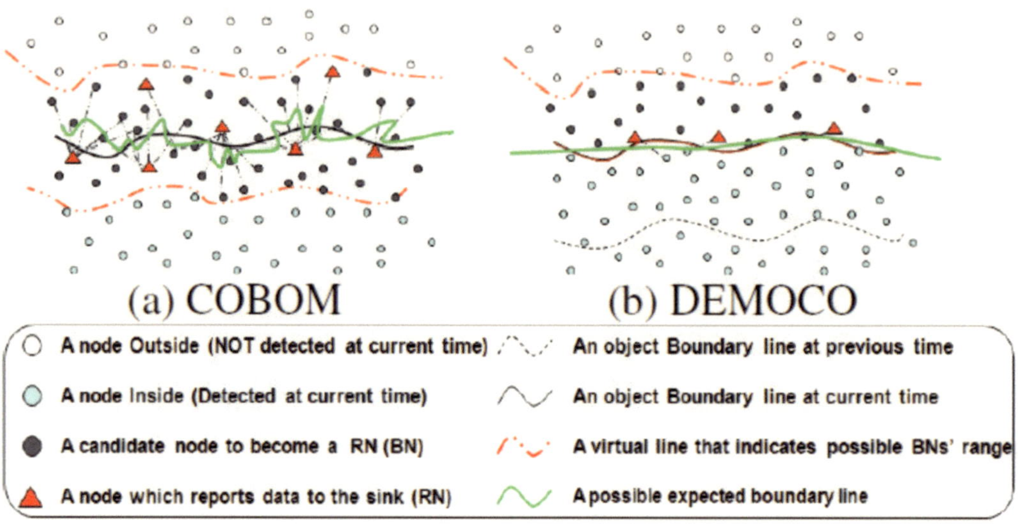

Fig. 2. Expected boundary lines (Kim, et al, 2008).

3. Problem

We first present the problems before outlining how our proposal can benefit dynamic boundary detection for 2D and 3D sensor networks in the coming sections. To generally analyze the existing problems for superior boundary detection in a 3D impediment scenario, sensor nodes in the network usually have slight mobility which makes it difficult to establish their locations. Figure 3 illustrates the possible boundary line changes in a 2D scenario when the object is shrunk or expanded which becomes a big problem that involves frequent inquiries among BNs and massive modifications to BN arrays.

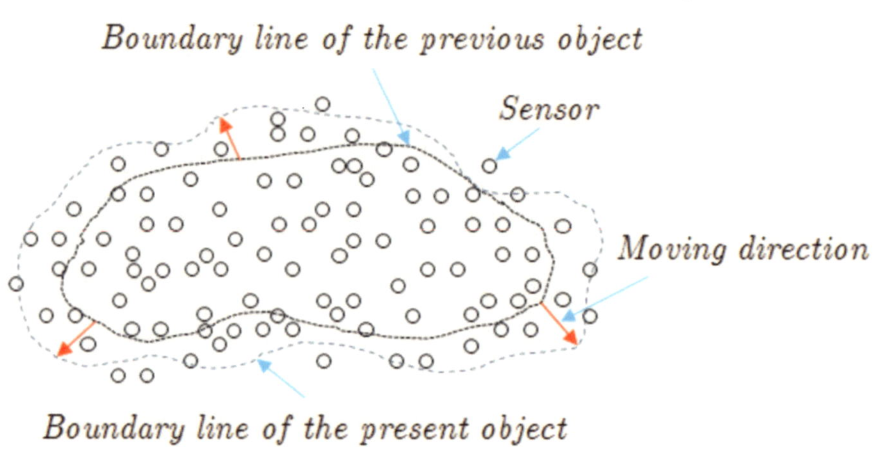

Fig. 3. Possible boundary line changes when the object shrunk or expanded.

4. Boundary Detection for a 3D Sensor Network (BD3D)

This section involves the main objective of achieving a flexible and energy-efficient 3D continuous boundary detection with a clear knowledge of EBN and non-EBN. Assume that sensor nodes are randomly deployed over 3-dimensional terrain. Each sensor node has limited resources (CPU, battery, etc), and is equipped with an omni-directional antenna. For the radio model, E_{elec} is for running the transmitter or receiver circuitry and $\mathcal{E}_{amp}$ is for the transmit amplifier. To transmit a δ -bit message a distance l using this radio model, the radio expends ($E_{elec} \times \delta + \mathcal{E}_{amp} \times \delta \times l^2$), to receive the message, the radio expands ($E_{elec} \times \delta$) (Heinzelman, et al, 2000). This energy model assumes a continuous energy consumption function. Moreover, we currently assume that sensor node failures are primarily caused by energy depletion. Note that in our model, no assumptions are made about (1) homogeneity of sensor node distribution; (2) network and BN density; (3) proximity of querying observers and sensor node synchronization.

Our major contribution could be creating a statistical property of the finite mixture model, especially the Gaussian mixture model (GMM) and adopting it to distributed sensing scenarios. Suppose that we have a set of data observations $\psi_i = \{\chi_1, \chi_2, \ldots, \chi_n\}$, $n \le N$ (N is the total number of sensor nodes in the network) with each χ_i representing a D-dimensional random vector. Assume that ψ_i follows a k-component finite mixture distribution (Mclachlan, et al,2000) as follows:

$$\mathcal{P}(\chi_i|\theta) = \sum_{j=1}^{k} \alpha_j \mathcal{P}(\chi_i \mid \theta_j),\ j = 1,2,\ldots,k;\ I = 1,2,\ldots,n. \quad (1)$$

subject to $\sum_{j=1}^{k} \alpha_j = 1$

where α_j is the mixing weight or sometimes called the prior weight and θ_j is the set of parameters of the jth mixture component $\mathcal{P}(\chi_i|\theta)$. Denote $\theta = \{\alpha_1, \theta_1, \alpha_2, \theta_2, \ldots, \alpha_k, \theta_k\}$. The objective function of estimating θ from ψ_i is to maximize the log-likelihood criterion as follows:

$$\text{Log} \prod_{i=1}^{N} \mathcal{P}(\psi_i|\theta) = \sum_{i=1}^{n} \log \sum_{j=1}^{k} \alpha_j \mathcal{P}(\chi_i|\theta_j) \quad (2)$$

Therefore, the maximum likelihood estimator of θ is:

$$\widehat{\theta}_{ML} = \arg \underbrace{\max}_{\theta} \{\log \prod_{i=1}^{n} \mathcal{P}(\psi_i|\theta)\} \quad (3)$$

Obviously, $\hat{\theta}_{ML}$ cannot be computed analytically from the above equation. Instead, GMM is applied as its general solver to iteratively find the maximum likelihood solution of $\hat{\theta}_{ML}$. GMM is the most important class of finite mixture densities. GMM is formulated by using a Gaussian density $\mathcal{G}(\chi_i \mid \mu_j, \Sigma_j)$ with its mean vector μ_i and covariance matrix Σ_j to replace the general probability density function $\mathcal{P}(\chi_i|\theta_j)$ in the finite mixture model:

$$\mathcal{P}(\chi_i|\theta) = \sum_{j=1}^{k} \alpha_j \mathcal{G}(\chi_i \mid \mu_j, \Sigma_j) \quad (4)$$

where a multi-dimensional multivariate Gaussian distribution is defined as:

$$\mathcal{G}(\chi|\mu, \Sigma) = \frac{1}{|\Sigma|^{\frac{1}{2}}(2\pi)^{\frac{D}{2}}} \exp\{-\frac{1}{2}(\chi - \mu)'\Sigma^{-1}(\chi - \mu)\} \quad (5)$$

The Bayesian Information Criterion (BIC) (Schwarz, et al,1978) is one of the most popular model selection criteria based on penalty terms of model complexity. In this paper, we use BIC for GMM model selection:

$$\text{BIC}(\theta) = -2\log(\mathcal{P}(\psi_i \mid \theta)) + K\log(m) \quad (6)$$

where, m is the data sample number, and K is the total number of parameters to be estimated in GMM.

In this paper, we provide an algorithm for classifying EBNs. Given a sensor network $\{S_i\}$, we assume that sensor nodes are deployed with moderate density in the spatial terrain. From a mathematical perspective, sensor readings provide a dense, but discrete sampling of the underlying continuous distribution. To check whether or not N_v is a sensor node lying on the boundary of an event, we put the data $\{\chi_n\}$ from readings of the sensor nodes in $\xi(N_v)$ and then build our best GMM based on $\{\chi_n\}$.

In more detail, we first set the upper bound of the mixture component number to be K. Then for each j = 1,2,…,K, the data set $\{\chi_n\}$is fed into (5) (6) for estimation of $\theta(J)$. Let BM denote the number of mixture components of the best model. We select BM where $\text{BIC}(\theta(\text{BM})) = \min_{j=1}^{K} \text{BIC}(\theta(J))$. Therefore our final is $\theta(\text{BM})$ or $\{\mu_j, \Sigma_j, \alpha_j\}_{j=1,2,\ldots,BM}$.

To classify if N_v is a EBN, the conditional probability for χ_i given model $\theta'(\text{BM})$ is computed by

$$\mathcal{P}(\chi_i \mid \theta'(\text{BM})) = \sum_{j=1}^{BM} \alpha_j \mathcal{G}(\chi_i \mid \mu_j, \Sigma_j) \quad (7)$$

then $\mathcal{P}(\chi_i \mid \theta'(\text{BM})) < \gamma$, N_v is classified as a EBN. Where γ is used as a threshold to measure EBN which has significantly low probability density values given the final model $\theta'(\text{BM})$. The threshold is set as $\gamma = 0.25$, the upper bound of the component number is set as K = 5. These parameters are used as the default in Sections 4 and 5, unless otherwise stated.

To dynamically update the estimates of observations by conducting (Zivkovic, et al,2004), we have the following dynamic evolvement and observation equations:

$$\chi_v^i = f(\chi_v^{i-1}) + w_v^i \quad (8)$$

$$\chi_v^i = g(\chi_v^{i-1}) + v_v^i \quad (9)$$

where f(.) is the linear or nonlinear state evolvement function and g(.) is highly nonlinear observation function. w_v^i and v_v^i are the standard deviation (noise sequences). For example in static sensor node location, where χ_v^i remain the same after deployment because of the governance of Equation (8). Therefore, we get the expression:

$$\chi_v^{i+1} = f(\chi_v^i) + w_v^i \quad (10)$$

where, w_v^i model the small position perturbation or other effects.

4.1. BD3D scheme in 2D model [D = 2 in (5)]

Suppose n total sensor nodes are randomly distributed in a 2D terrain, with network density μ enough to perform a boundary detection application. BD3D provides simultaneous selection of EBN and non-EBN during BN selection process by tactfully using the proposed BD3D BN Array (see Table 3) and GMM-based mathematic model.

Sensor reading of N_v(***head***)	Sensor readings of $\xi(N_v)$(***rear***)

Table 3. BD3D BN Array of N_v.

Note: $\xi(N_v) = \sum_{i=1}^{k} N_{u_i}$ (see Section 2) and both head and rear are initialized with "0"; Rule: a sensor node is EN if its own reading equals to "1" and vice versa.
Although the determination of sensor node status e.g., EBN or non-EBN etc. is practically meaningful, in literature fewer works are focusing on this issue. In BD3D, we tactfully utilize a BD3D BN array to adequately energy-friendly determine sensor node status (see Table 4).

BD3D BN Array			*BN*			
	EN	*Non-EN*	*EBN*	*Non-EBN*	*Non-BN*	
Head	1	0	1	1	0	1
HR	random	random	All 0 & random	All 1	All 0	All 1

Table 4. Head &HR based sensor node status determination.

*EBN ∈ EN, EBN ∪ non-EBN = BN (see Figure 4) and *random* means it is either all 1 or all 0.

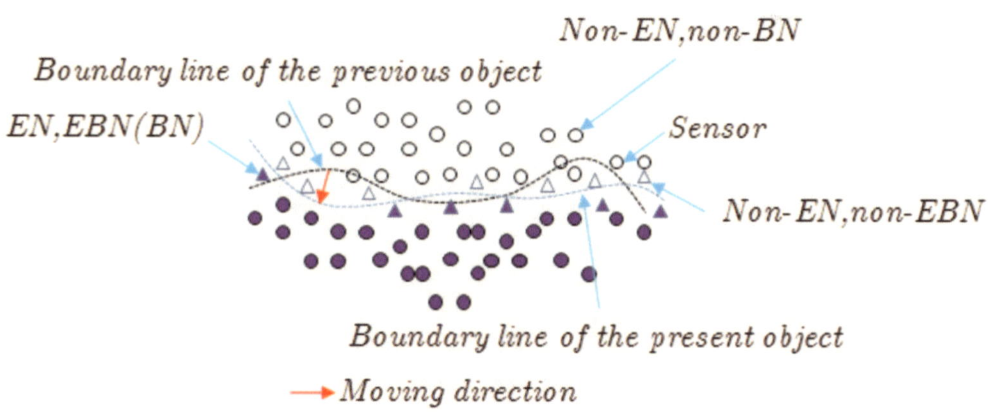

Fig. 4. EBN and non-EBN on BL in BD3D 2D model when object expanded or shrunk.

$$\text{head}=\begin{cases} 1 & \text{EN, EBN(BN), non}-\text{BN} \\ 0 & \text{non}-\text{EN, non}-\text{BN} \end{cases}$$

$$\text{HR}=\begin{cases} \text{all 1} & \text{non}-\text{EBN(BN), non}-\text{BN} \\ \text{all 0} & \text{EBN(BN), non}-\text{BN} \\ \text{random} & \text{EN, non}-\text{EN or EBN(BN)} \end{cases}$$

1	0	1	1	1	1	1	0	1

Table 5. Example of BD3D BN Array of N_v^i.

N_v^i is determined to be EN or EBN (BN) if head is 1 and HR is random based on the values in Table 5.

Expanding from 2D to 3D, we find that not only the sensing area of sensor node but the network topology is getting more complex, therefore when talking about the relative position of sensor nodes, we need 3D sense of space to construct the model.

4.2. BD3D 3D model [D = 3 in (5)]

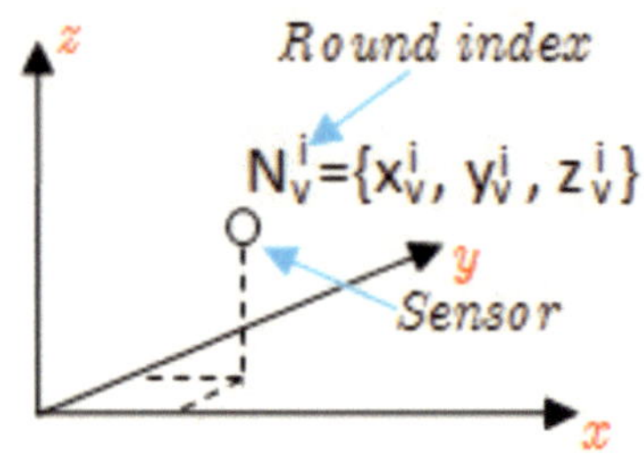

Fig. 5. Position of N_v in 3D co-ordinate.

Define the state variable as 3D position for a specific sensor node modeled in Figure 5.

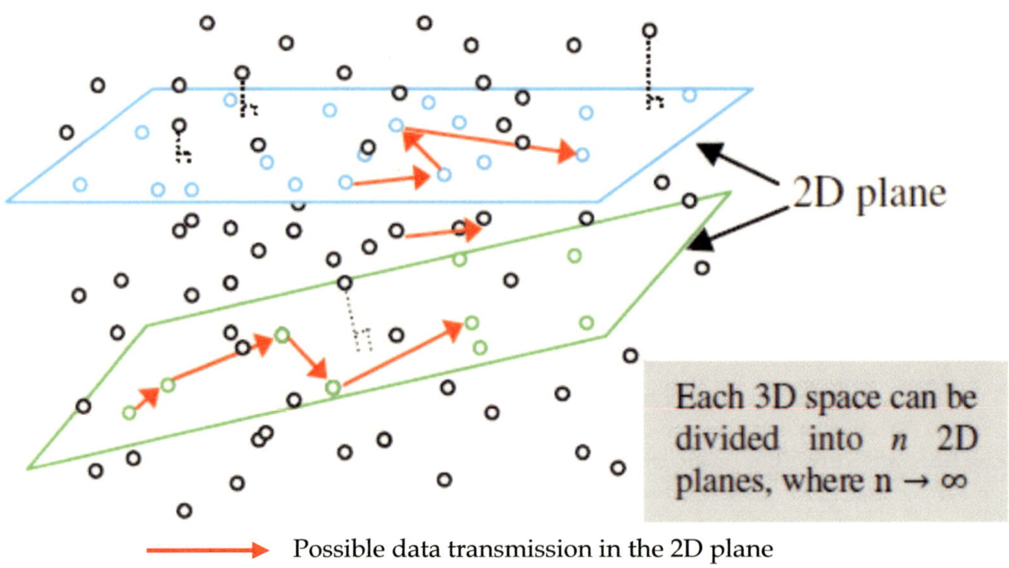

Fig. 6. Concept of 2D plane for 3D sensing space.

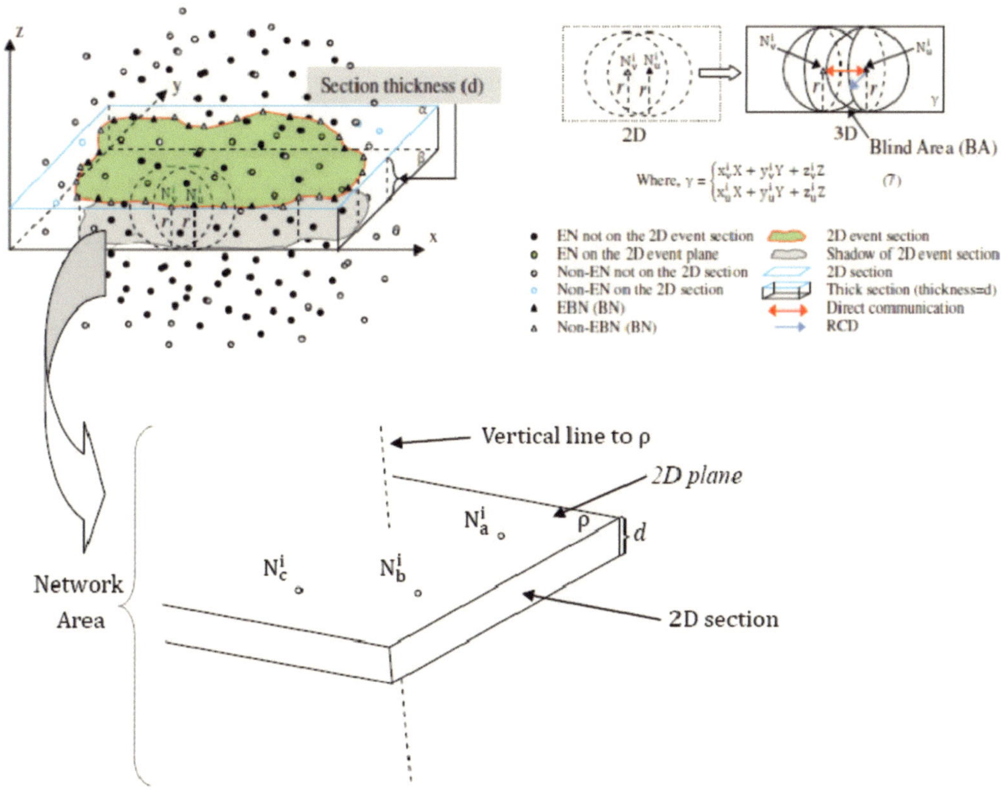

Fig. 7. TSM concept with $d = r$ for explanation simplicity.

In a 3D sensing space, sensor nodes are randomly distributed to form a network. To simplify the complicated operations in dealing with sensor node localization in 3D model, we apply a new concept of 2D plane that each 3D space can be divided into n 2D planes, where $n \to \infty$ (see Figure 6). The methodology of selection and representation of the 2D plane is described as:

- Randomly pick up to three sensor nodes $\{N_a^i\,(x_a^i, y_a^i, z_a^i),\ N_b^i\,(x_b^i, y_b^i, z_b^i),\ N_c^i\,(x_c^i, y_c^i, z_c^i)\}$(see Figure 7) from the 3D sensing space to form a 2D plane (either the 2D plane in blue or green in Figure 6).
- Suppose N_a^i N_b^i and N_c^i are arbitrary points (sensor nodes) on the formed 2D plane called "ρ", the plane representation is (see Figure 7).

$$\rho = \begin{cases} x_a^i X + y_a^i Y + z_a^i Z \\ x_b^i X + y_b^i Y + z_b^i Z \\ x_c^i X + y_c^i Y + z_c^i Z \end{cases} \tag{11}$$

Strictly speaking, a 2D plane is definitely as a 2D section (see Figure 7). Therefore, one 3D sensor network needs n 2D sections ($n \to \infty$) to reconstruct. Due to the impossibility of computing n in programming, we introduce a virtual Thick Section Model (TSM).

Figure 7 may help understand the concept of TSM. A 2D section is modeled as a thick plane (ρ) with section thickness (*d*) with a set of representative points { $N_a^i(x_a^i, y_a^i, z_a^i)$, $N_b^i(x_b^i\ y_b^i, z_b^i)$,$N_c^i(x_c^i\ y_c^i, z_c^i)$ }describing the elements of the section. In our model the boundaries are actually modeled as parametric line segments and points taking into account not only the position of the plane but also the uncertainties of the plane contour.
Suppose the 3D network area is ζ^3 cube (ζ is pre-determined in programming) and $d \neq r$. Therefore the selected 2D section could be those in Figure 8:

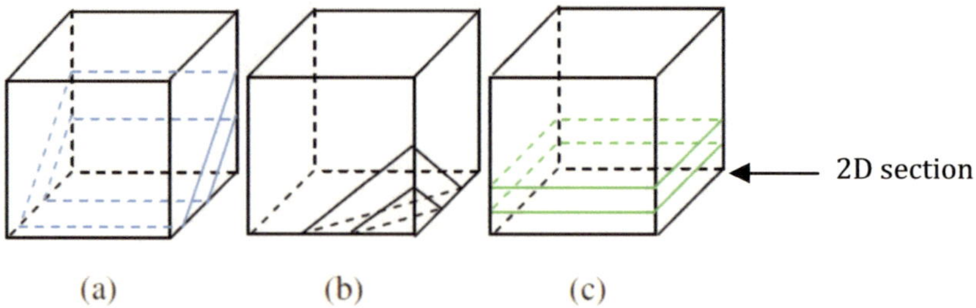

Fig. 8. Possible 2D sections in 3D network area and (c) is the model used in simulations.

The 2D section in our simulation model actually is a (ζ^2*d) area [see Figure 8 (c)]. (the section thickness (*d*) is determined a priori to programming).Thus, the simulator only need to perform TSM (ζ/d times that significantly improve the maneuverability.

5. Simulation

In this section, we evaluated the performance of BD3D 2D and 3D model implemented in Matlab respectively. The simulation parameters are given in the following table Sensor nodes make local observations every 2 time slots:

Parameter	Value
Network Area (2D,3D)	$(100\text{ m})^2(100\text{ m})^3$
Number of sensor nodes(2D,3D)	2,500,10,000
The sink (2D,3D)	(50,175), (50,175,50)
Transmission range(2D)	10 m
Time slots	100 seconds
Initial Energy	2J/battery
Message size	100 Bytes
E_{elec}	50 nJ/bit
E_{fs}	10 pJ/bit/m2
δ_{amp}	0.0013 pJ/bit/m4
E_{DA}	5 nJ/bit/signal

5.1. Simulation model

(2D model)

- Design a regular variation object: a circle initially centered at (50, 50) and continually expand it by increasing its radius by 10 meters every 10 time slots. (see Figure 9.)

- Design an irregular variation object: the initial ENs that adequately covers a area $\{(x-50)^2+(y-50)^2=R_{circle}^2\}$ to initiate the event. At every time slot, EN propagates by picking up a random number of neighbors to join the event. In this way, the network is guaranteed to be fully connected. (see Figure 9.)

(3D model)

- Design a regular variation 3D object: the object center is (50, 50, 50) and continually expand its radius by 10 meters every 10 time slots.
- Design an irregular variation 3D object: the initial ENs are within a spherical area $\{(x-50)^2+(y-50)^2+(z-50)^2 = R_{sphere}^2$. EN propagates in a similar way as that used for the irregular variation object in 2D model.
-

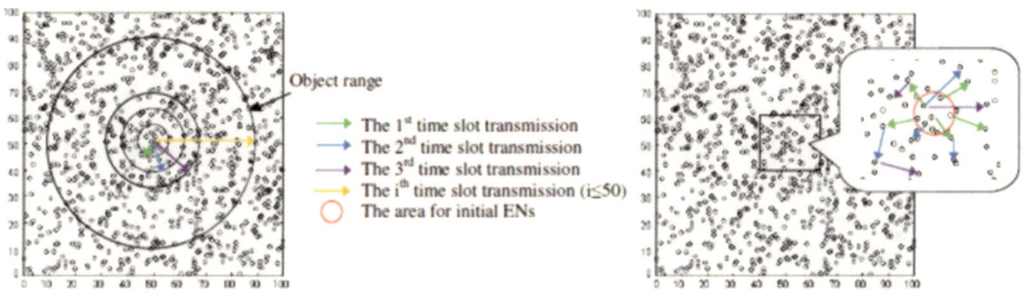

(a) Regular variation object (b) Irregular variation object

Fig. 9. Sample of BD3D 2D model with regular variation and irregular variation object.

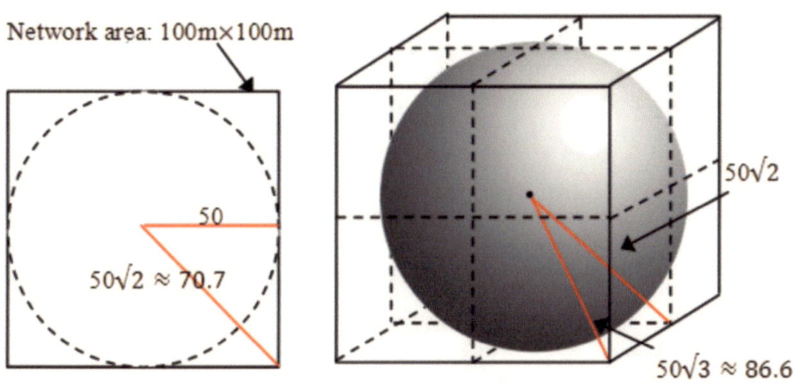

(a) 2D model
Rebound distance=50m,
Boundary distance=50√2m

(b) 3D model
Rebound distance=50m
Boundary distance=50√3m

Fig. 10. Rebound and boundary distances for BD3D.

The BD3D is flexible enough to be used in a clustered network or a non-clustered network since it does not put any constraints on cluster architecture. However, BNs are usually heavily utilized to send aggregated data associated with the object/network boundary information to cluster head (in clustered networks) or the sink (non-clustered networks), they would run out of energy more quickly. Therefore, achieving a reasonable amount of BNs (the less the better) benefits energy saving.

5.2. BD3D 2D model

This section discusses the performance evaluations based on BD3D 2D model. Figure 11 demonstrates the performances of the BD3D 2D model and DEMOCO and COBOM in terms of the average residual energy per sensor node at 50 and 100 time slots of operation respectively. Obviously, the performance of BD3D 2D is apparently better than DEMOCO and COBOM. However at the meanwhile, it shows the good stability of energy load balancing among the sensor nodes over the individual residual energy differences.

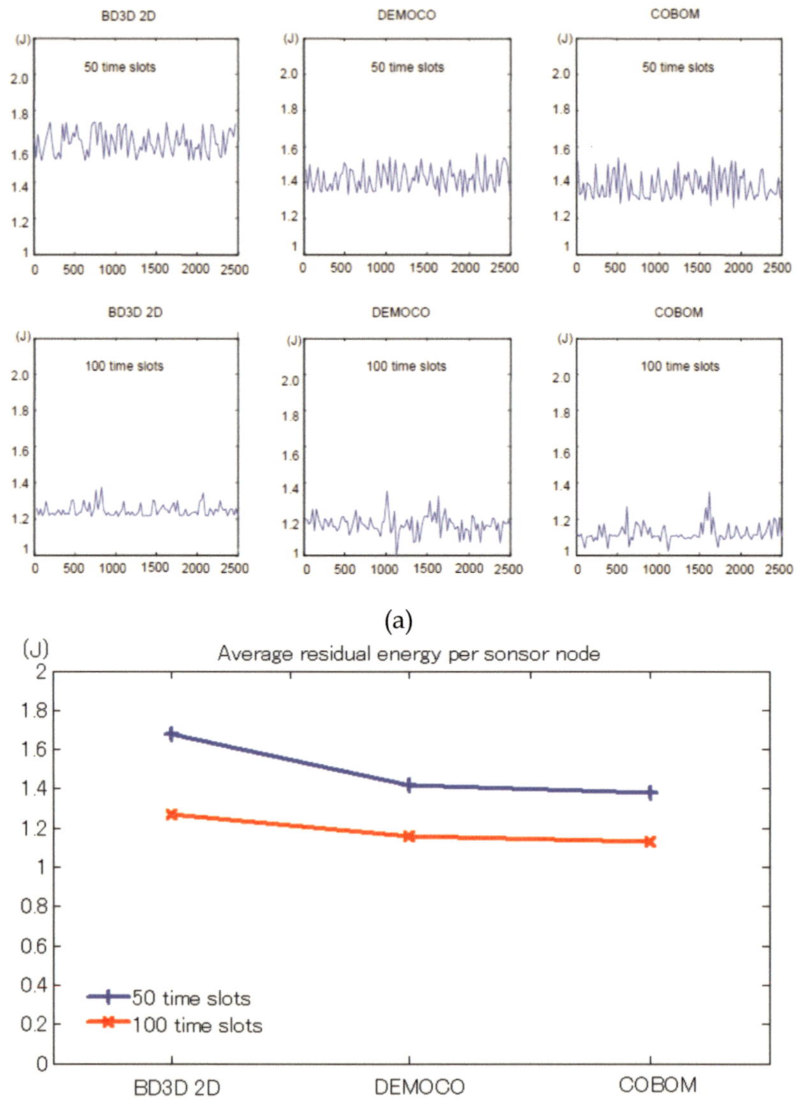

(b)

Fig. 11. Average energy level status of 2,500 sensor nodes after 50 and 100 time slots operation.

Comparison of the number of BNs for a regular variation object with COBOM and DEMOCO is shown in Figure 12 (a). To increase the comparability, the network is only operated during 50 time slots. Figure 12 (a) shows that the BD3D 2D model consistently provides less than half of the BNs selected by COBOM and reduces approximately by 1/3 those achieved by DEMOCO in the same environment. This could be due to the use of the BD3D BN array (see Section 2) and GMM that helps selecting potential BNs easier than the aforementioned COBOM and DEMOCO. Consequently, this avoids low data delivery rates and excessive energy consumption by frequent flooding of inquiring packets

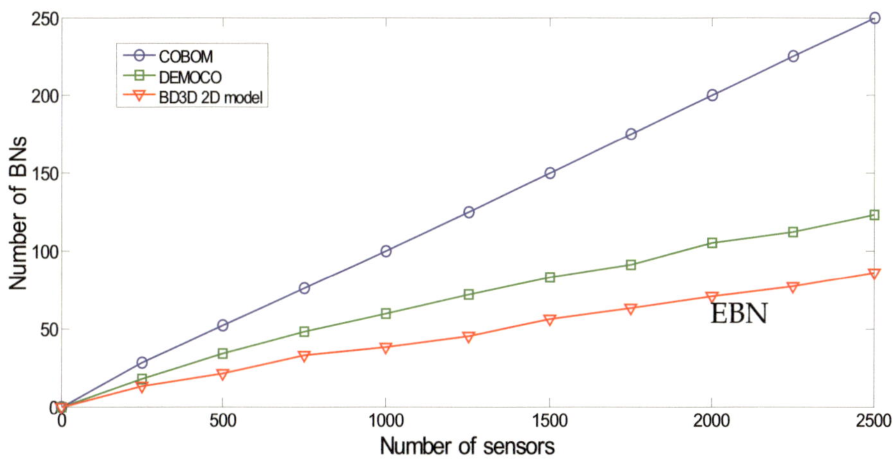

(a)

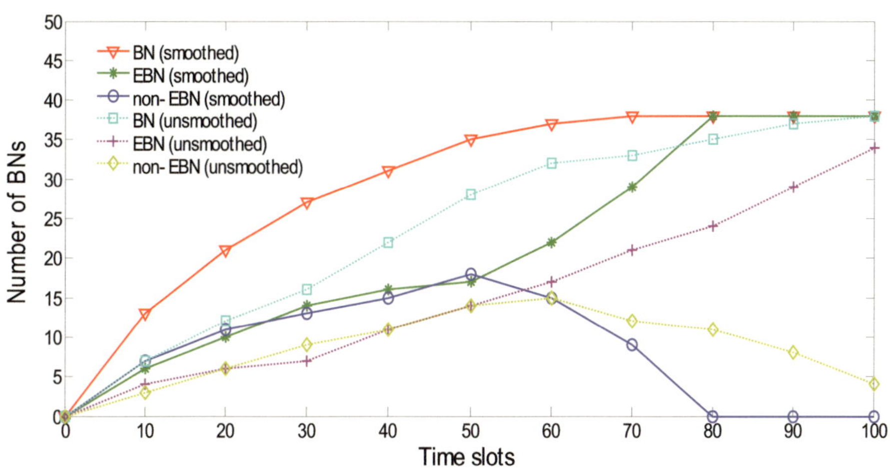

(b)

Fig. 12. Performance evaluation by using BD3D 2D when r = 10 m. **(a)** Comparison with DEMOCO and COBOM on the number of BNs based on the number of sensor nodes (regular variation object case for 50 time slots) **(b)** Number of BNs (EBNs and non-EBNs) based on time slots (both regular variation and irregular variation object case).

However, due to the elusive ways proposed to expand the irregular variation object, we can hardly do comparison with COBOM and DEMOCO anymore. Figure 12 (b) shows only the performance evaluation of the BD3D 2D model. As promised, EBNs and non-EBNs for regular variation and irregular variation object cases are clearly found. From our analysis, the value of BN (irregular variation) tends to be affected by irregular BL movements due to the elusive change of object compared to that of BN (regular variation) that looks more euphemistic. When the regular variation object expands over the *rebound distance* which indicates the end of saturated distribution around the BL, Figure 12 (b) shows a rapid increase of EBN (regular variation) and decrease of non-EBN (regular variation) until the object covers the whole network. On the other hand, due to the non-determinacy of irregular variation object shape changes, trajectories of both EBN (irregular variation) and non-EBN (irregular variation) are always difficult to size up. However, we find similarity or resemblance as to be essentially interchangeable for the first 50 time slots of operation between regular variation and irregular variation object cases. For the second 50 time slots, the performance of EBN (irregular variation) and non-EBN (irregular variation) are going to split up, but show no direct relationship with *rebound distance* and *boundary distance*.

5.3. BD3D 3D model

Figure 13 shows a vertical section view of 3D sensor network area using TSM—a combinational view of three conditions { $d < r$, $d > r$, $d >> r$ }。

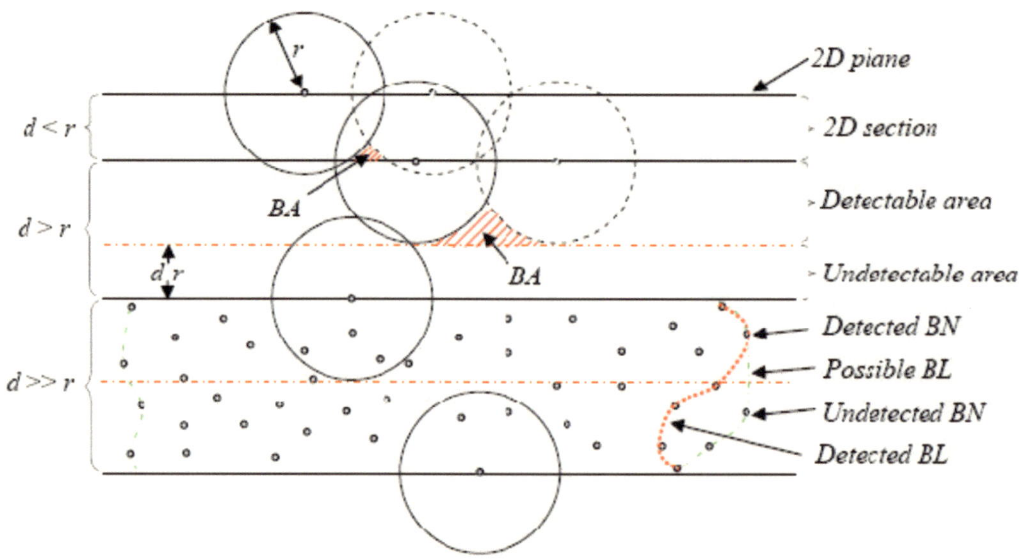

Fig. 13. A combinational vertical section view of 3D sensor network with {d < r, d > r, d >> r}

In this section, we modeled the BD3D 3D with different values of r and d by using TSM for regular variation and irregular variation objects, respectively. Figure 14(a) compares the number of BNs based on the value of d with r = 10m. As $d < r$, the values we got are approximately the same. Moreover, we varied the value of d ($d > r$) for simulating the cases with the significant existence of BA (see Figure 13), many BNs of highly possible BA got lost, resulting in decrease of the number of BNs. Figure 14(b) compares the number of BNs based on the value of r with d fixed at 8 m. As $d > r$, the performance shows the comparatively worst. By increasing r to meet $d < r$, it shows the significant improvement on the performances with very imperceptible distinctions. It is easy to guess that if communication range of a sensor node is large, there will be many neighbors that it can communicate with, which will result in more BNs.

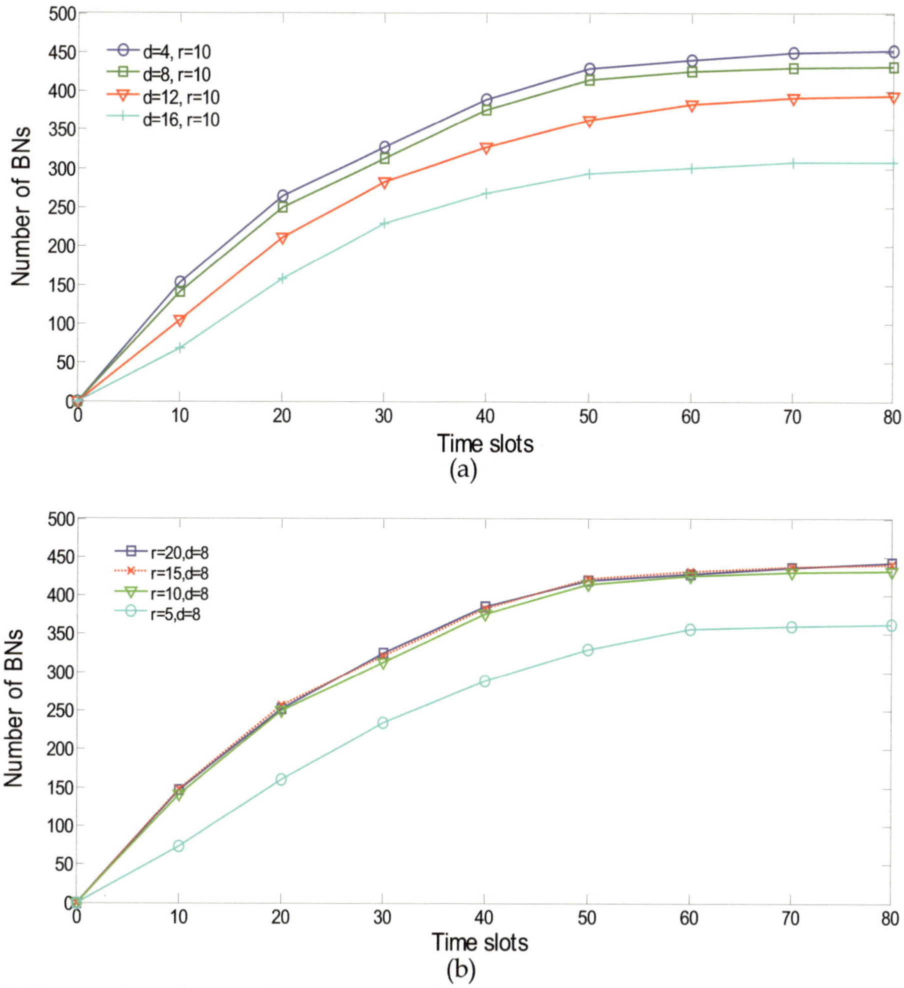

Fig. 14. Comparison for regular variation object case using BD3D 3D model. **(a)** Number of BNs based on time slots via varying d (r = 10 m), **(b)** Number of BNs based on time slots via varying r (d = 8 m).

Meanwhile, we set the same parameter environment in the BD3D 3D model for evaluating the number of BNs in the network in Figure 15. The most interesting feature is that the network based on $d < r$ apparently performs better than that with $d > r$. As a result, it can be alleged that there is no strong relationship between the number of BNs and the communication range (r) using TSM once $d < r$. This occurrence can be clarified by the analysis illustrated in Figure 13. Another interesting feature we can observe from Figure 15(b) is that when r = 5 and d = 8, it undergoes a very slow increase first and then experiences a sharp increase in the number of BNs after 40 time slots. This was caused by a phenomenon that the objects expand highly depending on the number of existing BNs. However, at network initialization, we have relatively fewer existing BNs. As the cardinal number designating the existence of BNs is over a special value (available at around 40 time slots), the performance miraculously achieves a sudden improvement.

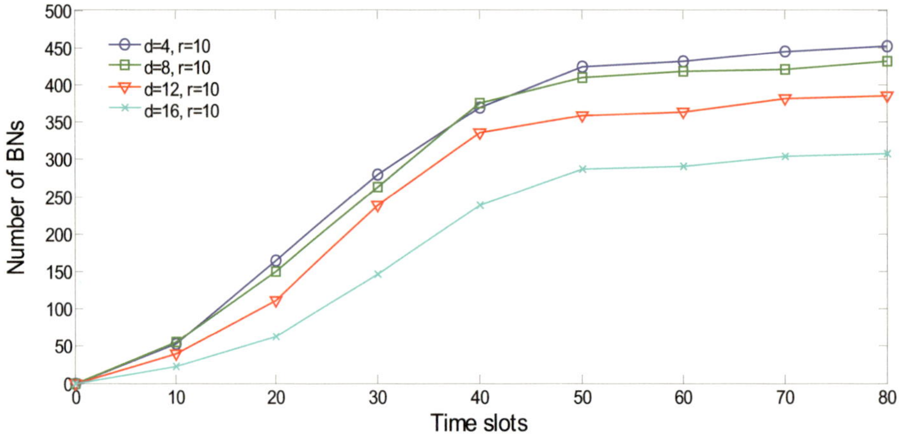

(a)

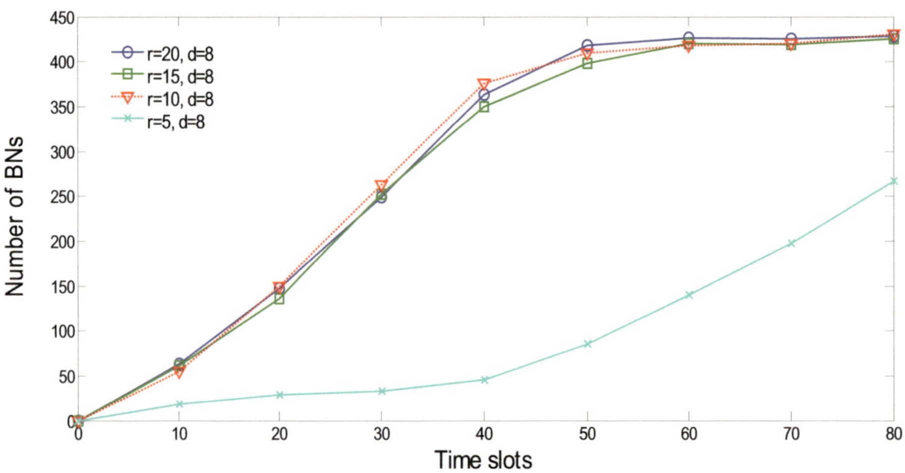

(b)

Fig. 15. Performance comparison for irregular variation object case using BD3D 3D model. **(a)**Number of BNs based on time slots via varying *d* (r = 10m); **(b)**Number of BNs based on time slots via varying *r* (d = 8m).

We hereby conclude that our BD3D for continuous boundary detection in 3D case works well especially when $d < r$ using TSM. An in depth study about the impact of localization impact on various routing protocols and its implications on design of location-dependent system are left as future work.

6. Conclusions

This paper has proposed a novel Gaussian Mixture Model-based BD3D scheme for boundary detection of continuously moving object in a 3D sensor network. We adequately presented the proposed protocol, and the simulation results shown support our allegation that the BD3D 2D model surely outperforms COBOM and DEMOCO in terms of average residual energy per sensor node and the number of selected BNs, and the BD3D 3D model achieves accurate boundary detections by soundly selecting EBN and non-EBN for both regular variation and irregular variation object cases. Our future work will include additional optimization desired to improve the performance of our algorithm and verification of the precision of the expected boundaries and invention of a new protocol that considers data losses and route failures due to unpredictable errors such as sensor node failures, contention, interference and fading (Woo, et al, 2003; Seada, et al, 2004). Moreover, the more accurate energy and mobility model will be addressed in future work.

Acknowledgements

This research was supported by Waseda University Global COE Program International Research and Education Center for Ambient SoC sponsored by MEXT, Japan.

7. References

Kim, J.H.; Kim, K.B.; Sajjad, H.C.; Yang, W.C.;&Park, M.S.(2008). DEMOCO: Energy-Efficient Detection and Monitoring for Continuous Objects in Wireless Sensor Networks. IEICE Trans. Com. 2008, E91-B, pp.3648-3656.

Zhong, C.& Worboys, M.(2007) Energy-efficient continuous boundary monitoring in sensor networks. Technical Report, 2007. Available online: http://ilab1.korea.ac.kr/papers/ref2.pdf/ (accessed on 31 July 2010).

Basu, A.; Jie, G.; Joseph, S.B.M.& Girishkumar, S.(2006) Distributed Localization by Noisy Distance and Angle Information. In Proceedings of ACM MOBIHOC'06, Los Angeles, CA, USA, 2006;pp. 262-273

Eren, T.; Goldenberg, D.K.; Whiteley, W.& Yang, Y.R.(2004). Rigidity, Computation, and Randomization in Network Localization. In Proceedings of IEEE INFOCOM'04,March 2004, Hongkong, China.

He, T.; Huang C.D.; Blum, B.M.; John A.S.& Tarek, A.(2003) Range-Free Localization Schemes for Large Scale Sensor Networks. In Proceedings of ACM MOBICOM'03, Annapolis, MD, USA, June 2003; pp. 81-95

Nissanka, B.; Priyantha Hari, B.; Erik, D.& Seth, T.(2003) Anchor-Free Distributed Localization in Sensor Networks. LCS Technical Report #892; MIT: Cambridge, MA, USA, April 2003.

Guo, Z.; Zhou, M.& Jiang, G.(2008) Adaptive optimal sensor placement and boundary estimation for dynamic mass objects. IEEE Trans. Syst. Man Cybern B. Cybern. 2008, 38, 222-32.

Olfati-Saber, R.(2007). Distributed tracking for mobile sensor networks with information driven mobility. In Proceedings of Amer. Control Conference, New York, NY, USA, July, 2007; pp. 4606-4612.

Funke, S. & Klein, C(2006). Hole Detection or: How Much Geometry Hides in Connectivity? In Proceedings of the Twenty-Second Annual Symposium on Computational Geometry, SCG '06, ACM Press: New York, NY, USA, 2006; pp. 377-385.

Funke, S.& Milosavljevic, N.(2007). Network sketching or: how much geometry hides in connectivity?–part ii. In Proceedings of the Eighteenth Annual ACM-SIAM Symposium on Discrete Algorithms (SODA2007), New Orleans, LA, USA, 2007; pp. 958-967.

Peng, R.& Sichitiu, M.L.(2006) Angle of Arrival Localization for Wireless Sensor Networks. In Proceedings of Third Annual IEEE Communications Society Conference on Sensor, Mesh and Ad Hoc Communications and Networks (Secon06), Reston, VA, USA, September 2006; pp. 25-28.

Lance, D.; Kristofer S.J.P.& Laurent EL G.(2001) Convex Position Estimation in Wireless Sensor Networks. In Proceedings of IEEE INFOCOM'01, Anchorage, April 2001, AK, USA.

Hu, L.X & David, E.(2004) Localization for Mobile Sensor Networks. In Proceedings of ACM MOBICOM'04, Philadelphia, PA, USA, September 2004; pp. 45-57.

Ji, X. & Zha, H.(2004) Sensor Positioning in Wireless Ad-hoc Sensor Networks Using Multidimensional Scaling. In Proceedings of INFOCOM'04, March 2004, Hongkong, China.

Yi, S.; Wheeler, R.; Zhang, Y.& Markus, P.J.F.(2003) Localization From Mere Connectivity, In Proceedings of ACM MOBIHOC'03, Annapolis, MD, USA, June 2003; pp. 201-212.

Yi, S. & Wheeler, R.(2004) Improved MDS-Based Localization. In Proceedings of IEEE INFOCOM'04, Hongkong, China, March 2004; pp. 2640-2651.

Andreas, S.; Park, H. & Mani, B.S.(2002) The Bits and Flops of the N-hop Multilateration Primitive for Node Localization Problems. In Proceedings of ACM WSNA02, Atlanta, GA, USA, September 28, 2002; pp. 112-121.

Zhang, L.Q.; Zhou, X.B. & Cheng, Q.(2006) Landscape-3D: A Robust Localization Scheme for Sensor Networks over Complex 3D Terrains. In Proceedings of 31st Annual IEEE Conference on Local Computer Networks (LCN), IEEE Computer Society Press: Tampa, FL, USA, November 2006;pp. 239-246.

Samitha, E. & Pubudu, P.(2010) RSS Based Technologies in Wireless Sensor Networks, Mobile and Wireless Communications Network Layer and Circuit Level Design, Fares, S.A., Fumiyuki Adachi, F., Eds.; INTECH Book: Vienna, Austria, 2010.

Bulusu, N.; Hohn, H. & Deborah, E.(2001) Density Adaptive Algorithms for Beacon Placement in Wireless Sensor Networks. In Proceedings of IEEE ICDCS'01; Phoenix, April 2001,AZ, USA.

Liu, L.; Wang, Z. & Zhou, M.(2009). An Innovative Beacon-Assisted Bi-Mode Positioning Method in Wireless Sensor Networks. In Proceedings of IEEE International Conference on Networking Sensing and Control (ICNSC09), Okayama, Japan, March 2009, pp. 570-575.

Liu, L.; Manli, E.; Wang, Z.G. & Zhou, M.C.(2009). A 3D Self-positioning Method for Wireless Sensor Nodes Based on Linear FMCW and TFDA. In Proceedings of IEEE International Conference on Systems, Man, and Cybernetics, San Antonio, TX, USA, October 2009; pp. 3069-3074.

Zhu, X.J.; Rik, S. & Gao, J.(2009). Segmenting a Sensor Field: Algorithm and Applications in Network Design. ACM Trans. Sensor Netw. (TOSN) 2009, 5, 1-31.

McLachlan, G. & Peel, D.(2000). Finite Mixture Models; John Wiley & Sons: New York: NY, USA, 2000.

Figueiredo, M. & Jain, A.K.(2002). Unsupervised learning of finite mixture models. IEEE Trans. Patt. Anal. Mach. Int. 2002, 24, 381-396.

Akaike, H.(1973). Information Theory and an Extension of the Maximum Likelihood Principle. In Proceedings of the Second International Symposium on Information Theory, Akadémiai Kiadó: Budapest, Hungary, 1973; pp. 267-281

Schwarz, G.(1978). Estimating the dimension of a model. Ann. Statist. 1978, 6, 461-464.

Solla, S.A.; Leen, T.K. & Muller, K.R.(2000). The Infinite Gaussian Mixture Model. In Advances in Neural Information Processing Systems; MIT Press: Cambridge, MA, USA, 2000; pp. 554-560.

Chintalapudi, K. & Govindan, R.(2003) Localized edge detection in sensor fields. IEEE Ad Hoc Netw. J. 2003, pp.59-70

Jin, G. & Nittel, S.(2006) NED: An Efficient Noise-Tolerant Event and Event Boundary Detection Algorithm in Wireless Sensor Networks. In Proceedings of the 7th International Conferences on Mobile Data Management, Nara, Japan, May, 2006; pp. 1551-6245.

Min, D.; Chen, D.; Kai, X. & Cheng, X.(2005). Localized Fault-Tolerant Event Boundary Detection in Sensor Networks. IEEE Infocom. 2005; Miami, FL, USA, March, 2005; pp. 902-913.

Heinzelman, W.R.; Chandrakasan, A. & Balakrishnan. H.(2000). Energy-Efficient Communication Protocol for Wireless Microsensor Networks. In the Proceedings of the Hawaii International Conference on System Sciences, Maui, Hawaii, USA, January 4-7, 2000; pp.3005-3014.

Schwarz, G.(1978). Estimating the dimension of a model. Ann. Stat. 1978, 6, pp.461-464.

Zivkovic, Z. & van der Heijden, F.(2004). Recursive Unsupervised Learning of Finite Mixture Models. In Proceedings of IEEE Transactions on Pattern Analysis and Machine Intelligence, Washington, DC, USA, May 2004; pp. 651-656.

Woo, A.; Tong, T. & Culler, D.(2003). Taming the underlying challenges of reliable multihop routing in sensor networks. In Proceedings of the 1st International Conference on Embedded Networked Sensor Systems, Los Angeles, CA, USA, 2003; pp. 14-27.

Seada, A.K.; Zuniga, M.; Helmy, A. & Bhaskar, K.(2004). Energy-Efficient Forwarding Strategies for Geographic Routing in Lossy Wireless Sensor Networks. In Proceedings of the 2nd International Conference on Embedded Networked Sensor Systems, Baltimore, MD, USA, 2004; pp. 108-121.

21

Monitoring Wireless Sensor Network Performance by Tracking Node operational Deviation

Yaqoob J. Y. Al-raisi[1] and Nazar E. M. Adam[2]
[1]*HIS Department, Sultan Qaboos University Hospital, Oman*
[2]*Computer Engineering Department, Fahad Bin Sultan University Saudi Arabia*

1. Introduction

Wireless Sensor Network (WSN) is a very powerful tool that enables its users to closely monitor, understand and control application processes. It is different from traditional wired sensor networks in that its characteristics make it cheap to manufacture, implement and deploy. However, this tool is still at an early stage and many aspects need to be addressed in order to increase its reliability. One of these aspects is the degradation of network performance as a result of network nodes deviation. This may directly reduces the quality and the quantity of data collected by the network and may cause, in turn, the monitoring application to fail or the network lifetime to be reduced.
Deviations in sensor node operations arise as a result of systematic or/and transient errors (Elnahrawy, 2004). Systematic error is mainly caused by hardware faults, such as calibration error after prolonged use, a reduction in operating power levels, or a change in operating conditions; this type of error affects node operations continuously until the problem is rectified. Transient errors, on the other hand, occur as a result of temporary external or internal circumstances, such as various random environmental effects, unstable hardware, software bugs, channel interface, and multi-path effects. This type of error deviates node operations until the effect disappears.
These two types of error may directly and indirectly affect the quality and the quantity of data collected by the WSN. They directly affect sensor measurements and cause drift by a constant value (i.e. bias); they change the difference between a sensor measurement and the actual value, (i.e. drift); and can cause sensor measurements to remain constant, regardless of changes in the actual value, (i.e. complete failure). In addition, they affect the communication and exchange of packets by dropping them. On the other hand, the above-mentioned errors can have an indirect effect on the network's collaboration function, the construction of routing tables, the selection of the node reporting rate, and the selection of data gathering points. Analysis of the data collected by the network (in some practical deployments, such as (Ramanathan, 2004), (Tolle, 2005)), shows that these error reduces the

quality of network collected data by 49%; and in some cases, the network had to be redeployed in order to collect the data because of the failure of the monitored application. Analysis also indicate that a 51% overall improvement of WSN functionality can be expected, as well as an improvement in the quality of the collected data, if real-time monitoring tools are used.

2. Motivations

To detect and isolate operational deviations in WSNs researchers proposed several data clearance, fault-tolerance, diagnosis, and performance measurement techniques.

Data cleaning techniques work at a high network level and consider reading impacts from a deviated sensor on multi-sensor aggregation/fusion such as in (Yao-jung, 2004). Such research proposes several methods that isolate deviated readings by tracking or predicting correlation between neighbour node measurements. Most of this research uses complex methods or models that need a high resource usage to detect and predict sensor measurements. Moreover, these techniques rectify deviated data after detecting them without checking their cause and their impact on network functionality.

Fault-tolerance techniques are important in embedded networks which are difficult to access physically. The advantage of these techniques is their ability to address all network levels; such as circuit level, logical level, memory level, program level and system level; but due to WSNs scare recourses these techniques have a limited usage. In general WSNs fault-tolerant techniques detect faults in fusion and aggregation operation, network deployment and collaboration, coverage and connectivity, energy consumption, energy event fault tolerance, reporting rate, network detection, and many others (Song, 2004, Linnyer, 2004, Bhaskar, 2004, Koushanfar, 2003, Luo, 2006). Faults are detected using logical decision predicates computed in individual sensors (Bhaskar, 2004), faulty node detection (Koushanfar, 2003), or event region and event boundary detection (Luo, 2006). These methods detect metrics either at high or low network level without relating them to each other and without checking their impact on network functionality. The main problem with these techniques is the impact of deviation on network functionality and collected data accuracy before it is detected.

Diagnosis techniques use passive or active monitoring to trace, visualize, simulate and debug historical network log files in real and non real time as discussed in (Jaikaeo, 2001). These techniques are used to detect faults at high or low network levels after testing their cause. For example, Nithya at (Ramanathan, 2005) proposed a debugging system that debugs low network level statistical changes by drawing correlations between seemingly unrelated, distributed events and producing graphs that highlight those correlations. Most of these diagnosis techniques are complex and use iteration tests for their detection. These techniques assume a minimal cost associated with continuously transmitting of debug information to centralized or distributed monitor nodes and send/receive test packets to conform the detection of a faultier.

Finally, performance techniques are similar to diagnosis techniques but without iteration tests and screw pack techniques. Unfortunately there is little literature and research on systematic measurement and monitoring in wireless sensor networks. Yonggang in (Yonggang, 2004) studied the effect of packet loss and their impact on network stability and network processing. He studied the effect of the environmental conditions, traffic load,

network dynamics, collaboration behavior, and constraint recourse on packet delivery performance using empirical experiments and simulations. Although packet delivery is important in wireless communication and can predict network performance, it can give wrong indications of network performance level due to collaboration behavior, and measurement redundancy which makes a network able to tolerate a certain degree of changes. Also, Yonggang proposed an energy map aggregation based approach that sends messages recording significant energy level drops to the sink.
The work in this paper has been motivated by the need to find a tool that uses a very low level of network resources and detects deviations in the network's operations that affect the quality and quantity of the data that are collected before they seriously degrade the network's overall functionality and reduce its lifetime.

3. Project Methodology

3.1 Layout of manuscript

The layout of this paper is organised as follows: Section 2 includes a discussion of related work on functionality degradation detection in WSNs, followed by an explanation of the algorithm's approach. The fourth section explains the practical implementation of the algorithm in a TinyOS *'Surge'* multi-hop application; results of experiments at the network level are then discussed. Finally, the paper ends with a conclusion and suggestions for future work.

3.2 Algorithm Approach

In order to overcome the above-mentioned drawbacks, the Voting Median Base Algorithm for Approximate Performance Measurements of Wireless Sensor Networks (VMBA) algorithm is proposed. This algorithm is a passive voting algorithm that collects its metrics directly from the application by utilizing the overhearing which exists in the neighbourhood. The algorithm requires only readings of neighbours' measurements and does not rely on any information regarding global topology. This makes it scalable to any network deployment size. The proposed algorithm uses parameters found in nodes for other networking and application protocols which makes it much cheaper in terms of resource usage. It uses only the transceiver to send warning messages if there is a network performance degradation or when the node disagrees with the warning messages of neighbours.
The algorithm is divided into four different modules; i.e. listening and filtering, data analysis and threshold test, decision and confidence control and warning packet exchange. In this section we give some definitions and then the VMBA functional algorithm is presented.

A. Listening and Filtering Module

The listening and filtering module is responsible for examining the validity of the received neighbour nodes measurements by filtering those readings beyond the range of the sensor's physical characteristics; as shown in the pseudo-code in Fig.1 . The module then constructs neighbour readings tables and builds statistics in the loss table for neighbour readings.

1: Each S_i senses the phenomenon and wait for
time T to receive N(S_i) readings
2: IF t > T THEN
3: For each unreceived x_j^i increment L_j^i;
4: IF $C_L > x_j^i > C_M$
5: Remove x_j^i from data set and increment D_j^i
6: Calculate med_i of the available S_i data set

Fig. 1. VMBA Algorithm Module 1

B. Data analysis and Threshold Test Module

The second module; i.e. data analysis and threshold test module; tests the content of these tables. This is done by evaluating the data with regard to assigned dynamic or static limits calculated from a reference value or median.

The proposed algorithm has followed a straightforward approach in calculating faulty deviations in sensor functionality. Its analysis assumes that true measurements of a phenomenon's characteristics, following a Gaussian pdf, centred on the calculated median of neighbourhood readings. Any deviation is controlled by the correlation expected at the end of the sensing range of a node, and the sensor nodes' measuring accuracy (where most of the physical processes monitored by WSNs are typically modeled as diffusion models with varying dispersion functions). This assumption is based on the fact that random errors are normally distributed with a zero mean and standard deviation is equal to the specification of the goals designed for the nodes and the network. Any sensor measurement that is not in this region is considered deviated to a degree equal to the ratio of the distance from the neighbourhood median value to the median value.

1: IF $| med_i - med_{i-1} | > \Delta med$
Increment M_i and let $med_i = med_{i-1}$
2: $d_j = | med_i - x_j^i |$
3: IF $d_j > \Theta_1$ and $| x_i^i - x_j^i | < \Theta_1$
4: Increment COV_j^i
5: ELSE increment R_i
6: IF $\frac{R_i}{k} > 40\%$
7: Increment N_i
8: IF $\frac{R_i}{k} * d_j > \Theta_1$
9: Increment D_j^i

Fig. 2. VMBA Algorithm Module 2

In addition, the second module tests the effect of losses on the reliability of the collected data by calculating the degree of distortion in the neighbourhood data that has occurred because of its affect on the collected data accuracy and network functionality. This is done by calculating the ratio of the number of healthy readings to the total number readings as shown in Fig 2 step 8.

C. Decision Confidence Control Module

The third module; i.e. Decision confidence control module; is concerned with tracking changes in the health of neighbour nodes in an assigned time window. This is set depending on the characteristics of the network application and the required response detection time. If exceeded, a request is sent to module four in order to send a detection message to the sink identifying suspected node number, the type of fault, the number of times it has been detected and the effect of the detection on the neighbourhood data and communication. The function of this module is shown in Fig 3.

1: Calculate ML_i
2: IF $ML_i > 60\%$
3: Send to module 4 a request to send an inefficient power consumption warning message
4: IF $M_i > \Theta_M$
5: Send to module 4 a request to send a neighbourhood malfunction due to losses warning message
6: IF $COV_j^i > \Theta_C$
7: Send to module 4 a request to send to detecting node j a coverage problem message
8: IF distortion $> \Theta_d$ & median of $L_j^i > 60\%$
9: Send to module 4 a request to send a degrade detection in network functionality message
10: IF $D_j^i > \Theta_w$
11: Send to module 4 a request to send a detection of node j malfunction message

Fig. 3. VMBA Algorithm Module 3

D. Warning Packet Exchange Module

When module four receives a send request, it checks its neighbours warning exchange memory to ensure that none of the neighbour nodes have reported the same fault in that monitoring window period. If none of the neighbours have so reported, it sends a message or it cancels the request. In addition, this module tests warning messages received from its neighbours with statistics from module three. If the suspected node flags up a counter indication smaller than a threshold, a message will be released indicating

'NO_FAULT_EVIDENCE' regarding the received warning message. On the other hand, if the threshold is higher or equal to the threshold, then the node cancels any similar warning message request from module three during that monitoring period. This is to ensure the reliability of the warning message detection and to correct any incorrect detection that may occur because of losses or other network circumstances. Moreover, module four reduces the algorithm warning packets released by checking if any of its neighbours sent the same message at that time interval. If it been sent the algorithm is going to discard module three requests as shown in Fig. 4 part 3.

1: Receiving neighbour warning
a) Check received warning with the same module 3 counter of reported node.
b) IF module 3 counter < 30%
c) Release 'NO-EVIDENCE-OF-FAULT' message
d) ELSE flag the stop sending of the same message from the node at this monitoring time.

2: Receiving module 3 request
a) Test stop flag of received request warning
b) IF flag = 1 discard message
c) IF send message repeated 3 times send stop reporting the fault message and flag stop fault counter.
d) ELSE send the requested message by module 3.

3: Testing warning packet release
a) IF detected fault returns to normal reset the same fault counters, send 'FAULT_CLEAR' message and recalculate protocol tables.
b) IF step 2 and 3-a alternate for the same fault three times in a predefined monitoring window, the module send s an 'UNSTABLE_DETECTION' warning message to report the detection and flags a permanent fault counter to stop reporting the same fault.
c) By the end of the predefined period reset all counters.

Fig. 4. VMBA Algorithm Module 4

4. Performance Evaluation

VMBA algorithm performance can be evaluate on eight different aspects: deviation detection in single and multi-hop levels, algorithm detection threshold, algorithm detection confidence, algorithm spatial and temporary change tracking for sensor nodes, the impact of packet losses on algorithm analysis, resource usage at node and network levels, the impact of algorithm programming location in the protocol stack, and algorithm released warning messages. In this paper, we considered the empirical performance evaluation of the algorithm at the network level.

4.1 Algorithm Programming in Protocol Stacks

The algorithm was implemented on a Berkeley (Crossbow) Mica2 sensor motes testbed that was programmed in nesC on TinyOS operation system. This is done by building the proposed algorithm on the TinyOS multi-hop routing protocol.

The TinyOS multi-hop protocol consists of MultiHopEngineM; which provides the over all packet movement logic for multi-hop functionality; and MultiHopLEPSM; which is used to provide the link estimation and parent selection mechanisms. These two TinyOS components were modified by added different functions from the proposed algorithm modules as shown at Figure 5.

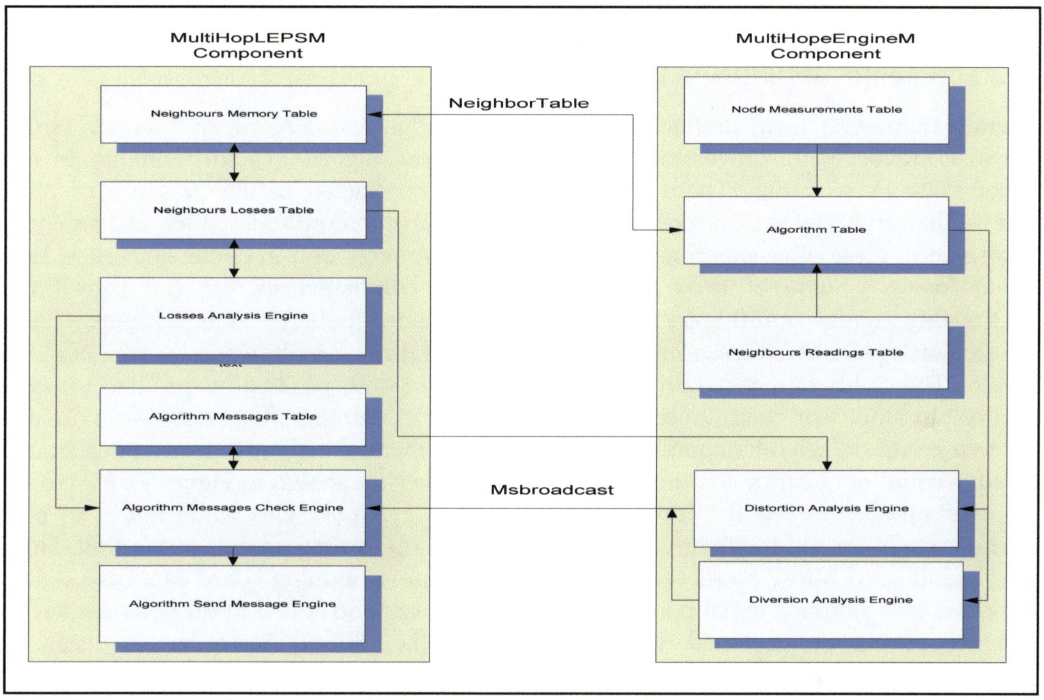

Fig. 5. Functions added to multi-hop components and links between the components

In order to send detected warning packets, a new packet type was constructed. This new packet carries the algorithm detection parameters; as shown at Figure 6. It has a total length of 20 bytes, the last 8 are used for algorithm detection, while the first 12 follow the multi-hop protocol configuration. This is to route the released warning packet in the network.

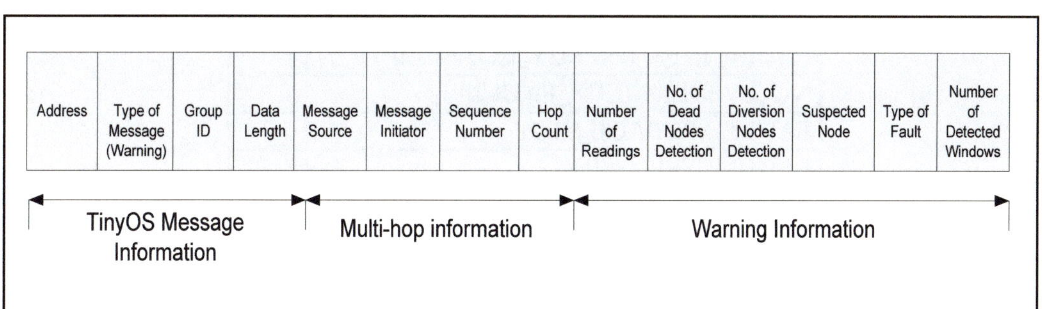

Fig. 6. Algorithm warning message packet

At the algorithm detection part, the first byte carries the total number of readings, that is the number of neighbour nodes in addition to the monitoring node. The next two bytes carry the number of neighbors detected by the node as dead and deviated respectively. This is followed by a byte that carries the identification number of the detected faulty neighbour node. The byte after this carries the type of fault codes; as shown in Table 1; and the final two bytes carry the number of times that the monitoring node detect the reported fault.

5. Experimental Setting and Evaluation Metrics

Several experiments were conducted indoors at the High Speed Network Research Group Lab in Loughborough University to test the proposed algorithm's functionality in real sensor network scenarios. These experiments were conducted in the presence of other devices that are able to interfere with the sensor transmission and reduce the antennae performance; these offer experiments in a dynamic topology and in circumstances of high packet losses. Some of these experiments were conducted to test the algorithm's functionality under multi-hop and highly dynamic topology configurations. These experiments used 13 Mica2 sensors, measuring temperature, distributed in an area of about 4mX5m. The nodes were programmed with the output power of -20 dBm and had top bent antennae to limit their communication range. In this configuration, the nodes were divided into two groups which overlapped in an area between them; thus, some of the nodes around the edge could not hear or communicate with each other (as shown in Figure 7). Moreover, this configuration forced the topology to be highly dynamic. This leads nodes to miss hearing each other and frequently change their multi-hop routing parents in the sink. These experiments used Mica2 nodes attached to a MIB510 programming board as a base station connected to a computer serial port. A snooping node was also added to the network setting with its power programmed to the maximum (i.e. 5dBm) in order to listen to communications among all the nodes within the network and to track packet exchanges in the multi-hop without increasing the usage of resources of the network's sensor nodes.

Fault Type	Code
TOPOLOGY_UNSTABLE	0
FAULT_TYPE_DEVIATION	1
FAULT_TYPE_COMMUNICATION	2
FAULT_TYPE_COVERAGE	3
FAULT_TYPE_ENERGY_CONSUMPTION	4
NO_EVEDENCE_OF_FAULT	5
FAULT_MESSAGE_STOP	6
FAULT_TYPE_DEID	7
FAULT_CLEAR	8
NEIGHBORHOOD_MULFUNCTION	9
PROTOCOL_EFFECT	10

Table 1. Codes of detected faults in algorithm warning messages

The metrics used to evaluate the results were, firstly, the percentage of incorrectly released dead node warnings. This is the ratio of the number of false dead node detections released by the algorithm as opposed to the total number of packets released by the application. This indicates the impact of high network dynamics on the algorithm's incorrect detection. The second metric was the percentage of 'NO-FAULT-EVIDENCE' messages released by the algorithm, which is the ratio of the number of 'NO-FAULT-EVIDENCE' messages to the total number of packets released by the application. This also indicates the impact of high network dynamics but on neighbours' passive tests of incorrect detections.

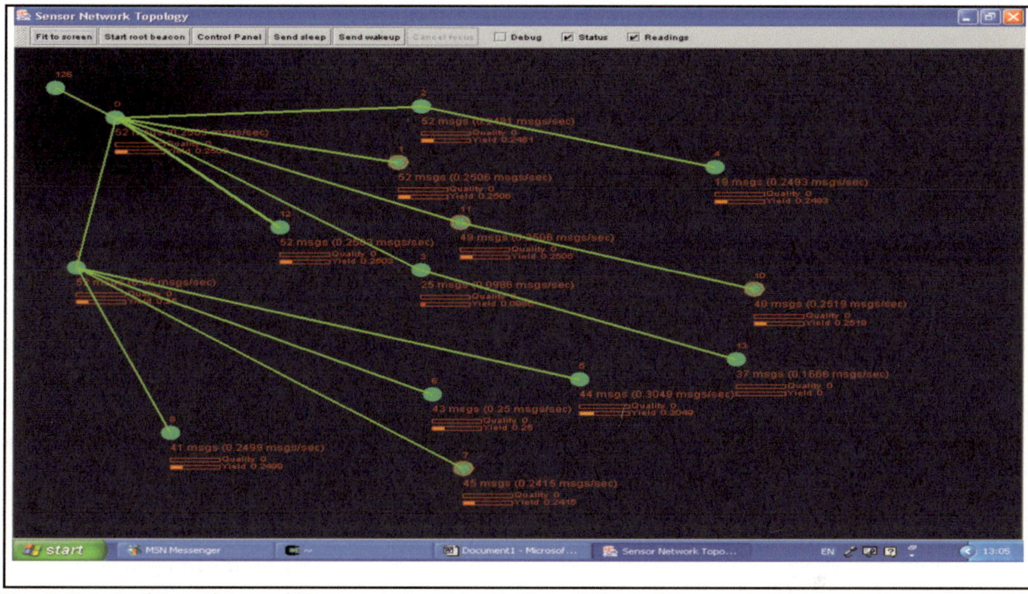

Fig. 7. Logical topology of the experiment at a time interval

These experiments tested the impact of the dead node window threshold, and monitoring window size on the algorithm's detection of dead nodes and the number of warning messages released by it in a highly dynamic network. The algorithm parameters that were tested, as shown in Table 2,and 3 were changed in different experiments to check their impact on the deductibility performance of the network and the exchange of warning packets.

Window Type	Small Monitoring window	Big Monitoring Window	Stop Reporting Window
Diversion	120 seconds (70% threshold)	480 seconds(8 minutes)	1920 seconds (32 minutes)
Distortion	60 seconds (84% loss threshold and larger than 25% accuracy of the two nodes)	240 seconds(4 minutes)	960 seconds (16 minutes)
Dead	60 seconds	240 seconds(4 minutes)	960 seconds (16 minutes)

Table 2. Sizes of monitoring windows in the experiments

Window	Small windows	Small window size	Size of Big window	Number of small window at the group	Total monitoring window size
1	Linear increased	240 seconds (4 minutes)	3 groups	4-8-12	48 minutes
2	Exponential increased			8-12-16	64 minutes
3				10-14-18	72 minutes
4				14-16-20	80 minutes

Table 3. Size of monitoring windows

5.1 Effect of Network Topology and Packet Losses on the Algorithm's Functionality

Figure 8 plots the relationship between the percentage of detected and 'No_Fault_Evidence' messages released from the algorithm for different application reporting rates. (Please note that reporting rates logs were used in the figure to plot these). The results of the experiments showed that at a 1 second reporting rate (a multi-hop protocol leads to congestion and an overflow of communication), a large amount of wrong suspected dead warnings occurred (around 3.2% of the total network packet exchange in the application). Furthermore, a large number of 'No_Fault_Evidence' replies were released from neighbour messages (i.e. around 0.5% of the total packets in the network application). Reducing the application's reporting rate to 2 seconds reduced the number of suspected dead messages; these decreased sharply to 0.5% of the total number of packets released by the network application. This happened alongside a reduction in 'No_Fault_Evidence' messages which reached around 0.01% of the total number of packets released. Thus, the number of suspected dead messages was reduced to almost 0% when the application's reporting rate was adjusted to 1 minute, along with a decrease in 'No_Fault_Evidence' messages released from neighbours. When the application's reporting rate was increased to 30 minutes, a sharp increase occurred in the number of suspected dead and 'No_Fault_Evidence' messages, as shown in the figure. Also, Figure 8 shows that, by increasing the application's reporting rate above 1 minute, the number of 'No_Fault_Evidence' messages increases so that it becomes higher than the number of suspected dead messages. This is as a result of the size of the monitoring windows and the highly dynamic network topology.

From these experiments, it can be concluded that dead node warnings will not disappear spatially in a monitored network when the network connections are highly dynamic. To reduce the number of wrong suspected dead messages, different window sizes and combinations were tested, as shown in Table 3. Figure 9 shows the relation between the percentage of correct, positive detected (wrong detection) by the algorithm, together with the negative false dead nodes for different sizes of large monitoring windows. The figure illustrates that, as the big monitoring window size increased, the confidence of the algorithm's detection of dead neighbour nodes increased, along with a decrease in the number of packets released by the algorithm. Although increasing window size will reduce the number of wrong messages, it also increases the response detection time and the probability of node failure occurring before releasing the warning message.

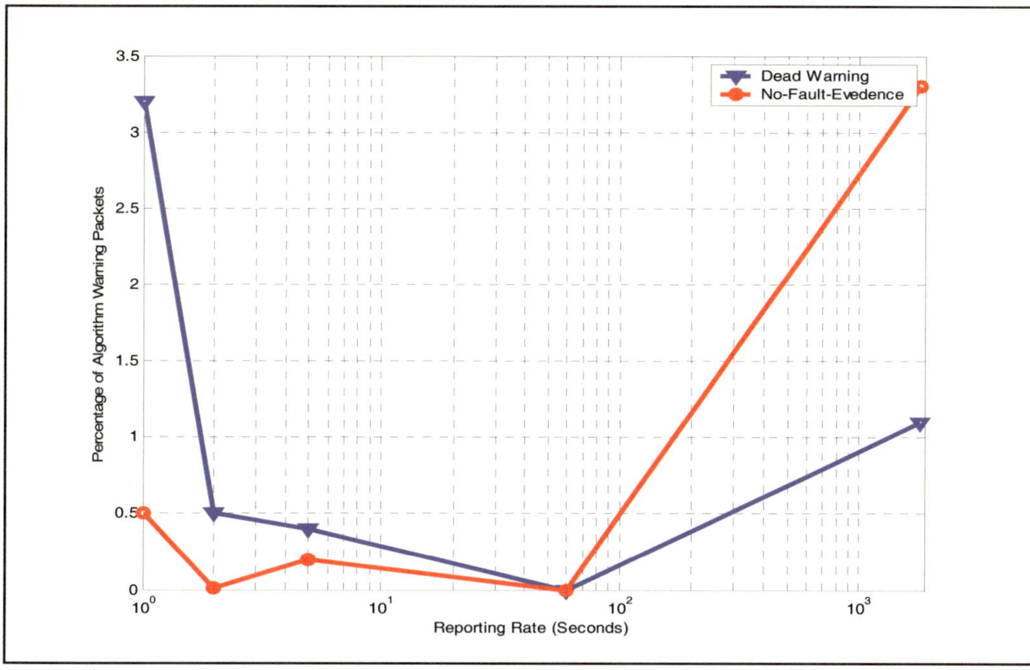

Fig. 8. Changing reporting rates with the percentage of warning messages released with the same window size

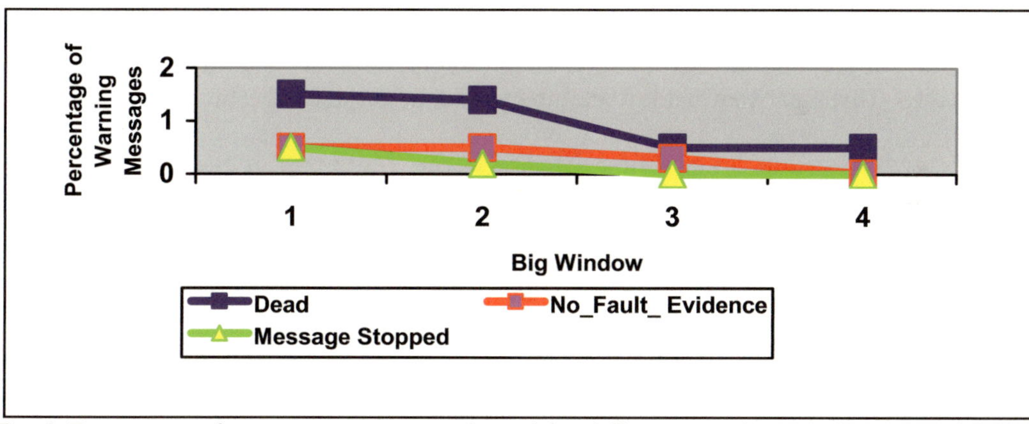

Fig. 9. Percentage of warning messages released for different window configurations

To solve this problem, the algorithm was programmed such that it would select the neighbours it would monitor; this selection depends on the amount of received packets. This configuration reduced the number of wrong packets reported by 80% and reduced 'No_Evidence_Fault' by 70%, as Figure 10 shows, but it also added additional complexity to algorithm's source code and its functionality. Moreover, there will be uncovered neighbour nodes in low density networks. In addition, the proposed algorithm was modified to send warning messages concerning the detection of connectivity problems between neighbour nodes. This makes the algorithm stop reporting a suspected node if the node is detected as dead and if 3 clear dead messages are detected at the stop reporting monitoring window. Figure 10 plots comparisons between the percentages of the algorithm's released dead and no evidence messages in a neighbourhood with and without the modification covering connectivity problems. The figure shows that there is a reduction of 20% in the number of 'No_Fault_Evidence' messages as a result of a 34% reduction in the detection of dead packets.

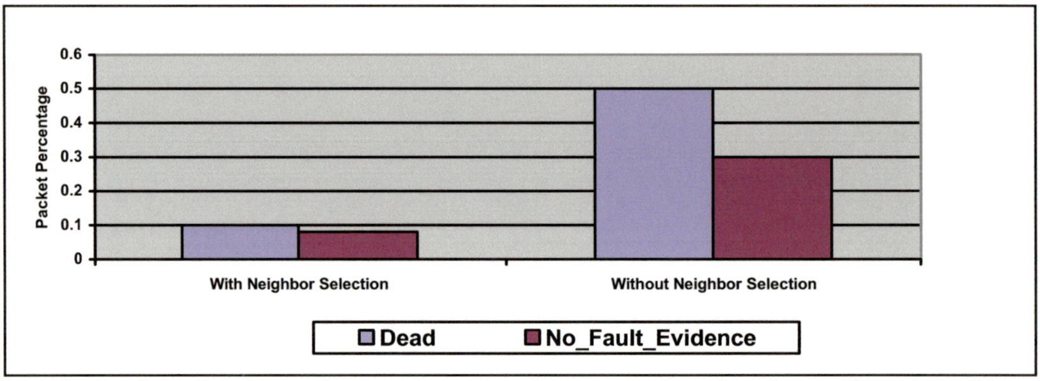

Fig. 10. Number of exchanged warning packets between selected and not selected neighbour nodes.

6. Conclusion and Future Work

We proposed a distributed performance algorithm that enables each sensor node at sensor network to detect the health of nodes at neighbourhood and their collaborative functionality. This algorithm sends a warning packet to the sink reporting any degradation detection.

The proposed algorithm tested using TinyOS 'Surge' multi-hop application on Berkely Mica2 sensor nodes testbed. These empirical experiments showed that the high loss in WSN causes proposed algorithm wrong detection of neighbour nodes aliveness and released more 'NO_EVIDENCE_FAULT' messages. This controlled by adjusting the monitoring window size and reduces the proposed algorithm wrong detection by 80% and the 'NO_EVIDENCE_FAULT' messages by 70%.

There are numerous aspects that can be considered in the future in order to extend this work and improve the algorithm's functionality, such as checking the impact of the mobility of sensor nodes on the algorithm's functionality. Also, it would be useful to study the impact of faulty data on individual WSN protocols and compare these results with the proposed approximate calculation that depends on the number of deviated nodes.

7. References

Elnahrawy Eiman and N. Badri, (2004). *Cleaning and Querying Noisy Sensors*, The First ACM Conference on Embedded Networked Sensor Systems (SenSys'03), pp. 78-87.

N. Ramanathan, T. Schoellhammer, D. Estrin, M. Hansen, T. Harmon, E. Kohler, and M. Srivastava,(2006). *The Final Frontier: Embedding Networked Sensors in the Soil, CENS Technical Report #68*, Center for Embedded Networked Sensing, UCLA, USA.

G. Tolle, J. Polastre, R. Szewczyk, D. Culler, N. Turner, K. Tu, S. Burgess, T. Dawson, P. Buonadonna, D. Gay, and W. Hong, (2005). *A Macroscope in the Redwoods,* ACM Conference on Embedded Networked Sensor Systems (SenSys'05), pp. 51-63.

W. Yao-jung , M. Alice Agogine and G. Kai. (2004). *Fuzzy Validation and Fusion for Wireless Sensor Networks,* in ASMEinternational Mechanical Engineering Congress and RD&D Expo (IMECE2004),Anaheim, California, USA..

H. Song and C. Edward. (2004). *Continuous Residual Energy Monitoring in Wireless Sensor Networks,* in International Symposium on Parallel and Distributed Processing and Applications (ISPA 2004), pp. 169-177.

Linnyer Beatrys Ruiz, Isabela G. Siqueria and Leonardo B. Oliveira. (2004). *Fault Management in Event-driven Wireless Sensor Networks,* in MSWiM'04, October 4-6, Venezia, Italy.

K. Bhaskar and S. S. Iyengar. (2004). *Distributes Bayesian Algorithms for Fult-tolerant Event Region Detection in Wireless Sensor Networks,*IEEE Transaction on Computers, vol. 53, pp. 421-250.

F. Koushanfar, M. Potkonjak and A. Sangiovanni-Vincentelli. (2003). *On-line Fault Detection of Sensor Measurements,* in Sensors. Proceedings of IEEE, 2003, pp. 974-979.

X. Luo, M. Dong and Y. Huang(2006). "On Distributed Fault-tolerant Detection in Wireless Sensor Networks," IEEE Transactions on Computers, vol. 55, pp. 58-70.

C. Jaikaeo, C. Srisathapornphat and C. Shen, (2001). *Diagnosis of Sensor Networks,* in Communications, 2001. ICC 2001. IEEE International Conference, pp. 1627-1632.

N. Ramanathan, K. Chang, R. Kapur, L. Girod, E. Kohler and D. Estrin, (2005). *Sympathy for the Sensor Network Debugger,* in The 3rd ACM Conf. Embedded Networked Sensor Systems (SenSys 2005), pp. 255-267.

Z. Yonggang, (2004). *Measurement and Monitoring in Wireless Sensor Networks,* PhD Thesis, Computer Science Department, University of Southern California, USA, June. 2004.

22

Building Context Aware Network of Wireless Sensors Using a Scalable Distributed Estimation Scheme for Real-time Data Manipulation

Amir Hossein Basirat and Asad I. Khan
Monash University
Australia

1. Introduction

Wireless sensor networks are an "exciting emerging domain of deeply networked systems of low-power wireless motes with a tiny amount of CPU and memory and large federated networks for high-resolution sensing of the environment" (Welsh et al., 2004). The capability to support plethora of new diverse applications has placed Wireless Sensor Network technology at threshold of an era of significant potential growth. The technology is advancing rapidly under the push of the new technological developments and the pull of vast and diverse potential applications. The near ubiquity of the internet coupled with recent engineering achievements, are opening the door to a new generation of low-cost and powerful sensor devices which are capable of delivering high-grade spatial and temporal resolution. In that regard, distributed estimation and tracking is one of the most fundamental collaborative information processing challenges in wireless sensor networks (WSNs). Moreover, estimation issues in wireless networks with packet-loss are gaining lion share of attention over the last few years. However, due to the inherent limitations of WSNs in terms of power and computational resources, deploying any distributed estimation technique within WSNs requires modifying the existing methods to address those limitations effectively. In fact, the problem with current approaches lies in the significant increase in the computational expenses of the deployed methods as the result of increase in the size of the network. This increase puts a heavy practical burden on deployment of those algorithms for resource-constrained wireless sensor networks.

Current decentralized Kalman filtering involves state estimation using a set of local Kalman filters that communicate with all other nodes. The information flow is all-to-all with approximate communication complexity of $O(n^2)$ which is not scalable for WSNs. In this chapter an attempt is made to explore new ways in provisioning distributed estimation in WSNs by introducing a light-weight distributed pattern recognition scheme which provides single-cycle learning and entails a large number of loosely coupled parallel operations. In fact, the focus of our approach would be on a novel scalable and distributed filtering scheme in which each node only communicates messages with its neighbours on a network to minimize the communication overhead to a high degree. In order to achieve higher level of simplicity to increase effectiveness of algorithm deployment in terms of utilizing limited available resources, a novel approach toward real-time distributed estimation is introduced, using a simplified graph-based method called Distributed Hierarchical Graph Neuron (DHGN). The proposed approach not only enjoys from conserving the limited power resources of resource-constrained sensor nodes, but also can be scaled effectively to address scalability issues which are of primary concern in wireless sensor networks. The proposed scheme is not intending to replace existing complex methods which are treated in the literature, but rather makes an attempt to reduce the computational expenses involved in the processing of the gathered in-situ information from sensor nodes. Strength of DHGN lies in the processing of non-uniform data patterns as it implements a finely distributable framework at the smallest (atomic) logical sub-pattern level. The results are easily obtained by summation at the overall pattern level. Hence, the algorithm is able to provide some sort of divide and distribute filtering process throughout the network in a fine-grained manner for minimizing the energy use. DHGN is also a highly scalable algorithm that allows real-time in-network data manipulation; essential feature for real-time data processing in WSNs. The outline for this chapter is as follows. Section 2 provides an overview of WSN technology and its current research trends. Section 3 provides a discussion on the limitations of existing approaches for data fusion and distributed estimation within WSNs. We will then introduce our proposed distributed event detection and pattern classification scheme in Section 4. Section 5 deals with performance metrics for our novel algorithm. Section 6 entails further discussion on our proposed scheme and future direction of this research. Finally, section 7 concludes the chapter.

2. WSN Overview

A wireless sensor network (WSN) in its simplest form can be defined as (Chong & Kumar, 2003; Akyildiz, Su, Sankarasubramaniam & Cayirci, 2002; Culler, Estrin & Srivastava, 2004) a network of (possibly low-size and low-complex) devices denoted as *nodes* that can sense the environment and communicate the information gathered from the monitored field (e.g., an area or volume) through wireless links; the data is forwarded, possibly via multiple hops relaying to a *sink* (sometimes denoted as *controller* or *monitor*) that can use it locally or is connected to other networks (e.g., the Internet) through a gateway. The nodes can be stationary or moving. They can be aware of their location or not. They can be homogeneous or not. A traditional single-sink WSN is illustrated in Figure 1. Almost all scientific papers in the literature deal with such a definition. This single-sink scenario suffers from the lack of scalability: by increasing the number of nodes the amount of data gathered by the sink increases and once its capacity is reached the network size can not be augmented. Moreover, for reasons related to medium access control (MAC) and routing aspects, network performance cannot be considered independent from the network size. A more general scenario includes multiple sinks in the network (see Figure 2). Given a level of node density, a larger number of sinks will decrease the probability of isolated clusters of nodes that cannot deliver their data owing to unfortunate signal propagation conditions.

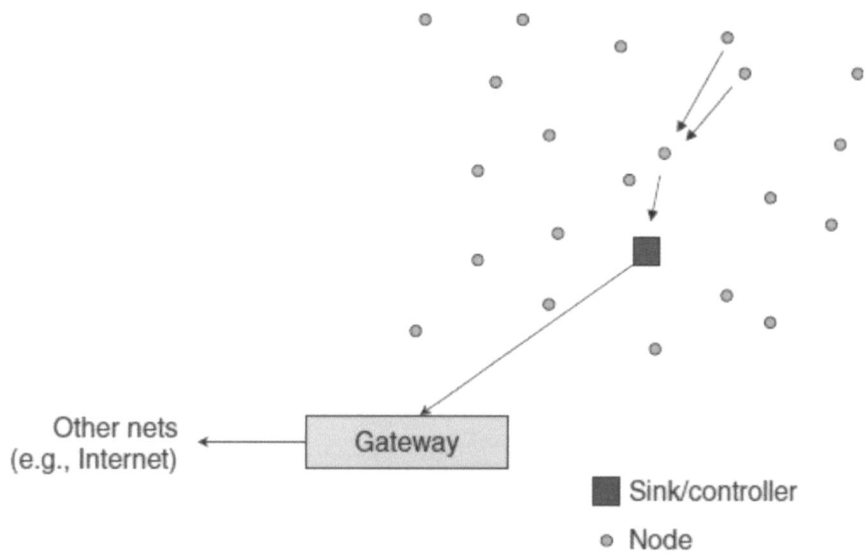

Fig. 1. Traditional single-sink WSN

In principle, a multiple-sink WSN can be scalable (i.e., the same performance can be achieved even by increasing the number of nodes), while this is clearly not true for a single-sink network. However, a multi-sink WSN does not represent a trivial extension of a single-sink case for the network engineer. There might be mainly two different cases: (1) all sinks are connected through a separate network (either wired or wireless), or (2) the sinks are disconnected. In the former case, a node needs to forward the data collected to any element in the set of sinks. From the protocol viewpoint, this means that a selection can be done based on a suitable criterion (e.g., minimum delay, maximum throughput, minimum number of hops, etc.). The presence of multiple sinks in this case ensures better network performance with respect to the single-sink case (assuming the same number of nodes is deployed over the same area), but the communication protocols must be more complex and should be designed according to suitable criteria. In the second case, when the sinks are not connected, the presence of multiple sinks tends to partition the monitored field into smaller areas; however from the communication protocols viewpoint no significant changes must be included, apart from simple sink discovery mechanisms. Clearly, the most general and interesting case (because of the better potential performance) is the first one, with the sinks connected through any type of mesh network, or via direct links with a common gateway. Both the single-sink and multiple-sink networks introduced above do not include the presence of actuators, that is, devices able to manipulate the environment rather than observe it. WSANs are composed of both sensing nodes and actuators (see Figure 3). Once more, the inclusion of actuators does not represent a simple extension of a WSN from the communication protocol viewpoint. In fact the information flow must be reversed in this case: the protocols should be able to manage many-to-one communications when sensors provide data, and one-to-many flows when the actuators need to be addressed, or even one-to-one links if a specific actuator has to be reached.

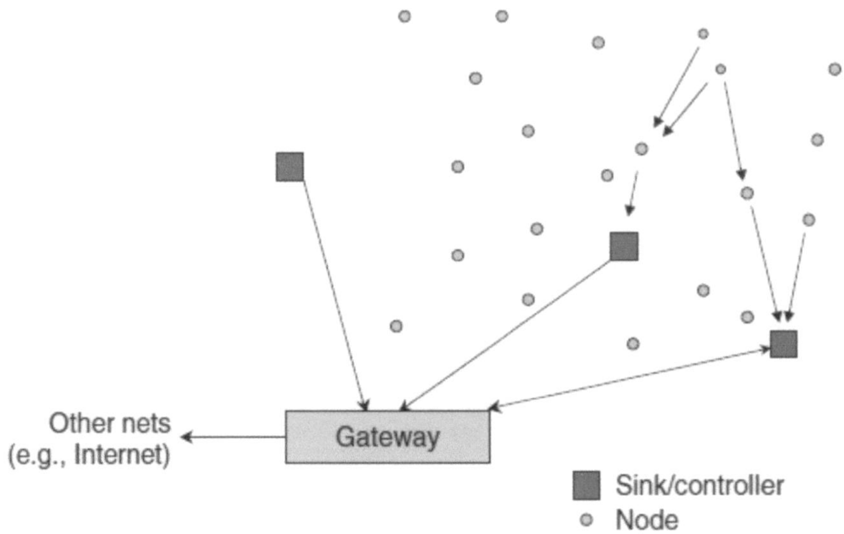

Fig. 2. Multi-sink WSN

The complexity of the protocols in this case is even larger. Given the very large number of nodes that can constitute a WSAN (more than hundreds sometimes), it is clear that MAC and the network layer are very relevant parts of the protocol stack. Tens of proposals specifically designed for WSANs have been made in the past few years. The communication protocols of a WSAN should also allow an easy deployment of nodes; the network must be able to self-organize and self-heal when some local failures are encountered.

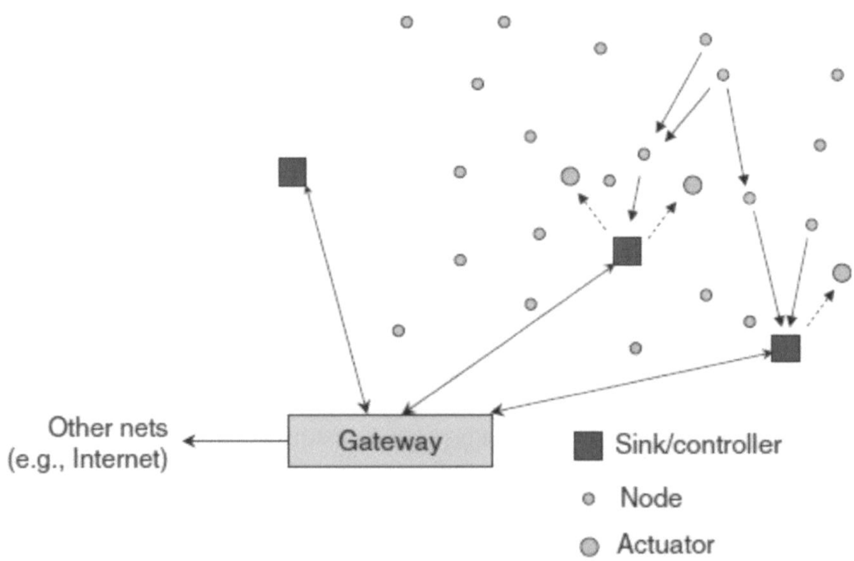

Fig. 3. Typical WSAN

2.1 Current and future research on WSANs

Many technical topics of WSANs are still considered by research as the current solutions are known to be non-optimized, or too much constrained. From the physical layer viewpoint, standardization is a key issue for success of WSAN markets. Currently the basic options for building HW/SW platforms for WSANs are Bluetooth, IEEE 802.15.4 and 802.15.4a. At least, most commercially available platforms use these three standards for the air interface. For low data rate applications (250 Kbits on the air), IEEE 802.15.4 seems to be the most flexible technology currently available. Clearly, the need to have low-complexity and low-cost devices does not push research in the direction of advanced transmission techniques. MAC and network layer have attracted a lot of attention in the past years and still deserve investigation. In particular, combined approaches that jointly consider MAC and routing seem to be very successful.

Topology creation, control and maintenance are very hot topics. Especially with IEEE 802.15.4, which allows creation of several types of topologies (stars, mesh, trees, cluster-trees), these issues play a very significant role. Transport protocols are needed for WSANs depending on the specific type of application. However, some of the most relevant issues investigated by research in WSANs are cross layer, dealing with vertical functionalities: security, localization, time synchronization. Basically, the research in the field of WSANs started very recently with respect to other areas of the wireless communication society, as broadcasting or cellular networks. The first IEEE papers on WSANs were published after the turn of the Millennium. The first European projects on WSANs were financed after year 2001. In the US the research on WSANs was boosted a few years before. Many theoretical issues still need a lot of investments.

The e-Mobility technology platform gathers all major players in the area of wireless and mobile communications. A strategic research agenda was released and updated in 2006. According to their views, by the year 2020 mobile and wireless communications will play a central role in all aspects of European citizen's lives, not just telephony, and will be a major influence on Europe's economy, wirelessly enabling every conceivable business endeavour and personal lifestyle. The aim of research in the field can be summarized as follows: The improvement of the individual's quality of life, achieved through the availability of an environment for instant provision and access to meaningful, multi-sensory information and content.

"Environment" means that the users will strongly interact with the environment that surrounds them, for example by using devices for personal use, or by having the location as a basis for many of the services to be used. This implies a totally different structure for the networks. Also, the context recognized by the system and it acting dynamically on the information is a major enabler for intelligent applications and services. This also means that sensor networks and radio frequency identifications (RFIDs) are increasingly important. "Multi-sensory" is related to all the users, devices, and also to the fact that the environment will be capable of sensing the users presence. Also, virtual presence may be considered, implying more sensory information being communicated, and an ideal of a rich communication close to the quality achieved in interpersonal communications or direct communications with another environment; this could also include non-invasive and context-aware communication characterizing polite human interactions. Therefore, this stretches mobile and wireless communications beyond radio and computer science into new areas of science, like biology, medicine, psychology, sociology, and nano-technologies, and

also requires full cooperation with other industries not traditionally associated with communications.
Finally, the information should be multi-sensory and multi-modal, making use of all human basic senses to properly capture context, mood, state of mind, and, for example, one's state of health. Clearly, the realization of this vision of mobile and wireless communications demands multi-disciplinary research and development, crossing the boundaries of the above sciences and different industries. Also, the number of electronic sensors and RFIDs surrounding us is quickly increasing. This will increase the amount of data traffic. The future system will be complex, consisting of a multitude of service and network types ranging across wireless sensor networks, personal area, local area, home networks, moving networks to wide area networks. Therefore, the e-Mobility vision emphasizes the key role played by WSANs as elements of a more complex system linking different types of access technologies.
ARTEMIS (advanced research & technology for embedded intelligence and systems) is the technology platform for embedded systems. The term 'embedded systems' describes electronic products, equipment or more complex systems, where the embedded computing devices are not visible from the outside and are generally inaccessible by the user. The sensor and actuator nodes of WSANs are embedded systems. According to the ARTEMIS strategic research agenda, intelligent functions embedded in components and devices will be a key factor in revolutionizing industrial production processes, from design to manufacturing and distribution, particularly in the traditional sectors. These technologies add intelligence to the control processes in manufacturing shop floors and improve the logistic and distribution chains, resulting in an increasing productivity in a wide range of industrial processes. The grand challenge in the area of sensors and actuators relates to the support of huge amounts of input and output data envisaged in the application contexts with minimal power requirements and fail-safe operation.

3. Data Fusion Techniques for WSANs

To fully exploit the potential of sensor networks, it is essential to develop energy-efficient and bandwidth-efficient signal processing algorithms that can also be implemented in a fully distributed manner. Distributed signal processing in a WSN has a communication aspect not present in the traditional centralized signal processing framework, thus it differs in several important aspects.

- Sensor measurements are collected in a distributed fashion across the network. This necessitates data sharing via inter-sensors communication. Given a low energy budget per sensor, it is unrealistic for sensors to communicate all their full-precision data samples with one another. Thus, local data compression becomes a part of the distributed signal processing design. In contrast, in a traditional signal processing framework where data is centrally collected, there is no need for distributed data compression.
- The design of optimal distributed signal processing algorithms depends on the models used to describe: the nodes connectivity, the nodes distribution, the knowledge of sensor noise distributions, the qualities of inter-sensors communication channels, and the underlying application metrics. Distributed signal processing over a wireless sensor network requires proper coordination and

planning of sensor computation as well as careful exploitation of the limited communication capability per sensor. In other words, distributed signal processing in sensor networks has communication aspects which are not present in the most of traditional signal processing frameworks.

- In a WSN, sensors may enter or leave the network dynamically, resulting in unpredictable changes in network size and topology. This can be due to failure between inter-sensors communication (propagation conditions, interference or non-available communication channels), duty cycling, drained batteries or nodes damages. This dynamism requires the necessity for distributed signal processing algorithms to be robust to the changes in network topology or size. These algorithms and protocols must also be robust to poor time synchronization across the network and to inaccurate knowledge of sensor locations.

There are many theoretical challenges such as establishing models, metrics, bounds, and algorithms for distributed multimodal sensor fusion, distributed management of sensor networks including auto-configuration, energy-efficient application-specific protocol designs, formal techniques for the study of architectures and protocols, representation of information requirements, and sensor network capabilities on a common mathematical framework that would enable efficient information filtering (Luo et al., 2006). From these aspects, it is clear that the design of sensor networks under energy, bandwidth, and application-specific constraints spans all layers of the protocol stacks and it is very important to have a common framework enabling all these points be taken into account even if with different approximations degrees.

In this view an example of cross-layer methodology of WSN design for environmental monitoring will be shown in the following with particular emphasis on the impact of distributed digital signal processing (DDSP) on the spatial process estimation error on one side and network lifetime on the other side. Depending on the process under monitoring and the goal of the WSAN, such as detection of distributed binary events and spatial process estimation, several techniques can be pursued with envisaging of centralized and distributed processing.

3.1 Distributed Estimation

Distributed estimation and tracking is one of the most fundamental collaborative information processing problems in wireless sensor networks (WSN). Multi-sensor fusion and tracking problems have a long history in signal processing, control theory, and robotics. Moreover, estimation issues in wireless networks with packet-loss have been the center of much attention lately. In most applications, the intelligent fusion of information from geographically-dispersed sensor nodes, commonly known as distributed data fusion, is an important issue. A related problem is the binary decentralized (or distributed) detection problem, where a large number of identical sensor nodes deployed randomly over a wide region, together with a global detector or fusion centre (FC), cooperatively undertake the task of identifying the presence or absence of a phenomenon of interest (PoI) (see Figure 4). Specifi cally, each node takes a local decision about the presence or absence of the PoI and sends its decision to the FC which is responsible for the final decision based on the information gathered from local sensors.

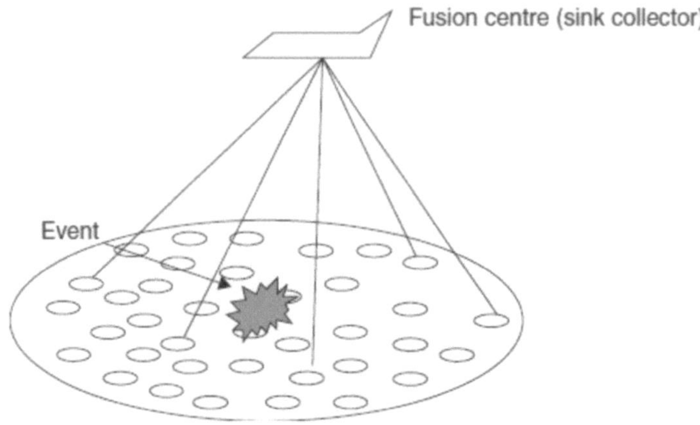

Fig. 4. Example of distributed detection scenario

Two problems have to be considered: the design of the decision rule at the FC and the design of the local sensor signal processing strategies. In the case of perfect knowledge of system parameters the design of the decision rule at the FC is a well-established task. The design of the local sensor decision rule, that is, the likelihood ratio test (LRT) threshold in binary detection, is more challenging due to the distributed nature of the system. In fact, the optimal choice of each sensor LRT threshold is coupled to each other node threshold, although nodes are not in general fully connected due to propagation effects or energy constraints. Recently it has been demonstrated that under the asymptotic regime (i.e., large number of nodes), the identical LRT threshold rule at the sensors provides the optimal error exponent if local sensor observations are independent and identically distributed (Sung, Tong & Swami, 2005). For a more complete overview of decentralized detection the reader is recommend to read Varshney, 1997; Signal Processing Magazine, Special Issue , 2006. Unlike in classical decentralized detection problems (Blum, Kassam & Poor, 1997; Varshney, 1997), greater challenges exist in a WSN setting. There are stringent power constraints for each node, and communication channels from nodes to the FC are severely bandwidth-constrained. In addition, the communication channels are no longer lossless (e.g, fading, noise and, possibly, interference are present), and the observation at each sensor node is spatially varying (Sung et al., 2005; Niu & Varshney, 2005). Recently, there has been great interest in cooperative communication (Winters, 1987; Sendonaris, Erkip & Aazhang, 2003). One may also exploit diversity associated with spatially distributed users, or simply cooperative diversity , in WSNs. In these networks, multiple sensor nodes pool their resources in a distributed manner to enhance the reliability of the transmission link. Specifi cally, in the context of decentralized detection, cooperation allows sensor nodes to exchange information and to continuously update their local decisions until consensus is reached across the nodes (Quek, Dardari & Win, 2006c; Quek, Dardari & Win, 2006b; Quek, Dardari & Win, 2006a). For example, cooperation in decentralized detection can be accomplished via the use of Parley algorithm (Swaszek & Willett, 1995). This algorithm has been shown to converge to a global decision after sufficient number of iterations when certain conditions are met. However, without a fully-connected network and given that the sensor observations are spatially varying, Parley algorithm may result in convergence to a wrong decision at most of the nodes.

3.2 Kalman Filtering

Although WSNs present attractive features, challenges associated with the scarcity of bandwidth and power in wireless communications have to be addressed. To perform state estimation, sensors may share these observations with each other or communicate them to a fusion center for centralized processing. In either scenario, the communication cost in terms of bandwidth and power required to convey observations is large enough to merit attention.. The Kalman filter is a very popular fusion method. Decentralized Kalman filtering (Speyer, 1979; Rao, Whyte & Sheen, 1993) involves state estimation using a set of local Kalman filters that communicate with all other nodes. The Kalman filter estimates the state x of a discrete-time controlled process that is ruled by the state-space model:

$$x\,[k]) = A.x\,[k-1] + w\,[k-1] \quad (1)$$

The system is influenced by process noise denoted w. The state dynamics determine the linear operator A. The state contributes to the observation y, which also includes a stochastic, additive measurement noise v:

$$y\,[k] = C.\,x[k] + v\,[k] \quad (2)$$

The process and measurement noises are assumed to be Normal processes with known variances W and V. The information flow is all-to-all with approxmiate communication complexity of $O(n^2)$ which can cause scalability issues for large-sclae WSNs.

4. Event Classification and Pattern Recognition

The reality in WSNs is that so far tremendous potential of the WSN has only been demonstrated for humble applications such as meter reading in buildings and basic forms of ecological monitoring. Reaping the full potential of this technology requires a second level of algorithms, which at present is missing. Current techniques solve the more immediate problem of conveying sensory data to a central entity known as the base station for most of the processing. This approach is not scalable and has potential to cast suspicions of big brother analysis. The WSN must therefore localize the computations within its monitored regions and communicate the computed results to small/roaming devices directly. Doing so will alleviate the scalability issue by localizing the computations and address the privacy and reliability concern through direct communications occurring within a decentralized model. Algorithms, such as the ones in this chapter, would help integrate vast networks of sensors into intelligent macro-scopes for observing our surroundings. These will bring unprecedented capabilities within our reach that transform the way we deal with phenomena occurring over large distances and inaccessible regions.

In order to achieve the broad objectives stated above, an initial step will be to develop a level of computability within the WSN whereby sensor readings can be instantly translated into event patterns and then rapidly analysed by the network (locally). This approach will entail two-fold benefit. On one hand it will enhance event detection e.g. surveillance, and target location capabilities and on the other hand it will aid in development of advanced threat detection systems for WSN, e.g. ones that can function like our biological immune system. The challenge is to evolve an approach, which can successfully detect complex real life patterns arising from heterogeneous data sets (generated by different types of sensors) in real-time. In this regard, events of interest need to be correlated to specific pattern classes of our definition. Furthermore the asymptotic limits of the approach need to be carefully examined. For instance how far will our scheme scale with increases in the complexity of the sensory patterns and the size of the network? What will be the trade off between the scheme's sensitivity and false alarms? Can the scheme degrade gracefully with disruptions to network nodes and wireless links?

4.1 Graph Neuron

DHGN extends the functionalities and capabilities of Graph Neuron (GN) (Khan & Mihailescu, 2004) and Hierarchical Graph Neuron (HGN) (Khan & Nasution, 2008) algorithms. Figure 5 depicts a GN array, which is capable of converting the spatial/temporal patterns into a simple graph-based representation, in which input patterns are compared with the edges in the graph for memorization or recall operations.

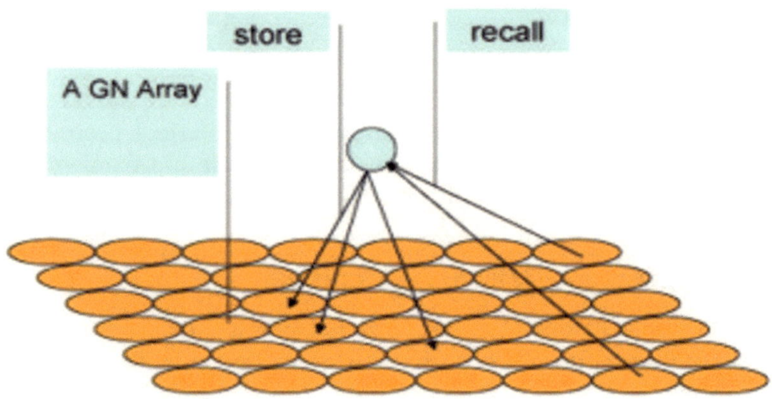

Fig. 5. Store and recall within a GN

GN is an associative memory algorithm, which implements a scalable AM device through its parallel in-network processing framework (Khan & Isreb, 2004). GN has been tested in pattern recognition applications within different types of distributed environments (Khan, Muhammad Amin, 2008a). GN makes use of graph-based model for pattern learning and recognition. One of the peculiarities of this technique is the employment of parallel in-network processing capabilities to address scalability issues effectively, which are of primary concern in distributed approaches. Graph neuron may be conceived as a directed graph in which processing nodes of the GN array are mapped to vertex set V of the graph, and links between nodes are mapped to the set of edges of the graph, E.

The communication is limited to the neighbouring nodes to decrease the amount of overhead involved and as a result, any significant increase in the number of nodes will not increase the communication overhead significantly. The information in the pattern space can be presented to each of the nodes in the graph in the form of (position, value) pair. Thus, GN array is capable of converting the spatial/temporal patterns into a simple graph-based representation in which input patterns are compared with the edges in the graph for memorization or possible recall operations (See Figure 6). Among various applications of the GN-based AM, an input pattern in GN pattern recognition may represent bit elements of an image (Khan & Muhammad Amin, 2007) or a stimulus/signal spike produced within a network intrusion detection application (Khan, Baig & Baqer, 2006).

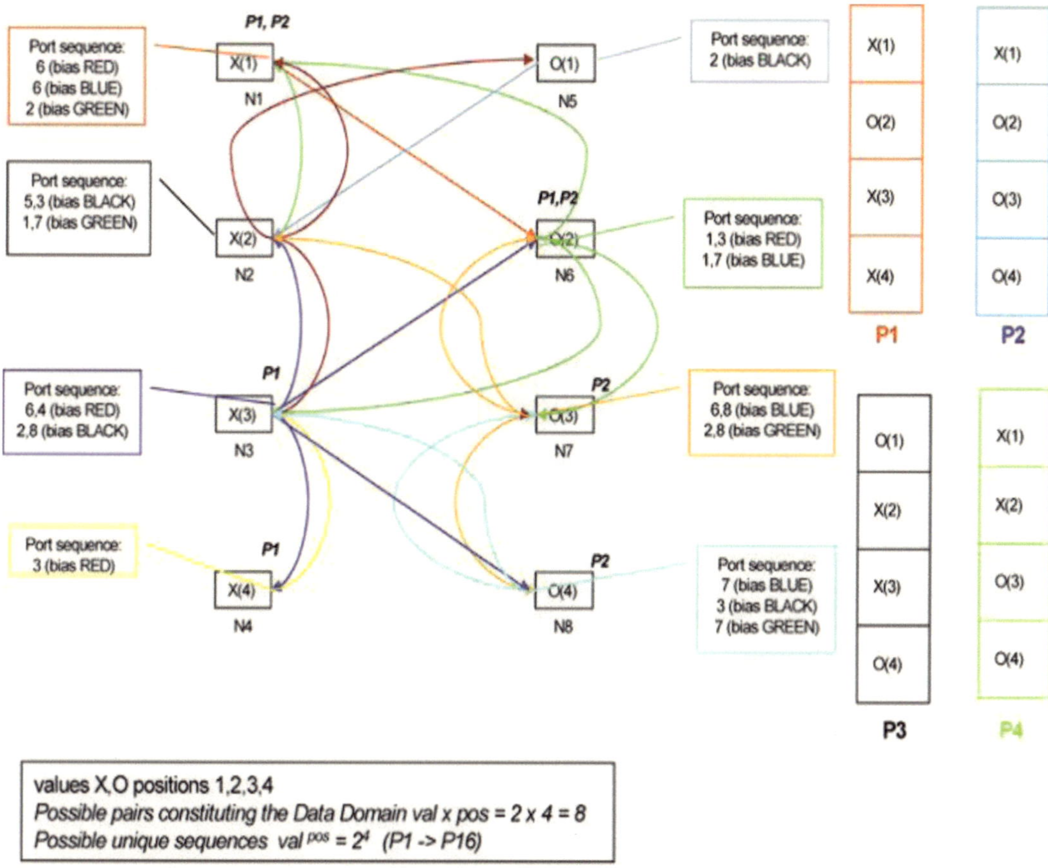

Note: The colouring scheme for interconnects is separate from the scheme used for the patterns

Fig. 6. An eight node GN network is in the process of storing patterns: P1 (RED), P2 (BLUE), P3 (BLACK), and P4 (GREEN)

4.2 Hierarchical Graph Neuron

In order to solve the issue of the crosstalk in GN model due to the limited perspective of GNs, the capabilities of perceiving GN neighbors in each GN was expanded in a new model called Hierarchical Graph Neuron (HGN) to prevent pattern interference (Khan & Nasution, 2008). Figure 7 illustrates a HGN scheme for pattern recognition with a pattern of size 7 and two possible values within the pattern being stored in the network.

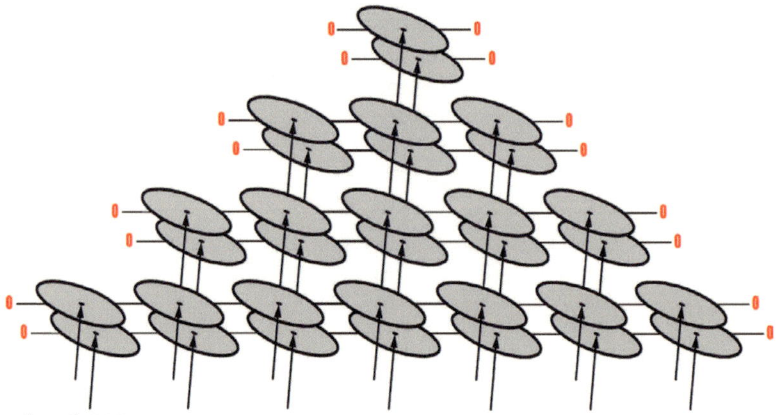

Fig. 7. Hierarchical GNs

The communication between HGN nodes is accomplished through different stages with the following fashion:

- Stage I. Each GN in the base layer will receive input pairs (value, command) sequentially. In this regard, all GNs are pre-programmed to receive these values. It should be mentioned that all the GNs in the same column will receive the same input pair, and all the GNs in the composition will receive the same command, whether store or recall.
- Stage II. Upon receiving the input, all active GNs will send a (column, row) pair to all the GNs on preceding and succeeding columns. It should be noted that, those GNs with matched IDs will issue the response; however other GNs enter the passive state, while still receiving reports from adjacent columns.
- Stage III. At this stage, all the active GNs at the edges have received one report message, while other active GNs would have received two reports from their adjacent columns. Then, each GN can determine the bias entry using input values, and report messages. Here, the GN can decide whether to store input in memory or declare it as a recall based on the command value.
- Stage IV. At this stage, after updating the bias array, each active GN will send its (row, bias index, command) value to all the GNs in the same column in the upper layer.

Finally, all the active GNs will return the (level, column, bias index) value back as the response. The process is continued by the next layer above the base layer. After repetitive execution of stages 2 to 4 by the remaining layers, the process will be stopped at one layer which is below the top layer. In general, the process consists of (S – 3) / 2 number of iterations for the last three stages, in which S denotes the size of the pattern space.

4.3 Distributed Hierarchical Graph Neuron

We then extended the HGN by dividing and distributing the recognition processes over the network. This distributed scheme minimizes the number of processing nodes by reducing the number of levels within the HGN. Figure 8 depicts the divide-and-distribute transformation from a monolithic HGN composition (top) to a DHGN configuration (bottom) for processing the same 35-bit patterns.

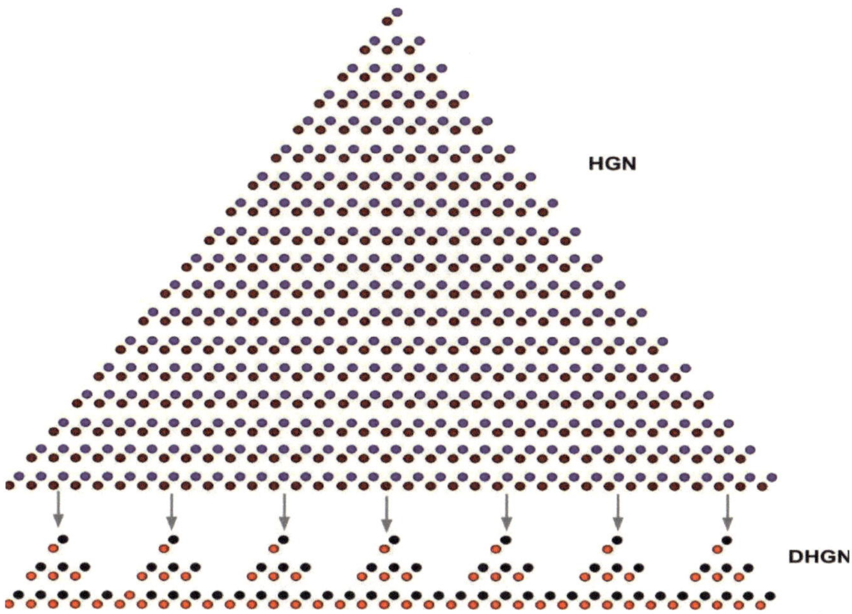

Fig. 8. Transformation of HGN structure (top) into an equivalent DHGN structure (bottom)

The base of the HGN structure in Figure 8 represents the size of the pattern. Note that the base of HGN structure is equivalent to the cumulative base of all the DHGN subnets/clusters. This transformation of HGN into equivalent DHGN composition allows on the average 80% reduction in the number of processing nodes required for the recognition process. Therefore, DHGN is able to substantially reduce the computational resource requirement for pattern recognition process - from 648 processing nodes to 126 for the case shown in Figure 8.

DHGN is in fact a single-cycle learning associative memory (AM) algorithm for pattern recognition. DHGN employs the collaborative-comparison learning approach (Khan & Muhammad Amin, 2009) in pattern recognition. It lowers the complexity of recognition processes by reducing the number of processing nodes (Khan & Muhammad Amin, 2007). In addition, pattern recognition using DHGN algorithm is improved through a two-level recognition process, which applies recognition at sub-pattern level and then recognition at the overall pattern level (See Figure 9). DHGN is already implemented in a grid environment (Khan & Muhammad Amin, 2008). We plan to adopt DHGN algorithm for our novel data estimation model in WSNs. DHGN allows the recognition process to be conducted in a smaller sub-pattern domain, hence minimizing the number of processing nodes which in turn reduces the complexity of pattern analysis. In addition, the recognition process performed using DHGN algorithm is unique in a way that each subnet is only responsible for memorizing a portion of the pattern (rather than the entire pattern). A collection of these subnets is able to form a distributed memory structure for the entire pattern. This feature enables recognition to be performed in parallel and independently. The decoupled nature of the sub-domains (subnets) is the key feature that brings dynamic scalability to data estimation within WSNs.

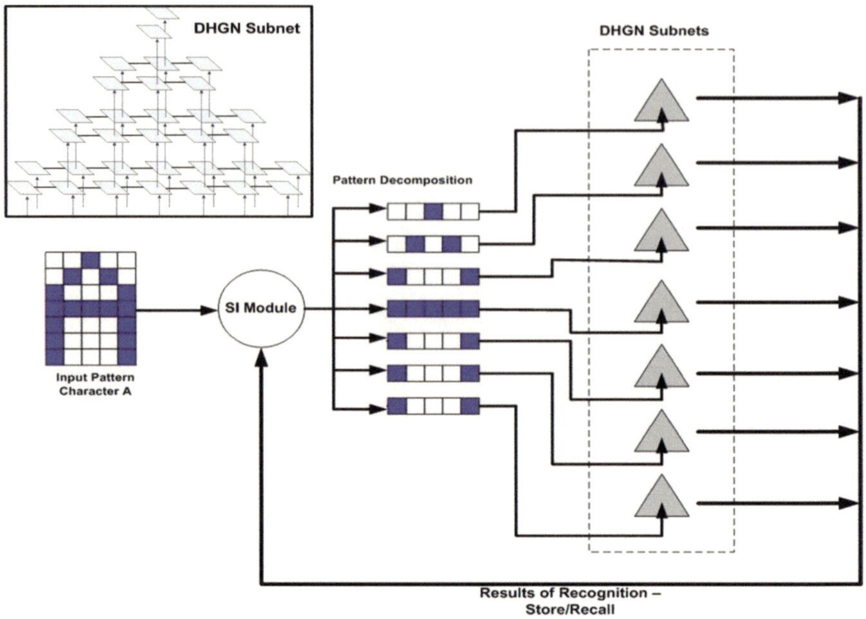

Fig. 9. DHGN distributed pattern recognition architecture

In fact, we envisage a vast associative memory network of wireless sensors to support user defined search and classification algorithms. Data in WSN may be viewed as uniform or non-uniform in nature. A uniform data block will incur the same computational cost as all the others. However in many instances, at least one data subset might be required to be known with a better quality of approximation or greater detail than the others, e.g. it may be invoked in the query process more often than the others. In this case the cost associated with data subset will be different. Larger variation can lead to inefficient use of the distributed resources. Strength of DHGN lies in the processing of non-uniform data patterns as it implements a finely distributable framework at the smallest (atomic) logical sub-pattern level. The results are easily obtained by summation at the overall pattern level. DHGN is also a highly scalable algorithm that incorporates content addressable memory within a clustered framework. Hence algorithmic strengths of the current estimation approaches can be investigated for the first time in combination with DHGN's single-cycle learning mechanism. This one shot approach will allow real-time in-network data manipulation; essential feature for data processing in WSNs.

5. Performance Metrics

By redesigning data processing models, data records are treated as patterns which enable data storage and retrieval by association over and above the existing simple data referential mechanisms. Hence, processing the input data and handling the dynamic load is handled by using a distributed pattern recognition approach that is implemented through the integration of loosely-coupled computational networks, followed by a divide-and-distribute approach that allows distribution of these networks within WSNs dynamically.

5.1 Scalable Simulation of the DHGN Application in WSNs:

Several tests are performed to ensure the accuracy of DHGN algorithm while evaluating its effectiveness in terms of complexity, timing and fault rate. It is essential that these software programs are fully integrated and tested as proof of concept that our approach will indeed work when applied to a dynamic WSN. In order to achieve this goal, we will formulate a wireless sensor network environment for executing our algorithms over very large numbers of GN nodes. Doing so will allow us to determine the asymptotical limits of our approach, find any problems that only arise when the algorithms are scaled-up, and test for robustness and scalability in face of dynamic changes to WSN configuration.
Our scheme relies on communications among adjacent nodes. The decentralized content location schemes will be implemented for discovering adjacent nodes in minimum number of hops. A GN based algorithm for optimally distributing DHGN subnets (clusters or sub-domains) among the WSN nodes will be provided to automate the boot-strapping of the distributed application over the network. Doing so will also provide means for investigating dynamic load balancing over the network. For this matter, a database of stored patterns is constructed. Patterns can be assumed as abstract notations for various parameters in the network. The user is able to configure the length and number of elements in the pattern, and based on these parameters, the program will generate a user-specified number of random patterns. At this stage, the user can manually enter patterns to be checked against the database. The program is not only capable of determining whether the pattern is matched with the previous stored ones or not, but is also able to determine the sub-patterns of the input pattern which have been visited before. This capability provides a tool for determining the level of distortion in the input pattern, which in turn adds to the robustness of the proposed algorithm significantly.

5.2 One shot pattern recognition within WSNs:

In contrast with hierarchical models proposed in the literature, DHGN's pattern matching capability and the small response time, that remains insensitive to the increases in the number of stored patterns, makes this approach ideal for WSNs. Moreover, the DHGN does not require definition of rules or manual interventions by the operator for setting of thresholds to achieve the desired results, nor does it require heuristics entailing iterative operations for memorization and recall of patterns. In addition, this approach allows induction of new patterns in a fixed number of steps. Whilst doing so it exhibits a high level of scalability i.e. the performance and accuracy do not degrade as the number of stored pattern increases over time. Its pattern recognition capability remains comparable with contemporary approaches. Furthermore all computations are completed within the pre-defined number of steps and as such the approach implements one shot, i.e. single-cycle or single-pass, learning. Tests have shown that the scheme has an acceptable level of pattern recognition accuracy in this context (Khan & Muhammad Amin, 2008b). As it is clearly depicted in Figure 10, the system's response (i.e. recognition times) does not trend upwards with the increase in the number of stored patterns. In other words, performance of the algorithm was not affected by the increase in the number of memorized patterns or the WSN network size.

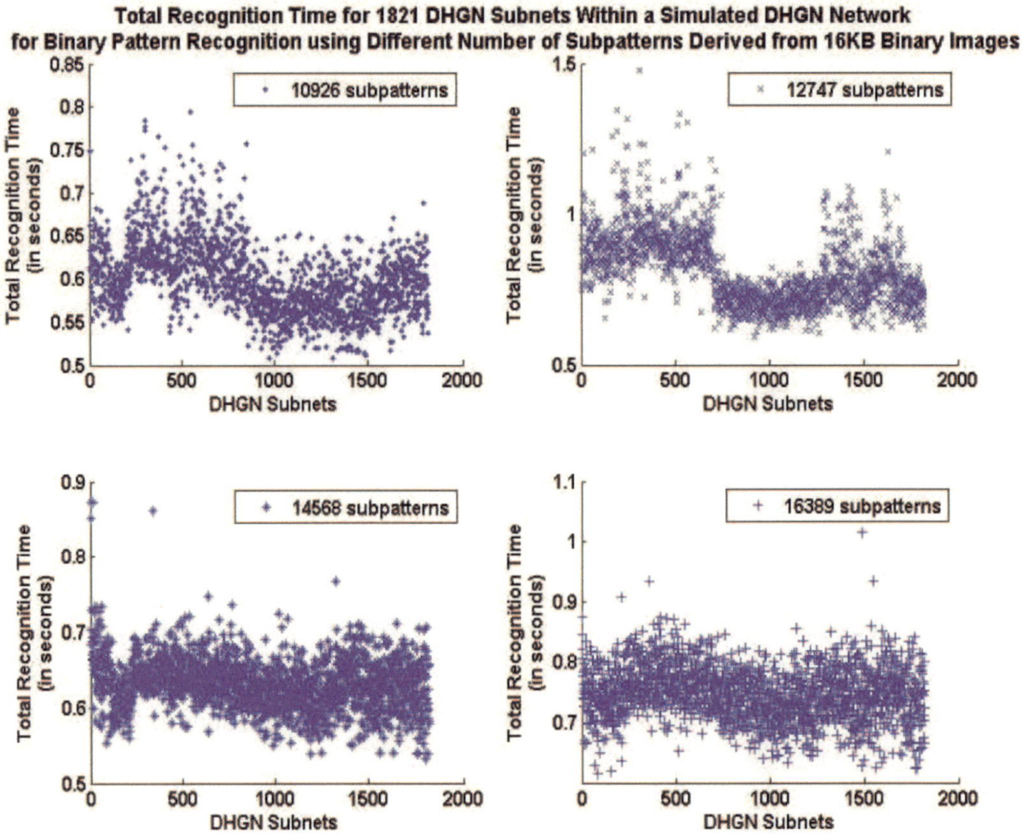

Fig. 10. Pattern recognition timings for storing a large number of 16KB images

5.3 One Shot Learning versus Back Propagation Network:

A comparative analysis with a Back-Propagation (BP) network layer and the input layer of the HGN highlights the ability of the GN layer to assimilate new patterns into the older patterns constructively (Khan & Nasution, 2008). The GN layer continues to improve its accuracy, as more and more patterns are stored, and exhibits better accuracy and performance as compared to the BP algorithm (See Figures 11 and 12). For performing a comparison approach, a BP perceptron layer was tested for single-cycle learning using a piecewise-linear activation function with a slope of 2.5 and a threshold of 0 for optimal results. These results were then compared with a single GN layer as shown in Figures 11 and 12 respectively. The perceptron layer and the GN layer were both trained once-only using up to 10,000 fully randomized patterns with a uniform distribution. Pattern sizes of 10 and 20 with random distortions (noise) were used in the tests. Each of these pattern sizes was associated with 16 possible values. Hence these pattern sizes corresponded to 16^10 or 40 bit (2^40) and 16^20 or 80 bit (2^80) patterns respectively. The accuracy was calculated as the percentage difference between the pattern size and the number of bits with recall errors.

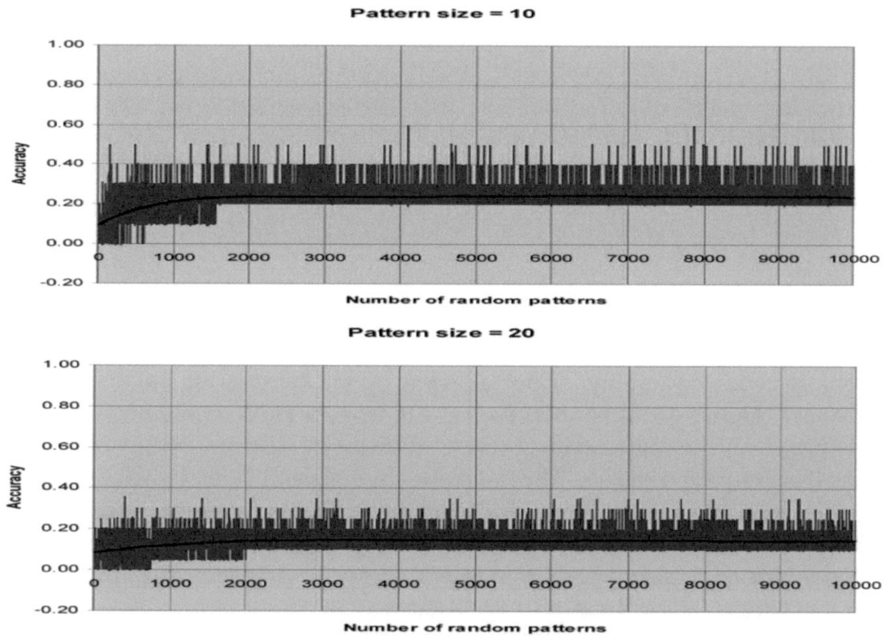

Fig. 11. Recall Accuracy of BP algorithm, after being fed with up to 10,000 fully randomized patterns

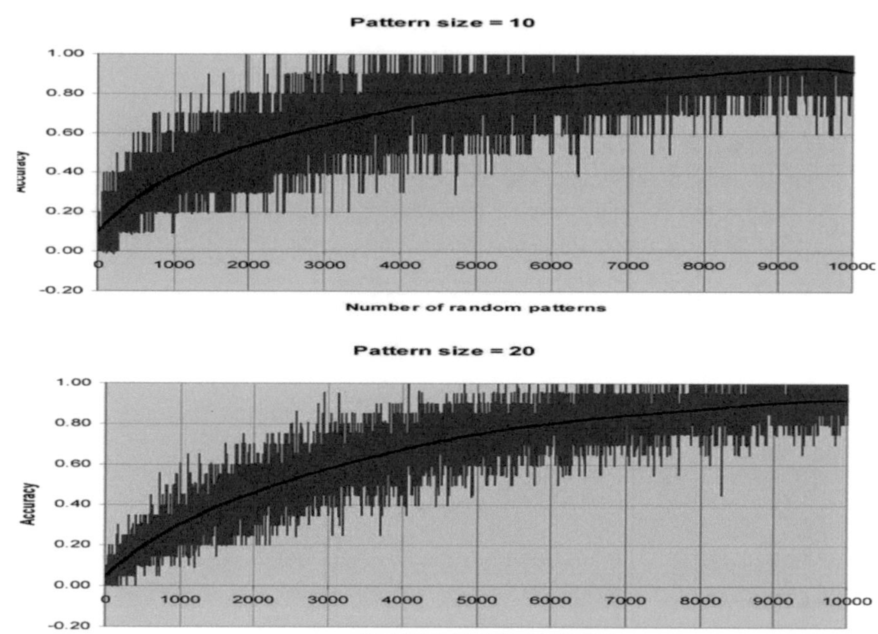

Figure 12. Recall Accuracy of single-cycle GN, after being fed with up to 10,000 fully randomized patterns (Adapted from [28])

It may be noted from Figures 11 and 12 that both the BP and the GN networks achieved comparable accuracy for the first 500 patterns. It can be seen from Figure 12 that the accuracy of GN in recognizing previously stored patterns increased as more patterns were stored (greater improvement with more one shot learning experiences). The GN in both the cases achieved above 80% accuracy after all the 10,000 patterns (with noise) had been presented. On the other hand, Figure 11 shows that the accuracy of BP network failed to improve after 2,000 patterns. Also, the recall accuracy of the perceptron layer noticeably deteriorated as the larger pattern size of 20 (80 bits) was introduced.

5.4 Superior Scalability:

Another important aspect of HGN is that it can remain highly scalable. In fact, its response time to store or recall operations is not affected by an increase in the size of the stored pattern database. Figure 13 illustrates the results from a HGN simulation in this regard. Figure 14 is based on the mathematical model derived in (Khan & Natusion, 2008), which corroborates the simulation results. The flat slopes in Figures 13 and 14, which show that the response times remain insensitive to the increase in stored patterns, represent the high scalability of the scheme. Hence, the issue of computational overhead increase due to the increase in the size of pattern space or number of stored patterns, as is the case in many graph-based matching algorithms will be alleviated in HGN, while the solution can be achieved within fixed number of steps of single cycle learning and recall.

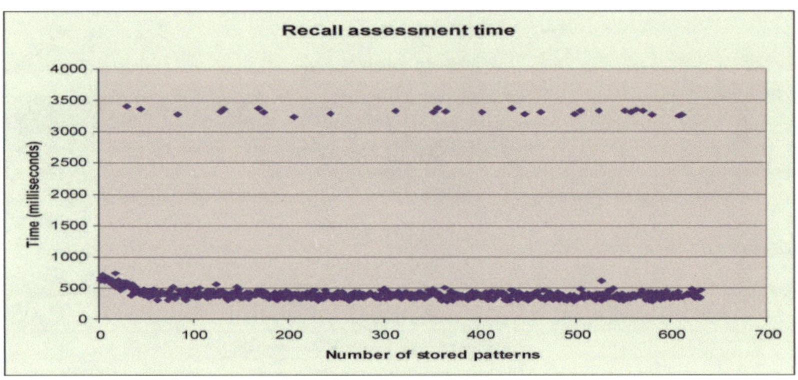

Fig. 13. Actual recall time for an analysed composition

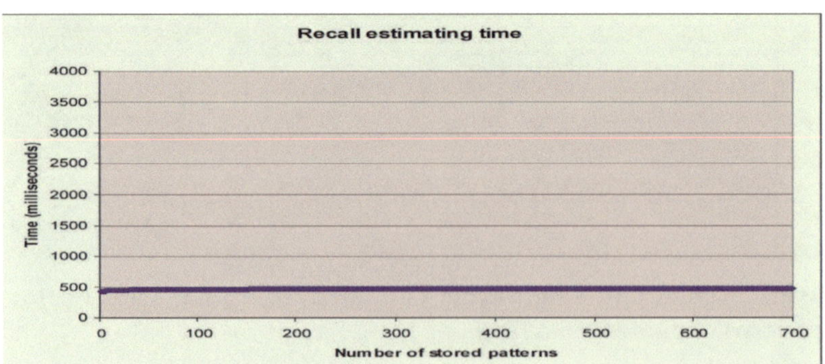

Fig. 14. Estimated recall time for a HGN composition

6. Discussion and Future Research

Our proposed approach is proven to work as a real-time pattern recogniser for the WSN. This is the only scheme which can operate within resource-constrained nodes of a WSN and still deliver speed and accuracies comparable to conventional pattern analysers. Furthermore our approach allows new patterns to be added at anytime without recourse to retraining. In contrast with contemporary distributed estimation approaches, our approach allows induction of new input patterns in a fixed number of steps. Whilst doing so it exhibits a high level of scalability i.e. the performance and accuracy do not degrade as the number of stored pattern increases over time. Its pattern recognition capability remains comparable with contemporary approaches. Furthermore all computations are completed within the pre-defined number of steps and as such the approach implements one shot, i.e. single-cycle or single-pass, learning. The one shot learning within this method is achieved by side-stepping the commonly used error/energy minimization and random walk approaches. The network functions as a matrix that holds all possible solutions for the problem domain. The network after applying our algorithm can find the solution in a single-cycle (i.e. fixed number of steps). Our approach finds and refines the initial solution by passing the results through a pyramidal hierarchy of similar arrays. In doing so it eliminates/resolves pattern defects, with distortions up to 20% being tolerated. Previously encountered patterns are revealed whilst new patterns are memorized without loss of stored information. In fact the pattern recognition accuracy continues to improve as the network processes more sensory inputs. HGN though highly suited for the WSN is limited by the well-known problem afflicting the hierarchical schemes i.e. large increases in network nodes as the pattern size and/or complexity increases. The distributed version of this scheme, DHGN, retains HGN's one-shot learning characteristic and reduces the computational complexity of our hierarchal scheme by distributing the recognition process into smaller clusters i.e. dividing-and-distributing simple pattern recognition processes across the network. The computational complexity of the scheme is significantly reduced and the accuracy is generally improved by using this strategy. The distributed scheme also lowers storage capacity requirements per node and incurs lesser communication cost thus improving the response-time characteristic. A comparison with the Hopfield network shows that our approach generally improves speed of recognition and accuracy. The distributed scheme also allows better control of the network resources - varying from coarse grained to very fine grained networks. This scheme compares well with contemporary approaches such as SOM and SVM in terms of speed and accuracy.

There are several benefits and advantages in our DHGN implementation for event detection and distributed estimation within WSN network. Our approach offers low memory consumption for event data storage using simple bias array representation. Furthermore, this scheme only stores sub-patterns or patterns that are related to normal event, rather than keeping the records of all occurring events. We have also shown that our approach is most effective for small sub-pattern size, since it uses only a small portion of the memory space in a typical physical sensor node in WSN network. In addition to this efficient memory usage, DHGN also eliminates the need for complex computations for event classification technique. With the adoption of single-cycle learning and adjacency comparison approaches, DHGN implements a non-iterative and lightweight computational mechanism for event recognition and classification. The results of our performance analysis have also shown that DHGN recognition time increases linearly with an increase in the number of processing elements

(PEs) within the network. This simply Lightweight Event Detection Scheme using Distributed Hierarchical Graph Neuron in Wireless Sensor Networks reveals that DHGN's computational complexity is also scalable with an increase in the size of the sub-patterns. DHGN is a distributed pattern recognition algorithm. By having this distributed characteristic, DHGN would be readily-deployable over a distributed network. With such feature, DHGN has the ability to perform as a front-end detection scheme for event detection within WSN. Through divide-and-distribute approach, complex events could be perceived as a composition of events occurring at specific time and location. Our approach eventually would be able to be used in event tracking as our proposed scheme has been demonstrated to perform efficiently within an event detection scheme such as forest fire detection using WSN. Despite all its benefits, DHGN has its own limitations. Firstly, DHGN simple data representation would requires significant advanced pre-processing at the front-end of the system. This might not be viable for strictly-resource constrained sensor nodes, where processing capability is very limited. In addition, DHGN single-hop communication for event detection scheme is not viable for large area monitoring, due to high possibility of communication error due to data packet loss during transmission. Our existing DHGN implementation has also been focusing on supervised classification. However, there is a need for unsupervised classification technique to be deployed for rapid event detection scheme. Overcoming the DHGN distributed event detection scheme limitations would be the path of our future research direction. We intend to look into event tracking scheme using DHGN distributed detection mechanism, as well as providing unsupervised classification capability for rapid and robust event detection scheme. Furthermore, we are looking forward into implementation of this scheme in large-area monitoring using multi-hop communication strategy.

7. Conclusion

The development of distributed estimation scheme within WSN has been made viable with the advancement in communication, computational, and sensor technologies. However, existing estimation/recognition algorithms fail to achieve optimum performance in a distributed environment, due to its tightly-coupled and computationally intensive nature. In this chapter, we have presented our readily-distributable distributed estimation scheme for WSN network which is known as Distributed Hierarchical Graph Neuron (DHGN). Throughout our studies, we discover that DHGN is able to perform recognition and classification processes with limited training data and within a one-shot learning. These DHGN features have given added-value for implementing this scheme within a lightweight distributed network such as WSN. Current implementation of DHGN in event detection using WSN has been focusing on the front-end processing, in which detection could be carried out earlier using the available wireless sensor nodes. Our approach differs from other existing estimation schemes in which major processing steps are conducted at the base station. By having a front-end detection, our proposed scheme is able to alleviate the computational costs experienced by the centralised-processing undertaken by the base station. In this chapter, we have also discussed the advantages and limitations of our proposed scheme. The future direction of this research lies in the development of a complete distributed estimation scheme that incorporates front-end detection and back-end complex event analysis. We foresee our DHGN distributed estimation scheme as a complete estimation scheme and recognition tool that is deployable over different types of event detection schemes on WSN networks.

8. References

Akyildiz , I. F. , Su , W. , Sankarasubramaniam & Cayirci , E. (2002). A survey on sensor networks. *IEEE Communications Magazin* , pp. 102 - 114 .

Blum , R. S. , Kassam , S. A. & Poor , H. V. (1997). Distributed detection with multiple sensors - Part II: Advanced topics. *Proceedings of IEEE*, pp. 64 - 79.

Chong , C.-Y. & Kumar, S. P. (2003). Sensor networks: evolution, opportunities, and challenges . *Proceedings of IEEE*, pp. 1247 - 1256 .

Culler , D. , Estrin , D. & Srivastava , M. (2004). Overview of sensor networks. *IEEE Computer*, pp. 41 - 49 .

Khan, A.I., Mihailescu, P. (2004), Parallel Pattern Recognition Computations within a Wireless Sensor Network, *In Proceedings of the 17th Intl. Conference on Pattern Recognition (ICPR'04)*, Cambridge, United Kingdom: IEEE Computer Society.

Khan A.I., Isreb M., Spindler R.S. (2004). A parallel distributed application of the wireless sensor network, *In Proceedings of the 7th Int'l Conf. on High Performance Computing and Grid.*

Khan, A.I., Baig Z.A., Baqer M., (2006). A pattern recognition scheme for distributed denial of service (DDOS) attacks in wireless sensor networks, *In Proceedings of the 18th International Conference on Pattern Recognition.*

Khan, A.I., Muhamad Amin, A.H. (2007). One Shot Associative Memory Method for Distorted Pattern Recognition, *In 20th Australian Joint Conference on Artificial Intelligence,* Gold Coast, Australia, pp. 705-709.

Khan, A.I., Muhamad Amin A.H., (2008a). Commodity-Grid Based Distributed Pattern Recognition Framework, *Sixth Australasian Symposium on Grid Computing and e-Research (AUSGRID 2008)*, Wollongong, NSW, Australia.

Khan, A.I., Muhamad Amin, A.H., (2008b). Single-cycle image recognition using an adaptive granularity associative memory network, *Lecture Notes in Artificial Intelligence*, vol 5360, Springer-Verlag, Berlin Germany, pp. 386-392.

Khan, A.I., Nasution, B. B. (2008) ,IEEE Transactions on Neural Networks, 212-229.

Khan, A.I., Muhamad Amin, A.H., (2009). Collaborative-Comparison Learning for Complex Event Detection using Distributed Hierarchical Graph Neuron (DHGN) Approach in Wireless Sensor Network, *In 22nd Australasian Joint Conference on Artificial Intelligence (AI'09)*, Melbourne, Australia.

Luo , Z-Q. , Gastpar , M. , Liu , J. & Swami , A. (2006). Distributed signal processing in sensor networks [from the guest editors] . I EEE Signal Processing Magazine, pp. 14 - 15.

Niu , R. , Chen , B. & Varshney , P. (2006). Fusion of decisions transmitted over Rayleigh fading channels in wireless sensor networks. *IEEE Transactions on Signal Processing*, pp. 1018 - 1026 .

Quek, T. Q.-S., Dardari, D. & Win, M. Z. (2006*a*). Cooperation in bandwidth-constrained wireless sensor networks. In *Proceedings of the IEEE International Workshop Wireless Ad Hoc Sensor Networks, New York, NY, USA* .

Quek, T. Q.-S., Dardari, D. & Win, M. Z. (2006 *b*). Energy efficiency of cooperative dense wireless sensor networks. In *Proceedings of the International Wireless Communication sand Mobile Computing Conference, IWCMC 2006, Vancouver, Canada* , pp. 1323–1330.

Quek, T. Q.-S., Dardari, D. & Win, M. Z. (2006*c*), Energy efficiency of dense wireless sensor networks: To cooperate or not to cooperate. In *Proceedings of the IEEE International Conference on Communications, ICC 2006, Istanbul, Turkey*, Vol. 10, pp. 4479–4484.

Rao, B. S. Y., Durrant-Whyte, H.F., and Sheen, J. A. (1993). A fully decentralized multi-sensor system for tracking and surveillance. *International Journal of Robotics Research*, 20–44.

Sendonaris , A. , Erkip , E. & Aazhang , B. (2003). User cooperation diversity - Part II: Implementation aspects and performance analysis. *IEEE Transactions on communications*, pp. 1939 - 1948.

Speyer. L. (1979). Computation and transmission requirements for a decentralized linear-quadratic-gaussian control problem. *IEEE Transactions on Automatic Control*, 266–269.

Sung , Y. , Tong , L. & Swami , A. (2005). Asymptotic locally optimal detector for large-scale sensor networks under the Poisson regime. *IEEE Transactions on Signal Processing*, pp. 2005 - 2017.

Swaszek, P. & Willett , P. (1995). Parley as an approach to distributed detection. *IEEE Transactions on Aerospace and Electronic Systems*, pp. 447 - 457.

Welsh, M. , Malan, D. , Duncan, B., Fulford-Jones, T., and Moulton, S. (2004). Wireless Sensor Networks for Emergency Medical Care", *GE Global Research Conference*, Harvard University and Boston University School of Medicine, Boston, MA.

Winters, J. H. (1987). On the capacity of radio communication systems with diversity in Rayleigh fading environment. *IEEE Journal Selected Areas in Communication*, pp. 871 - 878 . SAC

23

Multimedia Data Processing and Delivery in Wireless Sensor Networks

Javier Molina, Javier M. Mora-Merchan, Julio Barbancho and Carlos Leon
University of Seville (Department of Electronic Technology)
Spain

1. Introduction

In the last few years, multimedia processing in Wireless Sensor Networks (WSNs) has become a promising technology. It has the potential to enable a large class of applications, most of them related to surveillance and locating (i.e. target detection and tracking, border protection, patient and elderly assistance, people and object identification, environment monitoring, fire detection, industrial control, ...). To achieve an effective Quality of Service (QoS) in multimedia applications, special node and network capabilities are required. For example, compared to normal WSN nodes, multimedia nodes need additional hardware resources for memory, processing capability, transmission rate and energy. A Wireless Multimedia Sensor Network (WMSN) is a special WSN made up of several multimedia sensor nodes, specially designed to retrieve multimedia content such as video and audio streams, still images, and scalar sensor data from the environment.
In this paper, we focus on a different, but also very practical and common, sort of wireless network: the Heterogeneous Networks, where multimedia and non-multimedia nodes deliver data. In this scenario, the non-multimedia node constraints have to be taken into account to deliver multimedia data through multi-hop paths.
Generally, collaborative processing makes no sense in this case. Although in-sensor multimedia processing is a fundamental topic in order to obtain an effective data reduction.
Currently, research challenges in designing multimedia applications on WSNs include, but are not limited to the following:

- QoS requirements. Streaming media, system snapshots, audio/video store and playback applications have different requirements with respect to delay, jitter, and loss tolerance.
- Bandwidth. WMSNs require a bandwidth that is orders of magnitude higher than that supported by currently available sensors.
- Power. Compared to traditional WSNs, power consumption is greater in multimedia applications because of high volumes of data, high transmission rates, and extensive processing.
- In-network processing support, to efficiently extract relevant information from multimedia data (e.g. panoramic image fusion, target identification and location).

- In-node processing support, to compress data, signal analysis and features extraction (singular points, region segmentation, object detection, ...)
- Cross-layer design. An effective optimization of all the above parameters involve cross-layer protocol design ranging from Application to Physical Layer.
- New hardware design to better manage the energy with high QoS (i.e. power supply, microcontroller (MCUs) architectures and energy harvesting).

In this paper, we propose a new WMSN model that includes the Heterogeneous Network. For this reason, main topics of Wireless Sensor Networks are described in section 2, and the most relevant issues in multimedia networks are summed up in section 3. These issues are: standard network architectures, node constraints, compression techniques, memory, bandwidth, and energy consumption.
Different hardware structure alternatives for heterogeneous nodes are introduced in section 4. Moreover, a simple application algorithm is proposed to enable the low-end MCUs to capture multimedia data, or collaborate in delivering.

2. General vision on the design of Wireless Sensor Network Applications

Because of cost and size, the nodes of a WSN exhibit resource constraints in terms of CPU processing capacity, memory, bandwidth and energy. Figure 1 shows a node hardware structure, and table 1 sums up a comparative study of popular MCUs, RF chips, and platforms.

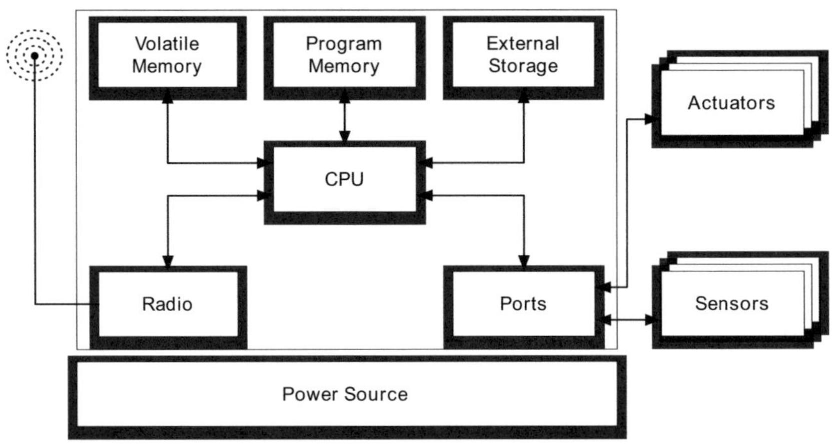

Fig. 1. Node hardware architecture (Soro & Heinzelman, 2009)

Many WSNs are designed for environmental sensing applications, acquiring data from scalar sensors (temperature, humidity, light, etc). In those cases where vector data sensing are required, the vector dimension is often low, for example movement sensing (accelerometers), wind data (anemometers), velocity and positioning (GPS sensors). In contrast, multimedia sensor nodes are characterized by audio and video streaming, and still image data.
Software development for WSNs nodes is a complex issue. Many researchers program the modes from scratch, using operating system components, specific middleware, or by higher programming abstractions (Mottola & Picco, 2010). Figure 2 summarizes a general software model on a WSNs node platform.

Sensor Node Name	Micro controller	Transceiver	Progra+DataMemory
BTnode	ATmega128L	CC1000	64k+180K
EPIC mote	MSP430	CC2420	10k
EyesIFX v2	MSP430F1611	TDA5250	10k+48k
FireFly	ATmega128L	CC2420	8k
GWnode	PIC18LF8722	BiM	64k
Imote 2.0	PXA271ARM	CC2420	256k
Mica2	ATmega128L	CC1000	64k+180k
TelosB	MSP430F1611	CC2420	10k+48k
TinyNode	MSP430	XE1205	8k
XYZ	ML67	CC2420	32k

Table 1. WSN Platforms

MAC- Medium Access Control is placed jointly to the operating systems because, in contrast to other wireless technologies where MAC protocol is done in hardware, in WSNs it is typically implemented mostly in software, using the low-level language associated to the operating system. MAC protocols must guarantee an efficient access to the communication media while carefully managing the energy budget available in the node. The latter goal is typically achieved by switching the radio to a low-power mode based on the current transmission schedule.

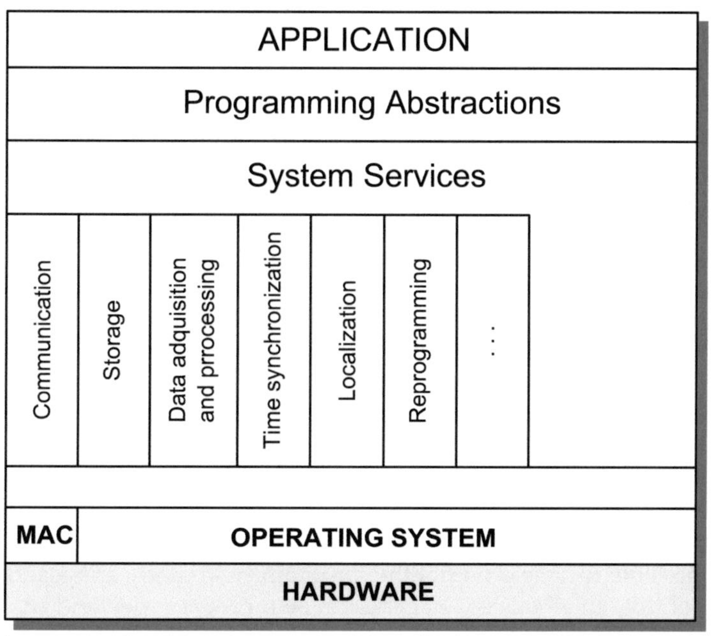

Fig. 2. Node software model

The operating system is essentially a library linked to the application code to generate an executable binary code. This way, the program memory resource is reduced to a minimum, and only necessary hardware specific routines are included. So far, many operating systems

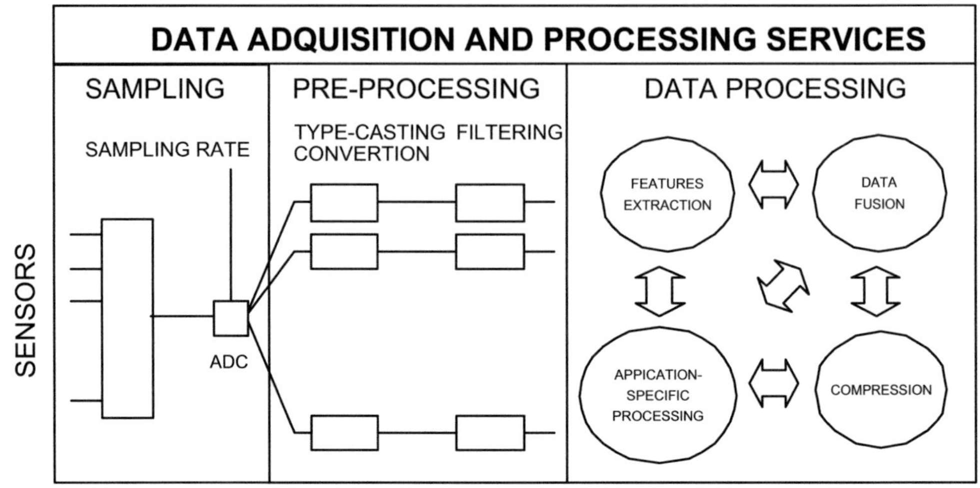

Fig. 3. Data acquisition and processing service

and specific program languages have been proposed. NesC language (Gay et al., 2003) for TinyOS (Hill et al., 2000) are the most popular. Alternatives include Contiki (Dunkels et al., 2004), SOS (Han et al., n.d.), Mantis (Abrach et al., 2003), RETOS (Cha et al., 2007), LiteOS (Cao & Abdelzaher, n.d.), t-Kernel (Gu, 2006), or NANOrk (Eswaran et al., 2005). Some of the above operating systems (e.g., SOS, LiteOS, and Contiki) also provide dynamic linking capabilities, i.e., new code modules can be added at run-time to the application running on a node. Dynamic linking is particularly important in supporting wireless reprogramming of the WSN.

System Services are programs designed to support applications. Typically, an application requires Location, Time Synchronization, Storage, Communication and Data acquisition/processing services. Location services estimate the own node position or the location of a target (He et al., 2003; Mao et al., 2007; Patwari et al., 2005). Time synchronization is usually required for data time-stamping, to measure the TOA (Time of Arrival) of a signal, to manage hibernate-wake up cycles, etc. Many algorithms have been proposed for this (Dai & Han, 2004; Elson & Römer, 2003; Maróti et al., 2004). Communication services deliver data reliably, manage the network and optimize QoS and energy savings. The researchers' community has developed many different algorithms for routing, cluster management, etc. Although, the industry has developed the Zigbee standard, it is probably the most widely accepted one so far.

In most applications, sensor networks are deployed once and intended to operate unattended for a long period of time. Reprogramming capabilities facilitate the management of application changes or updates (Rubio et al., 2007). Because of the data memory constraints in the nodes, it is not possible to save a monolithic program with many applications before being burned. Instead of this, small packets of code can be reprogrammed and can also be reliably delivered through the network. There are two basic schemes to reprogram sensor networks: code dissemination and code acquisition. The first one is initiated by the administrator to reprogram all the devices in the network. In the second one, the nodes request and reprogram a packet dynamically. Data acquisition is not always considered and programmed as a service. However, it is especially relevant in multimedia applications. Figure 3 shows a software

Phenomena	Sample rate (/Hz)
Very low frequency	
Atmospheric temperature	0.017–1
Barometric pressure	0.017–1
Low frequency	
Heart rate	0.8–3.2
Volcanic infra-sound	20–80
Natural seismic vibration	0.2–100
Mid frequency (100Hz – 1000 Hz)	
Earthquake vibrations	100-160
ECG (heart electrical activity)	100–250
High frequency (>1 kHz)	
Breathing sounds	100–5k
Industrial vibrations	40k
Audio (human hearing range)	15–44k
Audio (muzzle shock-wave)	1M
Video (digital television)	10M

Table 2. Sensor sampling rates of different phenomena.

model for data acquisition and processing service with three stages: Sampling, Pre-processing and Data processing.

In the sampling stage, a calibrated set of sensors are sampled, and the row data are time-stamped, if necessary. Sensor calibration is an important issue. Traditional methodologies can not be applied because of the cost of a manual and node-by-node adjusting. The sampling process is fired by either an event or periodically. Following the Nyquist information theory, sampling rate depends on phenomena dynamics. In Table 2 are shown data rates for different applications.

More modern theories, like compressive sensing, allow the reduction of the sampling rate for specific signals (e,g images).

Sensor calibration is an important issue. Traditional methodologies can not be applied because of the cost of a manual and node-by-node adjusting. Several collaborative techniques have been proposed to allow self-calibration (Feng et al., 2003; Ihler et al., 2004; Miluzzo et al., 2008).

Accommodation stage involves data transformation into engineering units, and data-type casting. To pick the right data type, value ranges and quantization errors should be taken into account. Although, using high precision arithmetic, like floating-point, makes the processing performance drastically fall in small CPUs (Zoumboulakis & Roussos, n.d.).

Filtering, if present, reduces or removes undesirable time tendencies, spurious interferences or frequency bands to enhance later processing algorithms. This stage is especially relevant in audio processing applications.

The application-specific processing stage often includes algorithms for data compression, data fusion and feature extraction. In-node data fusion allows the qualification and validation of sensor data, making it easier to associate different physical measures to complex phenomena. Feature extraction is highly application dependant. Its main goal is to obtain a pattern to better represent, detect, identify or predict events and complex phenomena.

Data storage strategies usually distinguish between short-term and long-term storage. Uncompressed, lossless compressed and labeled fused data are often necessary in short-term storage . Otherwise, lossy compression and ancient data loss (amnesic algorithms) are relevant topics in long-term storage (Girao et al., 2007; Sheng et al., 2006).
All the stages can interchange information for feedback and adaptive processing. In-network processing involves data exchange among several sensor nodes. At the sampling stage, for example, several collaborative algorithms have been proposed to calibrate the sensors (Feng et al., 2003; Ihler et al., 2004; Miluzzo et al., 2008). A smart analysis of fused data or features from different sensor nodes is necessary to process localized phenomena (i.e. target detection, tracking and positioning (Arora et al., 2004; Cao et al., 2005)). Collaborative in-network processing can also increase application reliability and performance (i.e. using voting schemes and/or analyzing different target features in different nodes (Bénézit, 2009)).

3. Wireless Multimedia Sensor Networks: specific characteristics

Classic Wireless Sensor Networks have been initially thought up to manage ambient data at low data rates. In general, for multimedia signals, high computing capabilities, high transmission rates, large memory and an unlimited power source supply are necessary. Wireless Multimedia Sensor Networks do not meet all of these conditions, though they do support ambient sensing and multimedia applications, but with a limited quality of service.

3.1 Architectures

The node model shown in figures 1 and 2 is also suitable to describe a node of a Wireless Multimedia Sensor Network (WMSN). However, multimedia signals require higher sampling rate and more complex data processing. These topics make multimedia applications high consumers of resources, in terms of memory, bandwidth, processing capabilities and energy. Limited resources in most hardware platforms are some of the more relevant constraints for multimedia applications. Other restrictions depend on the application. They can be described in terms of Quality of Service. For example, in image delivery, topics like transmission reliability or image quality can be relevant, and video streaming can require security, and low delay. (Faheem et al., 2010)
Overall network architecture is also another relevant topic in order to understand practical constraints. The research community has proposed three architectures, as described in the survey (Akyildiz et al., 2008). From this paper, figure 4 depicts these.
Elements of these architectures are described below:
Single-tier flat architecture is made up of homogeneous multimedia sensors connected to a sink that bridges data to a storage hub or a gateway. The function of these elements is described below:

- Standard Video and Audio Sensors capture sound, still or moving images of the sensed event.
- Scalar Sensors acquire data from physical variables such as temperature, light, humidity, etc. They are very resource-constraint devices.
- Multimedia Processing Hubs. These devices have larger resources compared to sensor nodes. They add data streams and reduce the volume of data delivered to the sink.
- Storage Hubs save still-images or multimedia streams for data-mining and feature extraction, even before the data is delivered to the end user.

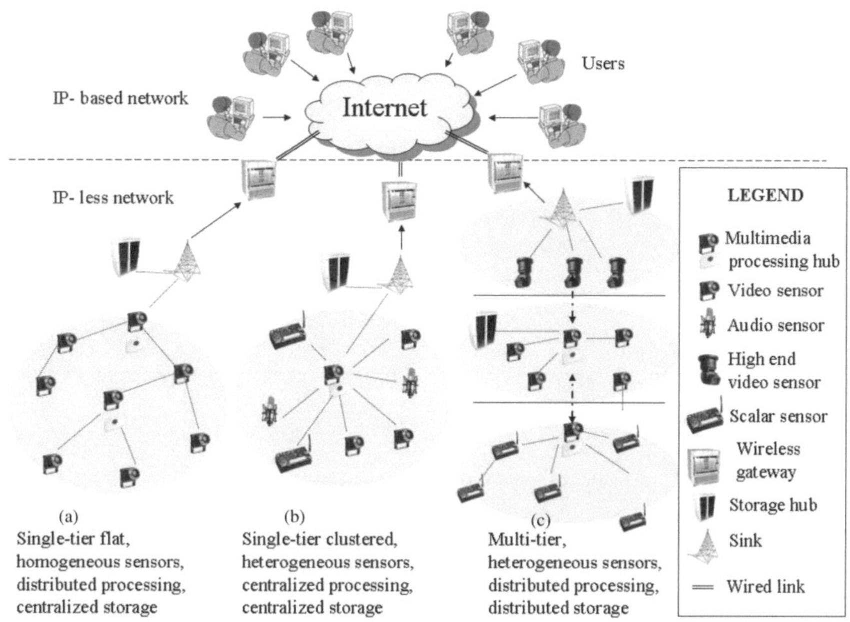

Fig. 4. Standard WMNs architectures (Akyildiz et al., 2008)

- The Sink is responsible for translating high level user queries to network-specific directives and returning the right chunk of the multimedia stream. Several sinks may be needed for large and heterogeneous networks.
- the Gateway bridges the sink to the Internet. IP is only assigned to the gateway, even with multiple sinks. It manages geographical information to allocate the right nodes and sinks for the selected area.

These architectures depict multimedia sensor networks where data flows between devices with the same hardware capabilities, or from lower-end to higher-end devices. Complex in-node and in-network processing is also performed in the multimedia processing hubs or high-end multimedia nodes. Scalar data acquisition and processing are managed in scalar nodes which are not involved in multimedia data processing or delivery. But, most environmental sensor networks manage scalar data and, currently it is becoming more frequent to include audio, sill-image and video delivery when an event has been taken place. This also means that image and audio processing capabilities are needed. This situation can be described as the fourth architecture: single-tier flat heterogeneous sensors (STFH), where low-end or high-end nodes must process and deliver data (streams are not often) through the deployed scalar sensor nodes (low-end devices). This contrasts with standard architectures, where each tier is in fact a sub-network with different performances.
Figure 5 describes the new architecture.
In the single-tier flat heterogeneous architecture, the constraints of the lowest-end nodes limit the performance of the overall network, no matter if it contains high-end video nodes. For this reason, only a limited set of services of image and audio processing make sense. This architecture will be the reference for this study, because it represents both: a very frequent situation in environmental applications, and the worst and most restrictive case for multimedia

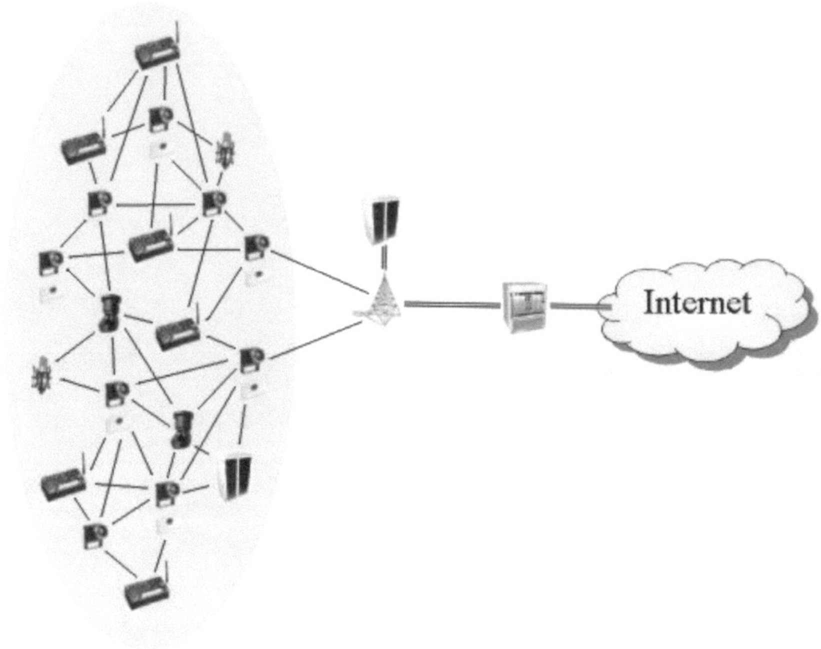

Fig. 5. Proposed WMNs architecture

capturing, processing and delivery. Any algorithm, specifically developed for it, will be easily and successfully applied to other architecture.
In the STFH architecture, any node can be involved in capturing, processing, delivering and the storage of multimedia data, the exact role depends on the application, but mainly on the node capabilities. In contrast to the multi-tier architectures, where nodes in the same tier have similar resources and functions, in STHF these functions can differ among equal nodes. For this reason, node constraints and applications requirements should be analyzed and compared.

3.2 Node constraints

Many commercial and research node platforms have been developed and proposed in scientific literature (Akyildiz et al., 2008). For Wireless Multimedia Sensor Network, these platforms can be classified into three categories that match node devices proposed in the general multimedia architectures (Akyildiz et al., 2008). Table 3 summarizes resources supported for these three node platform categories:

3.3 Compression

Multimedia data compression, which is one of the most relevant topics, is often present in any application with multimedia image storage and delivery. There are many image compression algorithms, but JPEG is in fact the most extensive. It is also included in most CMOS camera sensor chips. JPEG standard can easily reach from 10:1 up to 20:1 compression ratio without visible losses in color images. 30:1 to 50:1 compression is possible with small to moderate defects, while for very-low-quality purposes like indexing or previews, 100:1 compression

Low-end Category: Audio and video sensor					
Platform	Camera resolution	Memory	BW/Color	MCU	MIPS
MICA2 + Cyclops Polastre et al. (2004)Rahimi et al. (2005)	320x240 (CIF CMOS)	64Kb + 512Kb Flash	Color (8bits)	ATmega128L	16

Middle-End Category: Multimedia processing hub					
Platform	Camera resolution	Memory	BW/Color	MCU	MIPS
IMote2 + CMUCam2 *IMOTE2* (n.d.)Carnegie Mellon University (2007)	CIF (352x288) RGB	64Kb Ram + 128Kb Flash	Color (8bits)	ARM7TDMI	36
MeshEye Hengstler et al. (2007)	VGA CMOS (640x480)	64Kb + 256Kb Flash	Gray or Color (24bits)	Atmel AT91SAM7S	200
	30x30		Gray		
Panoptes Feng et al. (2005)	640x480	64Mb	Color	StrongARM 206 MHz	235

High-End Category: Video Processor					
Platform	Camera resolution	Memory	BW/Color	MCU	MIPS
Stargate X-scale + webcam *Stargate XScale Platform* (n.d.)	640x480	64Mb Ram + 32Mb Flash	Color	XScale PXA255 400MHz	370
SensEye Kulkarni et al. (2005)					
(Cyclops)	128x128	64Kb+512Kb	Color(8bits)	ATMEGA128	16
(CMUCAM)	352x288		Color(8bits)	SX52	80
(Webcam)	640x480		Color	Stargate	370
(SNC-RZ30N PTZ)	1024x768		Color	Embedded PC (P4 1.6 Ghz)	1237

Table 3. Constraints of several platforms

is quite feasible. Although, gray-scale images can not be compressed by such large factors because the human eye is much more sensitive to brightness variations than to hue variations. The threshold of visible loss using JPEG is often around 5:1 compression for gray-scale images. Table 4 shows comparative results for JPEG compression.

Image resolution	Format/Size	Size in memory		
		Raw	Low comp.	High comp.
Low	128 X 96 (SQCIF)	36K	3K	369
	176 x 144 (QCIF)	74K	7K	760
Med	320 x 240 (QVGA)	225K	22K	2K
	352 x 288 (CIF-PAL)	297K	29K	2K
High	640 x 480 (VGA)	900K	90K	9K
	704 x 576 (4CIF)	1M	118K	11K

Table 4. Size of images with several levels of compression

In many scientific papers, other compression schemes have been proposed, a great deal of them based on wavelets transform (Lewis & Knowles, 1992; Villasenor et al., 1995). Haar compression algorithms are very suitable for resource-limited devices like WSN nodes. Moreover, in image delivery, Haar compression allows variable data rates for adaptive transmission schemes, by assigning different priorities to the packets. In this way, most relevant coefficients will be reliably delivered and less relevant coefficients can be discarded if necessary.

Recently, compressions based on Wyner-Ziv encoder theory (Ziv & Lempel, 1978) are being studied (Aaron et al., 2003), especially for collaborative compressions (Chia et al., 2009). The main advantage is the simplicity and the light-weight computation of the encoder. For the same reasons, but with a fairly different perspective, Compressive Sensing (CS) is another promising technology. It exploits the sparse nature of an image to achieve very high compression rates.

A similar description can be done for audio data. Lossless compressions are more suitable for generic applications. In this sense, lossy encoders should be correctly picked to optimize audio data storage. Despite the fact that a vast amount of literature exists on lossless compression (Sayood, 2005), only few algorithms are shown to be feasible in WSNs due the strong HW constraints. For example: Delta Coding, Human Coding, Lempel-Ziv SS, and their variants: Delta Coding + Plain Huffman, Adaptive Huffman, Delta Coding+Adaptive Huffman, etc. The best compression rate results are about 65% (Alippi et al., 2007).

The application of lossy techniques requires specific signal degradation studies, and they should be picked and applied carefully. For example, many encoders take advantage of the human psychoacoustic hearing or human speech properties like the popular (mp3), and clearly, they are not directly applicable to compress vibrations or seismic data. Here again, Compressive Sensing is a promising technology because no previous signal characteristic is supposed, but sparsity is a very common property in nature.

3.4 Memory and bandwidth requirements

Memory supports compressed data storage and temporal raw data saving for multimedia processing and data transmission/relay. In this context, raw data means images or audio data without compression, with the same size or dimensions in which they have been captured, but in a suitable format. For example, 8-bit unsigned integer RGB or YCbCr matrices for images, and normalized floating-point (or 16-bit integer) arrays for audio. Memory occupation of multimedia data is depicted in 6. In an image or video application, compressed images or streams share memory with temporal images, packets for image delivery or relay, and local and global variables.

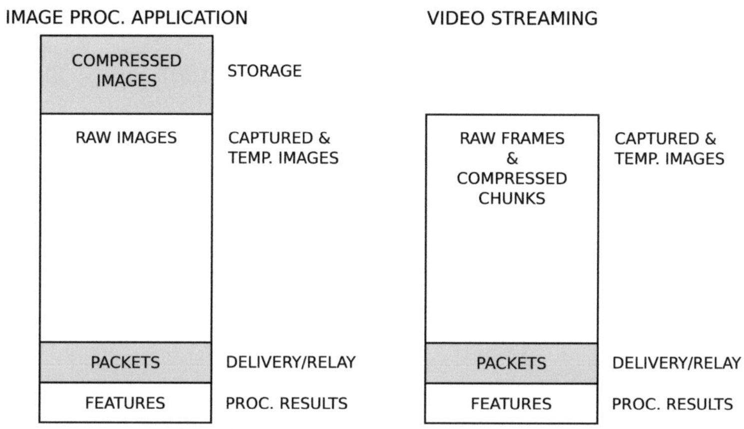

Fig. 6. Memory occupation of multimedia data

Compressed images memory occupation depends on the application. In general, considering the size of compressed images (table 4), it is not possible to save them into the microcontroller internal memory. If no processing or different compression format was necessary, the CMOS camera sensor could host the image in JPEG format in its internal memory. Otherwise, external RAM or FLASH memory would be needed.

Image transmission or relaying packet number, depend on the communication protocol, especially in the transport layer. By saving several packets, data can be retransmitted for reliability after an error occurs. But this, in general, just affects the source node, not the relaying nodes, except in the case where catching strategies were employed. It is always recommendable to use internal memory to quickly react to the communication events.

For image processing, temporal images must be allocated. Some special algorithms can be executed "on-the-fly" without saving the overall image. In latter sections, we will describe a basic model for that. But in general, for low-end platforms, an external RAM memory is required, normally connected by an SPI port. Computing performance will decrease if an intensive memory access is necessary. In that case, the algorithms must be released to support data pre-catching strategies.

The most successful communication standard for WSN is the IEEE 802.15.4. But for high-end nodes, Ultra Wide Band (UWB) is probably more suitable. Bandwidth for 802.15.4 compliant devices in short distance transmissions ranges from 224kbps to 20kbps for 10 and 75 meters, respectively.

Table 5 would clearly show that video streaming in WSN is only possible for medium or high-end platform, high compression, low frame rates and short distances. Fortunately, higher compression is possible, but with very specific algorithms.

3.5 Processing and energy consumption

In traditional WSN, energy consumption (fig. 8) comes mainly from radio transmissions. The energy required to process data is considered negligible if, as usual, external or timed events

Format/Size	Uncompressed		Low Compression		High Compression	
	25 FPS	15 FPS	25 FPS	15 FPS	25 FPS	15 FPS
128 X 96 (SQCIF)	7Mbps	4Mbps	720Kbps	432Kbps	72Kbps	43Kbps
176 x 144 (QCIF)	14Mbps	8Mbps	1Mbps	891Kbps	148Kbps	89Kbps
320 x 240 (QVGA)	43Mbps	26Mbps	4Mbps	2Mbps	450Kbps	270Kbps
352 x 288 (CIF-PAL)	58Mbps	34Mbps	5Mbps	3Mbps	594Kbps	356Kbps
640 x 480 (VGA)	175Mbps	105Mbps	17Mbps	10Mbps	1Mbps	1Mbps
704 x 576 (4CIF)	232Mbps	139Mbps	23Mbps	13Mbps	2Mbps	1Mbps

Table 5. Required Bandwidth for Video Streaming

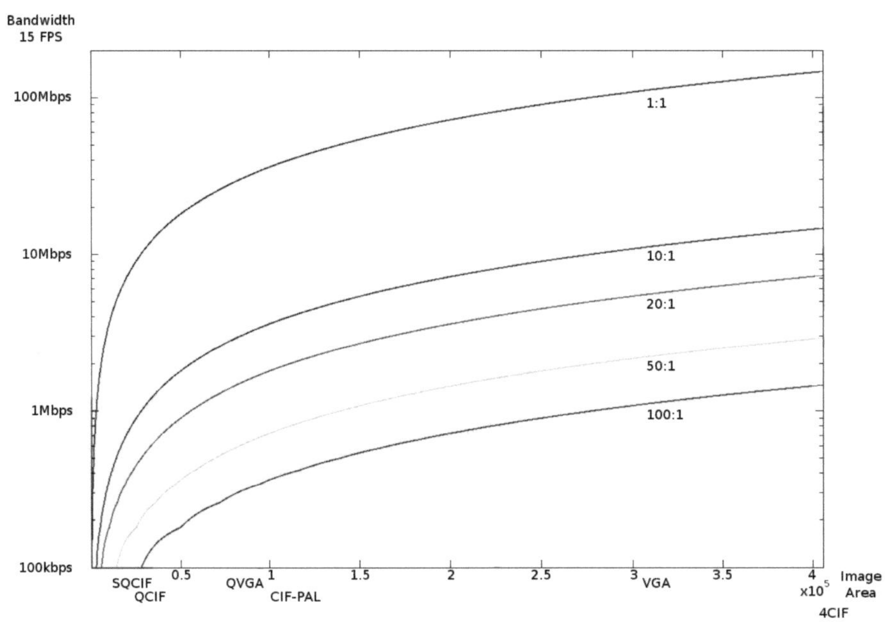

Fig. 7. Bandwidth usage in Streaming vs Image size and compression ratio

trigger the execution of an application, a task or a phase. Once completed, the node falls in a low power state (standby, hibernation, sleep...). For this reason, most research activities have been addressed to reduce radio transmissions, optimizing the protocol stack, compressing and processing data in the application level. Using these simple models, simulations allow the estimation of results for large networks (Hu et al., 2007; Zhong et al., 2004). Recently, different papers try to describe energy consumption more precisely. In (Kellner et al., 2008), energy states of different devices and OS activities are modeled as finite state machines.

In Wireless Multimedia Sensors, intensive computing keeps the MCU in the active state for a high percent of the working cycle. In this way, energy for signal processing is not negligible. Although, a high-end processor wastes more energy, intensive computing takes less time, and the overall energy can be lower. A new paradigm comes from this situation: finding the optimal MCU performance/consumption ratio (PCR) to minimize the computing energy. Influencing factors are clock frequency, internal architecture, the set of instructions, and low-

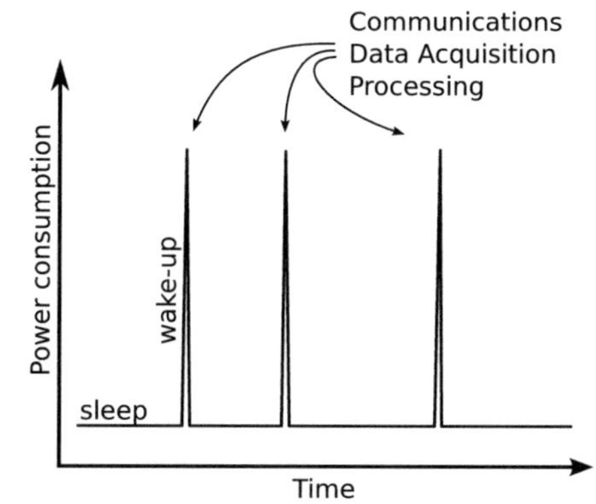

Fig. 8. Typical Power Consumption in WSN

power modes. There is no method to derive energy from these factors, so simulations and empirical tests are needed. But a sensor node must execute intensive and non-intensive tasks. Depending on the frequency of both, a high or low-performance processor will be necessary. In the next section we propose a general and flexible architecture that optimizes power consumption made up of a High-End MCU or DSP acting as a co-processor, and a Low-End MCU as the main processor. This architecture allows the adaptation of the performances and power to the instantaneous activity, enabling or disabling the coprocessor.

4. Hardware architecture alternatives and Application Optimization

Devices described in standard WMSN architectures are clearly divided into categories associated to different tiers. All the devices in the same tier are almost identical in resources. But in Single-tier Heterogeneous Architecture, high-end and low-end nodes work together to support multimedia streaming, still-image delivery, data relay, and multimedia processing. We propose a set of simple, also practical, hardware architectures that cover the functionality of all the categories. Moreover, we present multimedia manipulation schemes to increase acquisition and delivery performances in low-end nodes.

4.1 Image data: Split On the Fly algorithm

Proposed Split on the Fly (SoF) algorithm is designed for low-end platforms to increase the image resolution they can work with. These platforms consist of a resource limited MCU connected to a CMOS image sensor, as shown in figure 9. Most of these image sensors allow a command-based dialog using a serial interface (I2C or SPI). Typically, it is possible to select the image format (CIF, GCIF, JPEG), the data format (e.g. YCbCr 4:2:2, GRB 4:2:2 or RGB Raw Data), camera parameters (exposure, gain and white balance) and image quality parameters (brightness, contrast, saturation, sharpness gamma, and windowing). Other operations or algorithms must normally be executed in the microcontroller, and the original and processed images have to be saved in the internal memory. Only very low image resolutions (e.g. 30×30) can be processed in the internal memory.

This platform can support data relay, image capture, transmission and video streaming for very low resolutions. Compression algorithms, and transport protocols for reliable services, are very limited, or not possible because of memory constraints.

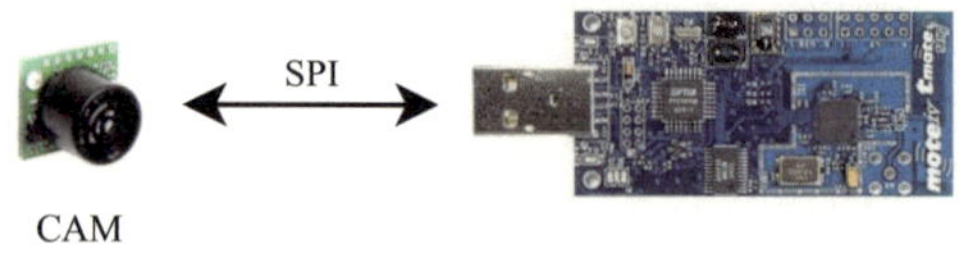

Fig. 9. Low-End node architecture

Delivery of higher image resolutions without compression can be supported if the camera sensor includes internal memory. In this case, nothing but performance limits the transmission. In any case, raw images can be delivered without processing if they are split *on the fly*, transmitting each row after being read from the camera. With this scheme, just one row needs to be saved in the memory before delivering.

Without transport protocols, a packet loss can degrade the image at the sink node. In this case a simple interpolation algorithm (e.g. bilinear) can be used to estimate the lost pixels and reconstruct the image. In a burst of lost packets, a region of the image will be completely lost and restoration will be impossible.

Reliability of this scheme and the quality of the restored image will increase if the rows are also split into two packets with the odd and even pixels, respectively. Choosing different delivery paths for both increases the probability of a packet loss, but one of them will also be more likely to reach the sink node.

Again, lost pixels can be estimated using an interpolation algorithm. But in this case, the real pixels are uniformly distributed throughout the image, so the quality of the image reconstructed by interpolation increases. See figure 10

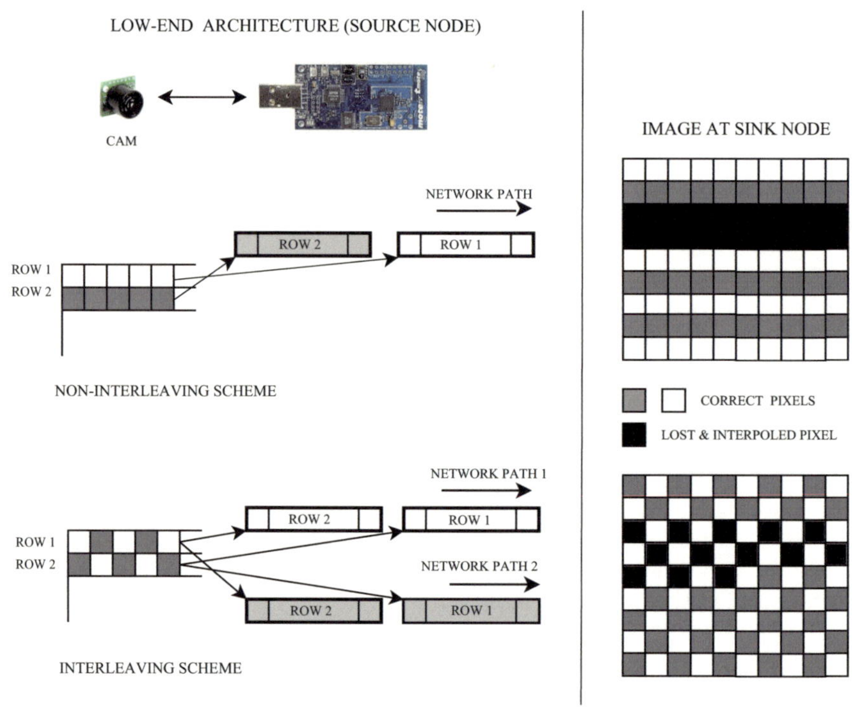

Fig. 10. Split on the Fly scheme (with and without interleaving)

4.2 Medium-End architecture

SoF is a suitable algorithm for MCUs with a lack of memory. But the energy required to deliver the image through the network is high. Let N be the size of a square image. Then,

$$E = O(N^2) \tag{1}$$

A compression algorithm can reduce this energy significantly, but no *splitting on the fly scheme* is possible, because the original and compressed images must be saved in the memory. An external memory (i.e. an I2C RAM) can solve the problem. But, the chip increases the energy consumption in three ways: its power supply, the time needed to transfer data from the camera (seconds), and a factor caused by the inefficient data access during the compression algorithm.

Fig. 11. MCU with external memory

As shown in (Molina et al., 2009) all of them are insignificant compared to the required processing time. Using Split on the Fly scheme as the reference, the energy saving of this solution can be approximated to:

$$\Delta E \approx N^2 \cdot E_{BB}\,(1 - R_C) + T_P \cdot PWC \tag{2}$$

where R_C is the compression rate, E_{BB} the energy to broadcast a byte, T_P the processing time, and PWC the CPU power consumption for the processing time. The first factor, R_C, is directly related to the image quality. The Haar transform (Raviraj & Sanavullah, 2007) is recommendable because it reduces the image size with a reasonable computation cost and high quality.

4.3 High-End architecture

To reduce the processing time, we propose a dual processor architecture: A low performance MCU managing the WSN services, and a high performance microcontroller acting as a *multimedia co-processor*. Even though the power consumption of this co-processor is much higher, the processing time is substantially lower. But co-processor selection is not trivial. If the option is another MCU or a processing card, a detailed analysis is necessary because it is possible not to achieve the desired energy reduction. This is the case of CMUCam (Carnegie Mellon University, 2007) . Multimedia co-processor should be optimized for data processing, and must necessarily contain an internal structure far different from the main MCU (e.g. PIC32XMX-460 with 32K RAM selected in (Molina et al., 2009)). A simple overclocked version of the main CPU reduces the processing time proportionally, but the power consumption increases the same rate, as shown following:
Using the general equation of power consumption of a CMOS integrated circuit (IC):

$$P_{IC} = KV^2 f \tag{3}$$

The energy in an overclocked CPU (E_{OV}) for a period T_{OV} is

$$E_{OV} = P_{OV} \cdot T_{OV} = KV^2 f_{OV} \cdot T_{OV} = KV^2 f_{OV} \cdot T \cdot \frac{f}{f_{OV}} = KV^2 fT = E \tag{4}$$

The camera sensor is connected directly to the co-processor, so no external memory is needed, and consequently, no delay time to transfer the pixels, no extra time to data access, etc.

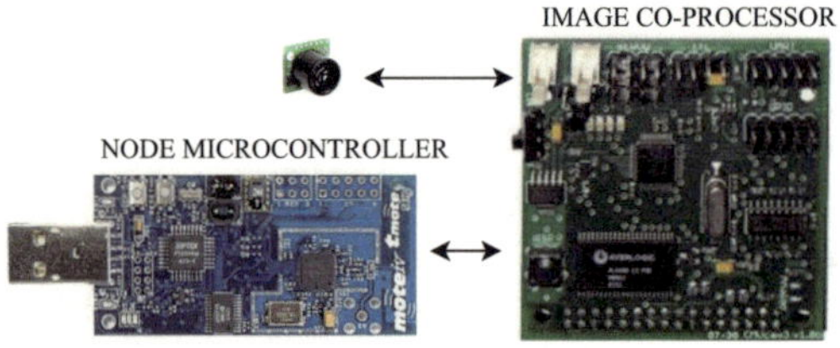

Fig. 12. High-End node architecture

5. Conclusions and future work

Extending proposed concepts and optimizations on any WMSNs standard architectures is obvious. Split on the Fly algorithm (SoF) is very easy to implement in any platform, and it can serve as a reference to compare it with.
Balancing the computing load between a dual processor architecture as proposed in the high-end one, allows the optimization of energy consumption, adapting the performance to any instantaneous processing profile.
Using FPGAs to implement the image co-processor is a promising research line. The capacity of a FPGA to execute parallel simple operations matches with the spatial redundancy of many image processing algorithms. It is reasonable to assume that the optimization of these algorithms for FPGAs, will reduce the processing time up to milliseconds, decreasing the processing energy significantly. Compared to ASICs, the power consumption of an FPGA is considerably higher. However, an FPGA can be reprogrammed to update it to a specific application.

6. References

Aaron, A., Rane, S., Zhang, R. & Girod, B. (2003). Wyner-ziv coding for video: Applications to compression and error resilience, *DCC '03: Proceedings of the Conference on Data Compression*, IEEE Computer Society, Washington, DC, USA, p. 93.

Abrach, H., Bhatti, S., Carlson, J., Dai, H., Rose, J., Sheth, A., Shucker, B., Deng, J. & Han, R. (2003). Mantis: System support for multimodal networks of in-situ sensors, *In 2nd ACM International Workshop on Wireless Sensor Networks and Applications (WSNA*, pp. 50–59.

Akyildiz, I. F., Melodia, T. & Chowdhury, K. R. (2008). Wireless multimedia sensor networks: Applications and testbeds, *Proceedings of the IEEE* **96**(10): 1588–1605.
URL: *http://dx.doi.org/10.1109/JPROC.2008.928756*

Alippi, C., Camplani, R. & Galperti, C. (2007). Lossless compression techniques in wireless sensor networks: Monitoring microacoustic emissions, *Robotic and Sensors Environments*.

Arora, A., Dutta, P., Bapat, S., Kulathumani, V., Zhang, H., Naik, V., Mittal, V., Cao, H., Demirbas, M., Gouda, M., Choi, Y., Herman, T., Kulkarni, S., Arumugam, U., Nesterenko, M., Vora, A. & Miyashita, M. (2004). A line in the sand: a wireless sensor network for target detection, classification, and tracking, *Computer Networks* **46**(5): 605 – 634. Military Communications Systems and Technologies.

URL: *http://www.sciencedirect.com/science/article/B6VRG-4CXKNBT-1/2/38791b700b4804d30e9df8aca23712e7*

Bénézit, F. (2009). *Distributed Average Consensus for Wireless Sensor Networks*, PhD thesis, École Polytechnique Fédérale de Lausanne.

Cao, Q. & Abdelzaher, T. (n.d.). The liteos operating system: Towards unix-like abstractions for wireless sensor networks.

Cao, Q., Yan, T., Stankovic, J. & Abdelzaher, T. (2005). Analysis of target detection performance for wireless sensor networks, *in* V. K. Prasanna, S. Iyengar, P. G. Spirakis & M. Welsh (eds), *Distributed Computing in Sensor Systems*, Vol. 3560 of *Lecture Notes in Computer Science*, Springer Berlin / Heidelberg, pp. 276–292.

Carnegie Mellon University (2007). Cmucam3 datasheet version 1.02. Pittsburgh, PA.

Cha, H., Choi, S., Jung, I., Kim, H., Shin, H., Yoo, J. & Yoon, C. (2007). Retos: resilient, expandable, and threaded operating system for wireless sensor networks, *IPSN '07: Proceedings of the 6th international conference on Information processing in sensor networks*, ACM, New York, NY, USA, pp. 148–157.

Chia, W. C., Ang, L.-M. & Seng, K. P. (2009). Multiview image compression for wireless multimedia sensor network using image stitching and spiht coding with ezw tree structure, *Intelligent Human-Machine Systems and Cybernetics, International Conference on* **2**: 298–301.

Dai, H. & Han, R. (2004). Tsync: a lightweight bidirectional time synchronization service for wireless sensor networks, *SIGMOBILE Mob. Comput. Commun. Rev.* **8**(1): 125–139.

Dunkels, A., Grönvall, B. & Voigt, T. (2004). Contiki - a lightweight and flexible operating system for tiny networked sensors.

Elson, J. & Römer, K. (2003). Wireless sensor networks: a new regime for time synchronization, *SIGCOMM Comput. Commun. Rev.* **33**(1): 149–154.

Eswaran, A., Rowe, A. & Rajkumar, R. (2005). Nano-rk: An energy-aware resource-centric rtos for sensor networks, *RTSS '05: Proceedings of the 26th IEEE International Real-Time Systems Symposium*, IEEE Computer Society, Washington, DC, USA, pp. 256–265.

Faheem, Y., Boudjit, S. & Chen, K. (2010). Wireless multimedia sensor networks:requirements & current paradigm, *L Update* .

Feng, J., Megerian, S. & Potkonjak, M. (2003). Model-based calibration for sensor networks, *International Conference on Sensors*.

Feng, W., Kaiser, E. & Le Baillif, M. (2005). Panoptes: scalable low-power video sensor networking technologies., *ACM Trans. Multimedia Comput. Commun. Appl.* **1**: 151–167.

Gay, D., Levis, P., von Behren, R., Welsh, M., Brewer, E. & Culler, D. (2003). The nesc language: A holistic approach to networked embedded systems, *PLDI '03: Proceedings of the ACM SIGPLAN 2003 conference on Programming language design and implementation*, ACM, New York, NY, USA, pp. 1–11.

Girao, J., Westhoff, D., Mykletun, E. & Araki, T. (2007). Tinypeds: Tiny persistent encrypted data storage in asynchronous wireless sensor networks, *Ad Hoc Networks* **5**(7): 1073 – 1089.

Gu, L. (2006). t-kernel: Providing reliable os support to wireless sensor networks, *In Proc. of the 4th ACM Conf. on Embedded Networked Sensor Systems (SenSys*, ACM Press, pp. 1–14.

Han, C.-c., Kumar, R., Shea, R., Kohler, E. & Srivastava, M. (n.d.). Abstract a dynamic operating system for sensor nodes.

He, T., Huang, C., Blum, B. M., Stankovic, J. A. & Abdelzaher, T. (2003). Range-free localization schemes for large scale sensor networks, *MobiCom '03: Proceedings of the 9th annual international conference on Mobile computing and networking*, ACM, New York, NY, USA, pp. 81–95.

Hengstler, S., Prashanth, D., Fong, S. & Aghajan, H. (2007). Mesheye: A hybrid-resolution smart camera mote for applications in distributed intelligent surveillance., *Int. Conf. Inf. Process. Sensor Netw. (IPSN), Cambridge, MA, USA.*, pp. 360–369.

Hill, J., Szewczyk, R., Woo, A., Hollar, S., Culler, D. & Pister, K. (2000). System architecture directions for networked sensors, *ASPLOS-IX: Proceedings of the ninth international conference on Architectural support for programming languages and operating systems*, ACM, New York, NY, USA, pp. 93–104.

Hu, P., Zhou, Z., Liu, Q. & Li, F. (2007). The hmm-based modeling for the energy level prediction in wireless sensor networks, *Industrial Electronics and Applications, 2007. ICIEA 2007. 2nd IEEE Conference on*, pp. 2253 –2258.

Ihler, A. T., Fisher, III, J. W., Moses, R. L. & Willsky, A. S. (2004). Nonparametric belief propagation for self-calibration in sensor networks, *IPSN '04: Proceedings of the 3rd international symposium on Information processing in sensor networks*, ACM, New York, NY, USA, pp. 225–233.

IMOTE2 (n.d.). http://www.xbow.com/Products/Imote2.aspx (accessed August 16, 2009).

Kellner, S., Pink, M., Meier, D. & Blass, E.-O. (2008). Towards a realistic energy model for wireless sensor networks, *Wireless on Demand Network Systems and Services, 2008. WONS 2008. Fifth Annual Conference on*, pp. 97 –100.

Kulkarni, P., Ganesan, D., Shenoy, P. & Lu, Q. (2005). Senseye: A multi-tier camera sensor network, *ACM Multimedia*.

Lewis, A. & Knowles, G. (1992). Image compression using the 2-d wavelet transform, *IEEE Transactions on Image Processing* **1**: 244–250.

Mao, G., Fridan, B. & Anderson, B. (2007). Wireless sensor network localization techniques, *Computer Networks. Elsevier* .

Maróti, M., Kusy, B., Simon, G. & Lédeczi, A. (2004). The flooding time synchronization protocol, *SenSys '04: Proceedings of the 2nd international conference on Embedded networked sensor systems*, ACM, New York, NY, USA, pp. 39–49.

Miluzzo, E., Lane, N., Campbell, A. & Olfati-Saber, R. (2008). Calibree: A self-calibration system for mobile sensor networks, *in* S. Nikoletseas, B. Chlebus, D. Johnson & B. Krishnamachari (eds), *Distributed Computing in Sensor Systems*, Vol. 5067 of *Lecture Notes in Computer Science*, Springer Berlin / Heidelberg, pp. 314–331.

Molina, F. J., Babancho, J., Mora, J. M. & Leon, C. (2009). Poster abstract: Practical issues in image acquisition and transmission over wireless sensor network, *IPSN '09: Proceedings of the 2009 International Conference on Information Processing in Sensor Networks*, IEEE Computer Society, Washington, DC, USA, pp. 395–396.

Mottola, L. & Picco, G. P. (2010). Programming wireless sensor networks: Fundamental concepts and state of the art, *ACM Computing Surveys* .

Patwari, N., Ash, J., Iii, A. O. H., Hero, A. O., Moses, R. M., Correal, N. S., Labs, M., Kyperountas, S., Correal, N. S. & (b, A. (2005). Locating the nodes: Cooperative localization in wireless sensor networks.

Polastre, J., Szewczyk, R., Sharp, C. & Culler, D. (2004). The mote revolution: low power wireless sensor network devices., *Hot Chips 16: A Symposium on High Performance Chips*.

Rahimi, M., Baer, R., Iroezi, O., Garcia, J., Warrior, J., Estrin, D. & Srivastava, M. (2005). Cyclops: In situ image sensing and interpretation in wireless sensor networks., *ACM Conf. Embed. Netw: Sensor Syst. (SenSys)*, San Diego, CA, USA, pp. 192–204.

Raviraj, P. & Sanavullah, M. (2007). The modified 2D-Haar transformation in image compression, *Middle-East journal of Scientific Research* **2**(2): 73–78.

Rubio, B., Diaz, M. & Troya, J. M. (2007). Programming approaches and challenges for wireless sensor networks, *ICSNC '07: Proceedings of the Second International Conference on Systems and Networks Communications*, IEEE Computer Society, Washington, DC, USA, p. 36.

Sayood, K. (2005). *Introduction to Data Compression, Third Edition (Morgan Kaufmann Series in Multimedia Information and Systems)*, Morgan Kaufmann Publishers Inc., San Francisco, CA, USA.

Sheng, B., Li, Q. & Mao, W. (2006). Data storage placement in sensor networks, *MobiHoc '06: Proceedings of the 7th ACM international symposium on Mobile ad hoc networking and computing*, ACM, New York, NY, USA, pp. 344–355.

Soro, S. & Heinzelman, W. (2009). A survey of visual sensor networks, *Advances in Multimedia* pp. 1–22.

Stargate XScale Platform (n.d.). http://blog.xbox.com/xblog/Stargate_Xscale_Platform/ (accessed August 16, 2009).

Villasenor, J., Belzer, B. & Liao, J. (1995). Wavelet filter evaluation for image compression, *IEEE Transactions on Image Processing* **4**: 1053–1060.

Zhong, L., Rabaey, J. & Wolisz, A. (2004). An integrated data-link energy model for wireless sensor networks, *Communications, 2004 IEEE International Conference on*, Vol. 7, pp. 3777 – 3783 Vol.7.

Ziv, J. & Lempel, A. (1978). Compression of individual sequences via variable-rate coding.

Zoumboulakis, M. & Roussos, G. (n.d.). Integer-based optimisations for resource-constrained sensor platforms.

24

Imaging in UWB Sensor Networks

Ole Hirsch, Rudolf Zetik, and Reiner S. Thomä
Technische Universität Ilmenau
Germany

1. Introduction

Sensor networks consist of a number of spatially distributed nodes. These nodes perform measurements and collect information about their surrounding. They transfer data to neighboring nodes or to a data fusion center. Often measurements are performed in cooperation of several nodes.
If the network consists of Ultra-Wideband (UWB) radar sensors, the network infrastructure can be used for a rough imaging of the surrounding. In this way bigger objects (walls, pillars, machines, furniture) can be detected and their position and shape can be estimated. These are valuable information for the autonomous orientation of robots and for the inspection of buildings, especially in case of dangerous environments (fire, smoke, dust, dangerous gases). Applications of UWB sensor networks are summarized in Thomä (2007).
In this article basic aspects of imaging in UWB sensor networks are discussed. We start with a brief description of two types of UWB radar devices: impulse radar and Noise/M-sequence radar. Network imaging is based on principles of Synthetic Aperture Radar (SAR). Starting from SAR some special aspects of imaging in networks are explained in section 3. Sections 4 and 5 form the main part of the article. Here two different imaging approaches are described in more detail. The first method is multistatic imaging, i.e. the measurements are performed in cooperation of several sensor nodes at fixed places and one mobile node. The second approach is imaging by an autonomous mobile sensor, equipped with one Tx and two Rx units. This sensor uses a network of fixed nodes for its own orientation.
Part of the described methods have been practically realized in a laboratory environment. Hence, practical examples support the presentation. Conclusions and references complete the article.

2. Ultra Wideband (UWB) Radar

2.1 Main Characteristics

The main characteristic of UWB technology is the use of a very wide frequency range. A system is referred to as UWB system if it operates in a frequency band of more than 500 MHz width, or if the fractional bandwith $bw_\mathrm{f} = 100\% \cdot (f_\mathrm{H} - f_\mathrm{L})/f_\mathrm{C}$ is larger than 25%. Here f_H, f_L, and f_C denote the upper and lower frequency limit and the centre frequency, respectively. For imaging applications the large bandwidth is of interest because it guarantees a high resolution in range direction, as explained in the next section. UWB systems always coexist with other radio services, operating in the same frequency range. To avoid intereference, a number of frequency masks and power restrictions have been agreed internationally. Current regulations are summarized in FCC (2002); Luediger & Kallenborn (2009).

A number of principles for UWB radar systems have been proposed (see Sachs et al. (2003)). In the rest of this section we briefly describe the two dominant methods 'Impulse Radar' and 'M-Sequence Radar'.

2.2 Impulse Radar

An Impulse radar measures distances by transmission of single RF pulses and subsequent reception of echoe signals. The frequency spectrum of an electric pulse covers a bandwidth which is inversely proportional to its duration. To achieve ultra-wide bandwidth, the single pulses generated in an impulse radar must have a duration of $t_{\mathrm{pulse}} \approx$1ns or even less. They can be generated by means of switching diodes, transistors or even laser-actuated semiconductor switches, see Hussain (1998) for an overview. Pulse shaping is required to adapt the frequency spectrum to common frequency masks. The principle of an impulse radar is shown in Fig. 1.

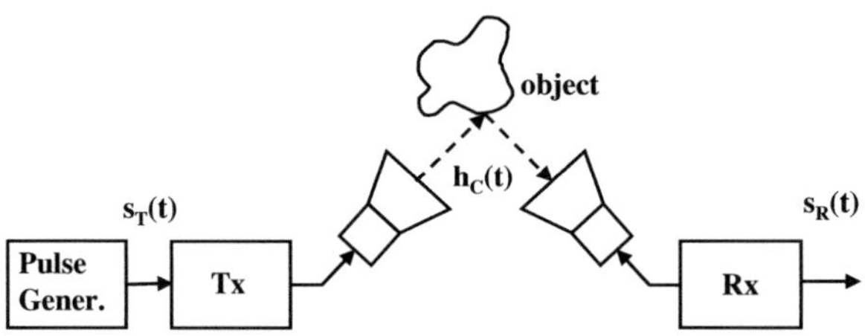

Fig. 1. Principle of an impulse radar.

The transmitter signal $s_{\mathrm{T}}(t)$ is radiated by the Tx-antenna. The received signal $s_{\mathrm{R}}(t)$ consists of a small fraction of the transmitted energy that was scattered at the object. $s_{\mathrm{R}}(t)$ can be calculated by convolution of $s_{\mathrm{T}}(t)$ with the channel impulse response $h_{\mathrm{C}}(t)$:

$$s_{\mathrm{R}}(t) = \int_{-\infty}^{\infty} h_{\mathrm{C}}(t') \cdot s_{\mathrm{T}}(t-t')\, dt'. \qquad (1)$$

Determination of the channel impulse reponse is possible via de-convolution, favourably performed with the Fourier-transformed quantities $S_{\mathrm{R}}(\omega), S_{\mathrm{T}}(\omega)$ in the frequency domain:

$$h_{\mathrm{C}}(t) = \mathcal{F}^{-1}\left\{\frac{S_{\mathrm{R}}(\omega)}{S_{\mathrm{T}}(\omega)} \cdot F(\omega)\right\}. \qquad (2)$$

$F(\omega)$ is a bandpass filter that suppresses high amplitudes at the edges of the frequency band, and $\mathcal{F}^{-1}$ symbolizes the inverse Fourier transform.
The minimum delay between two subsequent pulses (repetition time t_{rep}) is given by $t_{\mathrm{rep}} = d_{\mathrm{max}}/c$, where d_{max} is the maximum propagation distance and c is the speed of light. For smaller pulse distances no unique identification of pulse propagation times would be possible, especially in the case of more than one object. The necessity to introduce t_{rep} limits the average signal energy since only a fraction $t_{\mathrm{pulse}}/t_{\mathrm{rep}}$ of the total measurement time is used for transmission and the signal peak amplitude must not exceed the allowed power restrictions. Advantageously, the temporal shift between transmission and reception of signals reduces the problem of Tx/Rx crosstalk.

2.3 Noise Radar and M-Sequence Radar

Noise signals can possess a frequency spectrum which is as wide as the spectrum of a single short pulse. Because of random phase relations between the single Fourier components the signal energy of a noise signal is distributed over the entire time axis. Signals of this kind can be used in radar systems and an example is shown in Fig. 2.

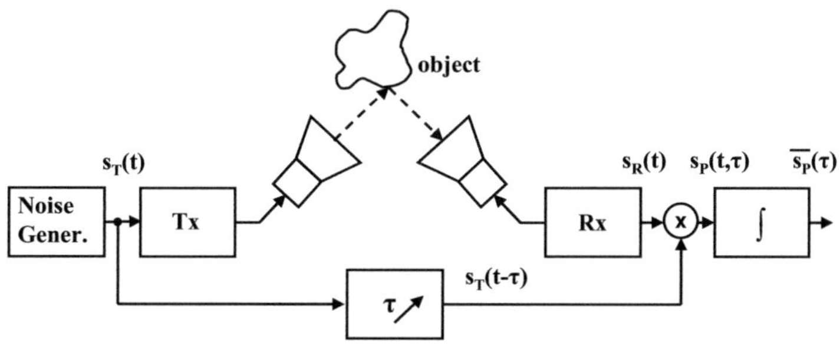

Fig. 2. Principle of a noise radar.

The relation between $s_R(t)$, $h_C(t)$, and $s_T(t)$ is of course the one already given in Equ. (1). In this kind of radar device, information about the propagation channel is extracted by correlation of the received signal with the transmitted $s_T(t)$. The correlator consists of a variable delay element introducing delay τ, a multiplicator that produces the product signal $s_P(t,\tau)$:

$$s_P(t,\tau) = s_R(t) \cdot s_T(t-\tau) \tag{3}$$

and an integrator that forms the signal average for one particular τ over all t. Introduction of the convolution integral (1) into (3) and averaging over a time interval that is long in comparison to usual signal variations gives the following expression for the averaged signal $\overline{s_P}(\tau)$:

$$\overline{s_P}(\tau) = \lim_{T\to\infty} \frac{1}{2T} \int_{-T}^{T} s_R(t) \cdot s_T(t-\tau)\, dt = \lim_{T\to\infty} \frac{1}{2T} \int_{-T}^{T} \int_{-\infty}^{\infty} h_C(t') \cdot s_T(t-t') \cdot s_T(t-\tau)\, dt' dt. \tag{4}$$

In case of white noise the average value of the product $s_T(t-t') \cdot s_T(t-\tau)$ is always zero, except for $t' = \tau$. This means the autocorrelation function of white noise is a δ-function. So we can perform the following substitution:

$$\lim_{T\to\infty} \frac{1}{2T} \int_{-T}^{T} s_T(t-t') \cdot s_T(t-\tau)\, dt \;=\; \delta(t'-\tau) \cdot \lim_{T\to\infty} \frac{1}{2T} \int_{-T}^{T} s_T^2(t)\, dt \;=\; \delta(t'-\tau) \cdot \overline{s_T^2(t)}. \tag{5}$$

Applying this result we see that the correlation of noise excitation $s_T(t)$ and receiver signal $s_R(t)$ delivers the channel impulse response $h_C(t)$ multiplied by a constant factor. This factor is the square of the effective value of $s_T(t)$:

$$\overline{s_P}(\tau) = \overline{s_T^2(t)} \int_{-\infty}^{\infty} h_C(t') \cdot \delta(t'-\tau)\, dt' \;=\; \overline{s_T^2(t)} \cdot h_C(\tau). \tag{6}$$

An M-sequnce radar is a special form of a noise radar, where $s_T(t)$ consists of a maximum length binary sequence, see Sachs (2004) for details. This pseudo-stochastic signal is generated in a shift register with feedback. Both noise radar and M-sequence radar use the full measurement duration for transmission and reception of signals, maximizing the UWB signal energy in this way. Decoupling between transmitter and receiver becomes more important since Tx and Rx operate at the same time.

3. Specifics of Imaging in Sensor Networks

3.1 Synthetic Aperture Radar (SAR)

Imaging in sensor networks is based on results of conventional microwave imaging, i.e. imaging with only one single Tx/Rx antenna pair. Especially the principles of "Synthetic Aperture Radar" (SAR) can be adapted to the special needs of sensor network imaging. Instead of using an antenna with a large aperture, here the aperture is synthesized by movement of antennas and sequential data acquisition from different positions. For an overview of SAR imaging see Oliver (1989), and for a typical UWB-SAR application see Gu et al. (2004). In 3.3 relations between the length of the scan path (aperture) and image resolution are explained. To achieve reasonable resolution, the antenna aperture of a radar imaging system must be significantly bigger than the wavelength λ. Processing of SAR data is explained in connection with general processing in 4.1 .

3.2 Arrangement of Network Nodes and Scan Path

The network consists of a number of nodes. These are individual sensors with Rx and/or Tx capabilities. Specialized nodes can collect data from several other nodes, and typically one node forms the fusion center, where the image is computed from the totality of the acquired data.

The network can be completed by so called 'anchor nodes'. These are nodes at known, fixed positions. Primarily they support position estimation of the mobile nodes, but additionally they can be employed in the imaging process.

The spatial arrangement of network nodes (network topology) strongly influences the performance of an imaging network. Together topology and scan path must guarantee that all objects are illuminated by the Tx antennas and that a significant part of the scattered radiation can be collected by the Rx antennas.

A number of frequently choosen scan geometries (node positions and scan paths) are shown in Fig. 3. At least one antenna must move during the measurement; or an array of antennas has to be used, as in Fig. 3(b). Two main cases can be distinguished with respect to the scan path selection:

1. The object positions are already known. In this case imaging shall give information on the shape of objects and small modifications of their position.
2. The object positions are entirely unknown. In this case a rough image of the entire surrounding has to be created.

The optimum scan geometry is concave shaped in case 1, e.g. Fig. 3(b) and (c). This shape guarantees that the antennas are always directed towards the objects, so that a significant part of the scattered radiation is received. If the region of interest is accessible from one side only, then semi circle or linear scan geometries are appropriate choices.

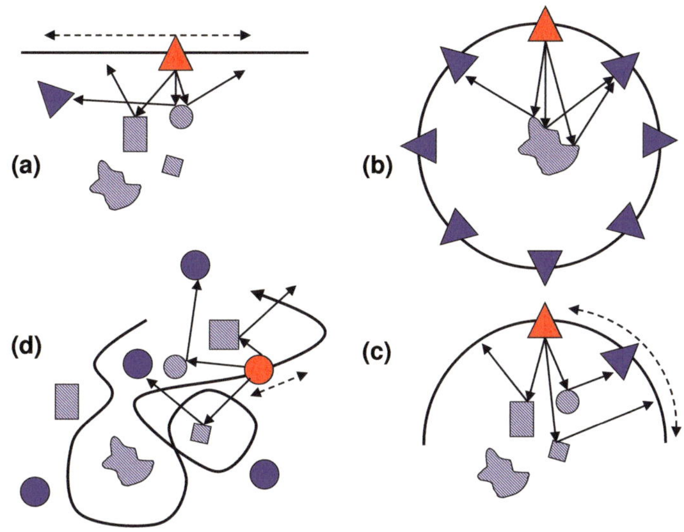

Fig. 3. Typical scan geometries in imaging sensor networks: (a) linear scan, (b) full circle, (c) semi circle, (d) arbitrary scan path. Filled triangles and circles: antennas, hatched figures: objects.

In entirely unknown environment a previous optimization of node positions is not possible. In this case the nodes are placed at random positions. They should have similar mutual distances. Node positioning can be improved after initial measurements, if some nodes don't receive sufficient signals. A network of randomly placed nodes requires the use of omnidirectional antennas, which can cause a reduction of the signal to clutter ration in comparison to directional antennas.

3.3 Resolution

Resolution is a measure of up to which distance two closely spaced objects are still imaged separately. In radar technique we must distinguish between 'range resolution' ρ_z (along the direction of wave propagation) and 'cross range resolution' ρ_x (perpendicular to the direction of propagation). An approximation for the former is:

$$\rho_z = \frac{c}{2bw}. \tag{7}$$

It is immediately understandable that ρ_z improves with the bandwidth bw because the speed of light c divided by bw is a measure for the width of the propagating wave packet in the spatial domain. The '2' results from two times passage of the geometrical distance in radar measurements.
A rough estimation of cross range resolution ρ_x can be derived by means of Fig. 4. d and d_1 are the path lengths to the end points of ρ_x when the antenna is at one end position of the aperture A. We assume the criterion that two neighbouring points can be resolved if a path difference $\Delta d = d - d_1$ of the order of half the wavelength λ appears during movement of the antenna along the aperture A, resulting in a signal phase difference of $\approx 2\pi$ (two way propagation). Typically the distances are related to each other as follows: $A \gg \rho_x$, $R \gg \rho_x$, $R > A$. Under

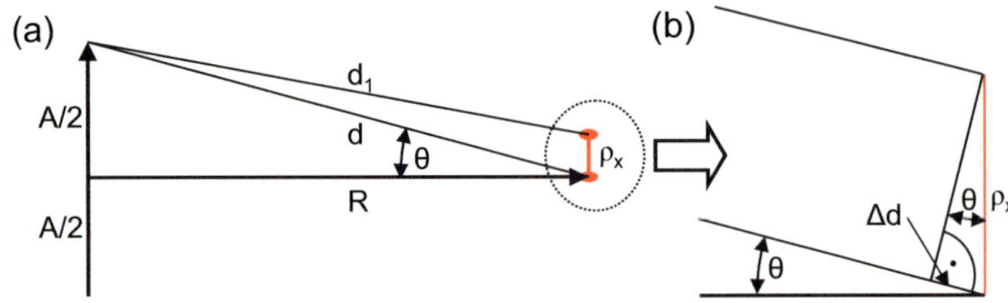

Fig. 4. Approximation of cross range resolution ρ_x. (b) is a zoomed version of the encircled section in (a).

these circumstances d and d_1 can be assumed as being parallel on short lengthscales. The angle θ appears both in the small triangle with sides ρ_x and Δd, and in the big triangle with half aperture $A/2$ and range R:

$$\sin\theta = \frac{\Delta d}{\rho_x}, \qquad \tan\theta = \frac{A/2}{R} \quad . \tag{8}$$

After rearranging of both equations, insertion into each other, and application of the relation $\sin(\arctan(x)) = x/\sqrt{1+x^2}$ we get an expression for ρ_x:

$$\rho_x = \frac{\Delta d \cdot 2R \cdot \sqrt{1+(A/(2R))^2}}{2A} \approx \frac{\lambda R}{2A} \quad . \tag{9}$$

Here Δd was replaced by $\lambda/2$. The extra '2' in the denominator results from the fact that the calculation was performed with only half the actual aperture length. With the assumed relation between A and R the square root expression can be set to 1 in this approximation. While range resolution depends on the bandwidth, cross range resolution is mainly dependent on the ratio between aperture and wavelength. In UWB systems resolution is estimated with an average wavelength. In imaging networks the two cases 'range' and 'cross range' are always mixed. For a proper resolution approximation the node arrangement and the signal pulse shape must be taken into account.

3.4 Localization of Nodes and Temporal Synchronization

Imaging-algorithms need the distance Tx $\rightarrow$ object $\rightarrow$ Rx at each position of the mobile nodes. This requires knowledge of all anchor node positions and continuous tracking of the mobile nodes. Time-based node localization is possible only with exact temporal synchronization of the singe nodes.

3.4.1 Localization of Nodes

Before we list the different localization tasks, we introduce abbreviations for the localization methods:

- TOA: Time of arrival ranging/localization
- TDOA:Time difference of arrival localization
- AOA: Angle of arrival localization
- ADOA: Angle difference of arrival localization

- RTT: Round trip time ranging
- RSS: Received signal strength ranging

It is not necessary to explain these methods here, because this subject is covered in the literature extensively. Summaries can be found in Patwari et al. (2005) and in Sayed et al. (2005). AOA and ADOA methods are explained in Rong & Sichitiu (2006), and TDOA methods are discussed in Stoica & Li (2006).
The single tasks are:

1. The positions of the static nodes (anchor nodes) must be estimated. If the network is a fixed installation, then this task is already fulfilled. Otherwise anchor node positions can be found by means of TOA localization (if synchronization is available) or by means of RTT estimations (synchronization not required).
2. The positions of mobile nodes must be tracked continuously. If the sensors move along predefined paths, then their positions are known in advance. In case of synchronization between mobile nodes and anchors, position estimation is possible with TOA methods. Without synchronization node positions may be found by TDOA, AOA, or ADOA methods. RSS is not very precise; RTT could be used in principle but requires much effort.

Methods that involve angle measurements (AOA and ADOA) can only be performed if the sensor is equipped with directional antennas or with an antenna array. Time-based methods require exact synchronization; in case of TDOA only on the individual sensor platform, for TOA and RTT within the network. The large bandwidth and good temporal resolution of UWB systems are huge advantages for time-based position measurements.

3.4.2 Temporal Synchronization of Network Nodes

Two main reasons exist for temporal synchronization of network nodes:

1. Application of time-based localization methods.
2. Use of correlation receivers in M-sequence systems.

Point 1 was alredy discussed. The necessity of synchronization in networks with correlation receivers can be seen from Fig. 5.

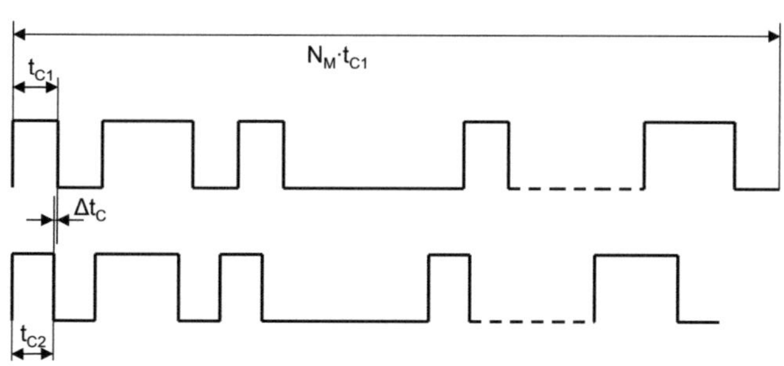

Fig. 5. Mismatch between a received M-sequence signal and the reference signal because of differing clock frequencies $1/t_{C1}$ and $1/t_{C2}$ of Tx and Rx. The total time shift is $N_M \cdot \Delta t_C$ (N_M: Number of chips; Δt_C: time difference per cycle).

Over the sequence duration of $N_M \cdot t_{C1}$ a maximum shift of $\approx \frac{1}{2}t_{C1}$ is tolerable. This corresponds to a maximum clock frequency difference of

$$\Delta f_C < \frac{1}{2\,N_M} f_{C1} \quad . \tag{10}$$

A comprehensive introduction into synchronization methods and protocols is given in Serpedin & Chaudhari (2009). Originally, many of these methods were developed for communications networks. The good time resolution of UWB signals makes them a candidate for synchronization tasks. An example is given in Yang & Yang (2006).

3.5 Data Fusion

Processing of data in an imaging sensor network is distributed across the nodes. Part of the processing steps are performed at the individual sensors while, after a data transfer, final processing is done at the fusion center. An example flow chart is shown in Fig. 6.

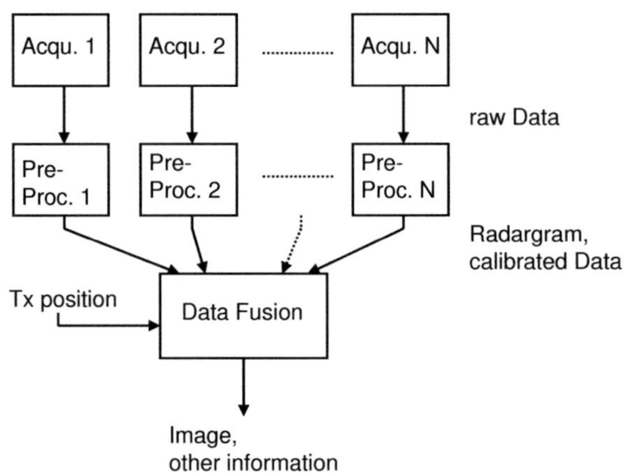

Fig. 6. Data processing in a network with one Tx and N Rx. The single steps are Data acquisition (Acqu.), Pre-processing (Pre-Proc.), and Data Fusion.

After transmitting a pulse or an M-sequence by the Tx, data are acquired by the Rx hardware. Typically the sensor hardware performs some additional tasks: analog to digital conversion, correlation with a known signal pattern (in case of M-sequence systems), and accumulation of measurements to improve the signal to noise ratio.

The next step is pre-processing of the raw data, usually performed in a signal processor at the sensor node. De-convolution of raw data with a measured calibration function can increase the usable bandwidth and in this way it can increase range resolution. In M-sequence systems data must be shifted to achieve coincidence between the moment of signal transmission and receiver time zero (Sachs (2004)). The result of pre-processing can be visualized in a radargram (Fig. 15). It displays the processed signals in form of vertical traces against the 'slow' time dimension of sensor movement. For some analyses only the TOA of the first echo is of importance. Then pre-processing includes a discrimination step, which reduces the information to a single TOA value.

Data fusion is a generic term for methods that combine information from the single sensor nodes and produce the image. While acquisition and pre-processing don't vary a lot between

the different imaging methods, data fusion is strongly dependent on network topology, sensor pathways, and imaging method. Examples are described in section 4.
Additional information, required for imaging, are the positions of mobile nodes. As long as the sensors follow predetermined pathways, this information is always available. In other cases the mobile node positions must be estimated by means of mechanical sensors or the position is extracted from radar signals.
Fusion is not always the last processing step. By application of image processing methods supplementary information can be extracted from the radar image.

4. Imaging in Distributed Multistatic Networks

4.1 Multistatic SAR Imaging

The multitude of different propagation pathways in a distributed sensor network can be used for rough imaging of the environment. A signal, transmitted by a Tx, is reflected or scattered at walls, furniture, and other objects. The individual Rx receive these scattered signals from different perspectives. The information about the object position is contained in the signal propagation times. The principle of this method is shown in Fig. 7. The propagation paths are sketched for a signal scattered at an objects corner. In principle, the positions of Tx and Rx could be swapped, but an arrangement with only one Tx and several Rx has the advantage of simultaneous operation of all Rx.

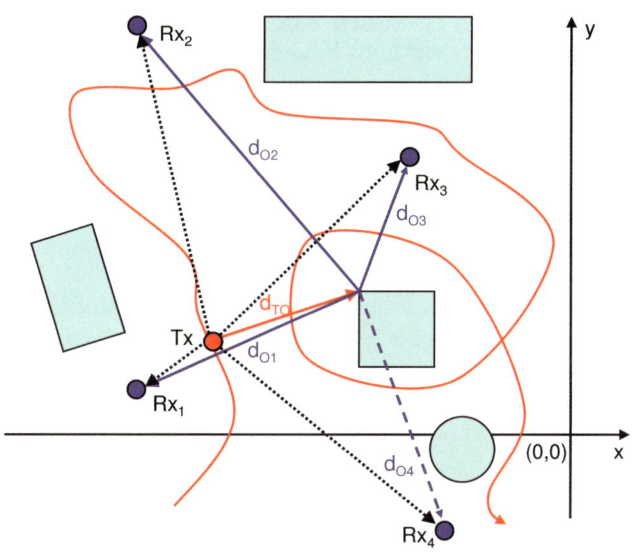

Fig. 7. Principle of imaging in a multistatic network. The four receivers (Rx_i) are placed at fixed positions. The Transmitter (Tx) moves along the curved pathway.

Prerequisites for application of this method are knowledge of the Rx positions, temporal synchronization of all nodes, and application of omnidirectional antennas. The synthetic aperture has the shape of an arbitrary path through the environment. Data acquisition is carried out as follows:

- The Tx moves through the region. It transmits signals every few centimeters.

- All Rx receive the scattered signals. From the totality of received signals a radargram can be drawn for each Rx.

The recorded data are processed in two ways:

- The Tx position at the individual measurement points are reconstructed from the LOS signals between Tx and Rx (dotted lines in Fig. 7).
- The image is computed by means of a simple migration algorithm.

Separately for each receiver Rx_i an image is computed. The brightness in one point $B_i(x,y)$ is the coherent sum of all signals s_r, originating from the scatterer at position (x,y), summarized along the aperture (n is the number of the measurement along the Tx path).

$$B_i(x,y) = \sum_{n=1}^{N} s_r(\tau_{in}, n)\,. \tag{11}$$

The delay τ_{in} is the time required for wave propagation along the way $Tx \rightarrow$ object point $(x,y) \rightarrow Rx_i$ with speed c:

$$\tau_{in} = \frac{1}{c}(d_{TO}(n) + d_{Oi}) = \frac{1}{c}\left(\sqrt{(x - x_{Tx}(n))^2 + (y - y_{Tx}(n))^2} + \sqrt{(x_{Rx_i} - x)^2 + (y_{Rx_i} - y)^2}\right). \tag{12}$$

The meaning of the used symbols can be seen from Fig. 7. This migration algorithm summarizes signals along ellipses, which have their foci at the respective Tx and Rx position. The ellipses for all possible Tx-Rx constellations have in common that they touch the considered object point.

For improved performance, migration algorithms, based on wave equations must be applied, see Margrave (2001). Stolt Migration, computed in the wavenumber-domain, is a fast migration method (Stolt (1978)). However, it requires an equally spaced net of sampling points. Therefore it cannot be applied in case of an arbitrary-shaped scan path.

4.2 Cross-Correlated Imaging

The summation, mentioned in the previous section, cumulates intensity of the image at positions, where objects, which evoked echoes in measured impulse responses, are present. However, this simple addition of multiple snapshots also creates disturbing artefacts in the focused image (see Fig. 9(a)). The elliptical traces do not only intersect at object's positions. They intersect also at other positions and even the ellipsis themselves make the image interpretation difficult or impossible.

In order to reduce these artefacts, a method based on cross-correlated back projection was proposed in Foo & Kashyap (2004). This method suggests a modification of the snapshot's computation. Instead of a simple remapping of an impulse response signal the modified snapshot is created by a cross-correlation of two impulse responses:

$$B(x,y) = \frac{1}{N}\sum_{n=1}^{N}\int_{-T/2}^{T/2} s_r\left(\frac{d_{TO}(n) + d_{Oi}}{c} + \xi\right) s_{ref}\left(\frac{d_{TO}(n) + d_{Oiref}}{c} + \xi\right) d\xi\,, \tag{13}$$

where s_{ref} is an impulse response measured by an auxiliary reference receiver at a suitable measurement position. Since two different delay terms $(d_{TO}(n) + d_{Oi})/c$ and $(d_{TO}(n) + d_{Oiref})/c$ have to match the actual scattering scenario in conjunction, the probability to add "wrong" energy to an image pixel (x,y), which does not coincide with an object, will be reduced. The integration interval T is chosen to match the duration of the stimulation impulse. Further improvement of this method was proposed in Zetik et al. (2005a); Zetik et al. (2005b); and Zetik et al. (2008). The first two references introduce modifications that improve the performance of the cross-correlated back projection from Foo & Kashyap (2004) by additional reference nodes. This drastically reduces artefacts in the focused image. In Zetik et al. (2008), a generalised form of the imaging algorithm, which is suitable for application in distributed sensor networks, is proposed:

$$B(x,y) = \frac{1}{N}\sum_{\mathrm{n}=1}^{N} W_{\mathrm{n}}(x,y) A[s_{\mathrm{rn}}(x,y), s_{\mathrm{ref1n}}(x,y), ..., s_{\mathrm{refMn}}(x,y)] \quad , \tag{14}$$

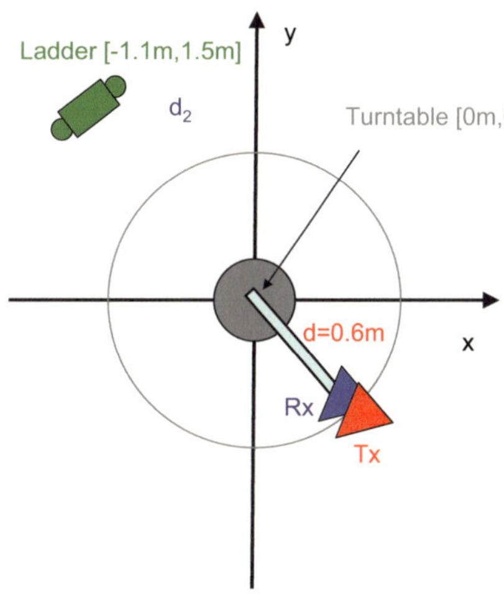

Fig. 8. Cross-correlated imaging with a circular aperture. The data measured at a certain position are multiplied with reference data, acquired at a position with 120° offset.

where $A[.]$ is an operator, which averages N spatially distributed observations and $W_{\mathrm{n}}(x,y)$ are weighing coefficients. It is assumed that all N averaged nodes can "see" an object at the position (x,y). In case of point-like objects, the incident EM waves are scattered in all directions. Hence, sensor nodes can be arbitrarily situated around the position (x,y) and they will still "see" an object (if there is one). However, extended objects, such as walls, reflect EM waves like a mirror. A sensor node can "see" only a small part of this object, which is observed under the perpendicular viewing angle. Therefore, the selection of the additional reference nodes must be done very carefully. A proper selection of sensor nodes for point like and also distributed objects is discussed in detail in Zetik et al. (2008). The weighting coefficients $W_{\mathrm{n}}(x,y)$ are inversely related to the number of nodes (measurement positions) that observe a specific part in the focused image. This reduces over and under illumination of the focused image by taking into account the topology of the network.
The following measured example demonstrates differences between images obtained by the conventional SAR algorithm (11) and the cross-correlated algorithm (14). The measurement constellation is shown in Fig. 8. The target - a metallic ladder - was observed by a sensor, which was moving along a circular track in its vicinity. The sensor comprised two closely

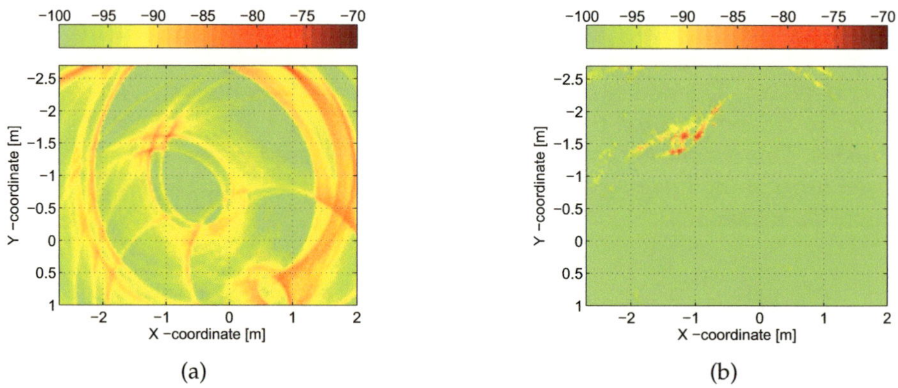

Fig. 9. (a) Image taken with the arrangement shown in Fig. 8 and processed with conventional migration algorithm. (b) Image processed with cross-correlated migration algorithm.

spaced antennas. One antenna was transmitting an UWB signal covering a bandwidth from 3.5 to 10.5 GHz. The second antenna was receiving signals reflected from the surroundings. Both antennas were mounted on the arm attached to the turntable. About 800 impulse responses were recorded. The origin of the local coordinate system was selected to be the middle of the turntable.

Firstly, the measured impulse responses were fused by the conventional SAR algorithm (11). The result of this imaging on a logarithmical scale is depicted in Fig. 9(a). The whole image is distorted by data fusion artefacts and is hard to interpret. The result can be improved by the generalised imaging algorithm (14). Here, the operator A[.] was replaced by the minimum value operator. It took the minimum magnitude from 3 observations s_{rn}, s_{ref1n} and s_{ref2n}. The position of two additional reference nodes Rref1n and Rref2n was computed adaptively for each pixel (x,y) of the focused image and to each measured impulse response Rn. The adaptation criterion was the 120° difference in the viewing angles of all 3 nodes. The reduction of disturbing artefacts is evident in Fig. 9(b).

4.3 Indirect Imaging of Objects

The procedure explained in 4.1 can be 'reversed'. Instead of measuring the signals reflected at objects, indirect imaging detects free LOS paths between objects within the area of interest.

An example is shown in Fig. 10. Generally a network of anchor nodes is required. First these anchors estimate their respective positions and, later on, they operate as Rx nodes at fixed positions. A mobile Tx moves around the area of objects and anchor nodes. The Tx emits UWB signals which are received at all Rx nodes. From the received signals two kinds of information are extracted: the propagation time, which is a measure for the current Tx-Rx distance, and the information about LOS or NLOS between Tx and Rx. The path of the Tx can be reconstructed from the totality of distance estimates. The second information allows creation of a map of LOS paths between the Tx path and the respective Rx position. An overlay of all individual LOS path maps reveals positions and approximate contours of the objects.

Diffraction at the edges of objects limits the performance of the described procedure and causes a underestimation of object dimensions. The method is explained in more detail in Hirsch et al. (2010).

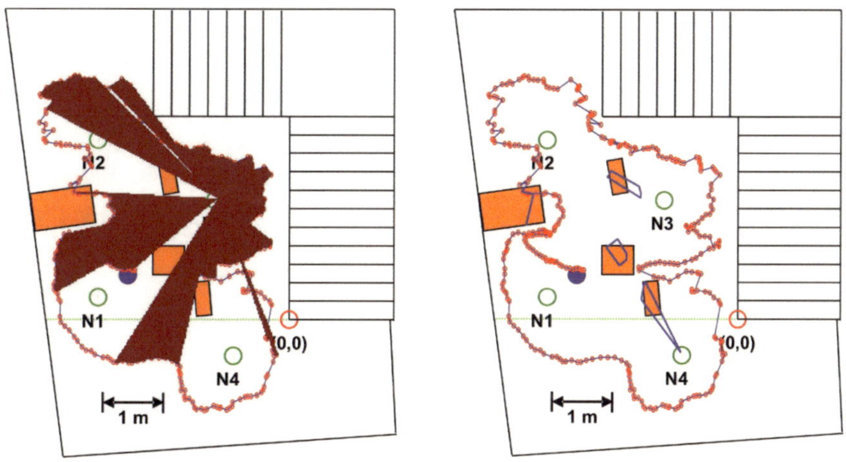

(a) LOS paths (b) Contours

Fig. 10. Indirect Imaging of objects. (a) LOS paths (dark regions) between the Tx pathway (small circles) and Rx node 3. (b) Position of objects (filled boxes) and estimated object contours (open boxes).

5. Imaging by Autonomous Rotating Sensors within a Network

5.1 Design

The networks presented in the previous sections consist of a number of nodes at fixed positions and one mobile node that moves along the imaging aperture. The imaging process requires cooperation of all nodes. Now we introduce a sensor that autonomously operates within a network of anchor nodes. It consists of a mobile platform equipped with one Tx and two Rx units and with the corresponding antennas. The sensor can move within the area of the network, varying its perspective in this way; and it can rotate to acquire 360° panoramic views. Because of the similarity to the ultrasound locating system of a bat we call it a bat-type sensor. By means of the anchor nodes the bat sensor can estimate its own position and its present orientation. Fig. 11 shows the geometry and a laboratory prototype. In principal the anchor nodes could be used as additional 'illuminators' or as additional receivers. These aspects were not investigated in the frame of this work.

5.2 Orientation within the Network

An image of the environment is typically assembled from several individual measurements performed with the bat-type sensor at different locations. For the correct assignment of these images the position and orientation of the sensor within the room must be estimated. As long as temporal synchronization exists between the network of anchor nodes and the bat-type sensor, a variety of time of arrival (TOA) localization methods can be applied for this purpose.

Here we present a method that neither requires temporal synchronization between mobile sensor and anchor nodes nor synchronization within the network of anchor nodes. It is based on angle measurements and can be classified as 'angle difference of arrival' (ADOA) localization. Line of sight from the mobile sensor to at least three anchor nodes is required. The basic idea consists in establishment of a system of two equations, where the input parameters

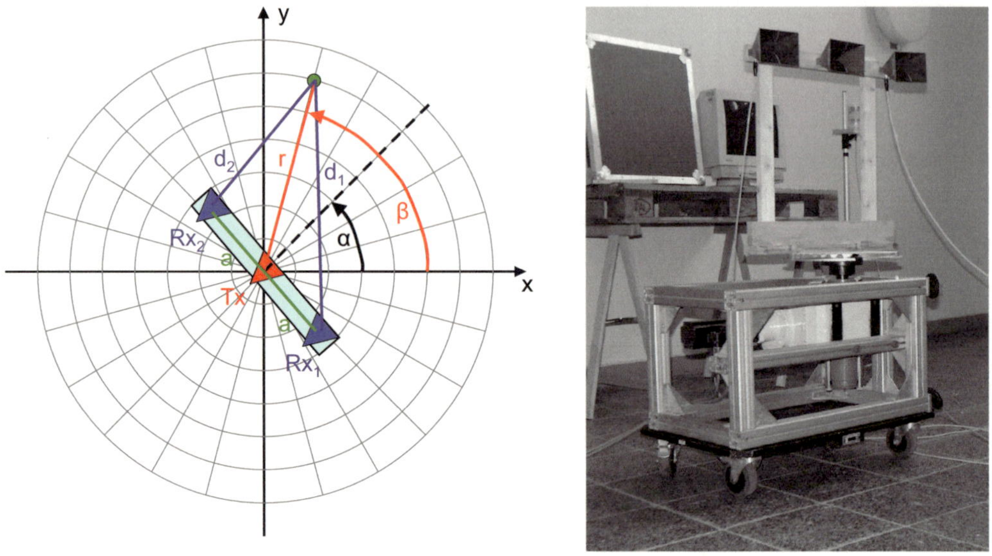

(a) Geometry (b) Prototype

Fig. 11. Geometry and laboratory prototype of a bat-type sensor. The point, where r, d_1, and d_2 come together, is an object point.

are the known positions of the anchor nodes and the difference angles between three anchor node directions, measured by the bat-type sensor. The solutions are the x- and y-position of the mobile sensor. Afterwards the sensor orientation can be estimated.

The situation is sketched in Fig. 12. We distinguish between the global coordinate system (x_g, y_g), defined by anchor nodes A_1, A_2, A_3, and the coordinate system of the bat-type sensor, which has its origin at coordinates (x_b, y_b) within the global system and which is rotated against that system by α_b. The mathematical effort is reduced if the global system is arranged in such a way that one node forms the origin and another node is placed directly on one coordinate axis. This choice does not reduce the generality of the method.

All anchor nodes operate in Tx mode. Then the bat sensor can easily estimate the direction angles to this nodes α_1, α_2, α_3 within its own coordinate system. During a sensor rotation the signal strength reaches a maximum at rotation angles where the sensor antennas are directed towards an anchor node. These angles can be extracted from the radargram with good accuracy because the TOA traces of left and right Rx antenna intersect at the transmitter positions, see Fig. 15. Because of the unknown α_b this information is not directly usable for position estimation but the angle differences $\alpha_{12} = \alpha_2 - \alpha_1$ and $\alpha_{31} = \alpha_1 - \alpha_3$ can be used, since they remain the same in both coordinate systems.

a_1, a_2, a_3 form the connection lines between sensor position (x_b, y_b) and the anchor nodes; and α_{12}, α_{31} represent the cutting angles between these lines. Now we establish the system of equations, which connects the cutting angles with the slopes m_i of the connection lines:

$$\tan \alpha_{12} = \frac{m_2 - m_1}{1 + m_2 m_1}, \quad \tan \alpha_{31} = \frac{m_1 - m_3}{1 + m_1 m_3} \quad . \tag{15}$$

The slopes follow from the anchor node coordinates and from the bat-type sensor position:

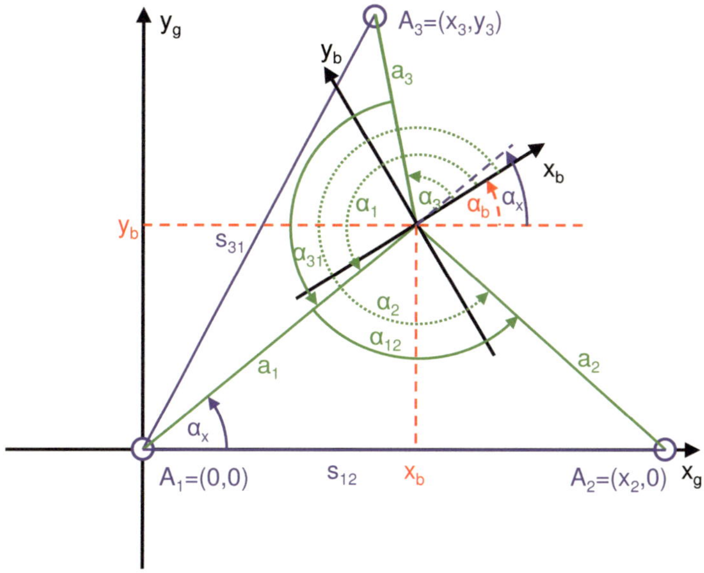

Fig. 12. Estimation of position (x_b, y_b) and orientation (α_b) of a bat-type sensor within the global coordinate system (x_g, y_g). See text for details.

$$m_1 = \frac{y_b}{x_b}, \quad m_2 = \frac{-y_b}{x_2 - x_b}, \quad m_3 = \frac{y_3 - y_b}{x_3 - x_b} \quad . \tag{16}$$

Insertion of (16) in (15), expansion of the resulting expressions, and summary of the coefficients of x_b and y_b, respectively, delivers the following system of equations:

$$x_b^2 + y_b^2 = x_2 x_b + \frac{x_2}{\tan\alpha_{12}} y_b, \quad x_b^2 + y_b^2 = \left(x_3 + \frac{y_3}{\tan\alpha_{31}}\right) x_b + \left(y_3 - \frac{x_3}{\tan\alpha_{31}}\right) y_b \quad . \tag{17}$$

Equalization of the two equations (17) gives the following expression, where we introduce the abbreviations S and T:

$$0 = \left(-x_2 + x_3 + \frac{y_3}{\tan\alpha_{31}}\right) x_b + \left(y_3 - \frac{x_3}{\tan\alpha_{31}} - \frac{x_2}{\tan\alpha_{12}}\right) y_b = S x_b + T y_b \quad . \tag{18}$$

In this way we have found expressions for y_b and y_b^2:

$$y_b = -\frac{S}{T} x_b, \quad y_b^2 = \left(\frac{S}{T}\right)^2 x_b^2 \quad . \tag{19}$$

Introduction of (19) in the first equation (17) gives a formula for x_b:

$$x_b = x_2 \left(1 - \frac{S}{\tan\alpha_{12} T}\right) \Big/ \left(1 + \left(\frac{S}{T}\right)^2\right) \quad . \tag{20}$$

Finally the orientation of the bat-type sensor is calculated. The angle between a_1 and the x-axis is $\alpha_x = \arctan(y_b / x_b)$. From Fig. 12 one can see that the orientation angle α_b must be:

$$\alpha_b = \alpha_x - \alpha_1 + \pi \quad . \tag{21}$$

An alternative angle of arrival (AOA) localization method, employing pairs of angle measurements between neighbor nodes, is described in Rong & Sichitiu (2006).

5.3 Calculation of Echo Profiles

In this section we calculate the signal propagation time for the paths from the Tx via the object to the Rx. The sensor acquires data during a full rotation. That's why the radar propagation time follows a systematic dependency. Knowledge of this dependency is required for the design of signal processing and focus algorithms. Additionally, the calculation delivers the angles of incidence both for the Tx and for the Rx antenna for each azimuth angle of the bat sensor. This allows inclusion of the antenna characteristics in the calculation of profiles of signal intensity vs. azimuth angle. Such profiles are valuable for classification of reflecting objects, for calculation of the field of view, and for the estimation of cross-range resolution.
The use of polar coordinates is convenient because of the rotational symmetry of the arrangement. The object coordinates are azimuth β and radius r. The bat sensor azimuth angle is α. In the inital sensor position the bar with the 3 antennas is aligned along the y-axis. In this position α is 0 and the sensor is looking towards the positive x-axis. The sensor rotates counter-clockwise (positive sense of rotation).
First we consider the simpler case of signal reflection at a point-like object, shown in Fig. 11(a). Because the Tx antenna is placed in the origin of the local bat coordinate system, r (the distance Tx-object) does not depend on α. The length of d_1 and d_2 (distances object-Rx_1, Rx_2) are determined by the oblique-angled triangles consisting of radius r, the arm of length a, and d_1 or d_2. The law of cosines is applied to find d_1 and d_2 in these triangles (in '$\pm$' or '$\mp$' the upper sign is always valid for index 1 while the lower sign is valid for index 2):

$$d_{1/2} = \sqrt{a^2 + r^2 - 2\,a\,r\,cos(\pi/2 \pm (\beta - \alpha))} \quad . \tag{22}$$

Then the total propagation times are:

$$\tau_{1/2}(\alpha) = (1/c) \cdot (r + d_{1/2}(\alpha)) \quad . \tag{23}$$

In radar imaging objects are often treated as if they were composed of single point scatterers. This is a simplification and fails in the case of bigger objects. In case of objects that are significantly bigger than the wavelength specular reflection dominates. From this reason we now consider the case of a reflection at a wall, which represents the extreme case of a spacious object and is more realistic than the point scatterer model in indoor imaging applications. A sketch of a bat sensor in front of a wall is shown in Fig. 13.
For convenience we use an azimuth angle δ that is $\pi/2 - \alpha$. In this way the sensor orientation is parallel to the wall at $\delta = 0$. In case of the wall reflection the point where the reflection occurs is not further fixed. Instead it moves during the rotation of the sensor. By means of the geometric construction in Fig. 13 those paths are to be found, which connect Tx and $Rx_{1/2}$ after a reflection at the wall and which fulfill the law of reflection at the same time. This law states that the angle of incidence is equal to the reflection angle. Then in our construction the perpendicular from the reflection point separates the x-axis in two regions of equal lengths s_{x1} and s_{x2}, respectively.
To determine the propagation path lengths $s_{T1/T2}(\delta)$ and $s_{R1/R2}(\delta)$ we first need to find the dependency of s_{x1} and s_{x2} on δ. We establish a linear equation of the general form $y = mx + n$ for the lines that contain $s_{R1/R2}$:

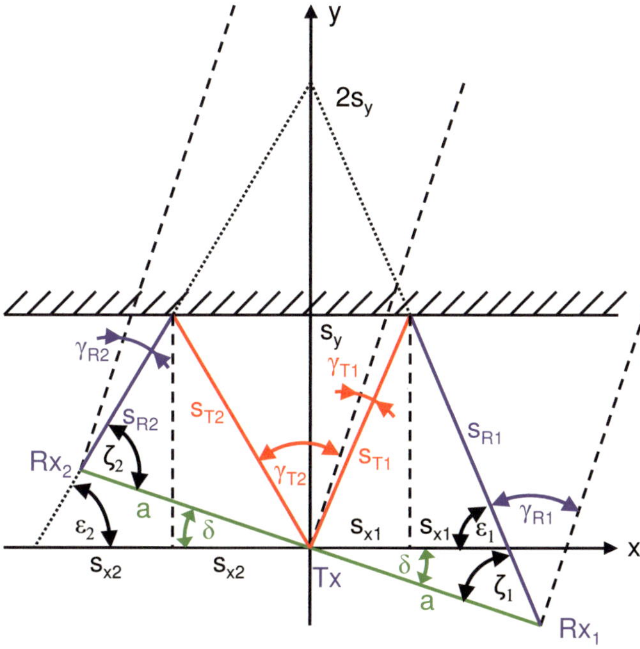

Fig. 13. Propagation paths in case of a bat sensor that rotates in front of a wall. The sensor is symbolized by the bar Rx_1-Tx-Rx_2. The wall goes parallel to the x-axis at a distance s_y. The antenna main lobe directions are given by the dashed lines. See text for details.

$$y = \frac{-a \sin\delta \mp 2\,s_y}{a \cos\delta} x + 2\,s_y \quad . \tag{24}$$

$a \cos\delta$ and $a \sin\delta$ are the projections of the bar a onto x- and y-axis. Setting $y = 0$ and rearranging Equ. (24) the intersection with the x-axis is found and in this way the length of $s_{x1/x2}$:

$$s_{x1/x2}(\delta) = \frac{s_y\, a \cos\delta}{a \sin\delta \pm 2\,s_y} \quad . \tag{25}$$

Using this result explicit expressions for the path lengths can now be given:

$$s_{T1/T2}(\delta) = \sqrt{s^2_{x1/x2}(\delta) + s^2_y} \quad . \tag{26}$$

$$s_{R1/R2}(\delta) = \sqrt{(a \cos\delta \mp s_{x1/x2}(\delta))^2 + (s_y \pm a \sin\delta)^2} \quad . \tag{27}$$

The total propagation times $\tau_{1/2}(\delta)$ are the quotients of the total propagation distances and the speed of light:

$$\tau_{1/2}(\delta) = \frac{s_{T1/T2}(\delta) + s_{R1/R2}(\delta)}{c} \quad . \tag{28}$$

Fig. 15 illustrates this dependency on a practical example. The result of Equ. (28) is shown for azimuth angles α from 0° to 359° (remember: $\delta = \pi/2 - \alpha$). When the bat sensor is facing the

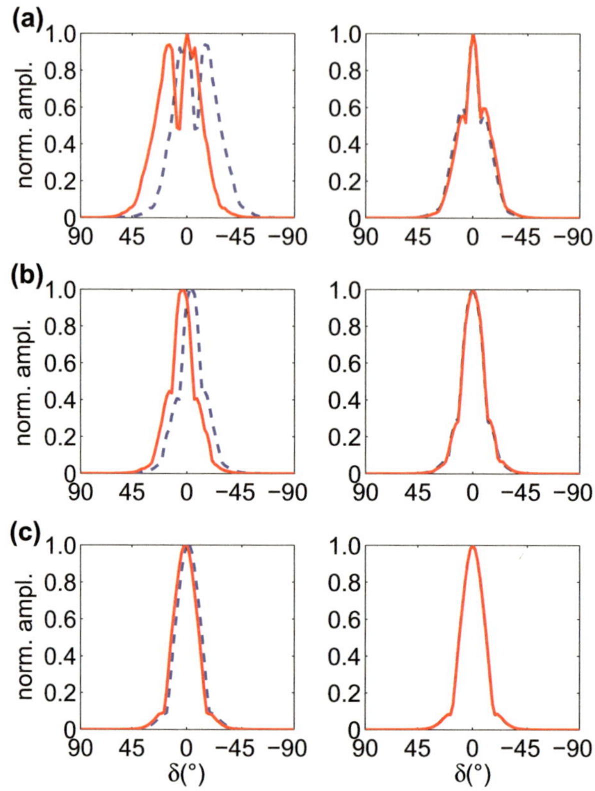

Fig. 14. Simulated amplitude vs. bat sensor rotation δ for point scatterer (left) and wall reflection (right). The length of the bat arm (variable a in Fig. 13) was 0.25 m. The distances between the center of rotation of the bat and the reflector were (a) 0.5 m, (b) 1.0 m, and (c) 5.0 m. Hatched line: signal at Rx_1, solid line: signal at Rx_2. The radiation pattern used in this simulation is the measured pattern of a double-ridge horn antenna described in Schwarz et al. (2010).

object both propagation times are equally long and the curves $\tau_{1/2}(\delta)$ intersect. The objects are visible only within a fraction of the rotation angle due to the antenna directivity.
For the calculation of the angles of incidence γ we first need to determine the auxiliary angle $\varepsilon_{1/2}$, which is part of the right-angled triangle with sides $s_{x1/x2}$, $s_{R1/R2}$, and s_y. It is

$$\sin\,\varepsilon_{1/2}(\delta) = \frac{s_y}{s_{T1/T2}(\delta)} \quad . \tag{29}$$

With this result we can give expressions for the angles of emergence $\gamma_{T1/T2}$ at the Tx antenna:

$$\gamma_{T1/T2}(\delta) = \frac{\pi}{2} \mp \delta - \arcsin\frac{s_y}{s_{T1/T2}(\delta)} \quad . \tag{30}$$

In a similar way we determine the angles of incience $\gamma_{R1/R2}$ at the receiver antennas. Here we have to use the auxiliary angle $\zeta_{1/2}$ that can be found from the sum of the internal angles of

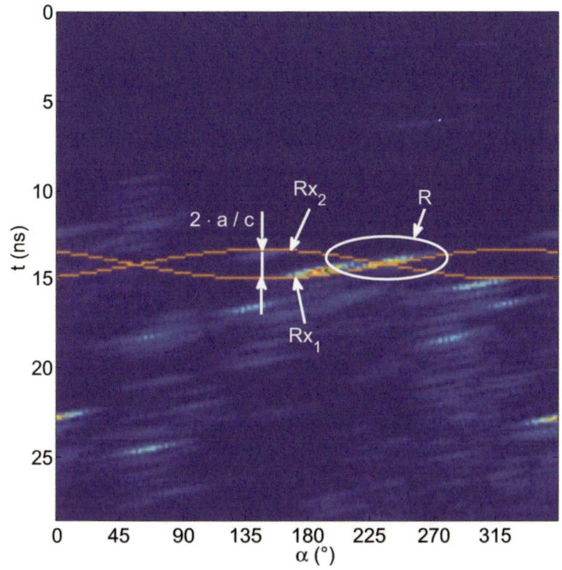

Fig. 15. Radargram of receiver Rx_1 with the curves of time of arrival (TOA) vs. bat azimuth angle α for Rx_1 and Rx_2. The Ellipse R marks the radar reflex for which the TOA curves have been calculated.

the triangles with corner points Tx, $Rx_{1/2}$, and the intersection point of $s_{R1/R2}$ with the x-axis. For Rx_1 and Rx_2 these sums are:

$$\pi = (\pi - \varepsilon_1) + \zeta_1 + \delta \qquad \pi = (\pi - \zeta_2) + \varepsilon_2 + \delta \quad . \tag{31}$$

By using these relations and Equ. (29) an expression for the angles of incidence can be given:

$$\gamma_{R1/R2}(\delta) = \frac{\pi}{2} - \zeta_{1/2}(\delta) = \frac{\pi}{2} \pm \delta - \arcsin \frac{s_y}{s_{T1/T2}(\delta)} \quad . \tag{32}$$

Now the amplitude $A_{1/2}$ of the received signal vs. the rotation angle δ can be computed. Given an antenna radiation patter $p(\gamma)$ the amplitude can be computed by a multiplication of weighting factors for the incidence at the Rx and Tx antennas and for the total propagation pathlength:

$$A_{1/2}(\delta) = p(\gamma_{R1/R2}(\delta)) \cdot p(\gamma_{T1/T2}(\delta)) \cdot \frac{s_{ref}}{s_{T1/T2}(\delta) + s_{R1/R2}(\delta)} \quad . \tag{33}$$

The reference distance s_{ref} is the total propagation distance at $\delta = 0$. In Fig. 14 results are plotted for the antenna DRH-90 (Schwarz et al. (2010)). At short distances (curves (a) and (b)) the successive appearance of the object first in the visibility range of Rx_2 and then of Rx_1 can clearly be seen. This effect is reduced in case of the wall reflection. The pattern at the largest distance (5 m) are practically identical with the square of the measured antenna radiation pattern. In this simulation the antennas are treated as point objects. This is still a simplification.

5.4 Detection of Individual Reflectors

Typically, a radar image is created from the acquired data by application of a migration algorithm. This procedure can as well be applied to the rotating bat-type sensor. In this case delays have to be calculated (similar to Equ. (12)) for both Rx separately. An example is shown in Fig. 17(b).

A disadvantage of this procedure is the appearance of parts of ellipses or circles in the final image, as they can be seen in the example. These are remains from the summation along curves. Furthermore, almost all migration algorithms were derived for point-like scatterers, but in indoor imaging most objects cause specular reflections.

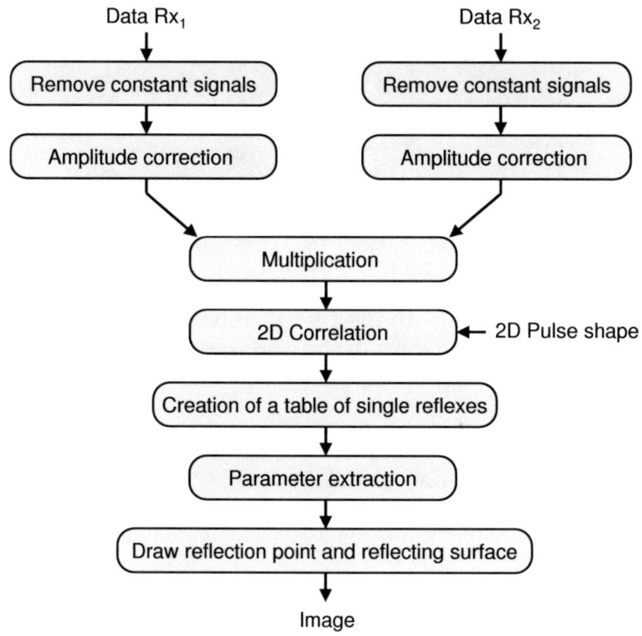

Fig. 16. Creation of an image from individual reflections at flat surfaces.

An attempt to create a more realistic final image is detection of individual flat reflectors from the radar returns. This is based on the following consideration: The reflection from a flat surface reaches a maximum when the irradiation occurs along the direction of the surface normal; from this it follows that during sensor rotation, signal maxima will occur when the sensor looks towards the surface normal of objects. Signal components that are reflected into directions far away from the surface normal cannot be detected from the bat-type sensor, because of the sensors small dimensions.

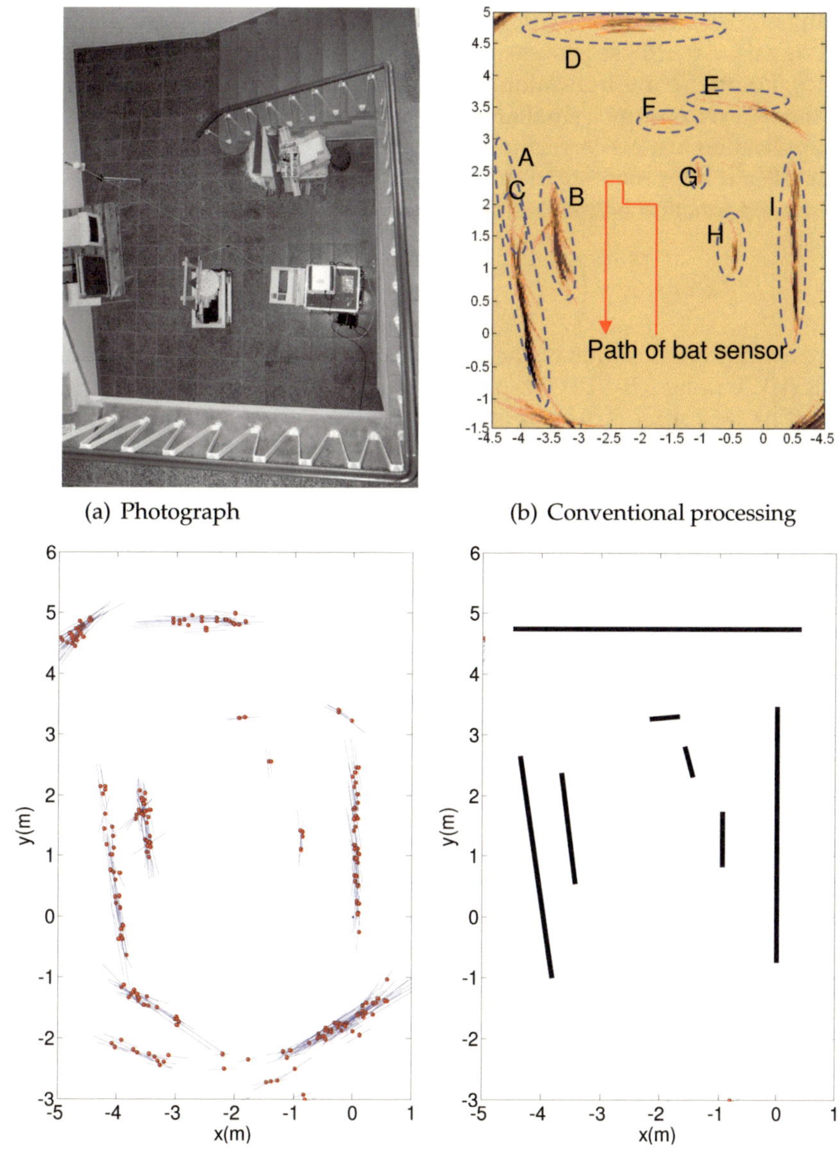

(a) Photograph
(b) Conventional processing
(c) Detection of single scatterers.
(d) Schematic representation of reflecting surfaces.

Fig. 17. Imaging with bat-type sensor in a stairwell. The images show an overlay of 15 measurement results, performed along the marked sensor path.

Now each reflex in the radargram can be interpreted as a flat surface fragment. The center of gravity of the reflex delivers the coordinates of the reflection point and the signal intensity is a measure for the reflector dimensions. Using this information one can construct an image consisting of individual flat surface fragments.
The method requires special processing of the Rx signals, schematically described in Fig. 16. Pre-processing includes the removal of constant signals (originating from cross-talk between

Tx and Rx antenna at the sensor platform) and distance-dependent amplitude correction of the received signals. After these processing steps a radargram can be drawn for each Rx separately (Fig. 15). Pointwise multiplication of both radargrams increases resolution, because the products of the reflexes occupy a smaller area. All product reflexes have a very similar shape. This allows application of a 2D correlation of the product-radargram with an example pattern, leading to significant improvement of the signal to noise ratio. Experiments have shown that a 2D Gauss-shaped function is a well suited pattern:

$$G(\alpha, r) = exp\left(-\left(\frac{\alpha - \alpha_0}{\Delta\alpha}\right)^2\right) \cdot exp\left(-\left(\frac{r - r_0}{\Delta r}\right)^2\right) \quad . \tag{34}$$

α_0 and r_0 are azimuth and radius of the reflex, respectively. For the equipment used in the practical tests (UWB radar 3.5-10.5 GHz, double-ridge horn antenna) the best parameters were $\Delta\alpha = 8.5°$ and Δr =55 mm. Afterwards a threshold detector isolates individual reflexes. A table of these reflexes is created. Then 3 parameters are extracted from each reflex : α_0, r_0, and the amplitude. Finally the image map is drawn (Fig. 17(c)). The detected reflex centers are marked with small points while the small bars are aligned along the surface orientation. The lengths of the bars are proportional to the signal amplitude. The image generated in this way comes already close to a schematic representation of the object surfaces (Fig. 17(d)).

6. Conclusions

UWB sensor networks can be used for rough imaging of the environment. Measurement results, presented e.g. in Fig. 9 and 17, demonstrate that the shape of simple objects or the structure of a room can be imaged with sufficient quality.
Methods which have been described elsewhere explicitely (e.g.4.3) were mentioned only briefly. New methods (especially chapter 5) were explained in more detail.
The methods presented in this article are especially useful for imaging of interiors. Special applications, e.g. through-wall imaging, have not been mentioned. 3D imaging in a sensor network doesn't seem to be a realistic issue. This would require time-consuming scanning of a 2D surface.
The subject is still under development. Some of the tasks that must be solved by future research are:

- Completely wireless operation of the imaging network must be achieved by introduction of wireless node synchronization.
- The time for data acquisition must be reduced, i.e. methods must be developed that generate an image from a smaller amount of raw data.
- Image artefacts must be reduced, e.g. by improved migration algorithms.

Imaging in UWB sensor networks can become part of surveillance systems with these improvements.

7. References

Federal Communications Commission, (2002). "Revision of part 15 of the Communication's Rules", *FCC 02-48*, April 2002, pp. 1-174. http://hraunfoss.fcc.gov

Foo, S. & Kashyap, S., "Cross-correlated back projection for UWB radar imaging", *Proceedings of IEEE Antennas and Propagation Society International Symposium*, pp. 1275-1278, ISBN: 0-7803-8302-8, Monterey, CA, USA, September 2004, IEEE Conference Publishing, Piscataway, NJ, USA.

Gu, K.; Wang, G. & Li, J., (2004). "Migration based SAR imaging for ground penetrating radar systems", *IEE Proceedings Radar Sonar & Navigation*, Vol. 151, No. 5, November 2004, pp. 317-325, ISSN: 1350-2395.

Hirsch, O.; Janson, M.; Wiesbeck, W. & Thomä, R. S., (2010). "Indirect Localization and Imaging of Objects in an UWB Sensor Network", *IEEE Transactions on Instrumentation and Measurement*, Vol. 59, No. 11, Nov. 2010, pp. 2949-2957, ISSN: 0018-9456.

Hussain, M. G. M., (1998). "Ultra-wideband impulse radar - an overview of the principles", *IEEE Aerospace and Electronic Systems Magazine*, Vol.13, No.9, September 1998, pp.9-14, ISSN: 0885-8985.

Luediger, H. & Kallenborn, R., (2009). "Generic UWB Regulation in Europe", *Frequenz, Journal of RF-Engineering and Telecommunications*, Vol.63, No.9-10, Sept./Oct. 2009, pp.172-174, ISSN: 0016-1136.

Margrave, F. G., (2001). *Numerical Methods of Exploration Seismology with algorithms in MATLAB*, http://www.crewes.org/ResearchLinks/FreeSoftware/EduSoftware/NMES_Margrave.pdf.

Oliver, C. J., (1989). "Synthetic-aperture radar imaging", *Journal of Physics D: Applied Physics*, Vol.22, No. 7, July 1989, pp. 871-890, doi: 10.1088/0022-3727/22/7/001.

Patwari, N.; Ash, J.N.; Kyperountas, S.; Hero, A.O., III; Moses, R.L.; Correal, N.S., (2005). "Locating the nodes", *IEEE Signal Processing Magazine*, Vol. 22, No. 4, June 2005, pp. 54-69, ISSN: 1053-5888.

Rong, P. & Sichitiu, M. L., (2006). "Angle of Arrival Localization for Wireless Sensor Networks", *Proceedings of SECON 2006*, pp. 374-382, ISBN: 1-4244-0626-9, Reston, VA, USA, September 2006, IEEE Catalog Number: 06EX1523.

Sachs, J.; Peyerl, P.; Zetik, R. & Crabbe, S., (2003). "M-Sequence Ultra-Wideband-Radar: State of Development and Applications", *Proceedings of Radar 2003*, pp. 224-229, ISBN: 0-7803-7871-7, Adelaide, Australia, September 2003, Causal Productions, Adelaide, Australia.

Sachs, J. (2004). "M-Sequence Radar", In: *Ground Penetrating Radar*, Daniels, D. J., (Ed.), 2nd ed., pp. 225-236, Inst. of Electrical Engineers, ISBN:0-85296-862-0, Stevenage, U.K. .

Sayed, A. H.; Tarighat, A.; Khajehnouri, N., (2005). "Network-Based Wireless Location", *IEEE Signal Processing Magazine*, Vol.22, No.4, July 2005, pp.24-40, ISSN: 1053-5888.

Schwarz, U.; Thiel, F,; Seifert, F.; Stephan, R. & Hein, M. A., (2010). "Ultra-Wideband Antennas for Magnetic Resonance Imaging Navigator Techniques", *IEEE Trans. on Antennas and Propagation*, vol. 58, No. 6, June 2010, pp.2107-2112, ISSN: 0018-926X.

Serpedin, E. & Chaudhari, Q. M. (2009). *Synchronization in Wireless Sensor Networks: Parameter Estimation, Performance Benchmarks and Protocols*, Cambridge University Press, ISBN: 978-0-521-76442-1, Cambridge, UK.

Stoica, P. & Li, J., (2006). "Lecture Notes - Source Localization from Range-Difference Measurements", *IEEE Signal Processing Magazine*, Vol.23, No.6, November 2006, pp.63-66, ISSN: 1053-5888.

Stolt, R. H., (1978). "Migration by Fourier Transform", *Geophysics*, Vol.43, No.1, February 1978, pp.23-48, ISSN: 0016-8033.

Thomä, R. S.; Hirsch, O.; Sachs, J. & Zetik, R., (2007). "UWB Sensor Networks for Position Location and Imaging of Objects and Environments", *Proceedings of the 2nd European Conference on Antennas and Propagation (EuCAP2007)*, pp. 1-9, ISBN: 978-0-86341-842-6, Edinburgh, U.K., November 2007, IET Conferences, London, U.K.

Yang, Y. & Yang, K., (2006). "Time Synchronization for Wireless Sensor Networks using the Principle of Radar Systems and UWB Signals", *Proceedings of IEEE International Conference on Information Acquisition 2006*, pp. 160-165, ISBN: 1-4244-0528-9, Weihai, China, August 2006, IEEE Conference Publishing, Piscataway, NJ, USA.

Zetik, R.; Sachs, J. & Thomä, R., (2005) "Modified Cross-Correlation Back Projection for UWB Imaging: Numerical Examples", *Proceedings of IEEE International Conference on Ultra-Wideband*, pp. 650-654, ISBN: 0-7803-9397-X, Zï£¡rich, Switzerland, September 2005, IEEE.

Zetik, R.; Sachs, J. & Thomä, R., (2005) "Modified Cross-Correlation Back Projection for UWB Imaging: Measurement Examples", *Proceedings of the 6th International Scientific Conference on Digital Signal Processing and Multimedia Communications - DSP-MCOM 2005*, pp. 56-59, ISBN: 80-8073-323-6, Kosice, Slovakia, September 2005, Technicka univerzita Kosice, Kosice, Slovakia.

Zetik, R.; Sachs, J. & Thomä, R., (2010) "Imaging of distributed objects in UWB sensor networks", In: *Short Pulse Electromagnetics 9*, Sabath, F.; Giri, D. ; Rachidi, F. & Kaelinpp, A. (Ed.), pp. 97-104, Springer-Verlag New York Inc, ISBN: 9780387778440, New York, NY, USA.